Springer Proceedings in Physics 28

Springer Proceedings in Physics

Managing Editor: H. K. V. Lotsch

Volume 1 *Fluctuations and Sensitivity in Nonequilibrium Systems*
Editors: W. Horsthemke and D. K. Kondepudi

Volume 2 *EXAFS and Near Edge Structure III*
Editors: K. O. Hodgson, B. Hedman, and J. E. Penner-Hahn

Volume 3 *Nonlinear Phenomena in Physics* Editor: F. Claro

Volume 4 *Time-Resolved Vibrational Spectroscopy*
Editors: A. Laubereau and M. Stockburger

Volume 5 *Physics of Finely Divided Matter* Editors: N. Boccara and M. Daoud

Volume 6 *Aerogels* Editor: J. Fricke

Volume 7 *Nonlinear Optics: Materials and Devices* Editors: C. Flytzanis and J. L. Oudar

Volume 8 *Optical Bistability III* Editors: H. M. Gibbs, P. Mandel, N. Peyghambarian, and S. D. Smith

Volume 9 *Ion Formation from Organic Solids (IFOS III)* Editor: A. Benninghoven

Volume 10 *Atomic Transport and Defects in Metals by Neutron Scattering*
Editors: C. Janot, W. Petry, D. Richter, and T. Springer

Volume 11 *Biophysical Effects of Steady Magnetic Fields*
Editors: G. Maret, J. Kiepenheuer, and N. Boccara

Volume 12 *Quantum Optics IV* Editors: J. D. Harvey and D. F. Walls

Volume 13 *The Physics and Fabrication of Microstructures and Microdevices*
Editors: M. J. Kelly and C. Weisbuch

Volume 14 *Magnetic Properties of Low-Dimensional Systems*
Editors: L. M. Falicov and J. L. Morán-López

Volume 15 *Gas Flow and Chemical Lasers* Editor: S. Rosenwaks

Volume 16 *Photons and Continuum States of Atoms and Molecules*
Editors: N. K. Rahman, C. Guidotti, and M. Allegrini

Volume 17 *Quantum Aspects of Molecular Motions in Solids*
Editors: A. Heidemann, A. Magerl, M. Prager, D. Richter, and T. Springer

Volume 18 *Electro-optic and Photorefractive Materials* Editor: P. Günter

Volume 19 *Lasers and Synergetics* Editors: R. Graham and A. Wunderlin

Volume 20 *Primary Processes in Photobiology* Editor: T. Kobayashi

Volume 21 *Physics of Amphiphilic Layers* Editors: J. Meunier, D. Langevin, and N. Boccara

Volume 22 *Semiconductor Interfaces: Formation and Properties*
Editors: G. Le Lay, J. Derrien, and N. Boccara

Volume 23 *Magnetic Excitations and Fluctuations II* Editors: U. Balucani, S. W. Lovesey, M. G. Rasetti, and V. Tognetti

Volume 24 *Recent Topics in Theoretical Physics* Editor: H. Takayama

Volume 25 *Excitons in Confined Systems* Editors: R. Del Sole, A. D'Andrea, and A. Lapiccirella

Volume 26 *The Elementary Structure of Matter* Editors: J.–M. Richard, E. Aslanides, and N. Boccara

Volume 27 *Competing Interactions and Microstructures Statics and Dynamics*
Editors: R. LeSar, A. Bishop, and R. Heffner

Volume 28 *Anderson Localization* Editors: T. Ando and H. Fukuyama

Volume 29 *Polymer Motion in Dense Systems* Editors: D. Richter and T. Springer

Anderson Localization

Proceedings of the International Symposium,
Tokyo, Japan, August 16–18, 1987

Editors: T. Ando and H. Fukuyama

With 187 Figures

Springer-Verlag Berlin Heidelberg New York
London Paris Tokyo

Professor Tsuneya Ando
Professor Hidetoshi Fukuyama

Institute for Solid State Physics, University of Tokyo,
7-22-1, Roppongi, Minato-ku, Tokyo 106, Japan

Organizing Committee:
W. Sasaki (Toho University, Chairman)
S. Kawaji (Gakushuin University)
Y. Nagaoka (Nagoya University)
S. Kobayashi (University of Tokyo)
A. Kawabata (Gakushuin University, Secretary)
H. Fukuyama (University of Tokyo)
T. Ando (University of Tokyo)
S. Ikehata (University of Tokyo)

International Advisory Committee:
E. Abrahams (USA)
B. L. Altshuler (USSR)
P. W. Anderson (USA)
G. Bergmann (USA)
Y. Bruynseraede (Belgium)
C. Di Castro (Italy)
R. C. Dynes (USA)
B. Kramer (FRG)
P. A. Lee (USA)
A. MacKinnon (UK)
N. F. Mott (UK)
G. A. Thomas (USA)
D. C. Tsui (USA)
R. A. Webb (USA)

ISBN-13: 978-3-642-73556-1 e-ISBN-13: 978-3-642-73554-7
DOI: 10.1007/ 978-3-642-73554-7

Softcover reprint of the hardcover 1st edition 1988

2154/3150-543210

Foreword

This volume contains the contributions presented at the University of Tokyo International Symposium on Anderson Localization, which was held at the University of Tokyo, August 16–18, 1987. The symposium was a satellite to the 18th International Conference on Low Temperature Physics and essentially followed the style of the International Conference on Localization, Interaction, and Transport Phenomena in Impure Metals (LIPTIM) held in Braunschweig in 1984 (the Proceedings of which were published as Vol. 61 of the Springer Series in Solid-State Sciences). The main purpose of the symposium was to review the present status and discuss various aspects of the rich phenomena related to Anderson localization. It was attended by about 100 participants from 15 countries.

This volume contains 25 invited papers, many of which not only present new results but also give a review of the respective topics. The 38 contributed papers were presented during lively poster sessions. The first three parts of this volume contain papers on the metal-insulator transition in doped semiconductors and disordered metals, which has long been an outstanding subject and which has recently achieved new and rapid development due to the scaling theory in the presence of interaction effects. Parts IV–VI deal with the weak localization phenomena in two- and three-dimensional dirty metals and semiconductor space-charge layers. Part VII contains several papers on the quantum Hall effect and localization in strong magnetic fields. Parts VIII–X deal with newer subjects, i.e., quasicrystals and mesoscopic systems. In the former the wave function is self-similar and marginally localized or delocalized, while in mesoscopic systems the conductance is no longer a self-averaged quantity and fluctuations play an essential role. This book provides an up-to-date overview of rapidly developing topics of Anderson localization and we are convinced that it will be of wide interest.

Finally, on behalf of all the participants we would like to express our sincere thanks for financial support from the Fund for Promotion of Science, the University of Tokyo, and the Nishina Memorial Foundation.

Tokyo, November 1987

W. Sasaki
T. Ando
H. Fukuyama

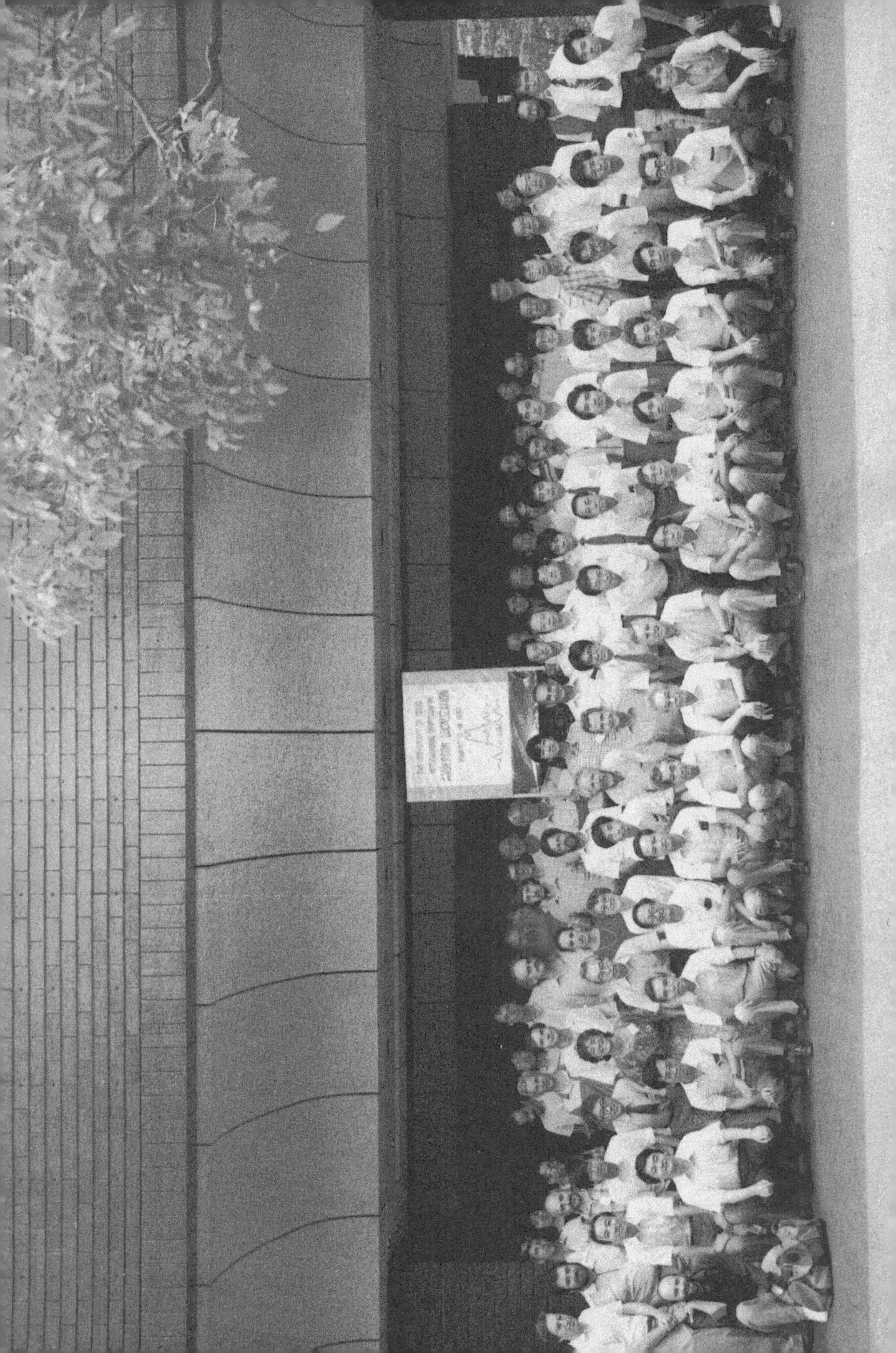

Contents

Part V **Weak Localization Effects: Space-Charge Layers**

Anderson Localization: An Introduction

H. Fukuyama and T. Ando

Institute for Solid-State Physics, University of Tokyo,
7-22-1 Roppongi, Minato-ku, Tokyo 106, Japan

A brief review is given of the present understanding of various phenomena related to Anderson localization, together with a historical survey.

Scaling Theory and Weak Localization Theory

In quantum mechanics electrons are described as waves. In perfect crystals they are propagating Bloch waves. In the presence of randomness, however, these waves are scattered and, if such randomness is strong, they are localized due to interference effects between scattered waves. This phenomenon is Anderson Localization, which was recognized first theoretically by P.W. ANDERSON in 1958 /1/. In spite of its importance as a basic notion in solid state physics, the microscopic understanding of Anderson localization has been hindered until recently by the intrinsic complexity due to randomness.

A key step was established in 1979, when ABRAHAMS, ANDERSON, LICCIARDELLO and RAMAKRISHNAN /2/ developed the scaling theory, which has been discussed earlier especially by WEGNER /3/. This scaling theory of the "gang of four" has provided a firm ground from which to investigate the phenomenon by use of similar methods developed for the many-body problem and critical phenomena. A surprising feature of the scaling theory is that it predicts a universal behavior in conductivity which depends only on the dimensionality and the symmetry of the system. (It turns out that the two-dimensional system is marginal and can have unique properties.) Moreover it has made it possible to think about metallic and insulating states on an equal footing.

Since the introduction of this scaling theory, an overwhelming amount of research on Anderson localization has been reported both experimentally and theoretically /4-10/. The first experimental verification of the theory was the logarithmic temperature dependence of conductivity in metallic films by DOLAN and OSHEROFF /11/.

These diverse investigations have established the validity of the scaling theory. Especially the weakly localized regime has played an essential role, since the theoretical predictions in this case can be made quantitative because of the validity of the perturbation with respect to the randomness parameter. In this regime, where the interference effect plays a vital role, effects of different symmetry of the system, i.e., orthogonal (normal impurity scattering), unitary (magnetic fields or magnetic impurities) or symplectic (spin-orbit scattering), have been worked out explicitly for two-dimensional systems by HIKAMI, LARKIN and NAGAOKA /12/ in the context of magnetoresistance. This theory has given explanations to the puzzling negative magnetoresistance, which had been observed before in the typical two-dimensional systems created by metal-oxide-semiconductor (MOS) structures /13/. Moreover, the application by KAWABATA /14/ of the theory to the three-dimensional case has resulted in an understanding of the long-lived puzzle of the negative magnetoresistance in doped semiconductors first observed by SASAKI in 1965 /15/. Not only they have provided explanations of old unresolved problems, they have also started to be used to extract more detailed information on the relatively weak scattering processes in solids, which had otherwise been difficult to obtain. For example, the characteristic times of inelastic and spin-orbit scattering and their subtle competition are determined through the analysis of magnetoresistance as first shown by KOMORI et al. for Cu films /16/ and by BERGMANN for Au coated Mg /17/ and elaborated to the level of an art by BERGMANN /18/. At the same time a remarkable oscillatory behavior of magnetoresistance in small cylinders has been predicted /19/ and observed actually /20/, which clearly demonstrates the importance of the interference effect.

Interaction Effects

In the course of study it was found by ALTSHULER, ARONOV and LEE /21/ and by FUKUYAMA /22/, independently, that the mutual interaction plays a vital role and can result in a logarithmic temperature (lnT) dependence in conductivity similar to the interference effects. This results from the direct interplay between randomness and the dynamical nature of interaction processes. These are so-called interaction effects /9,10/. This finding of interaction effects was surprising, since it was the convention that the mutual interactions are treated in the framework of Landau's Fermi liquid theory where electrons are treated as if free but with various parameters, e.g. the effective mass, being affected by the interactions and that these renormalized electrons are scattered by the randomness. This new understanding of interaction processes in random systems has disclosed various new features of interaction processes in solids. For example, even in ordinary bulk metals the same effect is shown to lead to lnT dependence of the conductivity, which is actually observed /23/. (This very effect but in a different context had originally been noted by ALTSHULER and ARONOV /24/, even before the appearance of the scaling theory, who examined the density of states at T=0.)

Once the interaction processes in the presence of randomness were known to have essentially new aspects, it was natural to study the same problem for the phase transition driven by mutual interaction between electrons. The typical phase transitions are superconductivity /25/ and itinerant magnetism /26/, the former of which has been examined so far in detail /8/. It is now confirmed by careful experiments on films /27,28/ that the critical temperature of superconductivity is a sensitive function of randomness as predicted by theories. Another typical interaction process of conduction electrons is the Kondo effect, which is also shown theoretically to be affected seriously by the randomness /29/, though there has not yet been any systematic experimental investigation on this problem.

Metal-Insulator Transition

Based on an understanding of the weakly localized regime the scaling theory has been extended to the region of strong disorder, especially to the transition between metallic and insulating states. This is now an area of active research. As in the weakly localized regime the Coulomb interaction plays a vital role for such a transition region and a new type of Fermi liquid theory has been searched for based on the renormalization group technique introduced by FINKELSTEIN /30/ and elaborated by DI CASTRO et al. /31,32/. Here the possibility of enhanced magnetic fluctuations near the transition is of central interest; the question is closely connected with the formation of local moments in itinerant systems.

One of the typical systems used to investigate this problem has been doped semiconductors. As regards the conductivity exponent, ν, theoretical studies have revealed that ν is dependent on cases, i.e. strong magnetic field, magnetic impurities and spin-orbit scattering /33,34/. Experimentally $\nu=1$ is rather widely observed, which is indicated to be consistent with theories if spin-flip or spin-orbit scattering is strong. The one exception is uncompensated Si:P where $\nu\simeq0.5$ is observed by the notable technique of stress tuning /35/. The real cause for this puzzling result is still not yet understood. Even in this case the duality relationship between the conductivity exponent and the dielectric constant exponent, ν', in the insulating phase is satisfied, i.e. $\nu'=2$, as indicated theoretically by IMRY et al. /36/ on phenomenological grounds and by KAWABATA /37/ based on a renormalization group method. The latter concludes the possible existence of interesting numerical differences in the prefactors, which depend on symmetry.

In this fascinating regime there are still many problems, both experimentally and theoretically, to be solved even in normal metals. The question can be more intriguing when the competition between superconductivity and the metal-insulator transition is to be addressed. One can think of two distinct cases for how the insulating state will overwhelm the superconducting state as the degree of randomness is increased; in one case the normal metallic states intervene between these two extremes whereas in another case the superconducting state is destroyed by the insulating state sharply. The mechanism resulting in such qualitatively different behavior is not fully explored.

Hence as regards the interaction effects there are still many interesting problems which wait for further investigation.

Quantum Hall Effect and Localization in Strong Magnetic Fields

The quantization of the Hall conductivity in two-dimensional systems in strong magnetic fields was first theoretically noted in 1975. ANDO, MATSUMOTO, and UEMURA calculated the Hall conductivity in different approximations and found that it is always given by integer multiples of e^2/h when the Fermi level lies in the energy gap between localized bound states associated with each Landau level /38/. After the experimental discovery of the quantum Hall effect /39,40/, various theoretical arguments appeared. Most of them have led to the conclusion that the Hall conductivity is quantized into integer multiple of e^2/h when the Fermi level lies in localized states. For example, LAUGHLIN has used a gauge-invariance argument using a cylindrical geometry /41/. On the other hand, AOKI and ANDO used Kubo's linear response theory /42/. Especially the latter work has proved that the Hall conductivity vanishes when all the states below the Fermi level are localized and that the Hall conductivity takes quantized values with the integer being the number of Landau levels below the Fermi level in sufficiently high magnetic fields. This has led to the important conclusion that there should be current-carrying extended states in each Landau level in high magnetic fields.

A more rigorous mathematical proof considers a two-dimensional system on a torus and the Hall conductivity averaged over two magnetic fluxes which penetrate inside and the opening of the torus /43-45/. The averaged conductivity becomes a topological quantity called winding number times e^2/h and is always integer multiples of e^2/h. Naturally, this does not necessarily mean that the observed Hall conductivity is always quantized. As a matter of fact, this integer varies rapidly if the Fermi level is swept across energy levels densely distributed in macroscopic systems. The observed Hall conductivity is an average over energy levels in the vicinity of the Fermi level and need not be an integer. When states in the vicinity of the Fermi level are localized, the average becomes also an integer multiple of e^2/h. It is clear that the localization effect is essential in the quantization of the Hall conductivity. Actually, current-carrying extended states are also necessary, since the integer vanishes when all the states below the Fermi level are localized.

There have been various investigations to clarify the localization problem /46/. Numerical studies have been quite effective and provided various useful results. Perturbation calculations have also been performed. As a result, it has now become clear that there exist extended states in each Landau level in high magnetic fields but only at a single energy in the vicinity of the center. The inverse localization length α, i.e., the decay rate of localized wave functions, approaches zero smoothly at the energy of extended states and the critical exponent s, defined by $\alpha(E)\propto|E|^s$, is close to 2. One of the important remaining questions is whether there is a critical magnetic field below which extended states disappear or whether they are always present and only disappear in the zero field limit. Another problem is the validity of a nonlinear σ model. Several authors proposed the nonlinear σ model with the diagonal and off-diagonal component of the conductivity tensor, σ_{xx} and σ_{xy}, as the two independent scaling variables /47/. The flow diagram in the σ_{xx}-σ_{xy} plane is periodic in the σ_{xy} direction with the periodicity e^2/h, which means that the critical behavior and the peak diagonal conductivity are independent of the Landau level index N. Numerical studies contradict the conclusion of this model, however /48/.

The quantized Hall resistance is now in actual use as the resistance standard. However, its application for the determination of the fine structure constant is still facing the severe difficulty of inaccuracies of existing resistors. It requires a reliable standard having accuracy one order of magnitude better than currently available.

Fractional Quantum Hall Effect

The fractional quantum Hall effect was first observed in two-dimensional systems in GaAs/AlGaAs heterostructures /49/. In the fractional quantum Hall effect, the Hall conductivity is quantized into fractions of $-e^2/h$ and the diagonal conductivity shows an activated temperature dependence. The observed fractions are 1/3, 2/3, 2/7, 2/5, etc. The fractional quantum Hall effect is related to the ground states of electrons interacting via the Coulomb repulsion in strong magnetic fields.

The basis wave function in which the angular momentum is diagonalized is

$$\varphi_m(x,y) = (2\pi 2^m m!)^{-1/2} z^m \exp\left[-\frac{|z|^2}{4}\right]$$

where $z=x+iy$ and the angular momentum is m with $m=0,1,..,M-1$ for a circular system with area $S=2\pi\ell^2(M-1)$. When N electrons are confined within this system, the wave function is given by a linear combination of Slater determinants, which are in general written as

$$\Psi(z_1,\ldots,z_N) \propto \prod_{i<j}(z_i-z_j) f(z_1,\ldots,z_N) \exp\left[-\sum_{i=1}^{N}\frac{|z_i|^2}{4}\right]$$

where $f(z_1,..,z_N)$ is a symmetric polynomial due to the antisymmetry of the wave function. The problem is to obtain the polynomial f which minimizes the total energy. The filling factor is $\nu=N/M$.

LAUGHLIN proposed possible polynomials which are expected to give the approximate ground state at desired fractional filling factors /50/. Especially for $\nu=1/p$ with p an odd integer, the pair distribution function approaches zero in proportion to r^{2p} near the origin. These wave functions describe in general (incompressible) liquid states without long range order. Low-energy excited states are described by quasi-holes or quasi-electrons with a fractional charge $\pm e/p$. Other higher order fractions are explained by hierarchy in which such quasi-holes or quasi-electrons form Laughlin states again and again with changing ν.

Justification of Laughlin's wave function is still unsatisfactory, however. The strongest support has been provided by numerical computer diagonalization of finite systems containing a few electrons /51/. The numerical study shows that the total energy has a small dip at 1/3 filling of the lowest Landau level and that the pair correlation function is close to but slightly different from Laughlin's wave function. The system is much too small to give any definite conclusions, however. There remains discrepancy between experiments and theory on excitation energies of quasi-electrons and quasi-holes /52/. The quantitative understanding requires detailed understanding of ground state properties in the presence of both Coulomb interactions and random potential fluctuations. A new and interesting problem is the recent observation of quantization into even denominator fractions for the excited Landau level N=1 /53,54/.

Mesoscopic Systems

Recent development of lithographic techniques has enabled fabrication of small systems in the submicron range. This has created a new field of physics called mesoscopic systems (much larger than atomic scales but not large enough to be regarded as macroscopic). The key feature of such mesoscopic systems is conductance fluctuations arising from the wave nature of quantum-mechanical electron motion.

Typical mesoscopic systems are metallic wires and quasi-one-dimensional (Quasi-1D) systems created by MOS structures with a very narrow gate. Metallic wires exhibit unexpected fluctuations as a function of magnetic field at low temperatures /55/. Quasi-1D MOS's in the metallic regime also show such aperiodic structure both in the magnetoresistance and in the resistance as a function of electron density /56,57/. These fluctuations are not noise in the usual sense but are time-independent changes in the resistance which vary between samples but are reproducible within a given sample. Such fluctuations disappear only slowly with increasing temperature.

This conductance fluctuation is inherent to the non-self-averaging nature of the transport coefficient and is a direct manifestation of coherent interference effects due to scatterings from different scatterers. In this sense it is quite similar to the Anderson localization phenomena. ALTSHULER /58/ and LEE and STONE /59/ have first shown that the fluctuation of the conductance in metallic systems is about e^2/h, independent of system size and randomness (the universal conductance fluctuation). One of the most striking features is the nonlocality in which a slight change of the positions of a single scatterer can cause measurable change in observed conductance /60/.

It has been found in the course of the development of the weak localization theory that elastic scattering processes alone do not destroy the phase memory of electrons and therefore the coherence of their wave functions. Interference of wave functions due to such long range coherence is the main origin of Anderson localization. The effective system size is the inelastic diffusion length or the phase coherence length since inelastic scattering destroys the phase coherence of wave functions. The inelastic diffusion length is much larger (typically 1μm and can be much larger at extremely low temperatures) than the elastic mean free path which is at most a few hundred angstrom. This is the main reason that the conductance fluctuation can be observed in mesoscopic systems whose size is still much larger than the elastic mean free path. With increasing temperature, the phase coherence length becomes smaller than the system size and a statistical average of the conductance within a given sample sets in and the fluctuation decreases slowly. The broadening of the Fermi distribution function gives rise to an additional statistical average /61/.

Quasi-1D MOS's in the nonmetallic regime, on the other hand, show giant fluctuations as large as 2 order of magnitude as a function of the electron density /62/. Proposed mechanisms for such giant fluctuations include resonant tunneling /63/ and variable range hopping /64/. In the former mechanism, the conductance exhibits a narrow resonance at the energy of a localized state present in the system. In the latter, the presence of a few important paths for hoppings gives rise to fluctuations. So far, experimental results seem to support the latter mechanism /65/.

References

1. P.W. Anderson: Phys. Rev. **109** (1958) 1492.
2. E. Abrahams, P.W. Anderson, D.C. Licciardello and T.V. Ramakrishnan: Phys. Rev. Lett. **42** (1979) 673.
3. F. Wegner: Z. Phys. **25** (1976) 327.
4. Anderson Localization, ed. by Y. Nagaoka and H. Fukuyama, (Springer, Berlin 1982)
5. Localization, Interaction, and Transport Phenomena, ed. by B. Kramer, G. Bergmann and Y. Bruynseraede, (Springer, Berlin, 1984).
6. Heavy Doping and the Metal-Insulator Transition in Semiconductors, ed. by P.T Landsberg (Solid-State Electronics **28** (1985))
7. P.A. Lee and T.V. Ramakrishnan: Rev. Mod. Phys. **57** (1985) 287.
8. Anderson Localization, Suppl. to Prog. Theor. Phys. **84** (1985)
9. Electron-Electron Interaction in Disordered Systems, ed. by A.L. Efros and M. Pollak (North-Holland, Amsterdam, 1985)
10. Localization and Interaction, ed. by D.M. Finlayson (Scottish Universities Summer School in Physics 31, 1986)
11. G.J. Dolan and D.D. Osheroff: Phys. Rev. Lett. **43** (1979) 721.
12. S. Hikami, A.I. Larkin and Y. Nagaoka: Prog. Theor. Phys. **63** (1980) 707.

13. Y. Kawaguchi, H. Kitahara and S. Kawaji: Surf. Sci. **73** (1978) 520.
14. A. Kawabata: J. Phys. Soc. Jpn. **53** (1984) 3540.
15. W. Sasaki: J. Phys. Soc. Jpn. **20** (1965) 825.
16. F. Komori, S. Kobayashi and W. Sasaki: J. Phys. Soc. Jpn **51** (1982) 3162.
17. G. Bergmann: Phys. Rev. Lett. **48** (1982) 1046.
18. G. Bergmann: Phys. Rept. **107** (1984) 3.
19. B.L. Altshuler, A.G. Aronov and B.Z. Spivak: JETP Lett. **33** (1981) 94.
20. D.Y. Sharvin and Y.V. Sharvin, JETP Lett. **34** (1981) 272.
21. B.L. Altshuler, A.G. Aronov and P.A. Lee: Phys. Rev. Lett. **44** (1980) 1288.
22. H. Fukuyama: J. Phys. Soc. Jpn. **48** (1980) 2169.
23. T.F. Rosenbaum, R.F. Milligan, G.A. Thomas, P.A. Lee , T.V. Ramakrishnan and R.N. Bhatt: Phys. Rev. Lett. **47** (1981) 1758.
24. B.L. Altshuler and A.G. Aronov: Solid State Commun. **30** (1979) 115.
25. S. Maekawa, and H. Fukuyama: Physica **107B** (1981) 123; J. Phys. Soc. Jpn. **51** (1982) 1380.
26. H. Fukuyama: J. Phys. Soc. Jpn. **54**, 2092 (1985)
27. H.R. Raffy, R.B. Laibowitz, P. Chaudhari and S. Maekawa: Phys. Rev. **B28** (1983) 6607.
28. J.M. Graybeal and M.R. Beasley: Phys. Rev. **B29**, 4167 (1984)
29. F.J. Ohkawa, H. Fukuyama and K. Yosida: J. Phys. Soc. Jpn. **52** (1983) 1701.
30. A.M. Finkelstein: Sov. Phys.-JETP **57** (1983) 97; Z. Phys. **56** (1984) 189.
31. C. Castellani, C. Di Castro, G. Forgacs and S. Sorella: Solid State Commun. **52** (1984) 261.
32. C. Castellani, C. Di Castro, P.A. Lee and M. Ma: Phys. Rev. **B30** (1984) 527.
33. S. Hikami: Phys. Rev. **B24** (1981) 2671 and p.15 in ref.4
34. C. Castellani, G. Kotliar and P.A. Lee: Phys. Rev. Lett. **59** (1987) 323.
35. M.A. Paalanen, T.F. Rosenbaum, G.A. Thomas and R.N. Bhatt: Phys. Rev. Lett. **48** (1982) 1284.
36. Y. Imry, Y. Gefen and D.J. Bergmann: p.138 in ref.4
37. A. Kawabata: J. Phys. Soc. Jpn. **53** (1984) 318; **53** (1984) 1429.
38. T. Ando, Y. Matsumoto and Y. Uemura: J. Phys. Soc. Jpn. **39** (1975) 279.
39. K. von Klitzing, G. Dorda and M. Pepper: Phys. Rev. Lett. **45** (1980) 494.
40. S. Kawaji and J. Wakabayashi: Physics in High Magnetic Fields ed. S. Chikazumi and N. Miura (Springer, Berlin, 1981), p.284.
41. R.B. Laughlin: Phys. Rev. **B23** (1981) 5632.
42. H. Aoki and T. Ando: Solid State Commun. **38** (1981) 1079.
43. Q. Niu, D.J. Thouless and Y.-S. Wu: Phys. Rev. **B31** (1985) 3372.
44. T. Ando: Statistical Physics and Condensed Matter Theory, ed. Xie Xide (World Scientific, Singapore, 1986), p.207; High Magnetic Fields in Semiconductor Physics, ed. G. Landwehr (Springer, Berlin, 1987), p.2.
45. H. Aoki and T. Ando: Phys. Rev. Lett. **57** (1986) 3093.
46. See, for examples, T. Ando: Prog. Theor. Phys. Suppl. **84**, 69 (1985), and references cited therein.
47. H. Levine, S.B. Libby and A.M.M. Pruisken, Phys. Rev. Lett. **51**, 1915 (1983); Nucl. Phys. B **240** [FS12] (1984) 30; 49; 71.
48. T. Ando: J. Phys. Soc. Jpn. **55** (1986) 3199.
49. D.C. Tsui, H.L. Stormer and A.C. Gossard: Phys. Rev. Lett. **48** (1982) 1559.
50. R.B. Laughlin: Phys. Rev. Lett. **50** (1983) 1395; Surf. Sci. **142** (1984) 163.
51. D. Yoshioka, B.I. Halperin and P.A. Lee: Phys. Rev. Lett. **50** (1983) 1219.
52. See, for example, S. Kawaji: Prog. Theor. Phys. Suppl. **84** (1985) 178.
53. R.G. Clark, R.J. Nicholas and A. Usher: Surf. Sci. **170** (1986) 141.
54. R. Willett, J.P. Eisenstein, H.L. Stormer, D.C. Tsui, A.C. Gossard and J.H. English: Phys. Rev. Lett. **59** (1987) 1776.
55. C.P. Umbach, S. Washburn, R.B. Laibowitz and R.A. Webb: Phys. Rev. **B30** (1984) 4048.
56. J.C. Licini, D.J. Bishop, M.A. Kastner and J. Melngailis: Phys. Rev. Lett. **55** (1985) 2987.
57. S.B. Kaplan and A. Hartstein: Phys. Rev. Lett. **56** (1986) 2403.
58. B.L. Altshuler: JETP Lett. **41** (1985) 648.
59. P.A. Lee and A.D. Stone: Phys. Rev. Lett. **55** (1985) 1622.
60. S. Feng, P.A. Lee and A.D. Stone: Phys. Rev. Lett. **56** (1986) 1960.
61. P.A. Lee, A.D. Stone and H. Fukuyama: Phys. Rev. **B35** (1987) 1039.

62. A.B. Fowler, A. Hartstein and R.A. Webb: Phys. Rev. Lett. **48** (1982) 196.
63. M.Ya Azbel, A. Hartstein and D.P. DiVicenzo: Phys. Rev. Lett. **52** (1984) 1641.
64. P.A. Lee: Phys. Rev. Lett. **52** (1984) 1641.
65. R.A. Webb, A. Hartstein, J.J. Wainer and A.B. Fowler: Phys. Rev. Lett. **54** (19 85) 1577.

Part I

The Metal-Insulator Transition in Doped Semiconductors

Critical Behavior Near the Metal-Insulator Transition in Doped Semiconductors

W. Sasaki

Department of Physics, Toho University, Miyama 2-2-1, Funabashi 274, Japan

Critical phenomena observed near the metal-insulator transition in 3-d disordered systems are reviewed with emphasis on Si:P. Physical quantities such as the conductivity are known to behave critically near the metal-insulator transition in various kinds of disordered materials. According to experiments, disordered systems seem to be divided into two classes depending on the value s of the conductivity exponent, 1 or < 1. The value of s is sensitive to compensation for multi-valley semiconductors, while it is usually 1 for other disordered systems, including amorphous alloys. Scaling theory of localization by Abrahams et al. leads to the exponent value 1 with a simple scattering mechanism, while the mechanism which leads to the exponent differing from 1 is as yet undetermined. The physical situations near the transition are considered with reference to observations of other quantities such as specific heat and magnetic properties.

1. INTRODUCTION

Recent developments in the field of electrical conduction in disordered materials owes a lot to the scaling theory of localization by Abrahams et al. [1]. Their argument was based on the scaling of the Thoules conductance, or the coupling to granularity ratio. The asymptotic behavior of the scaling function was considered and was shown to lead to important conclusions which enabled us to obtain an overview of electrical conduction in disordered systems. It was made clear that the dimensionality of the system plays an important role: no really metallic conduction is allowed in 1- and 2-dimensions, while in 3-d systems a critical conductance, an unstable fixed point which separates the metallic and insulator behaviors, is shown to exist. It was argued that the critical behavior of the electrical conductivity near the metal-insulator transition is correlated with the asymptotic form of the scaling function. The low temperature anomalies of electrical conduction observed in disordered metallic systems, doped semiconductors typically, were explained sucessfully in terms of a quantum correction to the Boltzmann conductance. Thus the theories for the transition and the precursive effects share the same framework of the asymptotic form of the scaling function. The development of disordered-system physics since 1979 has been summarized in review articles and is also contained in related publications[2,3].

In the next section the critical behavior of electrical conduction on the metallic side of the transition is reviewed. In Sect.3 that observed on the insulator side is reviewed. Because of the importance of the relation between the transition and the precursive state, thermal and magnetic properties, which do not behave critically near the transition, are reviewed in Sects. 4 and 5. Because of the variety of available experimental data, emphasis is given to the Si:P system.

2. CRITICAL BEHAVIOR OF METALLIC CONDUCTIVITY

A series of careful measurement of conductivity in uncompensated Si:P crystals has been carried out by Rosenbaum and coworkers[4]. The critical behavior was examined on the zero Kelvin conductivity, which was obtained by extrapolating the temperature dependence of electrical conductivity $\sigma(T)$ to zero Kelvin according to a theory by Altshuler and coworkers[5], which takes the effect of Coulomb interactions into account:

$$\sigma(T) = \sigma(0) + BT^{1/2}. \tag{1}$$

This relation was confirmed to be valid in practice over a wide temperature region above a few millikelvin. The change of $\sigma(0)$ follows a power law

$$\sigma(o) = \sigma_0(n/n_c - 1)^s \tag{2}$$

with n_c the critical concentration and s the critical exponent. The s value for uncompensated Si:P is estimated at 0.55 ± 0.05. A similar procedure[6] was carried out for compensated Si:P crystals, with compensation ratio ranging from 0.14 to 0.87, and for these cases s was estimated at 1. Figure 1 shows the change of $\sigma(0)$ as a function of (n/n_c). It should be noted that the power law (2) holds over a wide region covering n/n_c as large as 2.

The investigations along this line have been extended to other disordered systems. Except for uncompensated multi-valley semiconductors the majority of disordered systems, including amorphous alloys, are confirmed to have s = 1. Examples are as follows: $Al_xGa_{1-x}As$[7], Ge-Au[8], Nb:Si[9], $Gd_{3-x}V_xS_4$[10], and others[3]. Uncompensated multi-valley semiconductors are known to have s < 1; Si:As, s = 0.64[11], Si:P,As, s = 0.7[7], and Ge:Sb, s = 0.9[13]. Examining these results, it seems that the case s < 1 is rather marginal, being limited to uncompensated multi-valley semiconductors, and their s values are scattered over the range 0.5 to 0.9. Theoretical interpretations for these cases are not very convincing at present.

Another important point in these experiments is that the transition to the insulator occurs in a continuous way according to (2), in some cases crossing over the minimum metallic conductivity value of Mott σ_m. A jump of conductivity as suggested by Mott[14] has never been observed in 3-d systems. This fact supports the validity of the scaling approach by Abrahams et al.[1].

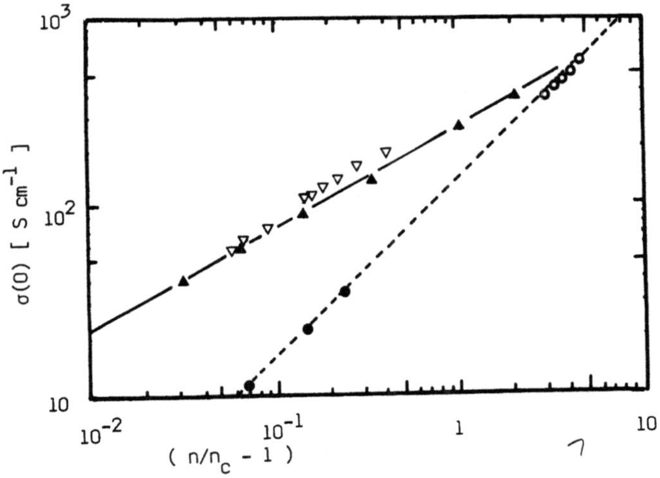

Fig.1 Zero temperature conductivity $\sigma(0)$ versus $(n/n_c - 1)$ for Si:P. ▽ ▲ for uncompensated and ● o for compensated samples. Straight lines indicate critical exponents s of 1/2 and 1

According to the scaling theory the value of s depends on the power n of the leading term of the $\varepsilon(=d-2)$ expansion of the scaling function

$$\beta = \varepsilon + a_n(1/g)^n + a_{n+1}(1/g)^{n+1} + \ldots.. \quad (3)$$

and

$$s = 1/n \, . \quad (4)$$

The form of β function is determined by the fundamental nature of the system: for the case of a disordered system a microscopic theory[15] predicts that s = 1 for the weak interaction limit, while no definitive theory which predicts s values smaller than 1 seems to be available.

3.CRITICAL BEHAVIOR ON INSULATOR SIDE

Physical quantities which characterize the critical behavior on the insulator side of the transition have been measured for Si:P systems: the dielectric susceptibility in uncompensated samples[16] and the localization length of the wave function in compensated samples[14]. These quantities diverge as the system approaches the transition from the insulator side, and the observed values of the critical exponent satisfy universal relations with those of metallic conductivity for uncompensated and compensated cases, respectively.

For uncompensated Si:P the dielectric susceptibility χ diverges with an exponent 1:

$$\chi = \chi_0(1 - n/n_c)^{s'}, \; s' = 1 \, . \quad (5)$$

This value of s' and that of s = 1/2 for the metallic conductivity satisfy a universal relation s' = 2s[30,17].

For compensated Si:P the localization length a of the wave function is estimated on the insulator side. a is obtained when the nonmetallic conduction is due to the variable range hopping, that is, the temperature dependence of the resistivity is described by the Mott formula

$$\rho = \rho_0 \exp(T_0/T)^{1/4}, \quad (6)$$

where the characteristic temperature T_0 is related to a by

$$T_0 = 18/(a^3 k_B N(E_F)). \quad (7)$$

Since the density of states $N(E_F)$ is known from the electronic specific heat, one can estimate a from the resistivity and specific heat data. Figure 2 illustrates the change of a as a function of n/n_c for compensated Si:P. a is shown to diverge with an exponent s" = 1 as the transition is approached from the insulator side. This value is also consistent with the conductivity exponent s = 1. Mention should be made of the temperature range where these data were obtained. A special temperature range 1.5 - 4 K was adopted where the Mott law (6) works satisfactorily and the electronic specific heat is well described by γT. In the lower temperature region the gradient of the Mott plot, lnρ versus $T^{-1/4}$, becomes steeper, which may be conjectured to indicate the opening of a gap with a width corresponding to the temperature where the plot shows a deviation from the Mott law. The specific heat in those crystals also shows an anomaly below around 1.5 K, which suggests the existence of low-lying excitations due to the gap or other origins.

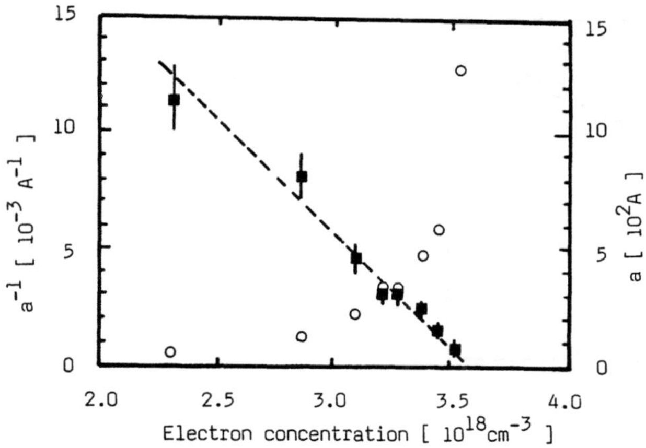

Fig.2 Decay length a of localized wave function (open circles) and its inverse (closed aquares) as function of excess electron concentration n. a is shown to diverge as $1/(1 - n/n_c)$

4. SPECIFIC HEAT

Low temperature specific heat is a useful tool to probe into the spectrum of collective excitations near the ground state. The information is straightforward when the excitations are described in terms of independent quasi-particles. The theory [5] which takes into account the effect of electron interactions in disordered systems is known to give the temperature dependence of conductivity correctly by (1). This theory predicts a dip in the one-particle excitation spectrum of electrons near the Fermi level. This is actually observed in an amorphous alloy system Nb:Si by tunnel spectroscopy[9]. If the thermal electronic excitation is given by a simple sum of one-particle excitations, the effect of such a dip must be reflected in the specific heat.

Specific heat measurements were done for both uncompensated and compensated Si:P[18,6]. Above 1.5 K the electronic specific heat is given by γT as is usual in normal metals, however, on lowering the temperature below 1.5 K, a rise is observed for specimens of lower electron concentrations, including the just metallic ones.

First, the data for γT are examined. The temperature dependence of the total specific heat in this region is described satisfactorily by

$$C = \gamma T + DT^3 . \tag{9}$$

This means that the specific heat consists of two terms corresponding to a continuum of electronic excitations and lattice vibrations. The coefficient γ of the electronic term is related to the density of states $N(E_F)$ by

$$\gamma = (\pi^2/3)k_B{}^2N(E_F). \tag{9}$$

In Fig. 3 the observed γ value is illustrated as a function of electron concentration, comparing it to the degenerate electron value calculated by (9) with the same mass parameters as the conduction band. It should be noted that no critical change is observed in the specific heat. For uncompensated samples the γ value follows the degenerate electron value in the high concentration region. At lower concentrations, γ shows a slight hump around the transition, but the change is continuous and a steep decrease is observed only after the concentration is

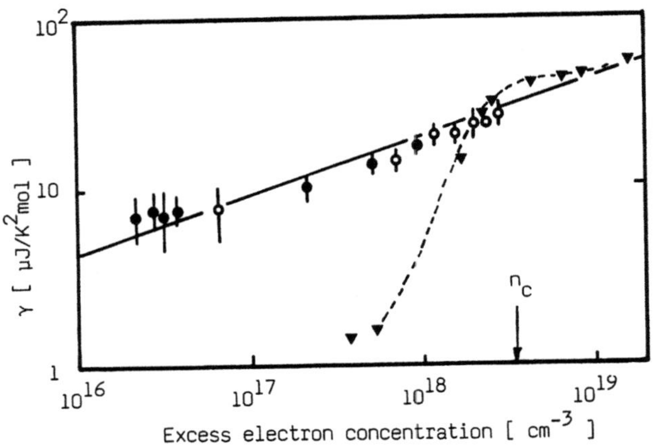

Fig.3 γ values of electronic specific heat as a function of excess electron concentration. ▼ for uncompensated and o , ● for compensated Si:P. The straight line is the value expected for degenerate electrons in the conduction band and the dashed curve is a guide to the eyes for uncompensated samples

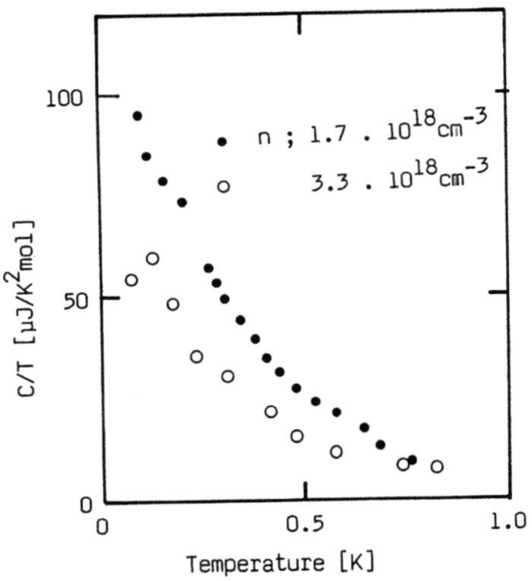

Fig.4 Anomalous specific heat of uncompensated Si:P observed in the lowest temperature region. ● nonmetallic, and o just-metallic samples

reduced below $0.5n_c$. Such a change of γ near the transition is consistent with a picture of Anderson localization in a sense that the mobility edge crosses the Fermi level at the transition, within a framework of a one-particle rigid band. Separation of the ground state from the band by a finite energy gap occurs when the concentration is reduced below 1/2 of the critical value. For compensated samples, the observed γ follows the degenerate value down to very low concentrations, suggesting that no separation of the impurity states occurs from the band. For the temperature and concentration regions where γT is valid, the electronic excitation is well approximated by non-interacting particles equivalent to bare electrons in the conduction band; no definite effect of disorder or interaction appears in the spectrum of electrons from the specific heat measurement with an energy window wider than 1.5 k_B.

Next, the rise of the specific heat below 1.5 K is considered. As mentioned above, such a phenomenon is observed in samples near the transition, and in both uncompensated and compensated samples. Figure 4 illustrates the case of uncompensated Si:P. It is considered that apart from the continuum there exist excitations with finite excitation energ smaller than 1.5 K. To correlate this phenomenon directly with the dip of the one-particle excitation spectrum, as predicted by the theory of interacting electrons[5], does not seem plausible, because what is observed [9] is the dip of tunneling current and the voltage width of the dip is as wide as 20 mV. It has already been pointed out that this low-temperature anomaly changes into a Schottky hump on applying a magnetic field, with conservation of the entropy of the specific heat anomaly. This fact suggests that the zero-field anomaly is due to the localized magnetic moments which are placed in an effective local field. The number density n_s of such excitations can be estimated from the excess entropy for uncompensated Si:P as follows[19]:

$n = 4.5 \times 10^{18} cm^{-3}$	$n_s = 1 \times 10^{17} cm^{-3}$
3.3 "	2 "
2.7 "	3 "
1.7 "	5 "
0.53 "	4 "

An important point is that such an anomalous rise of the specific heat below 1.5 K is observed even in metallic samples. These features are shared by both uncompensated and compensated Si:P.

5. MAGNETIC PROPERTIES

Static and ESR magnetic susceptibilities have been measured on Si:P. Earlier papers dealt with extremely metallic[20] and nonmetallic samples[21], and confirmed the Landau-Peierls-Pauli magnetism in the former and Curie-like paramagnetism of weakly interacting donors in the latter. Static susceptibility measurements covering the concentration region of the metal-insulator transition [22-24] have revealed that no critical behavior of susceptibility is observed by the transition. A Curie-Weiss paramagnetism, which is the leading term in the insulator region, becomes smaller with increasing electron concentration but stays still finite even in the metallic region. ESR susceptibility measurements [25] were useful to discriminate the electron spin contribution among the total static susceptibility. Measurements mentioned above were done over temperature regions above 1.2 K. Recently the temperature region was extended down to the dilution refrigerator region[26,27], and it was revealed that the magnitude of paramagnetism in uncompensated Si:P in the metallic region, with donor concentrations as high as twice the critical value, continues to increase with the decrease of temperature down to 20 mK. As was suggested in the preceding section, this Curie-type component must be related to the low temperature anomaly of the specific heat.

These observations are consistent with NMR[28]. In Si:P, ^{29}Si and ^{31}P are available for NMR. The Knight shift K gives a measure of the averaged local field seen by the nuclei,

$$K = (8\pi/3) \langle \chi_S |\psi(r_N)|^2 \rangle. \tag{10}$$

As was discussed in relation to the γT term of the specific heat, the majority of the electrons belong to a degenerate electron gas. If these electrons are free the magnetic susceptibility is given by the Pauli paramagnetism, which is proportional to the density of states $N(E_F)$, and the Knight shift and γ are expected to vary in parallel. This is actually the case for ^{29}Si NMR in metallic samples with donor concentrations higher than $2n_c$. However, when the concentration is reduced below $2n_c$ the shift shows a steep decrease with decreasing n, which means that the probability amplitude of the electron wave function at ^{29}Si nuclei becomes smaller. It is interesting to observe that contrary to the decrease of the shift of ^{29}Si the shift of ^{31}P NMR line becomes larger with decreasing n down to just above n_c. So, at these concentrations, the probability amplitude of the electrons begins to decrease at Si sites and increase at P sites. This situation can be regarded as a gradual localization of electrons with decreasing n down to $\sim n_c$. A recent study[29] shows in a quantitative way that a considerable number of ^{31}P nuclei are wiped out of the observable resonance line of ^{31}P because of the increasing contact field due to the progress of concentration of electrons at P sites. The number of the wiped out electrons is shown to increase with decreasing temperature from 4.2 to 1.65 K and to behave critically, $(n/n_c - 1)^s$, with an exponent of approximately 0.5. Combined with Curie-type electron paramagnetism observed in the same concentration region, it is plausible to consider that a localized character of magnetic moments is induced among the metallic electrons in these crystals. Nearly the same value of exponents for the conduction and wiped out number suggests that the change in the degree of concentration of electrons at P sites as observed through NMR is a direct manifestation of microscopic states of electrons which is described in terms of the scaling equation.

In summary, critical behaviors observed near the metal insulator transition in disordered systems were reviewed with emphasis upon Si:P. Critical indices of various physical quantities were examined. Thermal and magnetic properties were considered to suggest a microscopic picture included in the scaling equation in the just-metallic state.

References

1. E. Abrahams, P. W. Anderson, D. C. Licciardello, and T. V. Ramakrishnan; Phys. Rev. Lett. 42, 673 (1979)
2. Anderson Localization, ed. by Y. Nagaoka and H. Fukuyama (Springer, Berlin, Heidelberg 1982)
   Anderson Localization, Prog. Theor. Phys. Suppl. 84 (1985)
   P. A. Lee and T. V. Ramakrishnan: Rev. Mod. Phys. 52, 287 (1985)
   G. Bergmann: Phys. Rep. 107, 1 (1987)
3. Heavy Doping and the Metal-Insulator Transition, ed. by P. T. Landsberg: Solid State Electronics 28, No. 1/2 (1985)
   Nevil Mott Festschrift, ed. by E. A. Davis, Phil. Mag. 52, No.3(1985)
   Disordered Semiconductors, ed. by M. A, Kastner, G. A. Thomas, and S. R. Ovshinsky (Plenum, New York, London 1987)
4. T. F. Rosenbaum, K. Andres, G. A. Thomas, and R. N. Bhatt: Phys. Rev. Lett. 45, 1723 (1980)
   T. F. Rosenbaum, R.F. Milligan, M. A. Paalanen, G. A. Thomas, R. N. Bhatt, and W. Lin: Phys. Rev.. B27, 7509 (1982)
5. B. L. Altshuler and A. G. Aronov: Solid State Commun. 36, 115 (1979); 38, 11 (1981)
   B. L. Altshuler, A. G. Aronov, and P. A. Lee: Phys. Rev. Lett. 44, 1288 (1980)
   B. L. Altshuler, D. Kmelnitzkii, A. L. Larkin, and P. A. Lee: Phys. Rev. B22, 5142 (1980)
6. Y. Nishio, K. Kajita, T. Iwata and W. Sasaki: In Proc. 18th Intern. Conf. Phys. Semiconductors, ed. by O. Engstrom, 2, 1257 (World Scientific 1987)
   W. Sasaki, Y. Nishio and K. Kajita: In Disordered Semiconductors, ed. by M. Kastner, G. A. Thomas, and S. Ovshinsky, 37 (Plenum, New York, London 1987)
   M. J. Hirsch, U. Thomanschefsky, and D, F. Holocomb: Private communication
7. S. Katsumoto, F. Komori, N. Sano and S. Kobayashi: J. Phys. Soc. Jpn. 56, 2259 (1987)
8. B. W. Dodson, W. L. McMillan, J. M. Mochel, and R. C. Dynes: Phys. Rev. Lett. 46, 46 (1981)

9. G. Hertel, B. J. Bishop, E. G. Spencer, J. M. Rowell, and R. C. Dynes: Phys. Rev. Lett. 50, 743 (1983)
10. S. von Molnar, A. Briggs, J. Floquet, and G. Remenyi: Phys. Rev. Lett. 51, 706 (1983)
11. R. F. Newman and D. F. Holcomb: Phys. Rev.B28, 638 (1983)
12. R. F. Newman and D. F. Holcomb: Phys. Rev. Lett. 51, 2144 (1983)
13. Y. Ootuka, H. Matsuoka, and S. Kobayashi: Present Issue
14. N. F. Mott: Phil. Mag. 26, 1015 (1972)
15. C. Castellani, C. Di Castro, P. A. Lee, and M. Ma: Phys. Rev. B30, 527 (1982)
16. T. G. Castner, N. K. Lee, G. S. Cieloszyk, and G. Salinger: Phys. Rev. Lett. 34, 1627 (1975)
H. S. Tan and T. G. Castner: Phys. Rev. B23, 3983 (1982)
H. F. Hess, K. DeConde, T. F. Rosenbaum, and G. A. Thomas: Phys. Rev. B25, 5585 (1982)
17. A. Kawabata: J. Phys. Soc. Jpn. 53, 1429 (1984); Prog. Theor. Phys. Suppl. 84, 16 (1985)
18. N.Kobayashi, S. Ikehata, S. Kobayashi and W. Sasaki: Solid State Commun. 24, 67 (1977); Solid State Commun. 32, 1147 (1979)
19. W. Sasaki: In Proc. Intern. Conf. Phys. Semiconductors, J. Phys. Soc. Jpn. 49, Suppl., A31 (1980)
20. E. Sonder and D. K. Stevens: Phys. Rev. 110, 1027 (1958)
21. E. Sonder and H. Schweinler: Phys. Rev. 117, 1216 (1960)
22. W. Sasaki and J. Kinoshita: J. Phys. Soc. Jpn. 25, 1622 (1968)
23. S. Ikehata, T. Ema, S. Kobayashi and W. Sasaki: J. Phys. Soc. Jpn. 50, 3655 (1981)
24. A. Roy, M. Turner and M. P. Sarachik: Private communication, submitted to Phys. Rev. B
25. H. Ue and S. Maekawa: Phys. Rev. B3, 4232 (1971)
J. D. Quirt and J. R. Marco:Phys. Rev. B5, 1716 (1972); Phys. Rev. B7, 3842 (1973)
26. S. Ikehata; Present Issue
27. Y. Ootuka and N. Matsunaga: Present Issue
28. W. Sasaki, S. Ikehata and S. Kobayashi: J. Phys. Soc. Jpn. 36, 1377 (1974)
S. Ikehata, W. Sasaki and S. Kobayashi: J. Phys. Soc. Jpn. 39, 1492 (1975)
S. Kobayashi, Y. Fukagawa, S. Ikehata and W. Sasaki: J. Phys. Soc. Jpn. 45, 1276 (1978)
29. H. Alloul and P. Dellouve: Phys. Rev. Lett. 59, 578 (1987)
30. Y. Imry, Y. Gefen and J. Bergman: In Anderson Localization, ed. by Y. Nagaoka, H. Fukuyama (Springer, 1982) p. 138

Anomalous Spin Susceptibility of Si:P Near the Metal-Insulator Transition

S. Ikehata

Department of Physics, Faculty of Science, University of Tokyo, Hongo 7-3-1, Bunkyo-ku, Tokyo 113, Japan

1 Introduction

Since the scaling theory of ABRAHAMS et al./1/, great progress has been made in understanding the electronic properties of disordered systems. Especially, in the weak localization regime the recent theoretical progress in the Anderson localization has explained many problems unsolved for many years and predicted new phenomena/2/. Recently, successful attempts have been made to describe the metal-insulator transition itself for non-interacting disordered electrons by means of the renormalization group treatment/3/. But in the presence of the electron-electron interaction, there are still unsolved questions concerning the localization/4/. Particularly it is an open question whether the metal-insulator transition will be associated with enhancement of the spin susceptibility or not. Thus, at present it is very important and interesting to measure the spin susceptibility through the metal-insulator transition. For this purpose we have investigated the spin susceptibility of heavily phosphorus-doped silicon through the metal-insulator transition. We have performed many experiments on this system using the same samples, such as measurements of electrical conductivity, static magnetic susceptibility, NMR, ESR, specific heat and so on/5-11/. The critical concentration N_c for the metal-insulator transition is now well established to be $N_c = 3.2 \times 10^{18}/cm^3$.

2 Samples

The phosphorus concentration of the samples studied here ranges from $3.4 \times 10^{17}/cm^3$ to $2.4 \times 10^{19}/cm^3$. The ESR samples were in the form of grains with diameter smaller than the skin depth to minimize eddy current losses. The phosphorus concentration N_D was determined from the room temperature Hall coefficient R_H as

$$N_D = -\frac{1}{R_H e}, \qquad (1)$$

where e is the electron charge.

3 Experimental

The ESR and ^{29}Si NMR were observed from 4.2K to 8mK using the conventional Q-meter continuous wave spectrometer at 7.7MHz. The absolute spin susceptibility was determined by measuring the absorption intensity of ESR I_e relative to that of ^{29}Si NMR I_n at the same radio frequency/12/. The NMR absorption intensity I_n is represented by a Curie term for N nuclear spins and the nuclear spin susceptibility χ_n is expressed as

$$\chi_n = \frac{N\gamma_n^2\hbar^2 I(I+1)}{3k_B T}, \qquad (2)$$

where γ_n is the ^{29}Si nuclear gyromagnetic ratio and $I = \frac{1}{2}$. Experimentally we can determine the absorption intensity ratio I_e/I_n and using this ratio the absolute spin susceptibility χ_s is given as

$$\chi_s = \chi_n \frac{\gamma_e}{\gamma_n} \frac{I_e}{I_n}, \qquad (3)$$

where γ_e is the electron gyromagnetic ratio.

The resonance magnetic field of ESR is about 2.75 Oe at 7.7MHz. Then by this method we can get χ_s at very low magnetic field.

4 Results and Discussion

Figure 1 shows the typical ESR lines (absorption derivatives).

In samples with $N_D \leq 3.5 \times 10^{18}/cm^3$, the ESR linewidth and the resonance magnetic field increase with decreasing temperature. This behavior is quite similar to that observed for the sample with $N_D = 2.05 \times 10^{18}/cm^3$ by MURAYAMA et al./13/. On the other hand, in samples with $N_D \geq 4.5 \times 10^{18}/cm^3$ the ESR linewidth and the resonance magnetic field are almost independent of temperature over the temperature range from 4.2K to 8mK. This behavior is shown in Figs. 2 and 3. The absolute spin susceptibility is obtained by the method

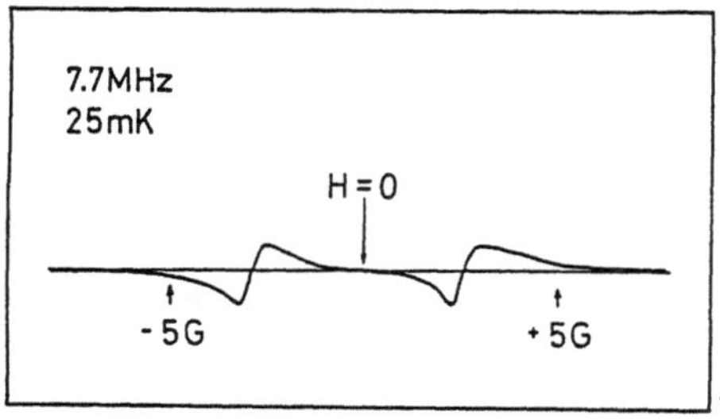

1

Fig. 1 Electron spin resonance signal of Si:P with $N_D = 4.5 \times 10^{18}/cm^3$. The magnetic field is swept from negative to positive through zero in order to get the absorption intensity.

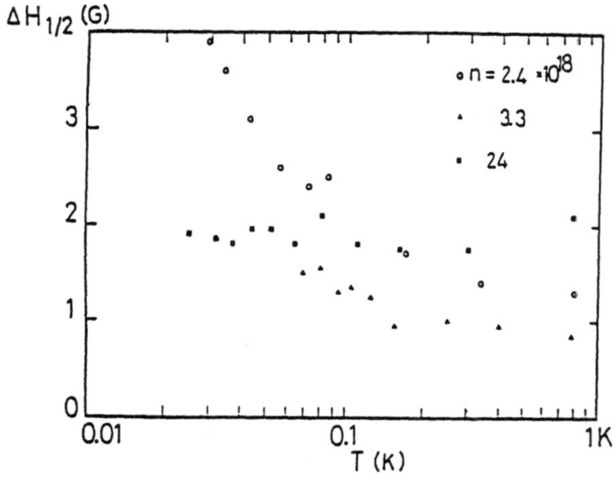

Fig. 2 Temperature dependence of ESR linewidth for three samples.

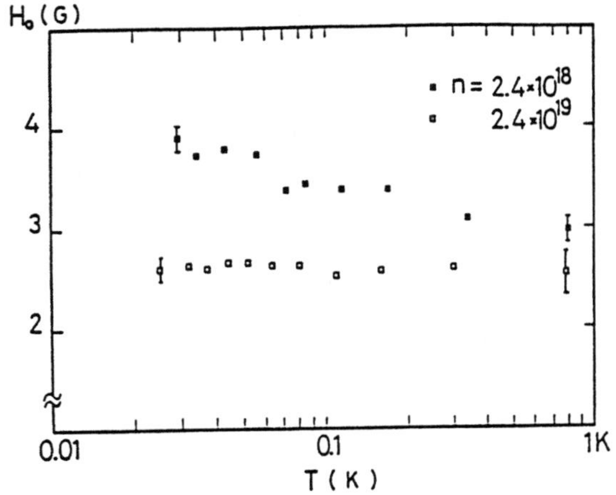

Fig. 3 Temperature dependence of the resonance magnetic field.

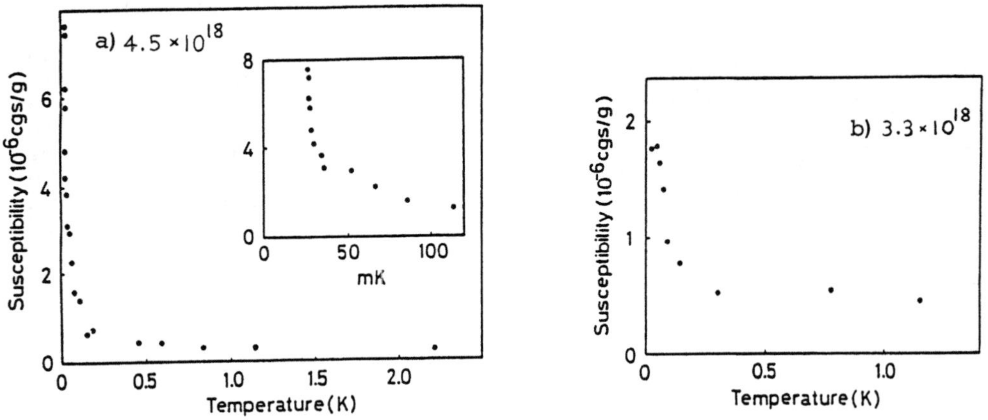

Figs. 4 a) and 4 b) Temperature dependence of the spin susceptibility of Si:P with a) N_D = 4.5 × 10^{18} /cm^3 and b) 3.3 × 10^{18}/cm^3. The low temperature range is shown on an expanded scale in the inset of Fig. 4 a). In the ordinate, there is an uncertainty ±40%.

mentioned here and its temperature dependence is shown in Fig. 4. The surprising feature is that there is a sharp increase of the spin susceptibility at very low temperatures in all samples studied here. The steep rise observed is quite different from that expected from the Curie-Weiss-type paramagnetic susceptibility observed in the earlier studies at higher temperatures/14/.

As regards the absolute value, we can compare our results with those obtained by the X band ESR measurements at about 1K. Even though there still exist large experimental errors at present, the spin susceptibility obtained here is substantially larger than those of X band ESR measurements/14/. This difference may be partly due to the difference in the strength of the magnetic fields between these two measurements. This fact suggests that there is some mechanism to enhance the spin susceptibility under low magnetic fields.

Recently PAALANEN et al./15/ measured the electron spin resonance of metallic Si:P near the metal-insulator transition down to 30mK using the conventional Q-meter continuous wave spectrometer operated at 18.4MHz. They also found the sharp increase of the spin susceptibility associated with the increase of the ESR linewidth for samples with $N_D = 1.09N_c$ and $1.25N_c$. Although a quantitative difference exists in the absolute value of the spin susceptibility, their results are similar to those for samples with $N_D \leq 3.5 \times 10^{18}/cm^3$ in our measurements. Their measurements are limited to the samples which have temperature dependence of the ESR linewidth at low temperatures. The interesting point of our measurement is that there still exists a sharp increase of the spin susceptibility at low temperatures in the real metallic samples which have no temperature dependence in the ESR linewidth.

Very recently OOTUKA and MATSUNAGA./16/ measured the static magnetic susceptibility of the same Si:P samples using a SQUID magnetometer under the magnetic field of 50 Oe. They also found the anomalous paramagnetic increase in the total susceptibility even in the most heavily doped sample with $N_D = 2.8 \times 10^{19}/cm^3$. Although it is not easy to deduce the spin susceptibility from the total susceptibility, the increment of the paramagnetic component at lowest temperatures seems smaller in the static measurement than in the ESR measurements. This may suggest the possibility of temperature dependence of the orbital diamagnetic susceptibility of disordered electrons with interaction/17/. At present , there seems to be no theory to explain the anomalous temperature dependence of the spin susceptibility in both the insulating and metallic phases. But the steep increase of the spin susceptibility observed here will be ascribed to the random nature inherent to this system.

Previously we have studied the NMR properties of Si:P at higher temperatures from 4.2K to 0.4K and have shown that in metallic Si:P the electron-electron interaction enhances the spin-lattice relaxation rate T_1^{-1} and the Knight shift (or the spin susceptibility)/8, 18/. This was ascribed to the effect of the spin fluctuations which will result in the anomalous temperature dependence of T_1^{-1} and the Knight shift (or the spin susceptibility).

Recently, FUKUYAMA/19/ has taken into account the effect of the spin fluctuations in the so-called weakly localized regime and has shown that the localization enhances the electron-electron interaction and the spin fluctuations.

According to FUKUYAMA, the spin susceptibility χ_s is expressed as

$$\left(\frac{\chi_s}{\chi_0}\right)^{-1} = 1 - u + \gamma_0(T) - 4u^2\left(1 - \frac{u}{2}\right)\lambda^2 \frac{3\sqrt{3}}{4}\pi^2\left(2 - 1.46\sqrt{\frac{2\pi\tau k_B T}{\hbar}}\right) , \quad (4)$$

where χ_0 is the Pauli susceptibility without interaction and u is the strength of the interaction. The parameter λ is a measure of the randomness and is defined by

$$\lambda = \frac{\hbar}{2\pi\varepsilon_F\tau} , \quad (5)$$

where ε_F is the Fermi energy and τ is the lifetime calculated from the resistivity. $\gamma_0(T)$ essentially represents the renormalization effect due to quantum fluctuations and is almost a positive constant at the temperature range studied here. At higher temperatures,

$$\frac{2\pi\tau k_B T}{\hbar} > 1 , \quad (6)$$

this equation leads to the so-called Stoner-type enhancement of the Pauli susceptibility due to the electron-electron interaction in the itinerant electron system.

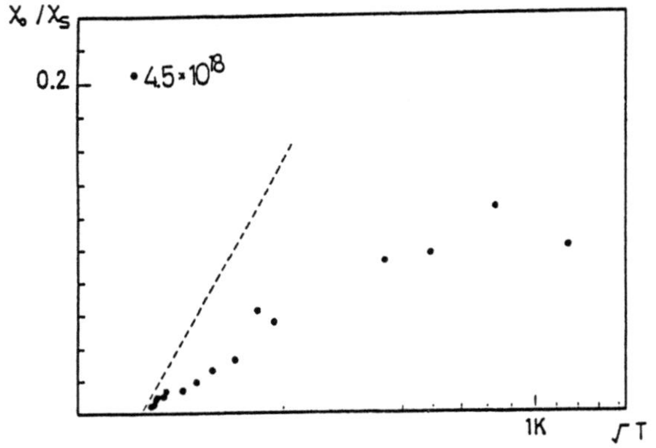

Fig. 5 Inverse of the spin susceptibility vs. square root of temperature for the sample with $N_D = 4.5 \times 10^{18}/cm^3$.

In Fig. 5, the inverse of the spin susceptibility is shown for the metallic sample against the square root of temperature $\sqrt{T}$. The linear dependence on $\sqrt{T}$ is clearly seen at low temperatures, as is predicted by (4). This fact indicates that the anomalous temperature dependence of the spin susceptibility observed here may be partly ascribed to the spin fluctuations in the disordered system. The dashed line in Fig. 5 is the theoretical prediction of (4). Considering experimental errors in χ_s, the agreement between theory and experiment is fairly good. For further comparison more precise determination of χ_s is needed.

5 Conclusions

We have reported the low field ESR data of phosphorus-doped silicon through the metal-insulator transition. An anomalous increase of the spin susceptibility was found at low temperatures on both sides of the transition. In the metallic phase near the transition, the temperature dependence of the spin susceptibility can be explained qualitatively by FUKUYAMA's theory, taking into account the spin fluctuations in the disordered electrons with interaction.

References

1. E. Abrahams, P.W. Anderson, D.C. Licciardello and T.V. Ramakrishnan: Phys. Rev. Lett. 42, 673 (1979)
2. For a review, see Anderson Localization, ed. by Y. Nagaoka and H. Fukuyama (Springer, New York, 1982)
3. A. Kawabata: Prog. Theor. Phys. Supplement No.84, 16 (1985)
4. A.M. Finkel'stein: Sov. Phys. JETP 57, 97 (1983)
5. C. Yamanouchi, K. Mizuguchi and W. Sasaki: J. Phys. Soc. Jpn. 22, 859 (1967)
6. S. Ikehata, T. Ema, S. Kobayashi and S. Sasaki: J. Phys. Soc. Jpn. 50, 3655 (1981)
7. W. Sasaki, S. Ikehata and S. Kobayashi: J. Phys. Soc. Jpn. 36, 1377 (1974)
8. S. Ikehata, W. Sasaki and S. Kobayashi: J. Phys. Soc. Jpn. 39, 1492 (1975)
9. S. Ikehata and S. Kobayashi: Solid State Commun. 56, 607 (1985)
10. N. Kobayashi, S. Ikehata, S. Kobayashi and W. Sasaki: Solid State Commun. 24, 67 (1977)

11. N. Kobayashi, S. Ikehata, S. Kobayashi and W. Sasaki: Solid State Commun. 32, 1147 (1979)
12. R.T. Schumacher and C.P. Slichter: Phys. Rev. 101, 58 (1956)
13. C.T. Murayama, W.G. Clark and J. Sanny: Phys. Rev. B29, 6063 (1984)
14. J.D. Quirt and J.R. Marko: Phys. Rev. B5, 1716 (1972)
15. M.A. Paalanen, S. Sachdev, R.N. Bhatt and A.E. Ruckenstein: Phys. Rev. Lett. 57, 2061 (1986)
16. Y. Ootuka and N. Matsunaga: in this volume.
17. B.L. Al'tshuler, A.G. Aronov and A. Yu. Zyuzin: Sov. Phys. JETP 57, 1525 (1983)
18. S. Ikehata, W. Sasaki and S. Kobayashi: Solid State Commun. 19, 655 (1976)
19. H. Fukuyama: J. Phys. Soc. Jpn 54, 2092 (1985)

Static Magnetic Susceptibility of Si:P Around the Metal-Insulator Transition

Y. Ootuka and N. Matsunaga

Cryogenic Center, University of Tokyo, Yayoi, Bunkyo-ku, Tokyo 113, Japan

1. Introduction

Recent theory of metal-insulator transition suggests that some magnetic instability takes place in disordered systems prior to the metal-insulator transition when the randomness is increased [1,2]. The ESR experiments of lightly metallic Si:P are reported to show anomalous behavior at very low temperatures, which might be ascribed to such an instability [3,4].

In spite of that the static magnetic susceptibility is one of the fundamental thermodynamic properties, that of Si:P in the metallic phase has been measured only above 1K. Recently, we have measured that of S:P in both insulating and metallic phase below 1K, and report the results in this paper.

2. Experiment

Six samples of Si:P with P concentration from 5.3×10^{17}/cc to 2.8×10^{19}/cc were studied. Some of them, 0.53, 1.62, 3.3, 4.5 $\times10^{18}cm^{-3}$, are identical with the samples in Ref.5. The donor concentration was determined from the Hall coefficient R_H at room temperature using the relation: $R_H=1/(eN)$. The metal-insulator transition takes place at about $3.1\times10^{18}cm^{-3}$ in this concentration scale. The samples were in the shape of cylinder of typical dimensions 5mm diameter x 6mm height with a hole for suspension at the center.

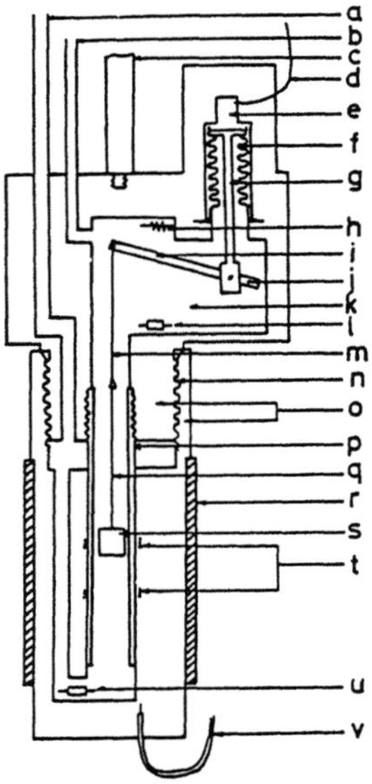

Fig.1: Magnetometer: (a) dilute solution outlet tube, (b) ^{3}He inlet tube, (c) support rod, (d) ^{4}He filling capillary, (e) ^{4}He chamber, (f) bellows, (g) connecting rod, (h) heater, (i) lever, (j) fulcrum, (k) mixing chamber of dilution refrigerator, (l) resistance thermometer, (m) copper wire, (n) soap-glycerin seal, (o) stycast 1266, (p) teflon tube, (q) nylon thread, (r) lead tube, (s) sample, (t) pick-up coils, (u) resistance thermometer, (v) solder capillary containing flux transfer lead.

The magnetism we are interested in is originated from the dilute impurity P atoms and the typical magnitude of the susceptibility is as small as 10^{-7}cgs. In order to detect such a small susceptibility below 1K free from the drift due to the magnetism of surrounding materials, we made a new magnetometer operating in a dilution refrigerator. The arrangement is shown in Fig.1. A sample is suspended inside a mixing chamber of the dilution refrigerator by a thin nylon thread. The end of the thread is fastened to a lever, to which a small bellows is connected. The deformation of the bellows caused by pressurizing the ^{4}He chamber is amplified by the lever and moves the sample. The detection system consists of a pair of astatically wound Nb pick-up coils, which are connected to a SQUID sensor, and a superconducting Pb cylinder, into which a field of 50G was frozen. The SQUID system is calibrated by using a pure Si whose susceptibility is known. The temperature is measured using a Matsushita carbon resistor which is calibrated against Ge thermometer and CMN magnetic susceptibility.

As the sample is immersed in liquid He in our arrangement, the output signal is proportional to the difference of the magnetic susceptibility of the sample and that of surrounding ^{3}He-^{4}He mixture. In the following analysis,we subtract from the data the calculated value of susceptibility of the mixture.

3. Results and Discussions

The absolute susceptibility of all samples below 4.2K is shown in Fig.2. The solid curves in the figure show the previous experimental results of the same sample above 1K and the susceptibility of pure Si.

All the samples are diamagnetic above 1K. In the samples with low concentration, the susceptibility changes to positive at lower temperature. Such a paramagnetic deviation is observed even in a most heavily doped sample although the magnitude is small.

The donor susceptibility χ_D ($=\chi - \chi_{pure\ Si}$) data of four samples are shown in Fig.3, where χ_D normalized to the calculated Curie susceptibility for free spins at nominal concentration at 1K is plotted against temperature in a doubly logarithmic plot. The reduction of χ_D from its free spin value is seen for all concentrations and it increases towards lower temperatures and higher concentrations.

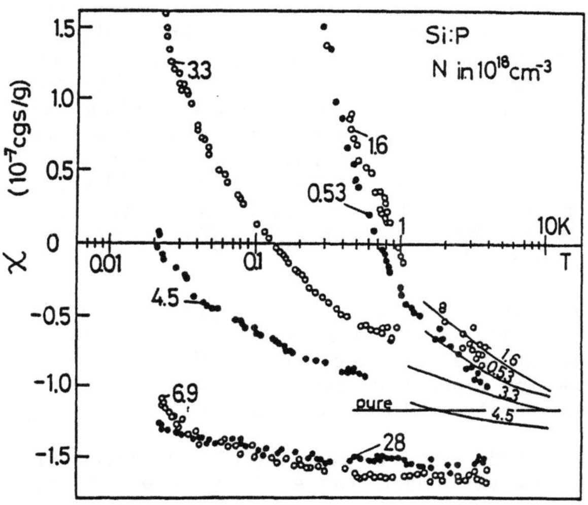

Fig.2: Absolute susceptibilities of Si:P samples whith various P concentrations. The solid lines indicate the data in Ref.5 for the same samples.

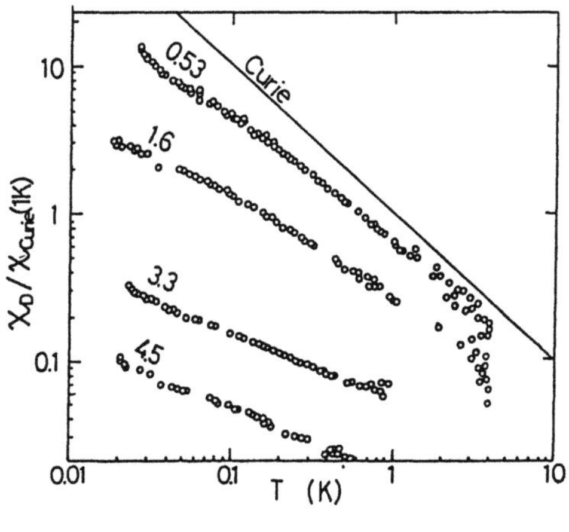

Fig.3: Donor susceptibility χ_D in Si:P normalized to the calculated Curie susceptibility for free spins at nominal concentrations at 1K. The solid line indicates the Curie susceptibility. The samples with N=3.3 and 4.5 $\times 10^{18}$/cc are in the metallic region of concentration.

The magnetic susceptibility of the insulating Si:P had been measured by ANDRES et al. [6] down to 3mK. They found the temperature dependence down to 30mK is experimentally fairly well described by a T^{-n} law, with the exponent n smaller than 1. Such a law can be seen also in this experiment. The temperature dependence of the sample 0.53×10^{18}/cc is described with n=0.88 and that of 1.6×10^{18}/cc with n=0.7. These values are somewhat larger than that obtained in Ref.6.

In Fig.3 , χ_D for lightly metallic samples are also shown. The orbital diamagnetism of conduction electrons really exists in these samples as seen at high temperatures in Fig.2, and such a plot may be less significant. Nevertheless, χ_D of these samples is described by the T^{-n} law very well (n=0.45 for N=3.3×10^{18}/cc and n=0.46 for N=4.5×10^{18}/cc) and the smooth cross-over from the insulating region seems to hold.

The data in Ref.6 level off from the T^{-n} law and flatten out in an unexpected fashion. The temperature at which this levelling off occurs is relatively independent of donor concentration. Unfortunately, the temperature range of our experiment is restricted above 20mK and the definite conclusion whether this occurs in our sample is not obtained.

The temperature dependence of susceptibility is manifest in the metallic samples including the most heavily doped one, although the change is smaller for samples with higher concentration. Recently several experiments reported the temperature dependent spin susceptibility in lightly metallic Si:P below 1K. The spin susceptibility was derived by integrating the ESR absorption line. In order to obtain the spin contribution to susceptibility, we have to subtract from the data the contributions of host Si susceptibility and of orbital diamagnetism due to conduction electrons. We obtain χ_s in the following way: We assume that the temperature variation arises only from χ_s and determine the temperature-independent base line using the data of χ_s at 1.1K by QUIRT and MARKO [7] which are derived from x-band ESR. The results are shown in Fig.4. Here, the spin susceptibility χ_s normalized to the Pauli value, $\chi_0=3n\mu_B^2/2k_BT_F$, is plotted against the temperature in a double logarithmic plot.

The experimental results of Ref.4 are also shown in the figure. The donor concentration in Ref.4 is given in the form of N/N_c and we estimate N assuming $N_c=3.1\times10^{18}cm^{-3}$. Although both experiments give enhanced susceptibility of the same order of magnitude,theirtemperature dependence seems to differ: the spin susceptibility in Ref.4 increases more steeply at lower temperatures.

The spin susceptibility derived from ESR has been reported also in Ref.3. The magnitude of χ_s in the literature is very much larger than others [4,7].

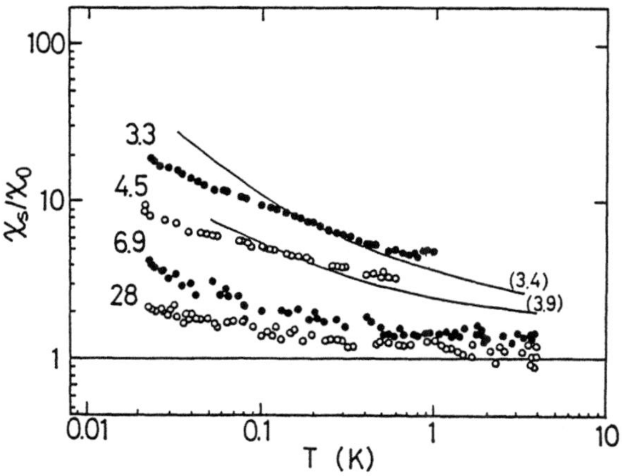

Fig.4: Spin susceptibility χ_s in S:P normalized by the Pauli susceptibility χ_0. The solid curves are data in Ref.4.

Concerning the temperature dependence, they found that the temperature dependence is described in the form of $\chi_0/\chi_s = A - BT^{1/2}$ fairly well [8] and the extrapolation brings the divergence of χ_s at finite temperature for lightly metallic samples. We find that the similar relation holds rather well even in the present data. The increase of χ_s is, however, weaker and the divergence, if it exists, seems to occur at much lower temperature.

In both cases, the temperature dependence of χ_s in this experiment is weaker than that in the ESR experiments. One of the possible reasons for this disagreement is in the evaluation of the diamagnetic susceptibility, which is necessary for the derivation of χ_s. In the above derivation, we assumed that the diamagnetism does not depend on temperature below 1K. The change of the diamagnetic susceptibility due to the Fermi degeneracy is surely negligible, because the Fermi temperature is as large as 100K for these samples. However, we now know that the electron-electron interaction brings the critical temperature dependence to the thermodynamic quantities such as orbital magnetic susceptibility at low temperature in disordered systems[9]. Although we have not evaluated the effects yet, this can be a reason for disagreements.

The measurement was made in the magnetic field of 50G in this experiment, while the ESR experiments were done in several G. The difference of the strength of the magnetic field might be another possible reason of the disagreement at lowest temperatures, although the data seem to differ in the whole range of temperature. The full discussions will appear in a following paper.

References

1. A. M. Finkelshtein: JETP Lett. **40** 796 (1984).
2. C. Castellani, C. Dicastro, P. A. Lee, M. Ma, S. Sorella and E. Tabet: Phys. Rev. B **33** 6169 (1986).
3. S. Ikehata and S. Kobayashi: Solid State Comm. **56** 607 (1985).
4. M. A. Paalanen, S. Sachdev, R. N. Bhatt and A. E. Ruckenstein: Phys. Rev. Lett. **57** 2061 (1986).
5. S. Ikehata, T. Ema, S. Kobayashi and W. Sasaki: J. Phys. Soc. Jpn. **50** 3655 (1981).
6. K. Andres, R. N. Bhatt, P. Goalwin, T. N. Rice and R. E. Walstedt: Phys. Rev. B **24** 244 (1981).
7. J. D. Quirt and J. R. Marko: Phys. Rev. B **5** 1716 (1972), Phys. Rev. B **7** 3842 (1973).
8. H. Fukuyama: J. Phys. Soc. Jpn. **54** 2092 (1985).
9. B. L. Altshuler, A. G. Aronov and A. Yu. Zyuzin: Sov. Phys. JETP **57** 889 (1983).

NMR in Doped Semiconductors

D.P. Tunstall

University of St. Andrews, St. Andrews, Fife KY16 9SS, Scotland

Doped semiconductors are an important experimental test-bed of ideas concerning the localization and interaction of electrons near the metal non-metal (mnm) transition, and conducting electron systems of one, two and three dimensions can be constructed and used to test theory. Many different aspects of the subject are covered, both theoretically and experimentally, in review articles published in the Proceedings of the 31st SUSSP [1].

This paper is concerned with reviewing the situation as to the magnetic properties of three-dimensional doped systems close to the metal non-metal transition. The motivation for doing nmr experiments in this area is based primarily on two series of measurements, and one theoretical problem.

1. MAGNETIC SUSCEPTIBILITY

A considerable quantity of good, ultra-low temperature, data has been published recently on the electronic magnetic susceptibility close to the mnm transition [2], [3], [4], [5].

Figure 1 summarizes the experimental situation here, with the QUIRT and MARKO [4] 1.1 K data showing a smooth transition from high density, Pauli-like, behaviour through to low density, Curie-like dependence. The more recent data at 30 mK is also shown since the deviations from Curie-like behaviour below 10^{18} cm^{-3} are well-understood in terms of a pair-wise, coupled anti-ferromagnetically, set of localized electronic spins [5]. The impact of Fig. 1 is that one needs to understand the magnetic behaviour of the donor electrons in a relatively narrow range of densities around the transition.

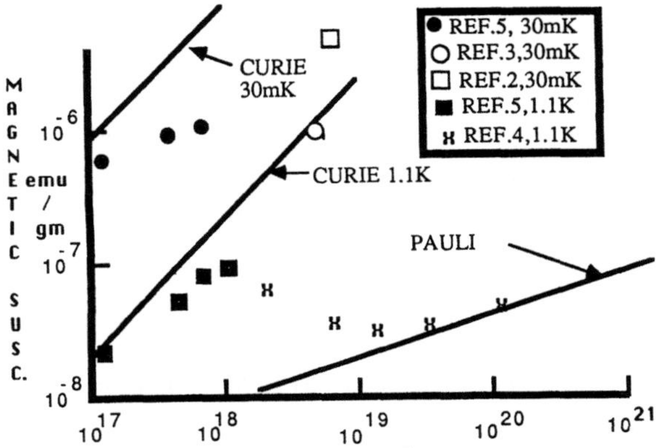

Fig. 1. The magnetic susceptibility against donor density, for two temperatures. The data from reference 2 has been scaled on the density axis by a factor of 1.3, to convert the Hall density to an Irvin curve density. Also shown on the graph are the theory curves for localized spins (Curie) and for delocalized spins (Pauli). The mnm transition in this Si:P system occurs at a density of $3.75*10^{18}$

Nmr measurements in this region hold out the promise of providing a more detailed breakdown of the electronic magnetic properties, since localized electronic moments perturb the nuclear resonance in a way quite different from the effect of itinerant, conduction electrons. Indeed, this promise is realised; nmr spectra in this mnm region do consist of a broad spectral response [6].

2. CONDUCTIVITY

An anomalous value of the critical exponent of the conductivity has emerged from the dilution refrigerator measurements of THOMAS et al [7] on metallic Si:P, and particularly from the elegant extension of these measurements into a region of density n an order of magnitude closer to n_c, the critical mnm density, by means of stress-tuning [8]. The focus of nmr measurements here is on providing information about any local electronic magnetic moments and on the existence of any electronic spin flip scattering mechanism, which theory indicates [9] could provide an explanation of the anomalous critical exponent.

3. THE IMPURITY BAND

In discussion of the mnm transition, another question for which nmr experiments provide at least a partial answer is that concerned with the character of the electronic wave-functions at the Fermi surface; do they look like the host conduction band wave-functions, or are they more properly described, and usually by tight-binding theory, as overlapping wave-functions centred on the impurities?

The target of this discussion is now to review a very restricted range of nmr experiments in the context of the above introduction, to see where and how they help in the interpretation of the mechanisms of the mnm transition.

In Si:P nmr, debate has centred on the ultra-low temperature Si^{29} data of PAALANEN et al [10], on the P^{31} data of DELLOUVE and ALLOUL [11] and on the re-publication of double resonance data by JEROME et al [12]. Using stress-tuning to explore the fine scale of densities around n_c, huge enhancements of Si^{29} spin-lattice relaxation rates, T_1^{-1}, have been observed at ultra-low temperature [10] and interpreted as evidence for a critical slowing down of electronic spin diffusion on the metallic side, orders of magnitude slower than charge diffusion. An alternative description, by GAN and LEE [13], in terms of localized electronic moments even in the metal, coupled pair-wise anti-ferromagnetically into singlet-triplet systems, is more palatable. The essential theoretical idea here is that an applied magnetic field creates singly occupied states in the metal just below the Fermi surface and the Hubbard U creates singly occupied states deep below the mobility edge (where the localization length becomes small enough for U to become large). These deep states are localized electron spins which couple to each other in a pair-wise fashion to form singlet-triplet systems. An applied magnetic field then splits the triplet, and when the lowest triplet level approximately coincides with the singlet a nuclear relaxation process occurs. The mechanism has the correct field dependence to explain the data, if taken in association with a standard Korringa mechanism [14] emanating from the electrons in the more extended states.

This new interpretation leaves localized electron spins above the mnm transition; the data of JEROME et al [12] provides evidence for Korringa relaxation [14] from weakly localized states below the transition. This latter experiment involved a Si:P system of density $n_d = 2.5 \times 10^{18}$ cm^{-3}, well below the value of $n_c = 3.75 \times 10^{18}$ cm^{-3}, and produced clear evidence for a Korringa behaviour [14] for Si^{29} nuclei close to the impurity site and for the P^{31} nuclei actually on the site. However, the direct nmr detection of P [11] makes clear that the active nuclei in the double resonance experiment [12] form only a very small fraction of the sample nuclei. Most of the nuclei must be seeing very large hyperfine fields from their nearest electrons and their resonance frequencies are spread over a sufficiently wide range as to render them unobservable. Both these experiments [11], [12] were restricted to pumped He^4 temperatures.

Other doped semiconductors have been studied. In Ge:As a series of experiments [15], [16], [17], has provided data, in the just-metallic regime, on the variation of the Knight shift with temperature and with concentration of donors, on the effect of the same parameters on spin-lattice relaxation times and on the uniaxial stress dependence of K and T_1. Briefly, in the concentration range 3.5 to 175 x 10^{17} of As donors cm^{-3} the Knight shift varies with concentration as $n_d^{1/3}$, a free-electron-like dependence [the critical concentration in this system n_c is 3.0 x 10^{17} cm^{-3}]. Similarly the nuclear relaxation rate T_1^{-1} varies as $n_d^{2/3}$, so that K^2T_1T, the Korringa constant, is independent of concentration and its value corresponds to that of non-interacting electrons. This data lends support to the thesis that the mnm transition occurs in the host conduction band; impurity band states which clearly exist at lower density have merged into the host electron structure. However the model foundations tremble when the experiment is repeated at a different magnetic field, since T_1^{-1} is shown to be field-dependent, and when uniaxial stress is applied. To understand this stress experiment, a quick review of the effect of stress is in order, and this effect can be explained from at least two different viewpoints:

(a) If the semiconductor is so heavily doped that a conduction band model is certainly correct, then, at zero stress, the donor electrons in doped germanium sit in four different k-space locations along <111> directions. Applied stress along one of these <111> directions will lower the energy of one k-space location with respect to the others, and donor electrons will migrate to the preferred location, eventually emptying the other three pockets of electrons. Consequential changes occur; the density of states at the Fermi surface is reduced, screening is reduced, the Lande' g-factor changes and the wave-function in real space becomes more anisotropic.

(b) Starting from the viewpoint that the semiconductor is so lightly doped that the wave-functions are isolated on donor sites, with appropriate valley-orbit splittings, then stress mixes in excited states with the ground state of the donor wave-function. This leads to an expansion of the wave-function and, with this model, one can envisage stress-tuning from a non-metal to a metal. The increasing anisotropy under stress however acts in the opposite sense to the expansion, and different materials will behave differently with respect to stress-tuning.

The experiments in Ge:As were at a doping density well above the metal non-metal transition (at 4.2 and 5.0 x 10^{17} cm^{-3}) so that viewpoint (a) is appropriate. The experimental result that the Knight shift of Ge^{73} does not change under saturation stress is disconcerting; host conduction band electrons in the stressed case would be expected to shift the resonance much less. On the other hand, the relaxation rate of Ge^{73} gets substantially longer under stress, and this is consistent with free-electron behaviour. However, a further additional feature creeps into the T_1^{-1} data under stress; it becomes non-exponential.

One may speculate that stress produces a radical change in correlation effects, because of the reduced screening, even allowing an impurity band to split off. The appearance of such a band would provide a natural explanation of the non-exponentiality of T_1^{-1}, since this would be associated with a developing inhomogeneity in the electron gas. The theoretical consensus [18], [19] is, however, that increased disorder should produce an enhancement of T_1^{-1}, and since the reduction of screening 'bares' the impurity potentials, stress clearly increases disorder. More data at different fields, and densities, is needed.

The last experiment to be discussed concerns nmr measurements on the field-driven transition in InSb. InSb has a low effective electron mass and a high dielectric constant. The mnm transition in zero magnetic field occurs at $n_c \sim 10^{14}$ cm^{-3} of doping density. However if a metallic sample of doping density

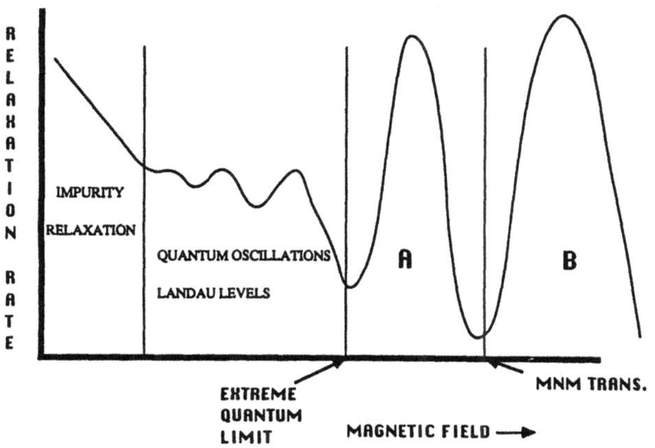

Fig. 2. A typical plot of the spin lattice relaxation rate against magnetic field, showing the various regimes of relaxation and the position of the metal non metal transition

say 10^{15} cm^{-3} is subject to a large magnetic field, then magnetic freeze-out is observed to take the sample back to a non-metallic state. Anomalous relaxation peaks occur on either side of the field at which freeze-out occurs [10].

Figure 2. The peak labelled B has lacked a fully satisfactory explanation up to now, but the new theory of GAN and LEE [13] may provide one. This peak occurs on the high field, non-metallic, side of the transition and the singlet-triplet mechanism of nuclear relaxation may be assumed to dominate. The effective Bohr radius in these high magnetic field conditions will be the cyclotron radius $\{\hbar/e\,B\}^{\frac{1}{2}}$. The relaxation rate is proportional to the number of pairs with $J \sim g\beta B$. Writing down the equation for the exchange interaction [13]

$$J_{\alpha\beta} = 1.636\,\frac{e^2}{\varepsilon a_B}\left\{\frac{R}{a_B}\right\}^{5/2}\exp\left\{\frac{-2R}{a_B}\right\}$$

the field dependence of a_B will cause the exponential prefactor to increase with B, whilst the exponential itself decreases with B. There exists therefore a maximum of J with R or B. As is pointed out [13], the number of such pairs is proportional to $\{dJ/dR\}^{-1}$; hence a relaxation peak occurs as the field is swept through this localized region.

There is some data [20] on the not inconsiderable temperature dependence of this B-peak, taken on a sample of density 1.2×10^{15} cm^{-3}; the peak position moves from 5T at 1.5 K to 7T at 4.2 K. In the helium temperature range the phonon energy is of the same order as the singlet-triplet and magnetic field induced splittings that the B peak is exploring. Consequently the magnetic environment of spin pairs is changing, as electrons from other pairs get thermally promoted to nearby states; the spectrum of exchange splittings as a function of distance between pairs will have some temperature dependence.

References

1. Proceedings of the 31st Scottish Universities Summer School in Physics, St. Andrews 1986, editor D.M. Finlayson, published by SUSSP, Physics Department, Edinburgh University
2. S. Ikehata, S. Kobayashi: Solid State Comm. 56, 607 (1985)
3. M. Paalanen, S. Sachdev, R.N. Bhatt: 18th International Conference on the Physics of Semiconductors, Stockholm, editor O. Engstrom, 1249 (1986)
4. J.D. Quirt, J.R. Marko: Phys.Rev.B. 7, 3842 (1973)
5. K. Andres, R.N. Bhatt, P. Gaolwin, T.M. Rice, R.E. Walstedt: Phys.Rev.B. 24, 244 (1981)

6. W. Sasaki, S.Ikehata, S. Kobayashi: J.Phys.Soc.Japan 36, 1377 (1974)
7. T.F. Rosenbaum, K. Andres, G.A. Thomas, R.N. Bhatt: Phys.Rev.Lett. 45, 1723 (1980)
8. M.A. Paalanen, T.F. Rosenbaum, G.A. Thomas, R.N. Bhatt: Phys.Rev.Lett. 51, 1896 (1983)
9. S. Hikami: Phys.Rev.B. 24, 2671 (1981)
10. M.A. Paalanen, A.E. Ruckenstein, G.A. Thomas: Phys.Rev.Lett. 54, 1295 (1985)
11. P. Dellouve, H. Alloul: Phys. Rev. Lett. 59, 578 (1987)
12. D. Jerome, Ryter, H.J. Schulz, J. Friedel: Phil.Mag.B. 52, 403 (1985)
13. Z-Z. Gan, P. Lee: Phys.Rev.B. 33, 3595 (1986)
14. J. Korringa: Physica 16, 601 (1950)
15. D.P. Tunstall, V.G.I. Deshmukh: J.Phys.C. : Sol.St.Phys. 12, 2295 (1979)
16. D.P. Tunstall, G.S. Sohal: J.Phys.C. : Sol.St.Phys. 16, L251 (1983)
17. D.P. Tunstall: J.Phys.C. : Sol.St.Phys. 17, L1003 (1984)
18. W.W. Warren: Phys.Rev.B. 3, 3708 (1971)
19. W. Goetze, W. Ketterle: Z.Phys.B. 54, 49 (1983)
20. D.P. Tunstall, R.E. Harper: Solid State Comm. 48, 507 (1983)

Magnetic-Field-Induced Metal-Insulator Transition in Degenerately Doped *n*-Type Ge

R.M. Westervelt[1], *M.J. Burns*[1], *P.F. Hopkins*[1], *A.J. Rimberg*[1], *and G.A. Thomas*[2]

[1]Division of Applied Sciences and Department of Physics, Harvard University, Cambridge, MA 02138, USA
[2]AT&T Bell Laboratories, Murray Hill, NJ 07974, USA

An experimental study is presented of the metal-insulator transition induced by strong magnetic fields at low temperatures in Ge degenerately doped with Sb donors to concentrations N_D up to twice the transition n_c in zero field. For unstressed samples with both the magnetic field and the electronic transport along <100> directions, the longitudinal and transverse resistivity increase strongly, by several orders of magnitude, above a characteristic field $H_c \sim 5$ Tesla for $N_D \sim 2n_c$ at temperatures $T < 1$ K. Below this field the conductivity changes slowly with temperature as $\Delta\sigma \propto T^{1/2}$, whereas above H_c the conductivity is thermally activated. The measured Hall coefficient changes relatively little through the transition, so that the apparent mobility is thermally activated in strong fields, while the apparent carrier concentration remains finite. By measuring the magnetoresistance at low temperatures ($T \sim 30$ mK to 0.5 K) and extrapolating to $T = 0$, we find that the functional dependence of the magnetoresistance with field near the transition is consistent with an exponent of one. These results agree with recent theory including both disordered Coulomb interactions and localization, which predicts that the transition in strong magnetic fields has the characteristics of localization.

1. INTRODUCTION

Strong magnetic fields are useful probes of localization and interaction effects in degenerately-doped semiconductors [1,2]. Magnetic freezeout, which has been studied for many years [3,4], occurs in non-degenerate semiconductors when compression of the electronic wavefunctions of isolated impurity atoms by the magnetic field increases the electronic binding energy and thereby decreases the thermally excited carrier concentration. Degenerately-doped semiconductors can be driven insulating by strong magnetic fields, but the nature of the transition in this case is more complex. Electron-impurity and electron-electron Coulomb interactions, and localization are all potentially important. Recent theory for metal-insulator transitions [5-7] includes both disordered Coulomb interactions and localization. For the case of a metal-insulator transition in strong magnetic fields [5-7] the qualitative type of transition is predicted to be localization. This is in contrast both to simple magnetic freezeout, and to the metal-insulator transition in low magnetic fields, which has been experimentally studied in Si [8] and Ge [9,10]. Possible collective electronic states including charge- and spin-density waves, and Wigner crystallization, have been theoretically predicted to occur [11-13] in strong magnetic fields when the impurity potential is weak. Experimentally this is seldom the case, and, and the interpretation of experiments [14-16] in $Hg_{1-x}Cd_xTe$ and InSb remains controversial.

In this paper we present an experimental study of the magnetic-field-induced metal-insulator transition in degenerately-doped n-type Ge. The data show a transition at low temperatures $T < 1$ K and strong magnetic fields $H \sim 5$ Tesla, for which the magnetoresistance increases by as much as a factor ~ 5000, while the Hall coefficient increases by only a factor ~ 3. The variation of the extrapolated zero temperature conductivity with field H is consistent with an exponent of one. Thus the transition in strong fields is of the localization type as predicted by recent theory [5-7] including both Coulomb interactions and localization and not simple magnetic freezeout. A preliminary account of some of these results has appeared elsewhere [17]. Experimental work along these lines in Si [18] was done at fields below the metal-insulator transition.

2. Ge SAMPLES AND APPARATUS

The material used for this study is uncompensated Czochralski-grown Ge degenerately-doped with Sb. Inhomogeneity of the impurity concentration can be a severe problem in heavily-doped crystals, both as slow variations in N_D across the crystal, and as narrow, planar impurity striations which form perpendicular to the growth direction, <100> for our crystals. We screened crystals for striations by cutting a thin, square sample with all faces perpendicular to <100> axes, and with the known growth direction along one side of the square. Contacts were attached to the four corners as described below, and the sample was driven through the metal-insulator transition at low temperatures $T < 1$ K by a magnetic field $\mathbf{H}$ oriented perpendicular to the square face. The isotropy of the dopant concentration was tested by comparing two sets of four-terminal resistance measurements with the leads rotated by 90°, which should be identical for this experimental geometry by crystal symmetry. For the two crystals discussed below, the anisotropy due to striations was typically ~10% in H at the transition. This method is very sensitive and easily detects striations not found by other techniques, such as spreading resistance measurements.

The two samples discussed below, 4H596 and EP2, were cut from two different Ge crystals. The donor concentrations determined by room temperature resistivity measurements [19] on each sample were $N_D \cong 1.9 \times 10^{17}$ cm^{-3} and $N_D \cong 2.7 \times 10^{17}$ cm^{-3} respectively; these values were checked by room temperature Hall measurements. Both samples were cut with all faces perpendicular to <100> crystal axes, lapped, and etched in 3:1 HNO_3 to HF to remove saw damage. Electrical contacts were made using Sn solder, either nominally pure, or doped with Sb; typical measured contact resistances were ~ 2-4 Ω at low temperatures. Sample 4H596 was prepared in a Hall bar geometry with dimensions 10x1.7x0.3 mm^3, two current contacts across opposite 1.7x0.3 mm^2 ends, and three voltage contacts arranged along the sides for virtual-contact Hall measurements, as well as four terminal resistance measurements. Sample EP2 was made in a van der Pauw geometry with outer dimensions 10x10x0.7 mm^3. Care was exercised to mount the samples stress-free.

Low temperature electrical measurements were made both in a He dilution refrigerator at Harvard to temperatures $T \sim 30$ mK and fields $H \sim 7$ Tesla, and in a pumped ^{3}He cryostat at the National Magnet Laboratory at M.I.T. to $T \sim 0.5$ K and $H \sim 20$ Tesla. Transverse and longitudinal magnetoresistance measurements were made with the field

H oriented along a <100> axis by rotating the sample through 90°. Because the magnetoresistance signal can be as much as a factor ~10^4 larger than the Hall signal, low-noise Hall measurements posed special difficulties. These were minimized by using a 5-contact Hall bar geometry with a virtual Hall contact [20] to null out unwanted magnetoresistance pickup, and by subtracting two data sets taken with the magnetic field reversed. Conventional low-noise a.c. lock-in amplifier detection was used for all electrical measurements, which were recorded on a microcomputer for analysis, and care was exercised to avoid heating and remain in the linear portion of the current-voltage characteristics.

3. EXPERIMENTAL RESULTS

3.1 Magnetoresistance

Figure 1 shows semilogarithmic plots of the transverse and longitudinal magnetoresistance ρ_{xx} and ρ_{zz} vs. field H for sample 4H596 at various temperatures between 30 mK and 400 mK. As shown, both ρ_{xx} and ρ_{zz} exhibit a strong increase in resistivity at low temperatures and high fields, by as much as a factor ≅ 5000 at 30 mK and 7 Tesla. Comparison of ρ_{xx} and ρ_{zz} in Fig. 1 shows that this resistivity increase is nearly isotropic. Furthermore, the raw data show evidence for a metal-insulator transition: at low fields H < 4 Tesla the magnetoresistance shows only a weak temperature dependence, whereas at high fields H > 4 Tesla this temperature dependence becomes activated. The transition sharpens somewhat at lower temperatures, but thermal rounding remains significant at the lowest temperature measured. The magnetoresistance of sample EP2 with higher donor concentration shows features very similar to Fig. 1, but shifted to higher fields by a factor ≅ 1.4, as discussed below.

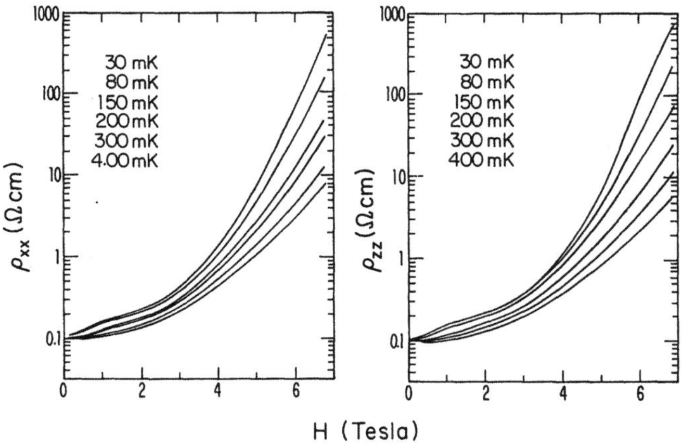

Fig. 1. Transverse and longitudinal magnetoresistance ρ_{xx} and ρ_{zz} for sample 4H596 at temperatures indicated

3.2 Hall Effect

Figure 2 shows a linear plot of the Hall coefficient R_H measured for sample 4H596 over the same range of temperatures and fields as for Fig. 1. The data is normalized to the Hall coefficient at zero field and T = 0.3 K as indicated, so that a constant apparent carrier concentration would correspond to a horizontal straight line in Fig. 2. The surprising feature of these data is that the Hall

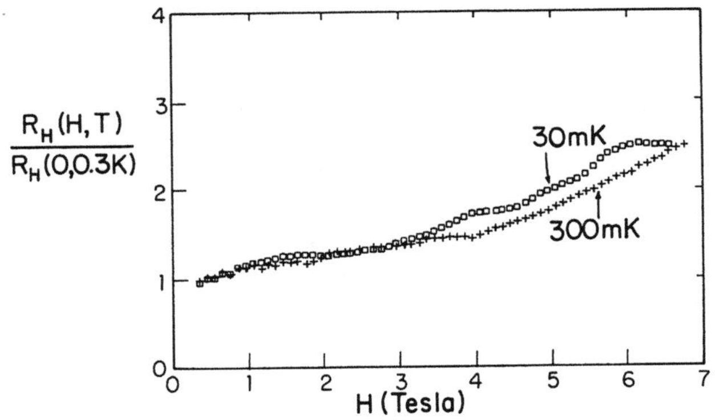

Fig. 2. Hall coefficient R_H normalized to the value at H = 0 and T = 0.3 K for sample 4H596 at the temperatures indicated

coefficient shows little evidence of the transition in Fig. 1: while the transverse magnetoresistance increases by a factor ≅ 5000 at 30 mK and 7 Tesla, the Hall coefficient increases by only a factor ≅ 2.5. Furthermore, the Hall coefficient shows little temperature dependence, both at low and high fields. As noted above, small Hall voltages are difficult to measure accurately in the presence of a large magnetoresistance signal, and the Hall voltage is also sensitive to inhomogeneity in dopant concentration. Thus the field dependence shown in Fig. 2 may be influenced by systematic errors, despite the precautions taken. Hall data from sample EP2 show a similar lack of field and temperature dependence on both sides of the metal-insulator transition.

The conclusion we draw from this data is that the magnetic-field-driven metal-insulator transition initially occurs as a reduction of the apparent carrier mobility, rather than the apparent carrier concentration, as for localization, and is thus different from simple magnetic freezeout. A localization-type transition is predicted in strong magnetic fields [5-7] although the theory of the Hall effect for this case has not been worked out. Qualitatively similar Hall data have been found for the two-dimensional electron gas in Si MOSFET's [21].

3.3 Exponent of the Transition

An important characteristic of the magnetic-field-induced metal-insulator transition is the exponent ζ describing the variation of the conductivity with magnetic field near the transition on the metallic side:

$$\sigma \propto (H_c - H)^{\zeta}. \tag{1}$$

The value of this exponent can help to distinguish different physical mechanisms which could be responsible for the transition. Recent theory [5-7] predicts $\zeta \cong 1$ for the transition in strong magnetic fields.

Because the amount of thermal rounding in our magnetoresistance data is significant even at 30 mK, we extrapolated the transverse conductivity to zero temperature in order to make a comparison with theory. Figure 3 shows a series of curves which illustrate the technique. Here the transverse conductivity σ_{xx} computed from magnetoresistance data for sample 4H596 and normalized to the value at H = 0 and T = 0.4 K, is plotted vs. $T^{1/2}$ on the metallic side of the

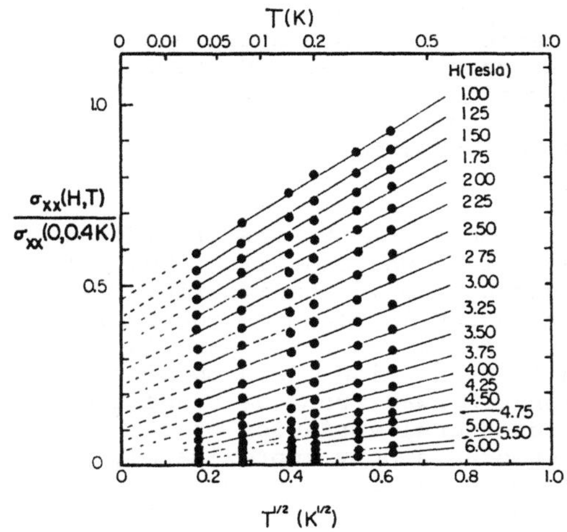

Fig. 3. Extrapolation of the normalized conductivity σ_{xx} for sample 4H596 to T = 0 K. The straight lines are least squares fits to the data at each field indicated

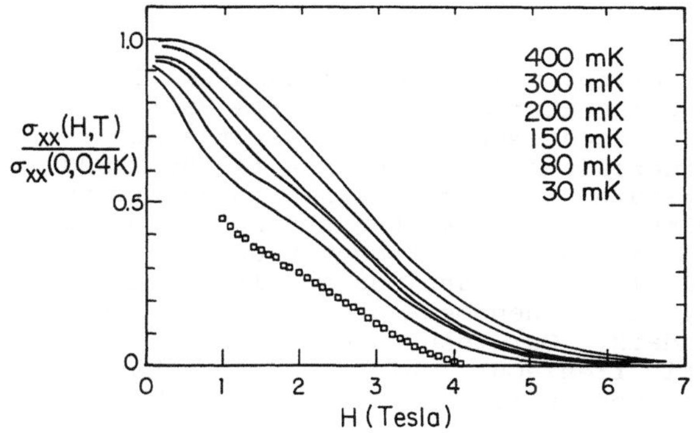

Fig. 4. Linear plots of the extrapolated zero-temperature conductivity σ_{xx} (squares) and finite temperature data (curves) vs. magnetic field H for sample 4H596 at the temperatures indicated

transition for various magnetic fields. This temperature dependence is experimentally found at zero field in degenerately-doped Ge [22], and is characteristic of corrections to the conductivity of three-dimensional samples from Coulomb interactions and from certain cases of weak localization; as shown in Fig. 3 it fits the variation of σ_{xx} with T quite well. Least squares fits to the data for each magnetic field, shown as the straight lines in Fig. 3, were used to find the zero-temperature intercepts.

Figure 4 shows a linear plot of the extrapolated zero temperature transverse conductivity σ_{xx} (shown as the square points) for sample 4H596 vs. field H, together with plots of the finite temperature conductivity (shown as curves) computed from magnetoresistance and Hall data at the temperatures indicated; the conductivity curves are all normalized to the value at H = 0 and T = 0.4 K as indicated. As shown in Fig. 4 the rounding of the transition is considerably reduced for the extrapolated data, and the zero-temperature conductivity approaches the transition in an approximately linear manner. Thus this data is consistent with the exponent $\zeta = 1$ predicted by recent theory [5-7] which includes both disordered Coulomb interactions and localization. Earlier experiments for degenerately-doped Si [18] in high magnetic fields were not able to drive the samples through the metal-insulator transition, and thus their estimate of the exponent is much more indirect. The critical field $H_C \cong 3.8$ Tesla for Ge sample 4H596 can be deduced from the extrapolated zero-temperature data of Fig. 4 by assuming a linear field dependence for the conductivity. Similar extrapolated data were obtained from sample EP2 with higher donor concentration, except that the critical field obtained by this procedure is larger by a factor $\cong 1.4$, as qualitatively expected for heavier doping.

A number of cautionary remarks regarding exponents are in order. The critical region very near the transition, where scaling theory is strictly valid, is obscured by small variations in the impurity concentration in material such as ours, which is doped by conventional methods. The small rounding evident in the extrapolated data in Fig. 4 is comparable to the variation in transition field obtained in our isotropy tests described in Section 2. Away from the transition where the data is most reliable, the functional variation of conductivity with field predicted by theory is only approximately correct. These problems are of course more severe in cases other than ours where the transition is not observed, and extrapolations from low fields are used to deduce the exponent.

4. CONCLUSIONS

A magnetic field-driven metal-insulator transition is observed in degenerately doped Ge:Sb at low temperatures $T < 1$ K. The Hall coefficient remains relatively unchanged while the transverse and longitudinal magnetoresistance increase by as much as a factor $\cong 5000$, and the extrapolated zero temperature transverse magnetoconductivity varies approximately linearly with field H near the transition on the metallic side. These data are consistent with recent theory [5-7] which includes both disordered Coulomb interactions and localization. Interpretations in terms of collective phenomena are also possible, but not required by the present data. Our results differ qualitatively from simple magnetic freezeout, which is driven by a reduction in the apparent carrier concentration.

5. ACKNOWLEDGEMENTS

We thank B.I. Halperin, P.A. Lee, A.B. Fowler, and P. Stiles for helpful comments. Part of this work was performed at the Francis Bitter National Magnet Laboratory, supported by the National Science Foundation at M.I.T. This work was supported by the National Science Foundation under grants DMR85-08733 and DMR-86-14003.

REFERENCES

1. B.I. Shklovskii and A.L. Efros: Electronic Properties of Doped Semiconductors, (Springer-Verlag, Berlin, 1984)
2. N.F. Mott and E.A. Davis: Electronic Properties in Non-Crystalline Materials, 2nd ed., (Oxford University Press, Oxford, 1979)
3. Y. Yafet, R.W. Keyes and E.N. Adams: J. Phys. Chem. Solids. **1**, 137 (1956)
4. R.W. Keyes and R. J. Sladek: J. Phys. Chem. Solids. **1**, 143 (1956)
5. A.M. Finkelstein: Sov. Phys. JETP, **57**, 97 (1983); Sov. Phys. JETP, **59**, 212 (1984)
6. C. Castellani, C. DiCastro, P.A. Lee, and M. Ma: Phys. Rev. **B30**, 527 (1984)

7. C. Castellani and C. DiCastro: Phys. Rev. B34, 5935 (1986)
8. Reviewed in G.A. Thomas: Phil. Mag. B52, 479 (1985)
9. G.A. Thomas, Y. Ootuka, S. Katsumoto, S. Kobayashi, and W. Sasaki: Phys. Rev. B25, 4288 (1982)
10. S.B. Field and T.F. Rosenbaum: Phys. Rev. Lett. 55, 522 (1985)
11. V. Celli and N.D. Mermin: Phys. Rev. 140, 839 (1965)
12. H. Fukuyama: Solid State Commun. 26, 783 (1978)
13. Z. Tesanovic and B.I. Halperin, preprint
14. G. Nimtz, B. Schlict, E. Tyssen, R. Dornhaus, and L. D. Hass: Solid State Commun. 32, 669 (1979)
15. T.F. Rosenbaum, S.B. Field, D.A. Nelson and P.B. Littlewood: Phys. Rev. Lett. 54, 241 (1985)
16. M. Shayegan, V. J. Goldman, H. D. Drew, D. A. Nelson and P. M. Tedrow: Phys. Rev. B32, 6952 (1986)
17. G.A. Thomas, M.J. Burns, P. Hopkins and R. M. Westervelt: Phil. Mag., in press
18. W.N. Shaferman, T.G. Castner, J.S. Brooks, K.P. Martin and M.J. Naughton: Phys. Rev. Lett. 56, 980 (1986)
19. S.M. Sze and J.C. Irvin: Solid State Electron. 11, 599 (1968)
20. H. Fritzche: In Methods of Experimental Physics, Vol. 6B: Solid State Physics, K. Lark-Horovitz and V.A. Johnson, eds. (Academic Press, New York 1959), ch. 8
21. C.J. Adkins: J. Phys. C 11, 851 (1978)
22. G.A. Thomas, A. Kawabata, Y. Ootuka, S. Katsumoto, S. Kobayashi, and W. Sasaki: Phys. Rev. B26, 2113 (1982)

The Metal-Insulator Transition in Ge:Sb

Y. Ootuka[1,*], *H. Matsuoka*[1,**], *and S. Kobayashi*[1]

Department of Physics, University of Tokyo, Hongo, Bunkyo-ku, Tokyo 113, Japan
*Present Address: Cryogenic Center, University of Tokyo, Yayoi, Bunkyo-ku, Tokyo 113, Japan
**Present Address: Central Laboratory, Hitachi Ltd. Kokubunji, Tokyo 185, Japan

1. Introduction

The critical behavior of the zero-temperature electrical conductivity $\sigma(0)$ near the metal-insulator transition point has been studied in various disordered systems. Among them, Si:P is peculiar in the point that the critical exponent s of conductivity as a function of donor concentration N;

$$\sigma(0) = \sigma_0 \, (N/N_c-1)^s, \tag{1}$$

is close to 0.5 [1], while most of others give 1 [2].

In this paper, we report the experimental results of another uncompensated semiconductor, Ge:Sb. The conductivity in Ge:Sb was measured across the metal-insulator transition in the magnetic fields as well as in zero field.

2. Experiment

A single crystal of Ge:Sb was prepared by the Chokralski method. Intentional compensation doping was not done. The crystal was a rod in shape with 20mm in diameter and 170mm in length, and the impurity concentrations are $3.5\times10^{16}\,cm^{-3}$ and $1\times10^{18}\,cm^{-3}$ at the both ends. In usual experiments, the crystal is cut into thin slices perpendicular to the crystal growth axis in order to obtain the best homogeneity of concentration. In this experiment, however, we adopted an alternative method: A sample with 1mmx2mmx70mm in size was cut from the center of the crystal in such a way that the long axis is parallel to the crystal growth axis. After the sample was etched chemically, Au:Sb wires were spot-welded at average interval of 2mm. As the donor concentration varied along the long axis, the resistance of Ge:Sb with finely varying impurity concentration could be obtained in a single run of the dilution refrigeration by measuring the electric potential at each electrode with the common current passed along the long axis.

The donor concentration was determined from the resistivity at 4.2K, whose relation to N had been established experimentally. At first, we expected that the donor concentration varies smoothly and monotonically in the sample. But, in fact, the distribution of the concentration is not monotonic and has fine structure. The change of the growth speed due to the temperature fluctuation during the crystal growth is thought to be an origin of such structure.

Experiment was done at temperatures down to 20mK using a dilution refrigerator. The magnetic fields up to 4T were applied in the same direction as that of the current. The inhomogeneity of the magnetic field was less than 2.5% over the sample.

3. Results and Discussions

Figure 1(a) shows the temperature dependence of conductivity below 1K in zero field. The abscissa is scaled in $T^{1/2}$. With decreasing temperature, the conductivity increases in highly metallic samples ($N>1.8\times10^{17}cm^{-3}$), while it decreases when N is low and the sample with $N=1.38\times10^{17}cm^{-3}$ behaves like an insulator.

It is known that T dependence of conductivity is described by the interaction theory[3]. Although the conductivity of Si:P shows clear $T^{1/2}$ dependence which the naive interaction theory predicts, that of Ge:Sb is not described by a simple $T^{1/2}$ law. This is interpreted to be an evidence of the renormalization of the coupling constant of the interaction: the renormalization of the interaction strength becomes important and causes an enhancement of the $T^{1/2}$ correction at low temperatures [4].

As we are interested in the zero-temperature conductivity, the data have to be extrapolated to T=0. In principle, we should analyze the data using the interaction theory which includes the effect of the renormalization of the coupling constant, but such a process is very much complicated and ambiguous. Here we use the following empirical fitting function to obtain $\sigma(0)$:

$$\sigma(T) = \sigma(0) + mT^{1/2} + BT. \qquad (2)$$

Such a function was used in a previous paper [5] and provides a rather good fit. Figure 2 shows $\sigma(0)$ as a function of N.

The temperature variation of conductivity in H=4T is shown in Fig.1(b). In contrast to the case of H=0, the conductivity exhibits clear $T^{1/2}$ dependence and can be explained with the simple interaction theory. The data are fit to (2) equating B=0 and we obtain $\sigma(0)$ plotted in Fig.2.

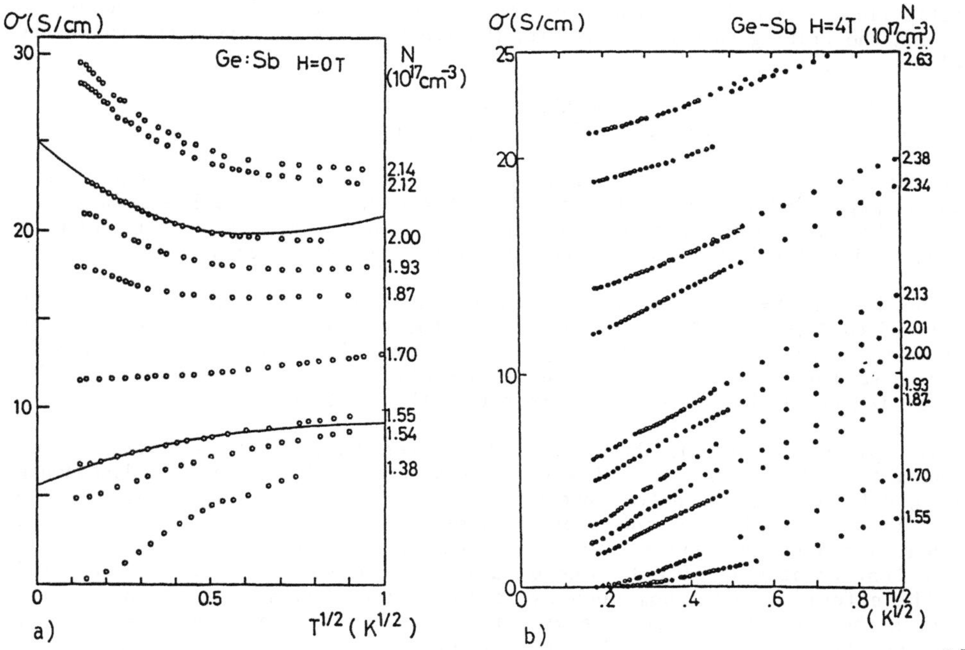

Fig.1: Conductivities of Ge:Sb samples plotted against $T^{1/2}$.
left:(a) H=0, right:(b) H=4T.

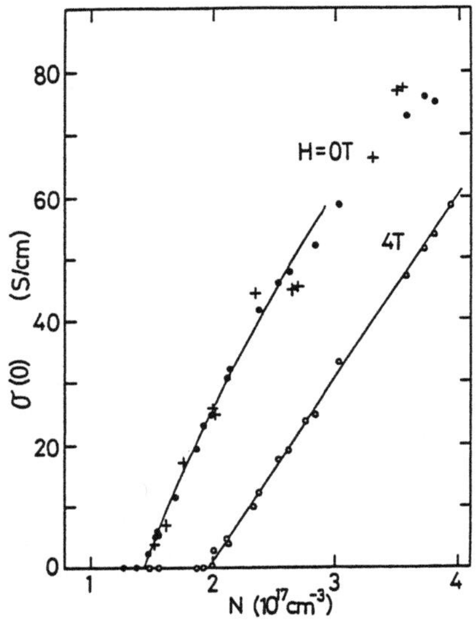

Fig.2: Zero-temperature conductivity in Ge:Sb as a function of donor concentration in H=0 and 4T. Solid curves indicate the results of fitting to (1). The symbols, ✝, indicate the previous results.

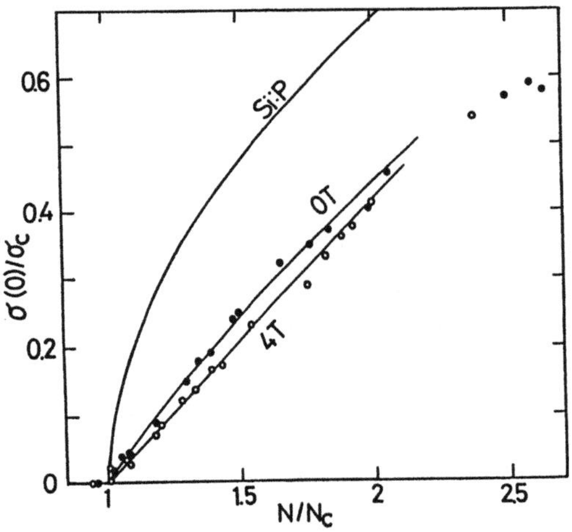

Fig.3: Reduced conductivity of Ge:Sb in H=0 and 4T as a function of reduced donor concentration. The solid lines are the results of fitting of (1). The result of Si:P is also plotted.

The critical concetration for the metal-insulator transition depends on the magnetic field strength. In each case, the conductivity decreases continuously to zero, and the fitting to (1) is possible. The least square fit gives the following results:

$$H=0T: \quad N_c=1.44\times10^{17}, \quad s=0.9, \quad \sigma_0=58\,S/cm,$$
$$H=4T: \quad N_c=1.98\times10^{17}, \quad s=1.0, \quad \sigma_0=59\,S/cm,$$

where the data of the samples with $N_c<N<2N_c$ is adopted for fitting. In a previous paper [6],we reported s in H=0 is 0.7 ± 0.2 for Ge:Sb. The samples in Ref.6 had been cut from other crystals made in a similar way. The data in Ref.6 is also plotted in Fig.2. They agree with the present data very well.

The data in H=0 and in 4T are compared in Fig.3. Here, the conductivity normalized to $\sigma_c=e^2N^{1/3}/\hbar$ is plotted against the reduced concentration N/N_c. The experimental result of Si:P in H=0 [1] is also shown. The data in different fields cannot be reduced to a single curve but apparently bring different value of s. The critical feature in H=0 is also different from that of Si:P.

The present system, Ge:Sb, and Si:P are both uncompensated doped semiconductors. The values of s, however, are different. The apparent N-dependence of $\sigma(0)$ differs from the actual one if the nonuniformity of donor concentration exists in the sample. As the present samples really have the gradient of concentration along the current axis and some distribution perpendicular to it, the evaluation of this effect is necessary. After a model calculation, we find that the effect of the inhomogeneity is important when a part of the sample ungergoes the transition, but otherwise it is not significant. The inhomogeneity in our sample is estimated to be less than 10% in both x- and y-directions. Therefore, we conclude that the effect of the inhomogeneity of donor concentration on the overall feature of $\sigma(0)$ vs N and also on the value of s is not important in the present experiment, as long as we deal with the rather wide range of concentration; $N_c < N < 2N_c$.

Within the existing theories of metal-insulator transition, only the unitary case (presence of either external magnetic field or of magnetic impurities) of non-interacting scaling theory leads to s=1/2 [7]. This is explained by supposing that the electron-electron interaction produce pseudo local magnetic moments which cause the spin scattering of electrons, but that the interaction affects the value of s little. To explain why the interaction is not important for s in Si:P, KAWABATA [8] presents the one-parameter renormalization theory including both the localization in the presence of magnetic scattering and the electron-electron interaction. He concluded that the critical exponent is determined by the relative importance of the localization and of the interaction and the latter effect is less important in multivalley semiconductors. The β -function in ν-valley case is given as:

$$\beta(g) = 1 - \frac{2}{\nu} g^{-1} - 3 g^{-2}. \tag{3}$$

As the valley number of Ge (ν=4) is smaller than that of Si (ν=6), the relative importance of the interaction is larger in Ge and s will be larger in Ge than in Si. Although this is consistent with the experiment qualitatively, the naive evaluation of (3) gives s = 0.55 for Si and 0.57 for Ge, which is much smaller than the experimental result.

The $\sigma(0)$-vs-N relation depends on the degree of compensation [6]. Recent experiment on Si:P compensated by the neutron irradiation suggests that the value of s is very close to unity even if the compensation is as small as 15% [9]. Although the intentional compensation doping was not done in our sample, some compensation effect by lattice imperfections may be an origin of the larger value of s.

The value of the prefactor σ_0 is discussed in Ref.10. They claimed that the relevant conductivity scale for localization in νvalley case is $0.18\ \nu^{2/3} e^2/\hbar N_c^{-1/3}$. This yields $\sigma_0/\sigma_c = 0.46$ for Ge:Sb, which is in agreement with the experimental results.

In conclusion, we have measured the conductivity of Ge:Sb across the metal-insulator transition. The conductivity extrapolated to T=0 decreases continuously to zero at N_c. The critical exponent s is about 0.9 in H=0. This value is smaller than unity but is rather large if it is compared with Si:P case, where s=0.55. The value of s varies in the magnetic field and is 1.0 in 4T.

Acknowledgements

We are grateful to Dr. C. Yamanouchi for providing the Ge:Sb crystal.

References

1. T.F. Rosenbaum, R. F. Milligan, M.A. Paalanen, G.A. Thomas, R.N. Bhatt and W. Lin: Phys. Rev. B**27** 7509 (1983).
2. For review, G.A. Thomas and M.A. Paalanen: Proc. Conf. LITPIM, ed. by Y. Bruynseraede (Springer Verlag,1985).
3. For review, Y. Ootuka and A. Kawabata: Prog. Theor. Phys. suppl.**84** 249 (1985).
4. C. Castellani, C. DiCastro, H. Fukuyama, P.A. Lee and M. Ma: Phys. Rev. B**33** 7277 (1986).
5. G.A. Thomas, A. Kawabata, Y. Ootuka, S. Katsumoto, S. Kobayashi and W. Sasaki: Phys. Rev. B**26** 2113 (1982).
6. G.A. Thomas, Y. Ootuka, S. Katsumoto, S. Kobayashi and W. Sasaki: Phys. Rev. B**25** 4288 (1982).

7. S. Hikami: Phys. Rev. B**24** 2671 (1981).
8. A. Kawabata: J. Phys. Soc. Jpn.: **53** 318 (1984).
9. W. Sasaki, Y. Nishio and K. Kajita: In Disordered Semiconductors, ed. by M.A. Kastner et al. (Plenum,1987) p.37, private communications.
10. R. N. Bhatt and T.V. Ramakrishnan: Phys. Rev. B28 6091 (1983).

The Metal-Insulator Transition in a Persistent Photoconductor

S. Katsumoto

Department of Physics, Faculty of Science, University of Tokyo, Hongo 7-3-1, Bunkyo-ku, Tokyo 113, Japan

ABSTRACT

The electric conductivity and the dielectric susceptibility of AlGaAs doped with Si were measured near the metal-insulator transition. The persistent photoconductivity effect was used to control finely the effective concentration of carriers in a single specimen. The transition was also brought about by the external magnetic field. In the metallic phase, the value of the critical exponent of the transition was determined to be 1. In the insulating phase, the data of the dielectric constant can be fitted with the values from 1.9 to 3.5 as the characteristic exponent. The critical behavior of the conductivity and the dielectric constant is compared with the predictions of the scaling theories with interacting electron models.

1. INTRODUCTION

The renormalization group treatment of the metal-insulator transition has enabled us to predict the critical behavior of physical quantities [1,2]. Recent theoretical efforts are focused on studying the characteristics of the solution for the renormalization group equations. On the experimental side, only a limited number of systematic works on the transition has been reported while that on the weakly localized regime is exhaustive.

One of the difficulties in the experimental studies of the metal-insulator transition is to tune a relevant parameter to the transition. For example, the the stress tuning technique has realized sophisticated works on the Si:P system [3].

Here we describe the systematic experiments on the metal-insulator transition in a new system, AlGaAs:Si (aluminum content 0.3). In the experiments, the effective electron concentration is finely controlled with the use of the phenomenon called persistent photoconductivity (PPC)[4]. AlGaAs:Si is a most popular material for PPC, and similar effects are observed in many species of semiconductors with a large number of deep states. Thus this method may be applied to such materials.

2. EXPERIMENT

2-1 Principle of the Experiment

Deep donors (or deep acceptors) with a high concentration (at least higher than n_c) whose capture rate is very small at low temperatures, are intentionally introduced into a semiconductor by doping or by some other techniques. Cooled slowly in the dark, the deep donors capture the electrons and the semiconductor becomes insulating at low temperatures. In the next step, the electrons are photo-excited from the deep states to the shallow states or to the conduction band. Because of the very small capture rate, even after the photo-irradiation, the electrons remain in the shallow states or in the conduction band and the conductivity remains finite. This is the PPC effect. When the temperature is so low that the relaxation time is much longer than the time for our measurements of the conductivity etc., we can ignore the relaxation and consider this PPC state to be stable and in equilibrium. In this state the concentration of the electrons free from the deep states (we call it the effective concentration) is determined by the total photo-exposure time.

This PPC system has several remarkable characteristics : 1) When the energy of the photons is well below the band gap so that the specimen is sufficiently transparent for the photons, this is a uniform three dimensional (3D) electronic system in which we can control the effective concentration by the photo-exposure time. 2) Because the deep states are strongly localized around the donors, a deep donor which has captured an electron does not produce a long range Coulomb potential. Therefore, in many cases it does not have an effective shallow state. In other words, only "photo-ionized" deep donors (actually they are no longer deep) have effective shallow states. The effective electron concentration is hence equal to the effective shallow donor concentration. The impurity band for this system is thus considered to be half filled. This system is different from compensation controlled systems [5] in this point. However, when the neutral deep donors work as scattering centers, the number of scattering center per electron is more than one. In this sense, the system is "compensated".

2-2 Sample Preparation

We used Si doped $Al_{0.3}Ga_{0.7}As$ mixed crystal which is one of well studied materials with PPC [6]. In this material all the doped Si becomes deep below 77K [7]. The samples were grown by molecular beam epitaxy (MBE) on (100) GaAs substrates. The MBE system and the growth conditions were the same as described elsewhere [8].

For the conductivity measurement, epitaxial layers were grown as follows. First, 0.98 μm thick, non-doped $Al_{0.3}Ga_{0.7}As$ layer was grown to avoid the generation of 2D electrons at the hetero-interface. Next, 2.97 μm thick, $Al_{0.3}Ga_{0.7}As$ doped with $1x10^{18}cm^{-3}$ Si was grown and covered with GaAs cap layer of 49 atomic monolayers which, after the growth, was removed by selective etching except in the vicinity of electric probe parts. Inhomogeneity of Si atom concentration in the layer is estimated to be less than 0.5%. The specimen was mesa-etched into a four-probe bridge shape, 200 μm by 900 μm in dimensions, by the photolithography. This is named Sample A.

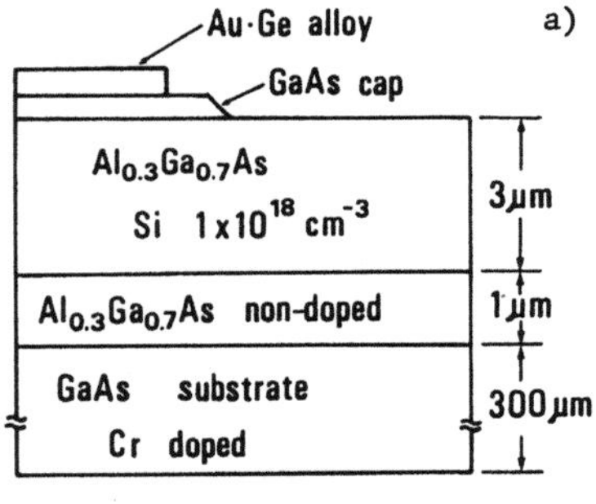

(a) Sample A for the conductivity measurement.

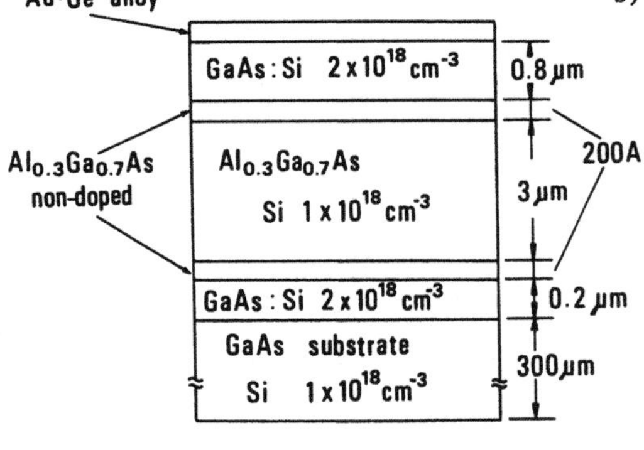

(b) Sample B for the dielectric susceptibility measurement.

Fig.1. Cross-sectional views of the epitaxially grown samples.

For the dielectric susceptibility measurement, a GaAs-AlGaAs-GaAs sandwich type structure was grown onto a Si doped n-type substrate. We used the n-GaAs layers as "metal electrodes", and measured the capacitance of the whole sample. Over the top GaAs layer, a Au/Ge alloy film was evaporated. Again by the photolithography, the alloy film was removed leaving a 1mm x 1mm square region. The epitaxial layers were also removed by mesa-etching leaving the square. The ohmic contact to the substrate is taken by an indium ball contact. This is named sample B. The cross sectional views of Samples A and B are shown in Fig.1.

2-3 Measurement

The linearity between the total exposure time and the effective electron concentration in the range up to $2x10^{17}cm^{-3}$ was confirmed by the measurement of the Hall coefficient as a function of exposure time at 77K. The electron mobility at 77K slowly increases around 800 cm^2/Vs with the increase of the electron concentration. Low mobility at 77K suggests that the neutral impurities act as scattering centers. This agrees with the recent calculation [9].

The specimens were installed in a mixing chamber of a dilution refrigerator faced to the bottom of the chamber. Cooled below 1K in the dark, they exhibited insulating manner ($\rho>10^{10}\,\Omega$ cm). They were illuminated by GaAs light emitting diodes (LEDs) placed at the bottom of a vacuum jacket. The emitted light passed through a Si doped GaAs cold filter, a diaphragm, and a wall of mixing chamber made of Stycast 1266, and was scattered by a diffusor plate in the mixing chamber. Stycast 1266 and $Al_{0.3}Ga_{0.7}As$ are transparent for the emitted light. Non-uniformity of the illumination in this configuration is calculated to be about 0.03% for Sample A and 0.05% for Sample B.

We call the illumination by LEDs during a certain period (three seconds) a "unit". To avoid the temperature increment of the LEDs which may result in some ambiguity in the light intensity, successive units of illumination are separated by more than two minutes. After some units of illumination, the samples attained finite conductance. Below 1K, the decay time of PPC was immeasurably long ($>10^9$ seconds). Variation of the refractive index of AlGaAs with the increase of the carrier concentration is estimated to be small and also supposed to be smooth at the metal-insulator transition [10].

The conductivity was measured by AC method at 63Hz in the temperature range from 800mK to 30mK. The capacitance was measured by a three-probe capacitance bridge at the frequency of 103kHz. At the lowest temperatures the Joule power loss in the samples was kept below 10^{-12}W.

3. RESULTS

3-1 Electrical Conductivity

The measured electrical conductivity without the magnetic field below 1K is plotted against square root of the temperature in Fig.2(a). Below 300mK, with the increase of the total exposure time, $\sqrt{T}$ dependence of the conductivity becomes clear. After a certain amount of exposure, the temperature dependence of the conductivity in the region from 300mK to 40mK is experimentally well represented as

$$\sigma(T,n) = \sigma(0,n)+m(n)\sqrt{T} \qquad (1)$$

where m(n) is a coefficient which is independent of the temperature and n the effective electron concentration.

The fitted value of $\sigma(0,n)$ changes its sign from negative to positive at a certain value of n. We define the critical concentration n_c as this value of n.

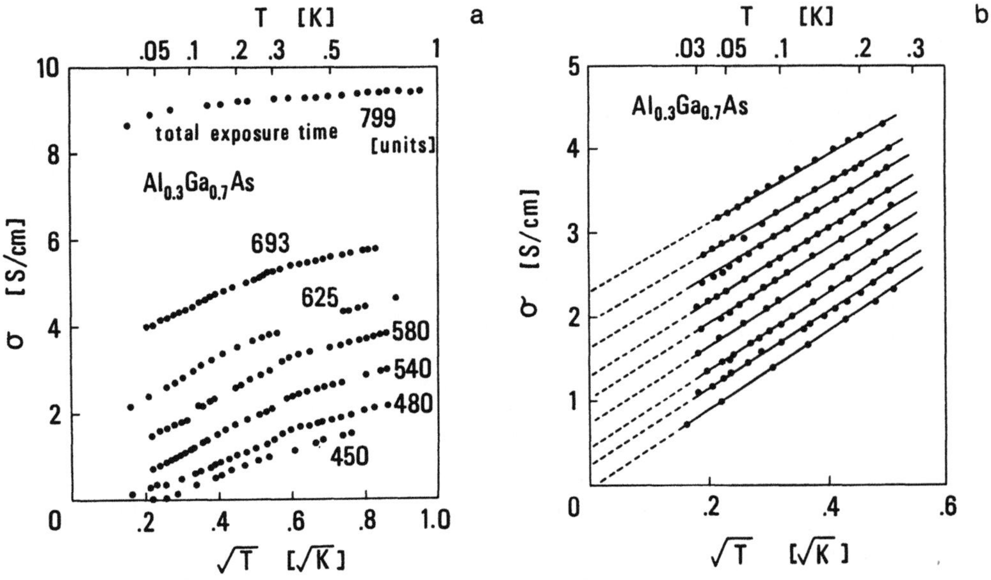

(a) Variation of the conducting behavior with the increase of the total exposure time.

(b) $\sqrt{T}$ dependence of the conductivity below 300mK in the vicinity of the metal-insulator transition.

Fig.2 Temperature dependence of the conductivity for $Al_{0.3}Ga_{0.7}As$:Si. Temperature is scaled in $\sqrt{T}$.

Here we calculate n using the results of the Hall measurement at 77K. The temperature dependence with negative $\sigma(0,n)$ must have an upward curvature below 40mK. Thus it is natural to replace these negative $\sigma(0,n)$ with zero. In the vicinity of the transition, m(n) is almost independent of n and is 4.5 $S/cmK^{1/2}$, as shown in Fig.2(b).

The critical exponent ν of the transition is defined as

$$\sigma(0,n) = \sigma_0(n/n_c-1)^{\nu}. \qquad (2)$$

Experimental values of the zero-temperature conductivity thus defined are plotted against the exposure time (i.e. the effective electron concentration) in Fig.3. With increasing n, $\sigma(0,n)$ increases almost linearly and the fitted value of the exponent is close to 1. This measurement resolves sub-one percent of the critical region as demonstrated in the inset of Fig.3.

Near the transition, even in the insulating phase as well as in the metallic phase, the magnetoresistance is negative up to 4T. In the vicinity of the critical region, the magnetic field changes only the value of $\sigma(0,n)$, i.e., the $\sqrt{T}$ dependence in the relation (1) and the value of m(n) are unchanged. The negative magnetoresistance is due to the delocalization of the wavefunction [11]. In the highly metallic region ($n>5\times10^{17}cm^{-3}$), the positive magnetoresistance appears in low magnetic field ($<3\times10^{-3}$T). This is presumably due to the spin-orbit scattering.

As the result of the delocalization of the wavefunction, the critical concentration n_c is lowered in the magnetic field. The behavior of the conductivity near the transition under the magnetic field of 4 Tesla is shown in Fig.2. We used again the relation (1) for the extrapolation. The values of m and

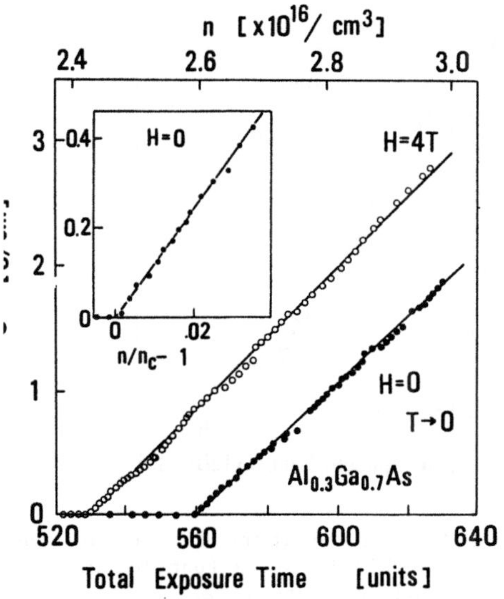

ig.3 Zero temperature conductivity near the metal-insulator transition in the magnetic field of 4T and 0T. The behavior very near the transition in zero field is shown in the inset.

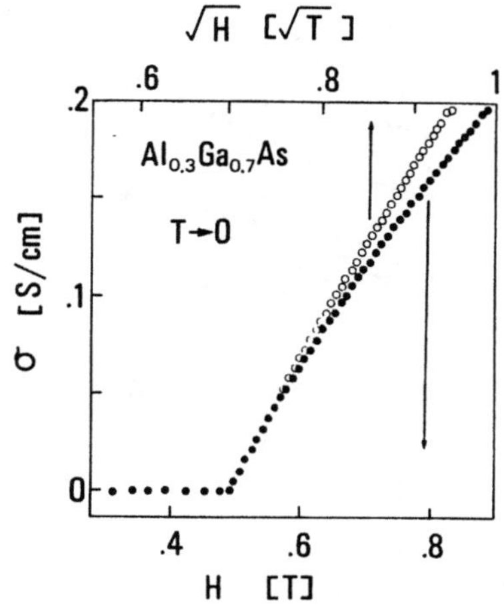

Fig.4 Zero temperature conductivity as a function of H and $\sqrt{H}$.

0 are the same as in zero magnetic field. The critical exponent ν again turned ut to be close to 1.

Adjusting the electron concentration just below the critical point by ontrolling the illumination, we carried out the magnetic field tuning of the ietal-insulator transition [12]. In this transition, we kept n constant and took he field strength H as the relevant parameter. As shown in Fig.4, $\sigma(0,H)$ ncreases linearly both with H and $\sqrt{H}$ near the critical point though the range of he linear dependence is wider in the $\sqrt{H}$ plot. The result again gives $\nu=1$.

-2 Dielectric Susceptibility

Before the illumination, measured value of the dielectric constant is about .3 times as large as the vacuum polarizability. This value is very close to that eported for pure GaAs. Thus, it is natural to define the intrinsic value of the lielectric constant of our sample as this value. Most of the frequency response of he circuit capacitance and the dissipation factor is explained by the equivalent ircuit shown as the inset of Fig.5.

The measured capacitance has strong temperature dependence in very low emperature region (<100mK) as shown in Fig.5. When the effective concentration is low and the capacitance is small, it varies as T^2. The coefficient of T^2 becomes larger with the increasing concentration. A similar T^2 dependence of the lielectric constant is observed by PAALANEN et al. [13] However, in our case, the lependence changes from T^2 to T^4. Thus the temperature dependence is fitted by the formula in these regions

$$\varepsilon(n,T)=\varepsilon(n,0)(1+a(n)T^2+b(n)T^4)\,. \qquad (3)$$

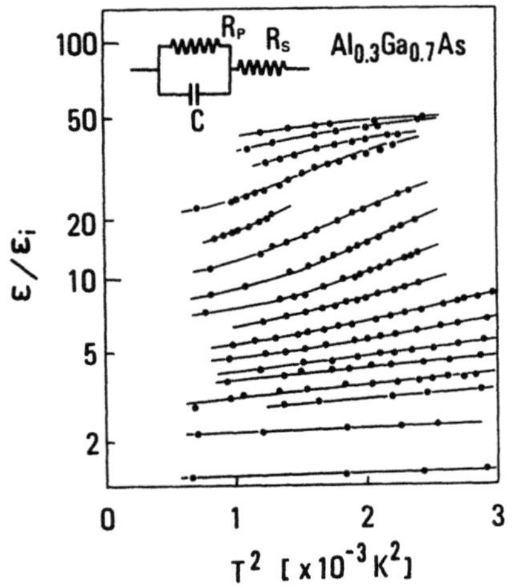

Fig.5 Temperature dependence of the dielectric susceptibility for $Al_{0.3}Ga_{0.7}As:Si$. The temperature is scaled in T^2. The susceptibility is calculated from the measured capacitance of Sample B assuming the equivalent circuit shown in the inset. $\varepsilon_i=13.1\varepsilon_0$.

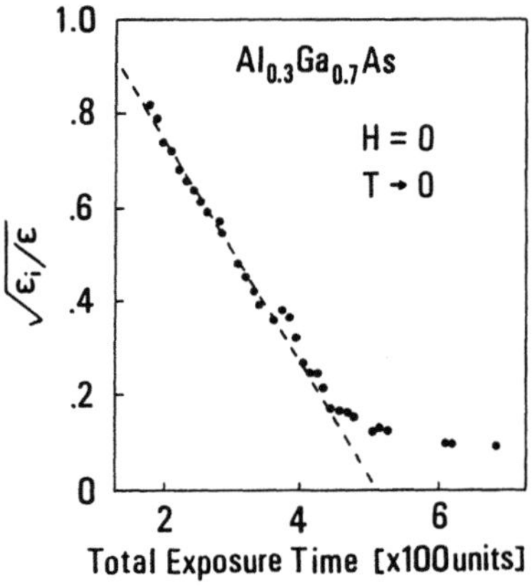

Fig.6 Zero-temperature dielectric susceptibility as a function of the total exposure time.

In contrast to the metallic phase, the lack of a reliable fitting formula brings in some systematic errors to our extrapolation.

The characteristic exponent ν' for the dielectric constant is defined as

$$\varepsilon(n,0)=\chi_0(1-n/n_c)^{-\nu'}. \quad (4)$$

We take ν' and n_c as fitting parameters. The best fit value of ν' is 2.3 which is close to 2. The statistical error is estimated to be +1.2 and -0.4. In Fig.6, $\sqrt{\varepsilon_i/\varepsilon(n,0)}$ is plotted against the total exposure time. As shown by the broken line, the data can be fitted by the relation (4) with $\nu'=2$ in a wide range. Some rounding is seen in the vicinity of the transition. In this region, the capacitance of the sample is very large and the rounding is presumably due to the inevitable series capacitance of the heterojunctions.

4. DISCUSSION

To compare the results with theories, we should clarify whether the spin-orbit scattering affects the critical behavior of the conductivity and the dielectric constant. The spin-orbit scattering time τ_{so} is calculated to be 4.6×10^{-10} seconds by using the positive magnetoresistance which appeared in the highly metallic region [14]. We can judge whether the spin-orbit scattering is effective or not by testing whether $\sqrt{D\tau_{so}}$ cuts off the characteristic length. Here the diffusion constant D is estimated from the conductivity at 77K based on the classical relation : $\sigma_{77K}=2e^2DN_F$. In the calculation of N_F, we adopted $0.092m_0$ for the value of the effective mass.

In the metallic phase, the characteristic length is the coherence length which is defined as $e^2/\hbar\sigma$. In the insulating phase it is the localization length which is represented as $\sqrt{\varepsilon}/\kappa$[15]. Here $\kappa=(8\pi e^2N_F)^{1/2}$ is proportional to the

Thomas-Fermi screening wavenumber. Hence, the conditions that $\sqrt{D\tau_{so}}$ is smaller than these values are represented by inequalities as $\sigma<10$ S/cm in the metallic phase and as $\varepsilon>2.5\times10^{-8}$ F/m in the insulating phase. Comparing these criteria with the experiment, it is concluded that the measured conductivity is affected by the spin-orbit scattering while the dielectric constant is not. On the other hand, in the magnetic field of 4T, the condition $g\mu_B H>\hbar/\tau_{so}$ holds and the effect of the spin-orbit scattering is reduced.

Firstly, we examine the critical exponent of the conductivity. The value of the critical exponent of the conductivity is predicted in the scaling theory with the interaction through particle-hole channel presented by FINKEL'STEIN [16] and by CASTELLANI et al. [17], to be unity both in the case of strong spin-orbit interaction and in the case of strong magnetic field. The prediction agrees with the present result.

Secondly, we discuss the temperature dependence of the conductivity. The theory with the interaction predicts that the conductivity should vary as $T^{2/5}$ when the spin-orbit scattering is frequent and should vary as $T^{1/3}$ in the strong magnetic field [17]. The theory which includes the effect of the Cooperon channel gives $\sqrt{T}$ dependence for the strong spin-orbit case which explains the data satisfactorily. Recently KAWABATA [19] calculated the coefficient of $\sqrt{T}$ for essentially the same model. According to the theory, the coefficient is nearly constant in the vicinity of the transition and given as

$$m = (12z_0/\pi)^{1/2}(e^2/\hbar)(k_B m^*/2\pi^2\hbar^2)^{1/2} \tag{5}$$

where m^* is the effective mass and z_0 is a quantity of order of unity which cannot be determined by the theory. This formula can be fitted to the data with $z_0=6$; the agreement is satisfactory. On the other hand, we cannot use the formula (5) in the magnetic field of 4T, because the spin-orbit scattering is reduced. Nevertheless the formula seems to be valid in the field up to 4T experimentally. The reason has not been clarified yet.

Lastly we discuss the results of the dielectric constant measurement. As mentioned above, the spin-orbit scattering is not effective in our measurement of the dielectric constant. However, the renormalization group equations for interacting electrons are difficult to solve in the absence of the spin-flip scattering or the magnetic field which reduce the spin fluctuation. Thus we cannot compare our results of the dielectric constant measurement directly with theories with the interacting model. Indeed the theories with the interaction and the strong spin flip scattering give 1.5 [17] or 1 [18,19] for the value of the critical exponent ν' which do not agree with our result.

It should be noted that our results satisfy the universal relations predicted in the non-interacting theory [15] and in the interacting theory with strong magnetic field [17] as

$$\nu'=2\nu \quad \text{and} \tag{6}$$

$$\sigma_0(\chi_0/N_F)^{1/2}=(e^3/\hbar)(8/\pi^5 A)^{1/2} \tag{7}$$

where A takes the value from 1 to 12 according to the universality class of the system. We can fit the data to the relation with A=8. Because of the ambiguities in the pre-factor and the value of N_F, more detailed analysis of the relation (7) is no more significant. Here we have assumed that the spin-orbit scattering does not change the value of σ_0 greatly. It remains as a future problem to explain why these relations hold in the absence of the magnetic field.

5. CONCLUSION

We measured the critical behavior of the conductivity and the dielectric constant of AlGaAs:Si. The effective concentration of the carriers was finely controlled in a single specimen by use of the persistent photoconductivity effect. The obtained results are:

1) The conductivity depends on the temperature as $\sqrt{T}$ in the vicinity of the metal-insulator transition.
2) The value of the critical exponent for the conductivity is 1 in the magnetic field of 0T and 4T.
3) The negative magnetoresistance near the transition also brings about the metal-insulator transition and the exponent is again 1.
4) The exponent for the dielectric constant is close to 2.

We have compared the above results with the scaling theories for the interacting electrons and clarified that:

1') The critical exponent and the temperature dependence of the conductivity are well explained by the theories except the $\sqrt{T}$ dependence in the magnetic field.
2') The critical behavior of the conductivity and the dielectric constant satisfies the universal relations predicted in the non-interacting theory and in the interacting theory with the magnetic field.

ACKNOWLEDGEMENT

I thank N. Sano and his colleagues in Kwansei Gakuin university for supplying the epitaxial wafers. I thank A. Kawabata, C. Di Castro, S. Kobayashi, F. Komori and Y. Ootuka for helpful discussions.

REFERENCES

1. E. Abraham, P. W. Anderson, D. C. Licciardello, and T. V. Ramakrishnan, Phys. Rev. Lett. 42, 673 (1979).
2. B. L. Altshuler, and A. G. Aronov, Solid State Commun. 46, 429 (1983).
3. M. A. Paalanen, T. F. Rosenbaum, G. A. Thomas, and R. N. Bhatt, Phys. Rev. Lett. 48, 1284 (1982).
4. R. J. Nelson, Appl. Phys. Lett. 31, 351 (1979).
5. W. Sasaki, and C. Yamanouchi, J. Non-Cryst. Sol. 4, 183 (1970).
6. D. V. Lang, R. A. Logan, and M. Jaros, Phys. Rev. B19, 1015 (1979).
7. M. O. Watanabe, K. Morizuka, M. Mashita, Y. Ashizawa, and Y. Zohta, Jpn. J. Appl. Phys. 23, L103 (1984).
8. M. Nakayama, K. Kubota, T. Kanata, H. Kato, S. Chika, and N. Sano, Jpn. J. Appl. Phys. 24, 1331 (1985).
9. T. J. Drummond and H. P. Hjalmalson, Appl. Phys. Lett. 48, 1144 (1986).
10. H. C. Casey, Jr., D. D. Sell, and M. B. Panish, Appl. Phys. Lett. 24, 63 (1974).
11. A. Kawabata, J. Phys. Soc. Jpn. 49, 628 (1980).
12. B. Shapiro, Phil. Mag. B50, 241 (1984).
13. M. A. Paalanen, T. F. Rosenbaum, G. A. Thomas, and R. N. Bhatt, Phys. Rev. Lett. 51, 896 (1983).
14. H. Fukuyama, and K. Hoshino, J. Phys. Soc. Jpn. 50, 2131 (1981).
15. A. Kawabata, Prog. Theor. Phys. Supple. 84, 16 (1985).
16. A. M. Finkel'stein, Sov. Phys. JETP57, 97 (1983).
17. C. Castellani, C. Di Castro, P. A. Lee, and M. Ma, Phys. Rev. B30, 527 (1984).
18. C. Castellani, C. Di Castro, Phys. Rev. B34, 5935 (1986).
19. A. Kawabata, preprint.

The Impurity Band Nature of the Metal-Insulator Transition in GaAs

D. Romero[1], *Ming-Way Lee*[1], *H.D. Drew*[1], *M. Shayegan*[2], *and B.S. Elman*[3]

[1]Joint Program for Advanced Electronic Materials: Department of Physics and Astronomy, University of Maryland, College Park, MD 20742, USA and Laboratory for Physical Sciences, College Park, MD 20740, USA

[2]Department of Electrical Engineering, Princeton University, Princeton, NJ 08544, USA

[3]GTE Laboratories Inc., Waltham, MA 02254, USA

1. Introduction

Despite much recent progress [1], there remain many unanswered fundamental questions about the electronic structure of doped semiconductors near the metal-insulator transition. In particular, whether the metal-insulator transition takes place in the conduction band or in an impurity band separated off from the conduction band has not been clearly established. Hartree-Fock cluster calculations [2] show that the impurity band merges with the conduction band for donor densities somewhat below the critical density, n_c. Other calculations [3] indicate that the impurity band persists up to densities of $n \simeq 10n_c$. Even the recent literature contains conflicting conclusions on this question based on the transport data. This issue relates to the question: Is the metallic state near the MI transition more correctly regarded as a tight binding amorphous metal or a dirty free electron metal?

In earlier work we presented evidence from transport [4] and far infrared cyclotron resonance [5] that the tight binding picture is appropriate near the magnetic field-induced MI transition in narrow gap semiconductors n-InSb and n-HgCdTe. Here we present transport and far infrared transmission data on GaAs doped with Si which give compelling evidence for the tight binding picture. We observe the impurity $1s\text{-}2p^{\pm}$ optical transition and, under the same conditions, metallic conductivity. This implies that the MI transition in this system occurs within the impurity band.

2. Transport Measurements

The GaAs samples were films typically 3 μm thick grown by molecular beam epitaxy on single crystal GaAs substrates. They were doped n-type by Si co-evaporation during growth. Transport measurements were made using a Hall bar configuration. For the three samples that we studied, we deduced carrier densities of 1.30×10^{16} cm^{-3} (sample A), 3.44×10^{16} cm^{-3} (sample B) and 5.48×10^{16} cm^{-3} (sample C). From the Mott relation ($n_c^{1/3}a^* \cong 0.25$), the zero magnetic field critical density for the MI transition, $n_c = 1.6 \times 10^{16} cm^{-3}$. From the Hall mobility data we estimate the compensation, $K = N_a/N_d \cong 0.3$ for these samples where $N_a (N_d)$ is the acceptor (donor) concentration.

Results of the low temperature transport measurements done on sample B are shown in Fig. 1. The longitudinal conductivity, σ_L, is proportional to $T^{1/2}$

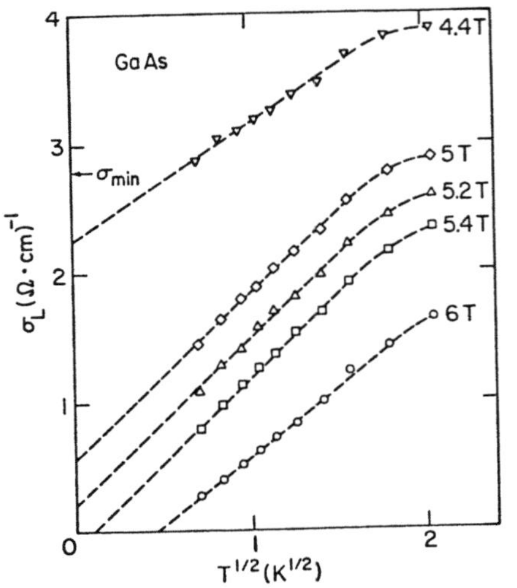

Fig. 1. Longitudinal conductivity of GaAs sample B for different values of the applied magnetic field. Mott's minimum metallic conductivity is indicated.

at low temperatures, consistent with the theory of disordered metals in high magnetic fields [6] and previous results reported for GaAs and other doped semiconductors [7]. This behavior is attributed to the effects of electron-electron interaction in disordered metals. The extrapolation of these curves to 0 K shows that the MI transition occurs at about 5.3T. Also, there is no "minimum metallic conductivity". The σ_{min} predicted by Mott [8] is indicated in Fig. 1. Although there is now general agreement that the minimum metallic conductivity concept is not appropriate for the zero field MI transition, the case of the magnetic field induced MI transition is still under debate [9].Similar transport behavior has been observed in sample C. The critical field for this case is 6.5 T.

3. Far-infrared Measurements

The Faraday geometry far infrared (FIR) measurements were made using a Specac Fourier transform spectrometer and a Ge composite bolometer detector. The transmission spectra for sample B at 4.2 K are shown for different magnetic fields in Fig. 2. Arrows in the figure mark the expected positions of cyclotron resonance and the 1s-2p$^{\pm}$ transitions from experiments [10] on low-doped samples and from theoretical calculations [11].

The data were least square fit to theoretical transmission curves based on the dielectric function, $\epsilon^{\pm}$, consisting of Lorentzian oscillators:

$$\epsilon^{\pm} = \epsilon_o + \frac{\omega_o^2 f^{\pm}}{\omega_{\pm}^2-\omega^2+i\omega\Gamma_{\pm}} - \frac{\omega_o^2 f_o}{\omega(\omega+i\Gamma_o)}$$

where the first oscillator models the *inter*-impurity band (1s-2p$^+$ and 1s-2p$^-$) transitions and the second is a zero frequency oscillator (Drude) term to model *intra*-impurity band transitions. The resulting curve fits, shown in Fig. 2, represent the data well except below $\nu\cong 15$ cm^{-1} and above $\nu\cong 160$ cm^{-1} where the low spectral intensity give large errors in the measured transmission ratios. Since it was necessary to include the Drude term to improve the fits at low frequency ($\nu\leq$ 25 cm^{-1}), we believe that the 1s-2p$^-$

interband absorption line is superimposed on a free-electron-like *intraband* absorption. This is corroborated by the fact that the Drude term reduces (≈ 0 at 9T)as the magnetic field is increased where the sample becomes insulating. Also, the resulting values of f^+ and f^- from these fits are consistent with the hydrogenic donor model. Furthermore, for increasing magnetic field f^+ increases and f^- decreases as is expected for isolated donors [12].This is a clear evidence that nearly all of the carriers are participating in the resonance since the oscillator line shape fits were made under the assumption that n is the Hall density.

The temperature dependence of these spectra also support the impurity transition hypothesis. At 40K the transmission at 4T (top trace in Fig. 2) shows a single resonance line at the position corresponding to the cyclotron resonance. As the temperature is lowered the amplitude of the cyclotron resonance decreases and that of the impurity absorption increases. The two resonances are nearly equal in strength at 20K.

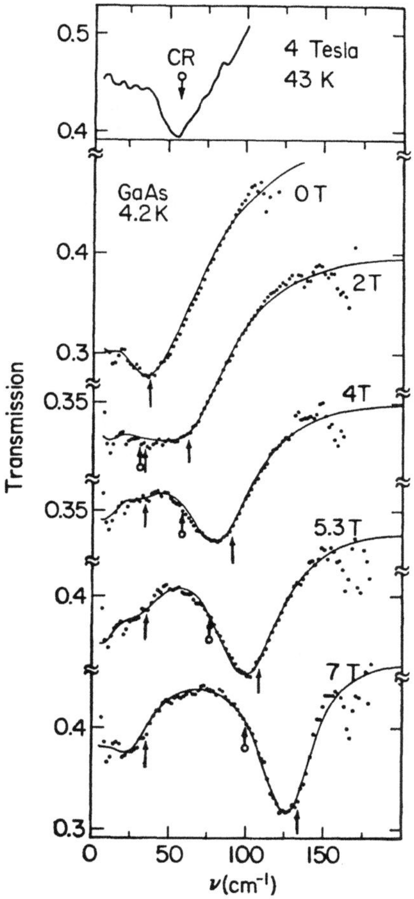

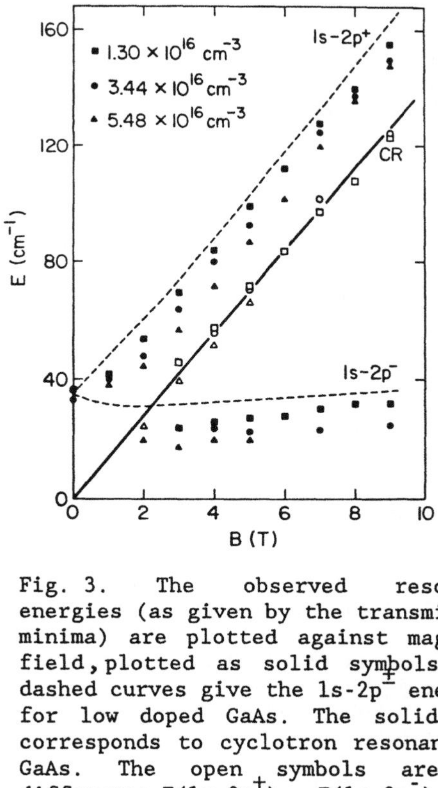

Fig. 3. The observed resonance energies (as given by the transmission minima) are plotted against magnetic field, plotted as solid symbols. The dashed curves give the 1s-$2p^{\pm}$ energies for low doped GaAs. The solid line corresponds to cyclotron resonance in GaAs. The open symbols are the difference $E(1s\text{-}2p^+) - E(1s\text{-}2p^-)$.

Fig. 2. The far infrared transmission at 4.2K of GaAs sample B vs. frequency for different magnetic fields. The dots are the data points and the solid line is an oscillator fit to the data. The plain arrows mark the positions of the 1s-$2p^{\pm}$ transitions for low doped samples. The dressed arrows mark the position expected for cyclotron resonance in GaAs. The top trace is a high temperature spectrum at 4T.

The field dependence of the measured resonance frequencies for the three samples are shown in Fig. 3 . Also shown are the energies for the isolated donors and the cyclotron resonance dispersion based on the conduction band effective mass (m^*=0.067) for GaAs. The resonance frequencies are shifted to lower frequencies than the isolated donor result. The magnitude of this shift increases with donor density. The 1s-2p resonance persists even down to zero magnetic field where samples B and C are metallic. It is also interesting to note that for each magnetic field the difference, $E(1s\text{-}2p^+) - E(1s\text{-}2p^-) \cong \omega_c$. This is clearly seen in Fig. 3 where all the open data points representing this difference lie very close to the cyclotron resonance line.

4. Conclusion

These spectroscopic results are surprising in view of the earlier work on Si:P on the insulating side of the MI transition and at zero magnetic field. For $n \approx n_c/4$ the 1s-2p transitions have disappeared and only a broad absorption band shifted to lower frequency by about 30% from the low doped 1s-2p resonance was observed [13]. This absorption was attributed to charge transfer transitions in donor pairs.

The data in Figs. 2 & 3 clearly demonstrate the persistence of the impurity 1s-2p related transitions on both the insulator and the metallic side of the MI transition at low temperatures. These results imply that the metallic conduction near the MI transition in this system occurs within the impurity band.

Acknowledgement

This work was supported in part by the National Science Foundation under grant Nos. ECS-8553110 and DMR-8704670. M. S. gratefully acknowledges an IBM Faculty Development Award.

References

1. See, for example, Proc. of the International Conference on Localization, Interaction and Transport Phenomena, ed. G. Bergmann and Y.Bruynseraede, (Springer-Verlag, New York, 1985).
2. K.A. Chao, R. Riklund, and A. Ferreira da Silva, Phys. Rev. B, 21, 5745, (1980).
3. N. F. Mott and E. A. Davis, Electronic Processes in Non-Crystalline Solids, (Oxford University Press, London, 1979)
4. V. J. Goldman, M. Shayegan, and H. D. Drew, Phys. Rev. Letters 57, (1986).
5. V. J. Goldman, H. D. Drew, M. Shayegan, and D. A. Nelson, Phys. Rev. Letters 56, 968 (1986).
6. B. L. Altshuler and A. G. Aronov,Sov. Phys. JETP 50, 968 (1979), and B. L. Altshuler, A. G. Aronov, P. A. Lee, Phys. Rev. Letters 44, 1288 (1980).
7. A. P. Long and M. Pepper, Solid- State Electronics 28, 61 (1985), and R. Mansfield, M. Abdul-Gader, and P. Fozooni, ibid, p.109.
8. N. F. Mott, Philos. Mag. 26, 1015 (1972).
9. N. F. Mott, Solid State Electronics 28, 57 (1985).
10. G. E. Stillman, C. M. Wolfe, and J. O. Dimmock, Proc. Int. Conf. on Photoconductivity at Stanford ed. E. M. Pell, (Pergamon, New York, 1971) p. 265.
11. D. Larsen, private communication.
12. H. Hasegawa and R. E. Howard, J. Phys. Chem. Solids 21, 179 (1961).
13. M. Capizzi, G. A. Thomas, F,. DeRosa, R. N. Bhatt, and T. M. Rice, Phys. Rev. Lett. 44, 1019 (1980).

Part II

The Metal-Insulator Transition in Metals

Interaction Effects Near the Metal-Insulator Transition in Semimagnetic Semiconductors

T. Dietl, M. Sawicki, T. Wojtowicz, J. Jaroszyński, W. Plesiewicz, L. Świerkowski, and J. Kossut

Institute of Physics, Polish Academy of Sciences,
Al. Lotników 32/46, PL-02-668 Warszawa, Poland

Experimental studies of electronic and magnetic properties of doped semimagnetic semiconductors (n-$Cd_{1-x}Mn_xSe$ and p-$Hg_{1-x}Mn_xTe$) are reviewed with a special emphasis on results yielding information on effects of electron-electron interactions near the Anderson-Mott transition.

1 INTRODUCTION

We discuss here results of recent studies of doped semimagnetic semiconductors [1] in the vicinity of the metal-to-insulator transition [2,3]. We shall focus on the low temperature data for the wide-gap n-$Cd_{0.95}Mn_{0.05}Se$ ($E_g \simeq 1.9$eV) [4-13] and the narrow-gap p-$Hg_{0.915}Mn_{0.085}Te$ ($E_g \simeq$ 50meV) [7,10,13]. It should be noted, however, that experimental results obtained for other Mn-based compounds [14,15] and/or different Mn mole fractions [5,8,16–21] are consistent with those discussed explicitly below.

In II-VI compounds the states derived from half filled d-shells of Mn substitutional atoms ($S = 5/2$, $g = 2.0$), are known to lie deeply below the Fermi energy [1]. The Mn s-electrons, on the other hand, contribute to the sp orbitals forming a wide valence and conduction bands. We limit our considerations to the paramagnetic region, where the d-spin magnetization can be described by the Brillouin function (phenomenologically modified [1] to account Mn-Mn antiferromagnetic interactions).

The most prominent feature of the materials in question is a strong exchange interaction, $-J\vec{S}\vec{s}$, between Mn d-electrons and effective mass sp electrons. Due to this interaction the doped semimagnetics exhibit a number of properties interesting from the viewpoint of the physics of disordered systems. First, it has recently been realized that the conductivity near the Anderson-Mott transition is very sensitive to symmetry lowering perturbations [2,3]. Two of these can be particularly large in magnetic systems: (i) the spin-flip scattering by magnetic ions and (ii) the spin-splitting of electron states which is known from magnetospectroscopic studies to be considerably enhanced by the sp-d interaction and which is proportional to the d-spin magnetization [1]. Second, because of sp-d coupling, the electron spin polarizes Mn spins within its localization orbit. Ferromagnetic clouds formed in this way (bound magnetic polarons [1]) manifest themselves in various experiments enabling a direct estimation of the s-spin concentration and their localization radius. Third, the materials in question, in contrast to nonmagnetic semiconductors, undergo the ***insulator-to-metal*** transition on ***increasing*** magnetic field. A remarkable aspect of the transition is that it takes place in the magnetic fields for which the Landau quantization of the density of states is insignificant while the spin-splitting is of the order of the Fermi energy, causing the majority of carriers to reside in one spin subband. For this spin-polarized case the renormalization group equations [22,23] can be solved making an experimental verification of the theory possible. Finally, we note that the p-d exchange interaction causes the valence band to be markedly anisotropic in the magnetic field [24]. It is possible then to test those predictions of the percolation [25] and scaling theories [26,27] which concern anisotropic systems.

Our paper is organized as follows. In Section 2 an anomalous temperature dependence of the conductivity observed below 0.5K in metallic n-$Cd_{0.95}Mn_{0.05}Se$ [6,9,12] is discussed in view of the current ideas concerning an enhancement of s-spin fluctuations [28] and persistence of quasi-static s-spins onto the metallic side of the transition [29]. Next, in Section 3, experimental results on the magnetoresistance of n-CdMnSe [4–13] are presented and discussed in terms of influence of the spin-splitting on the triplet channel of the electron-hole propagator [30] and of the redistribution of carriers between spin-subbands [31]. Finally, in Section 4, we turn to the temperature dependence of the conductivity near the magnetic-field driven transition in isotropic n-CdMnSe and anisotropic p-HgMnTe [7,10,13].

2 TEMPERATURE DEPENDENCE of CONDUCTIVITY

One of the open questions of the physics of the Anderson-Mott localization concerns spin properties on the metallic side of the transition. On the experimental side, an enhancement of the nuclear relaxation rate [32], magnetic susceptibility [33,34], and spin-resonance line width [34] has been detected in metallic Si:P at millikelvin temperatures. These findings are being described in two alternative ways. First is based on the Fermi liquid picture, with the interaction amplitudes depending on the length scale (temperature) [28]. The second approach involves a conjecture that the localized s-spins do not disappear at the insulator-to-metal transition (MIT) and reside, owing to the Hubbard repulsion, below the Fermi energy of the metal [29]. The latter model qualitatively accounts for a gradual (as opposed to critical) evolution of spin properties on proceeding from the insulator to the metal and relates magnetic properties on the both sides of the transition to the exchange interactions between s-spins. It is not clear at present whether the spin anomalies mentioned above occur in the case of any Anderson localization or rather are due to some characteristic features of doped semiconductors such as the 1Coulomb scattering potential and/or a strong energy dependence of the density of states.

Some light on this controversy may be shed by a recent observation by us [6] of a sudden and unexpected drop of the conductivity of the metallic n-$Cd_{0.95}Mn_{0.05}Se$ below ~0.5K (Fig.1). We have suggested [11] that this effect is a signature of the localized (quasi-static) s-spins present on the metallic side of the transition. Arguments supporting this conjecture are listed below.

As shown in Fig.1, the drop of the conductivity is hardly visible in the CdMnSe sample with the highest net donor concentration $n \simeq 2 \times 10^{18}\text{cm}^{-3}$, but then, on lowering n, and thus on approaching the MIT, its relative magnitude increases. The behavior of the samples with $n \leq 6 \times 10^{17}\text{cm}^{-3}$ can be viewed as the *temperature* driven MIT. These two facts imply strongly that the anomaly is associated with the proximity of the transition. Furthermore, it is seen in Fig.1 that the anomaly is absent in the case of non-magnetic n-CdSe. Hence, the anomalous dependence of σ on T in CdMnSe must be assigned to the s-d exchange interaction. It also turns out that $\sigma(T)$ in CdSe is correctly described by the current theory of the conductivity in the weakly localized regime (WLR) [3] assuming temperature *independent* Fermi-liquid amplitudes g_i. It is important to note that using the same values of g_i's one obtains a good description of the magnetoresistance in CdSe and CdMnSe [6].

A question arises whether the anomaly in $\sigma(T)$ is directly related to magnetic behavior of the system. Figure 2 presents magnetic susceptibility $\chi(T)$ for $Cd_{0.95}Mn_{0.05}Se$:In with $n \simeq 8\times10^{17}\text{cm}^{-3}$ [12]. The data are consistent with those reported for undoped $Cd_{0.95}Mn_{0.05}Se$ [35]. A spin-glass cusp is visible at $T_{sg} \approx 95$mK, while above 100mK $\chi(T) \sim T^{-\beta}$ with $\beta < 1$ ($\beta \approx 0.7$), a dependence characteristic for amorphous antiferromagnets [36]. Since above 100mK no magnetic long range order is observed and also because characteristic times of s-spin fluctuations are expected to be much shorter than those of d-spins we may disregard a direct influence of the s-d interaction on electron energies and conclude that the anomalous dependence of σ on T must be caused by spin-disorder scattering. In the vicinity of the MIT this scattering can be of particular importance because it destroys both the coherence of backscattering [3,37] and the effects of the interactions in the Cooper and electron-hole triplet channels [3] (the latter turns out to dominate quantitatively). Thus, in terms of Refs.22 and 23, the low-temperature drop of the conductivity can be regarded as a cross-over to the "singlet-only" universality class. Since in CdMnSe the s-d coupling is ferromagnetic [38], the Kondo effect is insignificant. Moreover, the exchange energy is smaller than the width of the conduction band. Therefore the subsystem

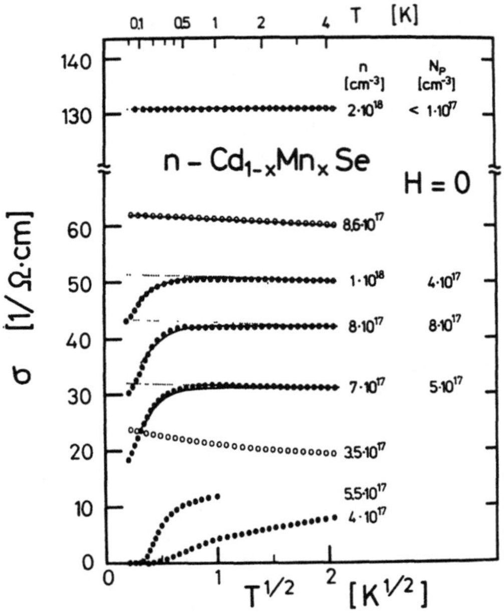

Fig.1 Temperature dependence of the conductivity in $Cd_{0.95}Mn_{0.05}$Se:In (●) and CdSe:In (○). The net donor concentrations n — from Hall data at 300 K. (—) calculated in the WLR ($k_F\ell_e > 1$) taking quantum corrections to the conductivity into account and treating the polaron concentration N_p (in n-CdMnSe) as an adjustable parameter. ($\cdots$) calculated for $N_p = 0$ (after [12])

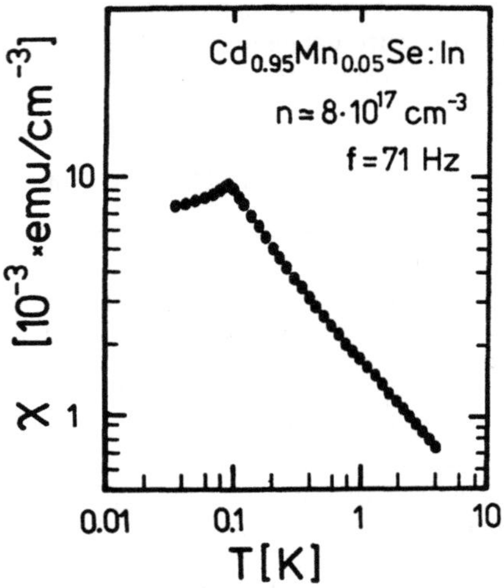

Fig.2 Magnetic susceptibility of n-$Cd_{0.95}Mn_{0.05}$Se measured by means of mutual inductance method using a.c. magnetic field of 1 Oe. An estimated error in absolute values of χ is about 5% above 1 K and perhaps 25% at the lowest temperature (after [12])

of d-spins can be described classically. In that case the inverse spin lifetime of s-electrons, τ_{si}^{-1}, is proportional to the correlation function of the d-spin magnetization, $\langle M_{\vec{q}}^{(i)}\, M_{-\vec{q}}^{(i)}\rangle$, averaged over the Fermi sphere, $0 < q < 2k_F$. We note also that a spin-glass transition is signalized by slowing-down of spin dynamics rather than by a change in spin configurations [39] and therefore is expected to have a minor influence on spin-disorder scattering.

In attempt to understand the anomalous drop of σ in terms of the spin-disorder scattering let us first assume that there are no localized s-spin excitations. Then the correlation function $\langle M_{\vec{q}}^{(i)}\, M_{-\vec{q}}^{(i)}\rangle$ is proportional to $\chi(T)\cdot T$ (for $T > T_{sg}$) and therefore its magnitude *decreases* with lowering temperature.The corresponding scattering rate is given by

$$\tau_{si}^{-1} = \frac{k_B T \alpha^2 k_F m^*}{4\pi\hbar^3 g^2 \mu_B^2}\mu\chi_i(T). \tag{1}$$

Here $i = x, y, z$ and α is the s-d exchange integral. The parameter μ stands for an enhancement factor brought about by a combined effect of the disorder and spin-fluctuations in the Fermi liquid [40]. While the former produces a temperature independent correction, the latter has been suggested [40] to scale with the triplet-channel interaction-amplitude and, therefore, to result in temperature dependent μ. We fitted the WLR theory to the experimental $\sigma(T)$ results treating μ as an adjustable parameter for each temperature. For $n = 8\times 10^{17}\mathrm{cm}^{-3}$ we obtained $\mu \simeq 8$ (i.e., $\hbar/k_B\tau_{si} \simeq 0.5$K) at 0.5K and $\mu \simeq 10^2$ (i.e., $\hbar/k_B\tau_{si} \simeq 4$K) at 0.1K. Thus our initial assumption led us to inconsistency of the model because we obtain $\hbar/\tau_{si} > k_BT$, and under such conditions a growth of the triplet amplitude on lowering temperature is precluded [22,23,28]. The above difficulty is not encountered within the model which allows the localized s-spins to be present on the metallic side of the MIT. Since such spins remain for a long time in a given region, they can polarize, via s-d interaction, the Mn spins inside their localization orbit ξ (there are about 200 Mn ions in the Bohr volume $\frac{4}{3}\pi a_B^3$ in $Cd_{0.95}Mn_{0.05}Se$). The degree of the Mn polarization (magnetization) produced by the s-spin is proportional to $\chi(T)$ [41] and thus strongly *increases* when the temperature is lowered. Ferromagnetic clouds (magnetic polarons) formed in this manner constitute centers of efficient spin-disorder scattering of delocalized electrons because $\xi \approx a_B$ is of the order of k_F^{-1}. After averaging over random positions of the polarons, the additional spin-disorder scattering rate assumes the form

$$\tau_{pi}^{-1} = f_i(T) N_p \mu \chi_i^2(T), \tag{2}$$

where N_p is the polaron concentration. Disregarding exchange interactions among s-spins, the explicit form of $\langle M_{\vec{q}}^{(i)} M_{-\vec{q}}^{(i)} \rangle$, and thus of the function f_i in Eq.2 , can be easily deduced from the expression for the free energy functional of the bound magnetic polarons [42]. Calculations represented by solid lines in Fig.1 were performed with τ_s containing scattering by both the fluctuations (Eq.1) and polarons (Eq.2). We assumed $\xi = a_B$ and treated N_p as an adjustable parameter. The enhancement factor μ in Eq.1 was tentatively approximated by [40] $\mu = (1 + 3\sqrt{3}/\pi k_F^2 \ell_e^2)(1 + F/2)$, while for long-range polaron scattering (Eq.2) by $[1 + 3\sqrt{3}(1 + k_F^2 a_B^2)/\pi k_F^2 \ell_e^2](1 + F/2)$. Here ℓ_e is the mean free path for elastic collisions and F is the triplet amplitude estimated from studies of the spin-splitting induced magnetoresistance (see next section) to be $F \simeq 3$ (the magnetoresistance is controlled by[3] $g_3 \simeq \frac{4}{3F}[4(1 + F/2)^{3/2} - 3F - 4]$). It can be seen in Fig.1 that our phenomenological model correctly reproduces experimental results and that estimated polaron concentrations are of expected order of magnitude.

3 MAGNETORESISTANCE

As shown in Fig.3, the magnetoresistance of n-CdMnSe near the MIT consists of positive at low fields and high-field negative components. We relate [6,9,11] the positive magnetoresistance (PM) to a destructive effect of the giant s-d exchange spin-splitting on the diffusive poles of the triplet particle-hole and singlet particle-particle channels [3,30] and a corresponding cross-over to the "high magnetic field" universality class [22,23]. This mechanism operates providing that the spin-splitting $\hbar\omega_s$ is larger than k_BT and than the spin-disorder scattering rate $\hbar/\tau_s$ [3]. A confirmation of the model comes from a good agreement between the theoretical predictions [3] and experimental results [6,9] in the WLR, as shown in Fig.4, where the solid lines represent magnetoresistance, $\rho(H) - \rho(0)$, calculated with one adjustable parameter, the coupling constant of electron-electron interactions, $g_3 + g_4$, chosen as 1.4. The calculations are limited to $T \geq 1.5$K below which the magnetization and hence $\hbar\omega_s$ are unknown. As can be seen in Fig.4,the low-field PM grows on lowering temperature reflecting an increase in $\hbar\omega_s$. However, at temperatures for which polaronic effects become significant, PM starts to diminish. This is an additional indication that polarons enhance the spin-disorder scattering rate. The theory referred to above neglects effects of redistribution of electrons between spin-subbands, since it has been derived under an assumption that $\hbar\omega_s$ is much smaller than both the Fermi energy E_F and the inverse momentum relaxation time, $\hbar/\tau$.This assumption is not well fulfilled in high fields because, e.g., $\hbar\omega_s$(20kOe) $\simeq$ 13meV, whereas $E_F(0) \simeq \hbar/\tau \simeq$ 25meV (for $n \simeq 8 \times 10^{17}$cm^{-3}). It has been suggested in Ref.31 that the spin-splitting causes the Fermi level to shift into the region of states having higher conductances. This, together with a destructive effect of the magnetic field on the spin-disorder scattering rate, appears to produce the negative magnetoresistance (NM) visible in Figs.3 and 4. NM is particularly strong in the case of samples with n in a immediate vicinity of the MIT. In particular, for $3 \times 10^{17} \leq n \leq 6 \times 10^{17}$cm^{-3} one observes, at sufficiently low temperatures, the insulator-to-metal transition driven by the magnetic field.

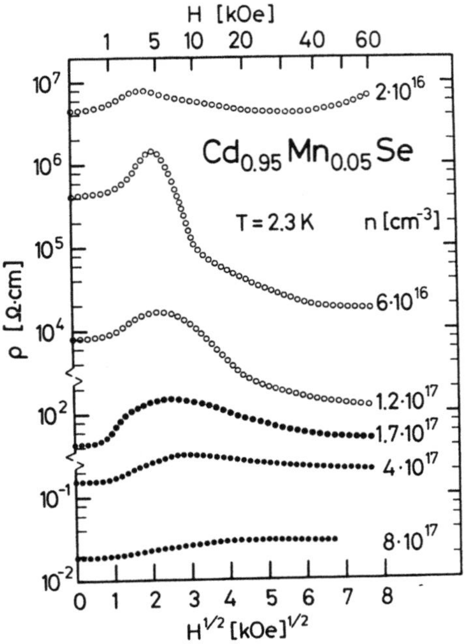

Fig.3 Magnetoresistivity of n-$Cd_{0.95}Mn_{0.05}Se$ for selected electron concentrations n. (●) In-doped crystals, (○) un-intentionally doped crystals. The Mott maximum metallic resistance is about 0.14 $(\Omega cm)^{-1}$ (after [11])

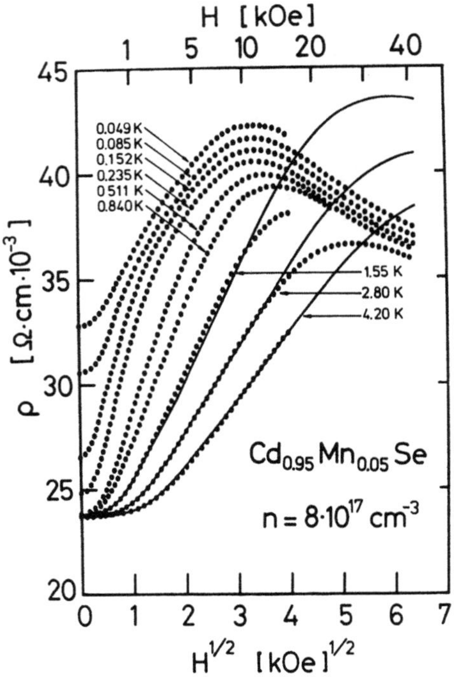

Fig.4 Magnetoresistivity of n-$Cd_{0.95}Mn_{0.05}Se$ at selected temperatures. (—) calculated neglecting the redistribution of electrons between spin subbands, which presumably affects the conductivity in high fields (after [9])

4 INSULATOR-TO-METAL TRANSITION

The experimental result showing that MIT takes place as the magnetic field is *decreased* in n-CdMnSe, is presented in Fig.5. We see that below a critical field $H_c \sim 13$kOe, the resistivity ρ strongly (exponentially) grows with decreasing temperature, and thus tends to infinity as $T \to 0$. Above H_c, the conductivity $\sigma \equiv 1/\rho$ can be described by [7] $\sigma(T,H) = \sigma_c(H/H_c - 1)^\nu + A(H)T^\delta$ (with $\nu \simeq 1$, $\delta \simeq 1/2$) which indicates a metallic behavior. To determine the universality class of this transition we note [7,10] that it occurs in weak magnetic fields from the viewpoint of the condition $\omega_c\tau \ll 1$ but in strong fields in the sense that the bare thermal diffusion length is longer than the magnetic length (for our estimates we take $\ell_e = k_F^{-1}$). This means that we can neglect Landau quantization of the density of states, as well as the interactions in the Cooper channel and the coherence of backscattering. At the same time, due to the large spin splitting, the distance between E_F and the bottom of the majority-spin subband is estimated to be at $H \simeq H_c$ about twice the distance to the minority spin subband. This means that the bare diffusion coefficient should be substantially different for these two subbands. Therefore we assume, following [5,8,17,31], that for $H \geq H_c$ the conductivity of the majority electrons scales with the system size to the metal,whereas that of the minority scales to the insulator. The localization of the minority electrons implies, in turn, the presence of a Coulomb gap [22,23,25] in their density of states. Presumably, it is the Coulomb gap which makes the resonant spin-flip scattering (tunneling) between the subbands of minor importance and allows us to consider the subbands to be decoupled. The above arguments lead to the conclusion that the nature of the magnetic-field driven insulator-to-metal transition in magnetic systems is entirely determined by electrons which reside in one spin-polarized subband. Electrons in the second subband

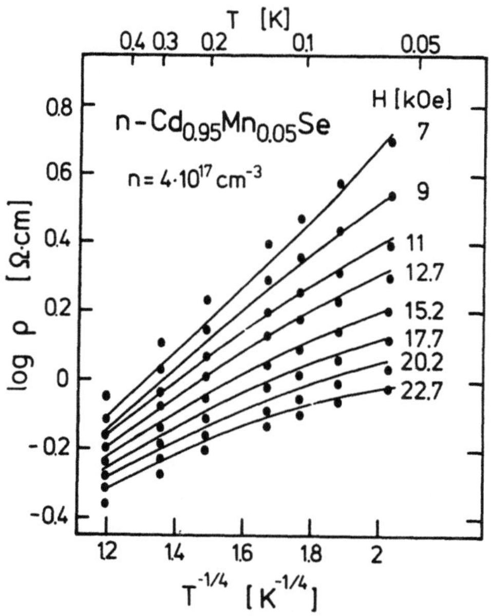

Fig.5 Resistivity of n-CdMnSe near the insulator-to-metal transition. (—) calculated from the renormalization group equations for the spin-polarized universality class and disregarding phonon-assistedprocesses (after [13])

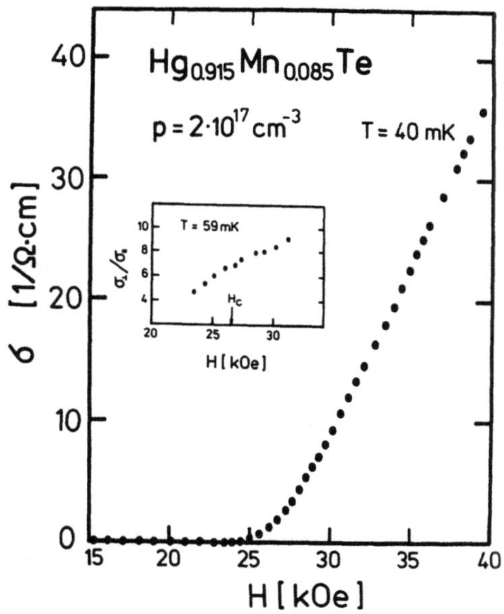

Fig.6 Conductivity in the transverse magnetic field in p-HgMnTe near the insulator-to-metal transition. The inset shows the ratio of the transverse and longitudinal conductivities in the same field range (after [13])

produce only an additional static random potential and serve as an electron reservoir which is depleted when ω_s increases. Recently, FINKELSTEIN [43] has derived the renormalization group equations (RGE) for the spin-polarized case. To the lowest order in G^{-1} and $\epsilon = d - 2$, where G is the dimensionless conductance and d is the space dimensionality, but to all orders in the Coulomb electron-electron interactions, he has obtained $d\ln G/d\ln L = d - 2 - G^*/G$ and $d\ln z/d\ln L = -1/2G$. Here $\sigma(L) = 8\sigma_N G(L)/L$, where $\sigma_N = e^2/2\pi^2\hbar$, $G^* = 1/\epsilon[1 + O(\epsilon^2)]$ and z is a function which relates the length and temperature scales according to

$$z(L)L^2 = \sigma(L)/\pi\sigma_N N_1 k_B T, \tag{3}$$

where N_1 is the thermodynamic density of states for the majority spin subband.

The $\sigma(T, H)$-values computed from the RGE were fitted to the experimental results [13] treating H_c, G^*, l and σ_c as adjustable parameters (l denotes the length at which the RGE begin to be valid and σ_c is given by $\sigma_c = 8\sigma_N l^{-1} H_c dG(l)/dH$ at H_c). This fitting procedure yields $H_c = 13^{+3}_{-1.5}$kOe, $G^* = 0.41^{+0.2}_{-0.1}$, $l = 260 \pm 50$Å, and $\sigma_c = 0.66 \pm 0.1(\Omega\text{cm})^{-1}$ [44]. As shown in Fig.5, the calculated conductivities are compatible with the experimental results on the both sides of the transition.

Turning to p-HgMnTe we recall that redistribution of holes between Γ_8 subbands, induced by the p-d exchange, is connected with a change of the transverse hole mass, changing from heavy mass to light mass [24], as the magnetic field is increased. This has two important implications. First, the MIT should be much sharper, and hence σ_c much larger, in p-HgMnTe than in n-CdMnSe. Second, the conductivity in p-HgMnTe is expected to exhibit strong anisotropy in the magnetic field. That this is indeed the case is shown in Fig.6, where the low-temperature values of $\sigma_\perp$ and $\sigma_\perp/\sigma_\parallel$ are plotted as a function of the magnetic field. These data give $\bar{\sigma}_c = (\sigma_c^{\perp 2} \cdot \sigma_c^{\parallel})^{1/3} \simeq 34(\Omega\text{cm})^{-1}$ [7], which is by a factor of ~ 50 greater than in n-CdMnSe. The sharpness of the transition causes that in the insulating phase, for $H_c - H \leq 0.9H_c$, one has $\xi \ll L$. Hence, the values of $\sigma(T)$ calculated from the RGE for this range of the magnetic fields are very small and in fact much smaller than those observed experimentally [13]. This indicates an onset of a phonon-assisted hopping. Nevertheless, $\sigma(T, H)$ in the metallic phase can be properly described by the RGE [7]. We have to mention here that the RGE which were derived [22,23] for $\hbar\omega_s \gg k_BT$ but, implicitly, for $\hbar\omega_s \ll E_F$ ("magnetic field" case in terms of Refs. 22,23), predict a dependence of σ on T in *disagreement* with our data for both n-CdMnSe and p-HgMnTe, as well as for other magnetic material, $Gd_{3-x}v_xS_4$ [45]. This substantiates the assignment of a separate universality class to systems with a large spin splitting, $\hbar\omega_s \sim E_F$ [7]. A smooth dependence of $\sigma_\perp/\sigma_\parallel$ on the magnetic field (Fig.6) demonstrates that in spite of a large anisotropy one observes the same critical field H_c and critical index ν in the transverse and longitudinal configurations. This confirms theoretical predictions of WÖLFLE and co-workers [26] but seems to contradict those of PENDRY [27].

5 CONCLUSIONS

The experimental results discussed above convincingly demonstrate a crucial role played by effects of electron-electron interactions in the temperature and field dependencies of the conductivity in the neighborhood of the MIT. Furthermore, these results and their interpretation help understand microscopic origins of the temperature induced localization and giant magnetoresistance effects, known for a long time to occur in magnetic semiconductors [45,46]. In particular, we hope to have shown that these effects reflect an interplay between two symmetry lowering perturbations: the spin splitting and the spin-disorder scattering, the latter being enhanced by formation of bound magnetic polarons persisting onto the metallic side of the MIT.

ACKNOWLEDGMENTS

We are grateful to R.R.Gałązka, A.Mycielski, and T.Skośkiewicz for continuous encouragement and support, as well as to A.M.Finkelstein, D.E.Khmel'nitskii, and Y.Ono for valuable discussions. This work was supported in part by Institute for Low Temperature and Structural Research, Polish Academy of Sciences.

REFERENCES

[1] for review of semimagnetics see, e.g., R.R. Gałązka: In *Proc. 18th Int. Conf. on The Physics of Semiconductors, Stockholm 1986*, ed. by O. Engström, (World Scientific, Singapore 1987) p.1727; J.K. Furdyna: *J. Vac. Sci. Technol.* **A4**, 2002 (1986); J. Mycielski: *Prog. Cryst. Growth Charact.* **10**, 101 (1984); N.B. Brandt, V.V. Moschalkov: *Adv. Phys.* **33**, 193 (1984).

[2] for review see, e.g., *Localisation and Interaction in Disordered Metals and Doped Semiconductors*, ed. by D.M. Finlayson, (SUSSP, Edinburgh1986); *Electron-Electron Interaction in Disordered Systems*, ed. by A.L. Efros and M. Pollak, (North-Holland, Amsterdam 1985); N.F. Mott, M. Kaveh: *Adv. Phys.* **34**, 329 (1985).

[3] B.L. Al'tshuler, A.G. Aronov, H. Fukuyama: In *Electron-Electron Interaction in Disordered Systems*, ed. by A.L. Efros and M. Pollak, (North-Holland, Amsterdam 1985); B.L. Al'tshuler, A.G. Aronov, D.E. Khmel'nitskii, A.I. Larkin: In *Quantum Theory of Solids*, ed. by I.M. Lifshits (MIR Publishers, Moscow 1982) p.130; P.A. Lee, T.V. Ramakrishnan: *Rev. Mod. Phys.* **57**, 287 (1985).

[4] T. Dietl, J. Antoszewski, L. Świerkowski: *Physica* **117-118B**, 491 (1983).

[5] Y. Shapira, D.H. Ridgley, K. Dwight, A. Wold, K.P. Martin, J.S. Brooks: *J. Appl. Phys.* **57**, 3210 (1985).

[6] M. Sawicki, T. Dietl, J. Kossut, J. Igalson, T. Wojtowicz, W. Plesiewicz: *Phys. Rev. Lett.* **56**, 508 (1986).

[7] T. Wojtowicz, T. Dietl, M. Sawicki, W. Plesiewicz, J. Jaroszyński: *Phys. Rev. Lett.* **56**, 2419 (1986).

[8] Y. Shapira, N.F. Oliveira,Jr., D.H. Ridgley, R. Kershaw, K. Dwight, A. Wold: *Phys. Rev.* **B34**, 4187 (1986).

[9] M. Sawicki, T. Wojtowicz, T. Dietl, J. Jaroszyński, W. Plesiewicz, J. Igalson: In *Proc. 18th Int. Conf. on Physics of Semiconductors, Stockholm 1986*, ed. by O. Engström, (World Scientific, Singapore 1987) p.1265.

[10] T. Wojtowicz, M. Sawicki, T. Dietl, W. Plesiewicz, J. Jaroszyński: In *High Magnetic Fields in Semiconductor Physics*, ed. by G. Landwehr, Springer Ser. Solid-State Sci., Vol. 71, 442 (Springer, Berlin, Heidelberg 1987).

[11] T. Dietl, L. Świerkowski, J. Jaroszyński, M. Sawicki, T. Wojtowicz: *Physica Scripta* **T14**, 29 (1986).

[12] M. Sawicki, T. Dietl, T. Wojtowicz, A. Lenard, W. Plesiewicz, L. Świerkowski: *The XVI International School on the Physics of Semiconducting Compounds, Jaszowiec, Poland 1987*, unpublished.

[13] T. Wojtowicz, T. Dietl, M. Sawicki, W. Plesiewicz: *The XVI International School on the Physics of Semiconducting Compounds, Jaszowiec, Poland 1987*, unpublished.

[14] D. Heiman, Y. Shapira, S. Foner: *Solid State Commun.* **45**, 899 (1983).

[15] M. von Ortenberg, W. Erhardt, A. Twardowski, M. Demianiuk: In *High Magnetic Fields in Semiconductor Physics*, ed. by G. Landwehr, Springer Ser. Solid-State Sci., Vol. 71, 446 (Springer, Berlin, Heidelberg 1987).

[16] T. Ichiguchi, H.D. Drew, J.K. Furdyna: *Phys. Rev. Lett.* **50**, 612 (1983).

[17] Y. Shapira, D.H. Ridgley, K. Dwight, A. Wold, K.P. Martin, J.S. Brooks, P.A. Lee: *Solid State Commun.* **54**, 593 (1985).

[18] J. Stankiewicz, S. von Molnar, W. Giriat: *Phys. Rev.* **B33**, 3573 (1986).

[19] T. Wojtowicz, A. Mycielski: *Physica* **117-118B**, 475 (1983).

[20] W.B. Johnson, J.R. Anderson, D.R. Stone: *Phys. Rev.* **B29**, 6679 (1984).

[21] A.V. Germanenko, L.P. Zverev, V.V. Kruzhaev, G.M. Min'kov, O.E. Rut, N.P. Gavalesho, W.M. Frasunyak: *Fiz. Tverd. Tela* **26**, 1754 (1984); A.V. Germanenko, V.V. Kruzhaev, G.M. Min'kov, O.E. Rut: *Fiz. Tekh. Poluprov.* **20**, 1662 (1986).

[22] A.M. Finkelstein: *Zh. Eksp. Teor. Fiz.* **84**, 168 (1983) [*Sov. Phys. JETP* **57**, 97 (1983)]; A.M. Finkelstein: *Zh. Eksp. Teor. Fiz.* **86**, 367 (1984), [*Sov. Phys. JETP* **59**, 212 (1984)].

[23] C. Castellani, C. Di Castro, P.A. Lee, M. Ma: *Phys. Rev. B* 30, 527 (1984).

[24] J.A. Gaj, J. Ginter, R.R. Gałązka: *Phys. Status Solidi (b)* **89**, 655 (1978).

[25] B.I. Shklovskii, A.L. Efros: *Electronic Properties of Doped Semiconductors* (Springer, Berlin 1984).

[26] P. Wölfle, R.N. Bhatt: *Phys. Rev. B* **30**, 3542 (1984); R.N. Bhatt, P. Wölfle, T.V. Ramakrishnan: *Phys. Rev. B* **32**, 569 (1985).

[27] J.B. Pendry: *J. Phys. C: Solid State Phys.* **19**, 3855 (1986).

[28] A.M. Finkelstein: *Pisma Zh. Eksp. Fiz.* **40**, 63 (1984) [*JETP Lett.* **40**, 796 (1984)]; C. Castellani, C. Di Castro, P.A. Lee, M. Ma, S. Sorella, E. Tabet: *Phys. Rev.* **B30**, 1596 (1984); **B33**, 6169 (1986); H. Fukuyama: *J. Phys. Soc. Jpn.* **54**, 2378 (1985).

[29] Zi-Zhao Gan, P.A. Lee: *Phys. Rev.* **B33**, 3595 (1986).

[30] A. Kawabata: *J. Phys. Soc. Jpn.* **50**, 2461 (1981); P.A. Lee, T.V. Ramakrishnan: *Phys. Rev.* **B26**, 4009 (1982).

[31] H. Fukuyama, K. Yosida: *Physica* **105B+C**, 132 (1981).

[32] M.A. Paalanen, A.E Ruckenstein, G.A. Thomas: *Phys. Rev. Lett.* **54**, 1295 (1985).

[33] S. Ikehata, S. Kobayashi: *Solid State Commun.* **56**, 607 (1985).

[34] M.A. Paalanen, S. Sachdev, R.N. Bhatt, A.E. Ruckenstein: *Phys. Rev. Lett.* **57**, 2061 (1986); in *Proc. 18th Int. Conf. on The Physics of Semiconductors, Stockholm 1986*, ed. by O.Engström, (World Scientific, Singapore 1987) p.1249.

[35] M.A. Novak, O.G. Symko, D.J. Zheng, S. Oseroff: *Phys. Rev.* **B33**, 6391 (1986).

[36] R.N. Bhatt, P.A. Lee: *Phys. Rev. Lett.* **48**, 344 (1982).

[37] Y. Ono, J. Kossut: *J. Phys. Soc. Jpn.* **53**, 1128 (1984).

[38] M. Dobrowolska, A. Witowski, J.K. Furdyna, T. Ichiguchi, H.D. Drew, P.A. Wolff: *Phys. Rev.* **B29**, 6652 (1984).
[39] S.F. Edwards, P.W. Anderson: *J. Phys.* **F5**, 965 (1975).
[40] S. Sachdev: *Phys. Rev.* **B34**, 6049 (1986).
[41] A proportionality between the local magnetization and $\chi(T)$ is not obeyed at low temperatures in the case of deep donors with small ξ since then the molecular field of s-spin saturates Mn-spins.
[42] T. Dietl, J. Spałek: *Phys. Rev. B* **28**, 1548 (1983).
[43] A.M. Finkelstein: private communication.
[44] The present value of l differs from that quoted previously [7] since in [7] a factor of 2π was omitted in Eq.3. The present form of Eq.3 together with $G^* = 0.4$ assures a quantitative agreement between $\sigma(T)$ calculated from the RGE and that from the perturbation calculation for the spin-polarized case and $d = 3$.
[45] S. von Molnar, J. Flouquet, F. Holtzberg, G. Remenyi: *Solid State Electron.* **28**, 127 (1985).
[46] J. Stankiewicz, S. von Molnar, F. Holtzberg: *J. Mag. Mag. Mat.* **54-57**, 1217 (1986).

Monotonic Increase of Magnetoresistance in Amorphous Mo_xGe_{1-x} Through the Metal-Insulator Transition

S. Yoshizumi[1], *T.H. Geballe*[2], *M. Kunchur*[3], *and W.L. McLean*[3]

[1]Department of Materials Science & Engineering, Stanford University, Stanford, CA 94305, USA*

[2]Department of Applied Physics, Stanford University, Stanford, CA 94305, USA†

[3]Serin Physics Laboratory, Rutgers University, Piscataway, NJ 08855-0849, USA††

Magnetoresistance measurements on amorphous Mo_xGe_{1-x} as x is decreased from the metallic to insulating values show a different behavior from that anticipated from earlier studies. In contrast to the granular aluminum system in which the magnetoresistance MR = $[\rho(T,H)-\rho(T,0)]/\rho(T,0)$ at high fields is negative and shows a maximum magnitude at the metal insulator transition [1], in amorphous Mo-Ge MR is positive, increases monotonically with decreasing x through the metal-insulator transition, and appears to saturate in the insulating regime. There is as yet no theory that explains these results. An additional experimental result for which there is no existing theory is that the magnetoresistance of the insulating sample (x = 0.083) was negative at low magnetic fields and positive at higher fields. It appears that different classes of metal-insulator transition occur in the two systems.

Three-dimensional films (thickness 2500Å) of amorphous Mo-Ge were deposited on to single-crystal sapphire substrates by cosputtering from pure Mo and Ge targets in Ar at about 10^{-2} torr. Compositional inhomogeneity was decreased by rotating the substrates at 300 rpm during deposition. The films were later patterned photolithographically for four-terminal resistance measurements which were made in the liquid ^{4}He temperature range and in magnetic fields up to 9 tesla.

All of the Mo_xGe_{1-x} samples had positive magnetoresistances at high fields. Fig. 1 shows how MR varies with x and with temperature. These results should be contrasted with those in granular aluminum $Al\text{-}Al_2O_3$ in which the magnetoresistance at high fields was negative [2]. In addition at low temperatures MR in granular aluminum passed through a maximum magnitude at the metal-insulator transition [1], rather than increasing monotonically with increasing resistivity ratio $\rho(4.2K)/\rho(300K)$, where $\rho(4.2K)$ is the resistivity in the normal state at 4.2K. In Mo_xGe_{1-x} the resistivity ratio increases monotonically with decreasing x as it does in $Al_x(Al_2O_3)_{1-x}$. We return later to a discussion of the differences between the magnetoresistance behaviour in the two systems.

The Mo-Ge samples discussed here all had values of x smaller than the minimum value (x_m = 0.13) for which superconductivity could be observed. The metallic samples had positive MR at all fields and at the largest fields the reduction in conductivity was proportional to $H^{1/2}$. At fixed values of the magnetic field the conductivity was proportional to $T^{1/2}$ plus a constant, as also were the extrapolated values of the high-field conductivity for $H^{1/2} \rightarrow 0$. All of these aspects are in accordance with the electron-electron interaction theory of Lee and

*Present address: Research Laboratories, Eastman Kodak Co., Rochester, NY 14650

†Work supported by NSF MRL grant.
††Work supported in part by NSF DMR-81-18062.

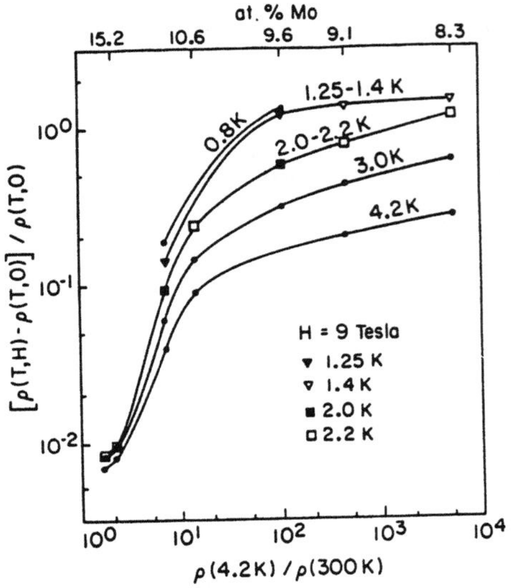

Fig. 1.
Magnetoresistance MR = $[\rho(T,H)-\rho(T,0)]/\rho(T,0)$ versus $\rho(4.2K,0)/\rho(300K,0)$ at different temperatures.

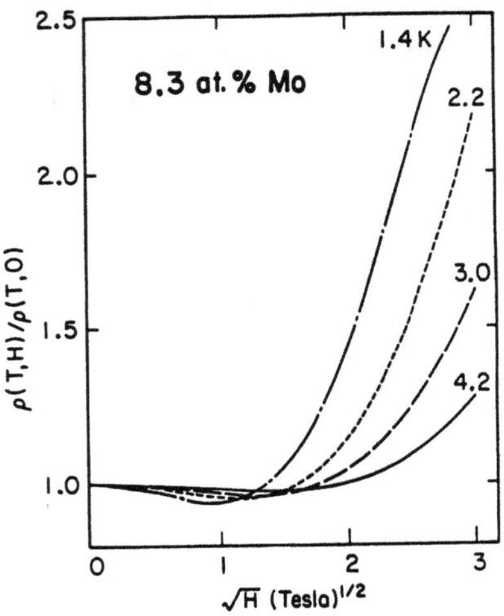

Fig. 2.
The ratio $\rho(T,H)/\rho(T,0)$ plotted against $H^{1/2}$ for $Mo_{0.083}Ge_{0.917}$ at T = 1.4, 2.2, 3.0, and 4.2K. Note values below unity at low fields.

Ramakrishnan [3,4]. However, as we show in Fig. 2, MR of the insulating sample (x = 0.083) was negative at low magnetic fields and positive at higher fields. This behaviour is opposite to that expected from localization effects in metallic samples with strong spin-orbit scattering [4] and is not yet understood in terms of existing theory.

We now return to a discussion of the monotonic behaviour exhibited in Fig. 1 which contrasts with the maximum in the negative MR observed in granular aluminum. It has been pointed out [1] that in the scaling theory of localization [5] the field-dependent length scale $L_H = (c\hbar/eH)^{1/2}$ leads to an increase of the magnitude of the negative MR at low temperatures as the metal-insulator transition is approached from the metallic side. At the conductance threshold MR = -1. This treatment cannot be used on the insulating side where electrical conduction is thermally activated. However, the location of the mobility edge is sensitive to magnetic field [6] and it has been shown that this field-dependence can lead to a magnetoresistance in variable-range hopping conduction given by [7,8]

$$MR = \{[\rho(T,0)/\rho(\infty,0)]\exp[B(\xi/L_H)^{1/\nu}]\} - 1$$

where ν is the critical exponent for the localization length $\xi \propto (n_c-n)^{-\nu}$, n_c is the value of the concentration n at the mobility edge in zero magnetic field, the variable-range hopping law is of the form $\rho(T,0) = \rho(\infty,0)\exp[(T_0/T)^{1/a}]$, and the constant $B = -A\nu(1-a)/a$. For a system in which MR is negative B and hence A are negative for a>1. As the metal-insulator transition is approached from the insulating side the magnitude of the negative MR again increases and at the conductance threshold MR = -1 (if we assume $\rho(T,0)/\rho(\infty,0)$ remains finite as expected [7] from the n-dependence of T_0).

The predominance of the positive MR in Mo-Ge suggests that electron-electron interactions dominate the localization effects arising from quantum interference. The scaling theory treatment given above is therefore inapplicable. There is also a lack of convergence of the treatment on the insulating side if the constant B is positive.

A complete understanding of the MR in strongly disordered systems near the metal-insulator transition thus awaits a comprehensive theory that includes both the effects of quantum interference as well as electron interactions.

References

1. H.K. Sin, P. Lindenfeld, and W.L. McLean, Phys. Rev. B 30, 4067 (1984); H.K. Sin, P. Lindenfeld, and W.L. McLean, Proc. LITPIM Suppl. (PTB-PB-1)p.287 (PTB Braunschweig, 1984).
2. T. Chui, P. Lindenfeld, and W.L. McLean, Phys. Rev. Lett. 47, 1617 (1981).
3. P.A. Lee and T.V. Ramakrishnan, Phys. Rev. B 26, 4009 (1982).
4. P.A. Lee and T.V. Ramakrishnan, Rev. Mod. Phys. 57, 287 (1985).
5. E. Abrahams, P.W. Anderson, D.C. Licciardello, and T.V. Ramakrishnan, Phys. Rev. Lett. 42, 673 (1979); Y. Imry, Phys. Rev. Lett. 44, 469 (1980).
6. D.E. Khmel'nitskii and A.I. Larkin, Solid State Commun. 39, 1069 (1981); B. Shapiro, Phil. Mag. B 50, 241 (1984).
7. B. Al'tshuler, A.G. Aronov, and D.E. Khmel'nitskii, Pis'ma Zh. Eksp. Teor. Fiz. 36, 157 (1982) [Sov. Phys. JETP Lett. 36, 195 (1983)].
8. We believe that the factor ξ in the exponent of this equation for MR was incorrectly omitted from reference 7. (W.L. McLean, unpublished)

Universal Behavior of Hall Conductivity Near the Metal-Insulator Transition in Disordered Systems

M. Rohde and H. Micklitz

Experimentalphysik IV, Ruhr Universität,
D-4630 Bochum, Fed. Rep. of Germany

Although there is a large number of articles dealing with the behavior of the dc conductivity σ near the metal insulator transition (MIT) in disordered systems due to Anderson localization /1/ there is only one theoretical /2/ and one experimental /3/ approach concerning the behavior of the Hall constant R_H near the MIT. A pure localization scaling theory /2/, excluding the influence of Coulomb interactions, predicted for the Hall conductivity $\sigma_H \propto (x-x_c)^{\nu_H}$ with $\nu_H=2\nu=2$. Field and Rosenbaum /3/, on the other hand, showed within their experimental work on the system Ge:Sb that $\nu_H \approx \nu \approx 1$ in contradiction to theory. This fact raises the question whether the observed behavior of σ_H is specific to the system Ge:Sb or the experimental value of ν_H is an universal one. For that reason we have measured $R_H(\sigma_H)$ in the amorphous metal rare gas mixture Bi_xKr_{1-x} which is known to show a MIT due to Anderson localization /4/, but is physically a quite different disordered system compared to crystalline, doped Ge.

Measurements of the Hall voltage U_H and the conductivity σ were performed on films with a thickness of ≈ 1000 Å obtained by the codeposition of Bi and Kr onto a cold (≈ 4.2 K) sapphire substrate. More experimental details are given elsewhere /5/. The measurements of U_H and σ were made in situ at 4.2 K wich is low enough to satisfy the low temperature condition $kT \lesssim 10^{-4}E_F$ (Fermi energy E_F of Bi ≈ 5 eV) /6/. U_H shows linear behavior with increasing magnetic field up to a maximum of 3 T indicating that the low field condition also is fullfilled, i.e. that the magnetic field is not able to tune the MIT. This fact is supported by the experimental observation that the magnetoresistive effect is small ($\Delta\rho/\rho \lesssim 10^{-3}$).

The conductivity σ and the Hall coefficient R_H are shown in fig. 1 as a function of the relative distance from the critical concentration $x_c=0.55$. The data points marked with triangles in fig. 1 and also in fig. 2 are obtained from a measurement of Bergmann /8/ on amorphous Bi. The solid lines are results of least squares fits to the data with $R_H \propto ((x-x_c)/x_c)^{-g}$ for the Hall constant and $\sigma \propto ((x-x_c)/x_c)^{\nu}$ for the conductivity, $g=0.69\pm0.1$ and $\nu=0.93\pm0.1$. The critical exponent for the Hall conductivity $\sigma_H=R_H \cdot \sigma^2$ is $\nu_H=2\nu-g=1.19\pm0.1$. The

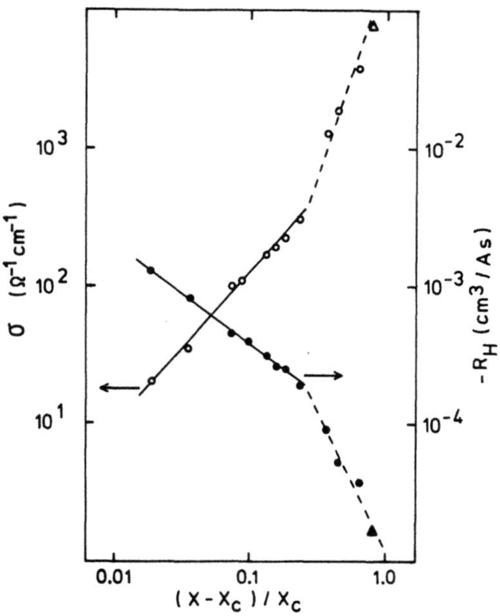

Fig. 1: Longitudinal conductivity σ (open symbols) and Hall coefficient R_H (filled symbols) as a function of relative distance from the MIT. The solid lines are least squares fits as described in the text. The dashed lines are only guides to the eye. The data values marked with a triangle are obtained from ref. 8

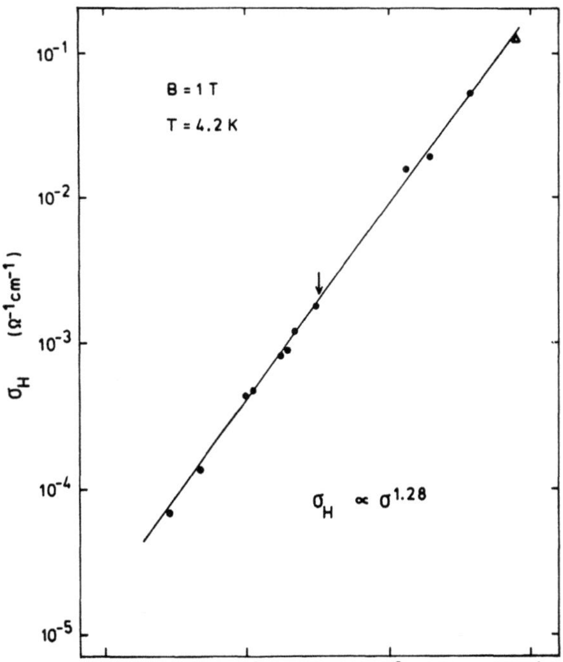

Fig. 2: Double log plot of the Hall conductivity σ_H as a function of the longitudinal conductivity σ at a field of 1 T and a temperature of 4.2 K. The arrow indicates the concentration above which the deviation from the critical behavior is observed. The triangle marks a data point which is obtained from ref. 8

deviation from the critical behavior at high x values, as indicated by the dashed lines in fig. 1, is assumed to be caused by bandstructure effect, i.e. the density of states probably varies rapidly in this region. However, the system is still in the region of incipient localization, since all our conductivity values are below the Mooij value of about $5\cdot10^3$ $(\Omega\,cm)^{-1}$ which marks the upper limit of the critical region /7/.

The plot of σ_H versus σ (see fig. 2) shows a unique relationship between these quantities. The arrow in fig. 2 indicates the concentration above which deviation from critical behavior is observed. The data can be fitted by the relation $\sigma_H \alpha\, \sigma^{1.28\pm0.05}$ or $\nu_H/\nu = 1.28\pm0.05$. This is a most surprising result. Bandstructure effects seem to cancel in this representation and the relation given above is fullfilled over a range of more than three orders of magnitude in σ_H and more than two orders of magnitude in σ. We want to point out that the representation given in fig. 2 is independent of the experimental determined value of x as well as the choice of the critical concentration x_c. Thus the above given ratio for ν_H/ν is quite reliable.

The critical exponents of the magnetotransport parameters found in the amorphous metal rare gas mixture Bi_xKr_{1-x} are the same as those found in Ge:Sb /3/. The amorphous metal rare gas mixture Bi_xKr_{1-x} and the highly doped semiconductor Ge:Sb are physically quite different systems concerning the absolute carrier density (five orders of magnitude difference in critical electron density n_c), mobility and kind of disorder. The agreement in the critical behavior of the Hall conductivity in such different systems indicates the universality of the critical exponents. The discrepancy between theory /2/ and experiment, therefore, seems to be of fundamental nature. It is probably caused by the neglection of contributions due to electron-electron interactions.

Finally we want to mention that the Hall conductivity in a classical, percolating system, namely a granular Sn-Ar mixture /9/, shows a quite different critical behavior (ν_H=2.7±0.1) which, however, agrees reasonably well with theoretical calculations /10/.

This work was supported by the Deutsche Forschungsgemeinschaft.

References:

1. See for instance the review article of P.A. Lee and T.V. Ramakrishnan, Rev. Mod. Phys. 57, 287 (1985) and references therein.
2. B. Shapiro and E. Abrahams, Phys. Rev. B24, 4025 (1981)
3. S.B. Field and T.F. Rosenbaum, Phys. Rev. Lett. 55, 522 (1985)

4. R. Ludwig and H. Micklitz, Solid State Commun. 50, 861 (1984)
5. M. Rohde and H. Micklitz, to be published in Phys. Rev. B
6. G. Thomas : In Localization and Interaction, ed. by D.M. Finlayson, Proceedings of the 31. Scottish Universities Summer School in Physics 1986 (Edinburgh University Press, 1986), p. 178
7. J.H. Mooij, Phys. Status Solidi (a)17, 521 (1973)
8. G. Bergmann, Z. Physik 255, 76 (1972)
9. M. Rohde and H. Micklitz, to be published
10. D.J. Bergman, Y. Kantor, D. Stroud and I. Webman, Phys. Rev. Lett. 50, 1512 (1983)

A Strong Voltage Dependence on the Conductivity in Granular Aluminum

A.J.G. Schellingerhout[1], *H.M. Appelboom*[1], *J.A. Kokkedee*[1], *J.E. Mooij*[1], *and T.M. Klapwijk**

[1]Department of Applied Physics, Delft University of Technology, Delft, The Netherlands

[2]Department of Applied Physics, Groningen University, Groningen, The Netherlands

1. Introduction

There have been many investigations in the transport properties of granular aluminium near the metal-insulator transition. Our intention is to investigate the electrical conduction of this material on a short length scale in the range of 20 to 1000 Å. This makes it possible to explore the conduction when the length scale of the sample is of the same order as the diverging length scale near the metal-insulator transition. When granular aluminium of this thickness is enclosed between two superconductors, a superconductor-insulator-superconductor (SIS) or a superconductor-normal metal-superconductor (SNS) junction is created and the superconducting properties can be explored, similar to experiments performed by BARRERA and BEASLEY [1] on NbSi. We report results on a length scale of a 1000 Å to investigate the properties of granular aluminium in this regime.

2. Experimental

The investigated samples consist of a thin granular aluminium layer sandwiched between two good conducting aluminium layers, which serve as electrodes. These layers are evaporated in an UHV-system.The substrate, held at room temperature during deposition, is (012) sapphire. The partial pressure of oxygen and the aluminium deposition rate are controlled with a feedback system. The oxygen pressure is 5.10^{-4} Pa and the aluminium rate is 46 Å/min. The trilayer is evaporated in one continuous run by changing the aluminium rate and the oxygen pressure. Small squares are patterned by a selective anodization process similar to SNAP used by KROGER et al. [2]. The completed structure is drawn in Fig.1.

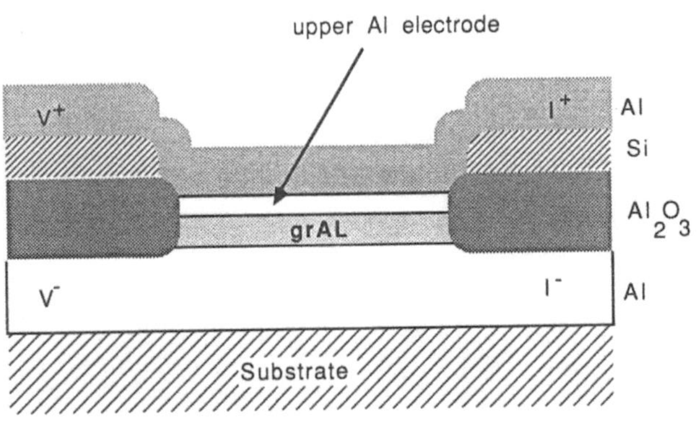

Fig.1 Cross-sectional view of an Al-grAl-Al junction. The barrier thickness is 1000 Å. The area of the junction is 8x8 μm^2. The silicon layer is a protection against shorts.

With a second shutter which was opened during the deposition of the 1000 Å granular aluminium layer, we made a layer of the same composition as the barrier on a separate substrate. It is used as a reference for the barrier composition.

3. Measurements

Resistance versus temperature measurements were performed on the junction and on the reference layer. The resistance of the reference layer follows the known relation

$$R(T) = R_o \exp((T_o/ T)^{0.5}) \qquad (1)$$

For this composition T_o was 220 K ,with $\rho_{4.2\,K}$ of 40 Ω cm.Similar results have been obtained by SIN et al. [3]. The granular aluminium is clearly in the insulating regime . The resistance of the junctions showed the same temperature dependence (1), with an identical T_o , although the measured $\rho_{4.2\,K}$ was surprisingly high : 400 Ω cm.

We observed strong changes in the conductance at higher voltages. The measured I(V) characteristics are shown in Fig.2 on logarithmic scales. Changing the measurement voltage for example from 0.1 to10 mV, leads to conductance changes of three orders of magnitude. At high voltages all the curves seem to converge to the same curve, independent of temperature .

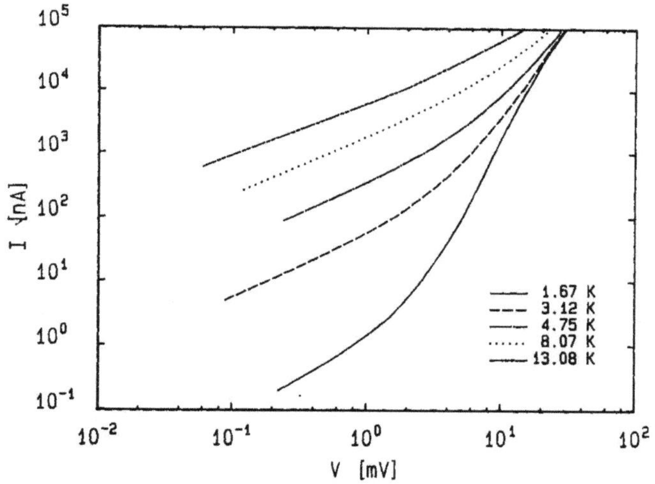

Fig.2 I(V) characteristics of a 8x8 μm^2 Al-grAl-Al junction.

4. Discussion and conclusion

A possible explanation for the non-ohmic conduction is the high electric field in the granular aluminium . Strong field effects have been observed by others (see MOTT and DAVIS[4] and EFROS and POLLAK [5]). We compare the thermal energy k_BT to the energy gained by an electron in an electric field eEL_E, with L_E a characteristic length. We propose the following qualitative model. At low field the conduction is governed by the temperature according to (1), at high fields the conduction is dominated by the electric field and hence independent of temperature.

We can define a cross-over field

$$e E_c L_E \approx k_B T \tag{2a}$$

and hence a cross-over voltage

$$V_c = E_c d \tag{2b}$$

with d the thickness of the barrier. In Fig. 3 we have plotted the derivative dV/dI of the I(V) curves. We have found that the high voltage behaviour can be fitted to the relation

$$dV/dI = R_0 \exp((V_0 / V)^{0.5}) \tag{3}$$

with R_0 and V_0 arbitrary constants. To obtain a cross-over voltage V_c from the experimental data in Fig.3 we used the voltage where the dV/dI at zero voltage equals (3). In Fig.4 the values of V_c are plotted versus the temperature.

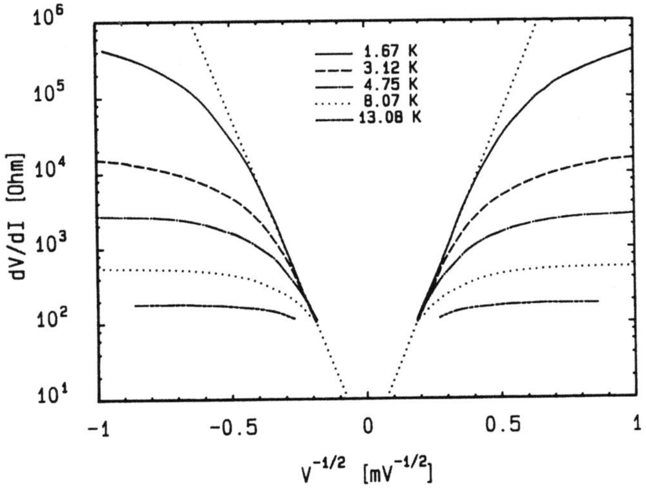

Fig.3 Differential resistance dV/dI versus the inverse voltage across an 8x8 μm^2 junction. The extra dotted straight line is a least squares fit to (3).

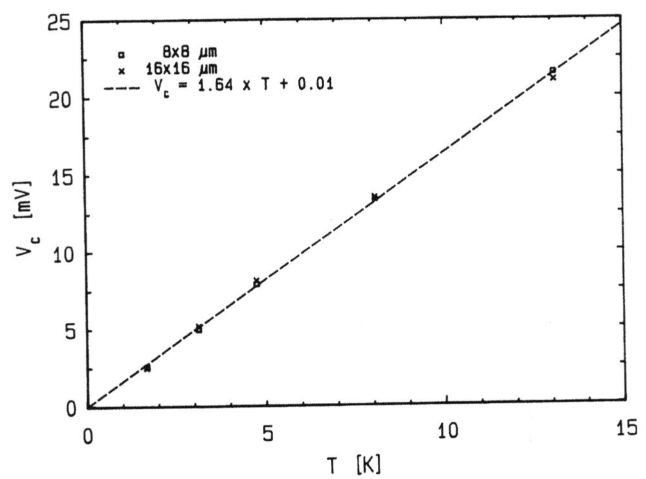

Fig.4 Cross-over voltage V_c versus temperature for two samples (8x8 μm^2 and 16x16 μm^2). The line is a least squares fit for all the data.

With Fig.4 and (2) it is possible to make an estimate for L_E which is about 50 Å. This value is compatible with distances between localized states or different grains.

In conclusion, we have observed a strong voltage dependence of the resistance in granular aluminium, which can be explained with an electric field effect. This effect should also be observable in lines of granular aluminium although it might be difficult to distinguish the electric field effect from a heating effect. We realize the necessity for junctions with a different barrier thickness to confirm that the electric field is causing the non-ohmic behaviour.

Acknowledgment
This project is supported by the Foundation for Fundamental Research (F.O.M.).

References

1. A.S. Barrera and M.R. Beasley, Applied Superconductivity Conference 1986, Baltimore IEEE trans. mag. to be published
2. H.Kroger, L.N. Smith and D.W. Jillie Appl. Phys.Lett.39(3),(1981),280
3. H.K. Sin, P. Lindenfeld and W.L. McLean Phys.Rev.B30(1984),4067
4. N.F. Mott and E.A. Davis , Electronic processes in non-crystalline materials (Clarendon Press, Oxford,1979)
5. A.L. Efros and M. Pollak , Electron-electron interactions in disordered systems (North-Holland,Amsterdam,1985)

The Metal-Insulator Transition in Disordered Metals

S. Okuma, F. Komori, and S. Kobayashi*

Department of Physics, Faculty of Science, University of Tokyo,
Hongo 7-3-1, Bunkyo-ku, Tokyo 113, Japan
*Department of Physics, Faculty of Science, Tokyo Institute
of Technology, 2-12-1, Ohokayama, Meguro-ku, Tokyo 152, Japan

Critical behavior of the metal-insulator transition is studied in several kinds of highly disordered metal films prepared with the newly developed method[1]. This enables us to examine systematically the roles of the various material parameters, such as magnetic scattering, spin-orbit scattering and interaction of superconductivity in the transition. According to the scaling theory[2], zero temperature conductivity $\sigma(0)$ with increasing disorder is expressed as

$$\sigma(0) = \sigma_0 \left(x / x_c - 1 \right)^{\nu} \tag{1}$$

where x is a controllable parameter which varies smoothly with the microscopic conductance and x_c is its critical value. The critical exponent ν, which reflects the form of the scaling β (g) function near the transition, is calculated not only in ordinary case but in the presence of those scattering mechanisms[3]. In our previous experiments the role of magnetic scattering in the transition was studied in Cu films with Mn (Cu:Mn)[1] and that of spin-orbit scattering was studied in Bi films[4].

With increasing disorder, superconducting ground state changes to insulator in two ways: One is S(superconductor)-N(normal metal)-I(insulator) transition, and the other is S-I transition. It was theoretically pointed out that the spin-orbit[5] and the short-range Coulomb interaction[6] are important for the appearance of S-N-I transition. Roles of these interactions on the phase transition are examined in Al films with small amount of gold (Al:Au) and palladium (Al:Pd)[7]. Gold has strong spin-orbit interaction, while the palladium is a typical transition metal with strong Coulomb interaction and yet not intrinsically ferromagnetic.

The disordered metal films are prepared by repeating a small amount of vacuum deposition of metal on glass substrate and the oxidation. The films are composed of small metallic grains coupled through thin oxide layer. To study the critical region of the transition, we prepare the films whose resistivity varies slowly through this region by introducing a temperature gradient on the substrate during the oxidation process. In Cu:Mn, Al:Au and Al:Pd films, 12 at% of Mn, 1.4 at% of Au and 2.2 at% Pd are mixed into pure Cu and Al as impurities.

In Figs.1, the conductivity σ_{LT} of Cu, Cu:Mn and Bi films at the lowest temperature is plotted against that at 77K. For Bi films, the result in the magnetic field of 4T is also shown. We simply use σ_{LT} instead of $\sigma(0)$ because the conductivity cannot be extrapolated to 0K without ambiguities. The σ_{77K} is adopted for x because it corresponds to the classical residual conductance. The σ_{LT} decreases

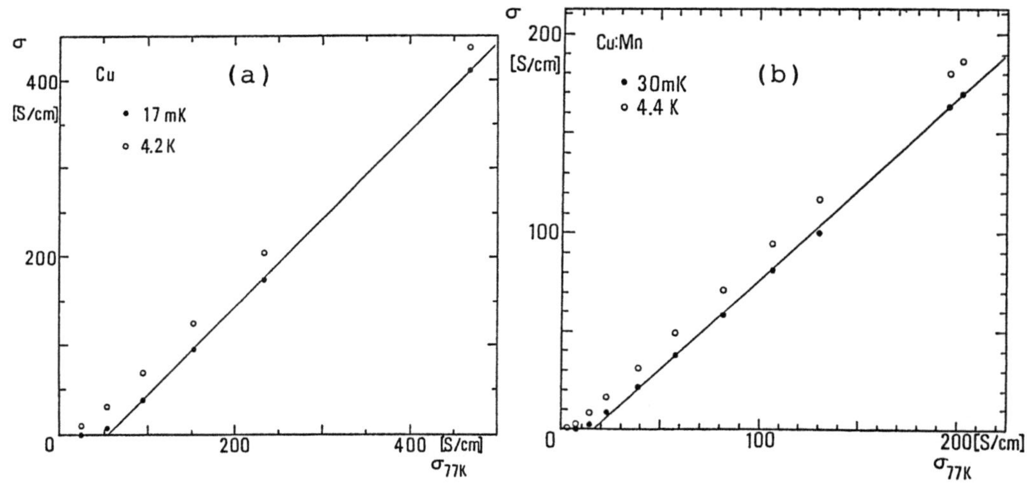

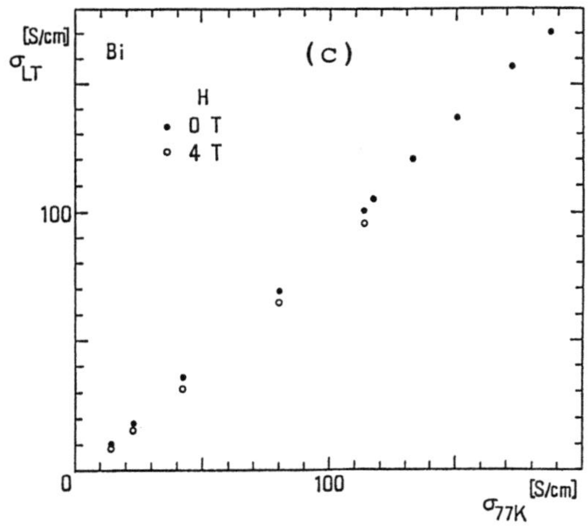

Fig.1 Metal-insulator transition observed in (a) Cu (b) Cu:Mn and (c) Bi films

Table 1 Parameters used to fit the data in Figs.1 to Eq.(1)

	Cu	Cu:Mn	Bi(0T)	Bi(4T)
ν	$\sim$1	$\sim$1	$\sim$1	$\sim$1
σ_c	56S/cm	14.2	2.2	4.2
σ_0	55S/cm	12.7	2.0	3.6

almost linearly with σ_{77K} from the metallic region and eventually tends to zero at a certain value of $\sigma_{c,77K}$. The data in Figs.1 are well reproduced by Eq.(1) with the parameters listed in Table 1.

The scaling theories with and without electron-electron interaction predict that the critical exponent ν is unity in ordinary case. Within one-body approximation, ν is calculated to be 1/2 when the magnetic scattering is strong and 0.57 in case of large spin-orbit scattering[8]. On the other hand, when the electron-electron interaction is considered, ν is expected to be unity in the presence of magnetic and/or spin-orbit scattering. The present results in Figs.1 demonstrate that the critical exponent is unchanged both by the magnetic scattering and the spin-orbit scattering. The values of the exponents ν =1 are consistent with the prediction of the scaling theory including the electron-electron interaction.

In the presence of strong magnetic field or magnetic impurities, localization effect is suppressed and the interaction mainly contributes to the change in conductivity. In Bi films, the contribution of localization is known to be negative(anti-localization). Therefore we attribute the smaller critical conductivity in Cu:Mn films than that in Cu films to the delocalization by strong magnetic scattering. Moreover, the enhancement of the critical conductivity observed in Bi films in the field of 4T compared to that in zero field is due to the suppression of the anti-localization effect.

Figure 2(a) shows temperature dependence of the conductivity of Al films near the N(S)-I transition. At low temperature and in zero field samples #2 and #3 are superconducting, while Sample #1 is insulating. Under the magnetic field of 4T, all these films become insulating instead of normal metallic. The insulator is determined from the definition that $\sigma(0)=0$ because in this case we can obtain $\sigma(0)$ by simple extrapolation. Results of Al:Au films are essentially the same as those of Al films. However, in the results of Al:Pd films, as shown in Fig.2(b), we note the following remarkable differences compared with those of Al films: (1) In the field of 4 T, all the samples which show the superconductivity are metallic. (2) The T_c's are reduced to less than 1 K. (3) The high temperature conductivity of Sample #2 of Al film, which is superconducting, is slightly smaller than that of Sample #4 of Al:Pd film, which is insulating.

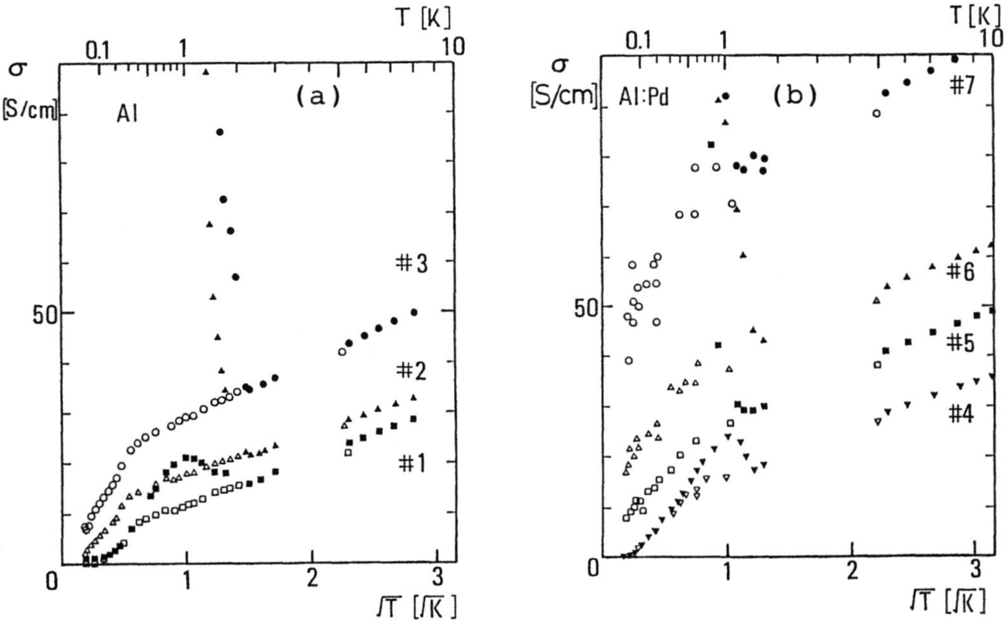

Fig.2 Conductivity of (a) Al and (b) Al:Pd films against $\sqrt{T}$. The solid symbols correspond to the data in 0T and the open symbols correspond to those in 4 T

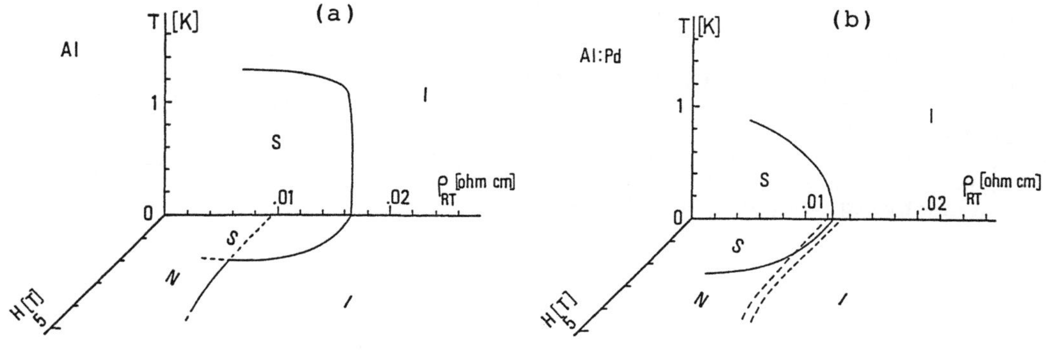

Fig.3 The S-N-I phase diagrams of (a) Al and (b) Al:Pd films

The S-N-I phase diagram for Al films in the space with magnetic field(H), temperature(T) and disorder(ρ_{RT}) axes is schematically illustrated in Fig.3(a). As a degree of disorder, we use resistivity ρ_{RT} at room temperature. The phase boundary shown with the broken line is approximative. It is noted that the Al films exhibit the S-I transition with increasing magnetic field and/or disorder. In Al:Au films, the S-I and N-I boundaries are essentially the same as those of Al films, although the same amount of Au in Cu film is known to increase the spin-orbit scattering rate more than fifty times[9]. Thus we conclude the spin-orbit interaction does not affect the S-I transition at all. This result definitely disagrees with the recent theoretical prediction that three-dimensional materials with strong spin-orbit scattering undergo S-N-I transition with increasing disorder[5]. On the other hand, the S phase in Al:Pd films is significantly suppressed especially in the highly disordered region as shown in Fig.3(b). Although both the spin-orbit and the Coulomb interaction increase by mixing Pd into Al, we have confirmed that the former does not affect the transition from the results of Al:Au films, whose spin-orbit interaction is estimated to be several times stronger than that of Al:Pd films. Consequently, we attribute the trend toward the S-N-I transition to the strong Coulomb interaction. This result is qualitatively explained by the recent theory on disorder-enhanced spin fluctuation[6].

In conclusion, the critical exponents in Cu, Cu:Mn and Bi films are found to be commonly unity. These are consistent with the scaling theory including electron-electron interaction. Effects of magnetic and spin-orbit scattering appear in $\sigma_{c,77K}$. The Al and Al:Au films exhibit almost the same S-I transition, while in Al:Pd films a suppression of the S phase is observed particularly in the highly disordered region. This tendency for the appearance of the S-N-I transition is ascribed to the strong Coulomb interaction.

References

1. S. Okuma, F. Komori, Y. Ootuka and S. Kobayashi: J. Phys. Soc. Jpn. **54**, 2382 (1985)
2. E. Abrahams, P. W. Anderson, D. C. Licciardello and T. V. Ramakrishnan: Phys. Rev. Lett. **42**, 673 (1979)
3. A. Kawabata: Prog. Theor. Phys. Suppl. **84**, 16 (1985)

4. F. Komori, S. Okuma and S. Kobayashi: J. Phys. Soc. Jpn. **56**, 691 (1987)
5. M. Ma and E. Fradkin: Phys. Rev. Lett. **31**, 1416 (1986)
6. H. Fukuyama: Physica **135B**, 458 (1985)
7. S. Okuma, F. Komori and S. Kobayashi: J. Phys. Soc. Jpn. **56**,1936 (1987)
8. S. Hikami: J. Physique Lett. **46**, L-719 (1985)
9. F. Komori, S. Okuma and S. Kobayashi: Proc. Int. Conf. on LITPIM, Braunschweig, Suppl, 147 (1984)

Part III

Theory of the Metal-Insulator Transition

Numerical Studies of Anderson Localization

B. Kramer

Physikalisch-Technische Bundesanstalt Braunschweig, Bundesallee 100,
D-3300 Braunschweig, Fed. Rep. of Germany

Numerical methods used for the investigation of the Anderson localization have included the Lanczos method, the equation of motion method, the transfer matrix method as well as real space scaling methods. They are briefly described, and their main results are summarized. Particular emphasis is on the comparison between numerical scaling studies and the one-parameter scaling theory.

1. Introduction

Numerical methods have been used extensively in the study of the problem of localization in disordered solids. The reason is that due to the randomness of the Hamiltonian the applicability of analytical techniques is limited. Except in one dimension there are no exactly solvable models. Approximate treatments (as the coherent potential approximation, for instance) are often uncontrollable.

Three regions of interest are distinguishable: (i) weak localization, (ii) strong localization, (iii) the Anderson transition. Weak and strong localization denote the asymptotic regimes of weak and strong disorder, respectively. The Anderson transition (AT) is the transition between non-localized and localized states upon changing the disorder and/or the energy. It corresponds to a disorder induced metal-insulator transition (MIT) [1].

The theoretical approaches may be divided into three groups. Renormalized perturbation expansions [2] usually work well within the asymptotic regions. Information about the AT has then to be obtained by interpolation. Renormalization group techniques [3] are especially tailored to work in the transition region. They yield information about the critical behaviour. Approximative methods like the mode coupling theory [4] treat the whole range of disorder between the asymptotic regimes and yield experimentally accessible information.

Numerical methods, on the other hand, may be considered as "ideal experiments" dealing with well-defined systems. They allow to test the assumptions made in the analytical methods, within computational errors. Limitations arise only from the finiteness of the available computer storage and the available computer time.

Problems which can be attacked numerically in the field of localization are (i) test of exponential localization in the localized region, (ii) test of the scaling theory, (iii) calculation of the conductivity, (iv) calculation of the distribution function of physical quantities as the conductivity, (v) determination of the critical behaviour. (i) and (ii) are perhaps the simplest and by now the most advanced. The numerical calculation of the conductivity is still a problem for $d>1$ (d: dimensionality) [5,6], especially for the dc case. The same holds true for the investigation of the statistics of experimentally accessible quantities. Most crucial is the determination of the critical behaviour. There are many experimental results available, but not all of them are consistent with the results of the analytical theories [7].

In this paper we concentrate on the description of the numerical methods which are available, and on results concerning (i), (ii) and (v).

2. Characterization of Localized States

Mathematically, one may classify the quantum states in an infinite disordered system by their asymptotic behaviour, their average size, their fractal dimensionality, and/or their transport properties.

The asymptotic behaviour of a random state may be described by an exponential decay length, the localization length λ'. It is a measure of the asymptotic exponential decay of the envelope of the wave function, i. e.

$$\psi(r) = f(r)\exp(-r/\lambda'), \qquad (r\to\infty) \tag{1}$$

$f(r)$ is a randomly varying prefactor. $\lambda'\to\infty$ corresponds to an extended state.

A measure of the average size of a state may be obtained by considering the inverse participation number P^{-1} defined by

$$P^{-1} = \sum_r |\psi(r)|^4 \qquad (\|\psi\| = 1)\ . \tag{2}$$

In general, we have the "diameter" $P^{1/d} \neq \lambda'$. Examples for this case are the states in the Lifshitz tails of the band [8] which are localized within macroscopically large accidental potential wells. Their asymptotic behaviour is given by the exponential decay outside the well.

The fractal dimensionality d^* may be obtained from P according to

$$\lim_{L\to\infty} P \sim L^{d^*}\ . \qquad \text{(L: system diameter)} \tag{3}$$

d^* tells us how space-filling a state is. Localized states have $d^* = 0$, extended states $d^* = d$.

The character of the states near a given energy determines their transport properties. In particular, one has a vanishing dc-conductivity in the limit of zero temperature if there are only localized states near the Fermi energy [9]. The transport properties are connected with the two-particle spectral function [10]

$$S^{(2)}(r_1,r_2,r_3,r_4;E_1E_2) = \overline{\langle r_1|\delta(E_1 - H)|r_2\rangle\langle r_3|\delta(E_2 - H)|r_4\rangle}. \tag{4}$$

$\overline{\ldots\ldots}$ denotes the configurational average. On the other hand, $S^{(2)}$ reflects the localization properties of the states.

Writing the δ-functions in $S^{(2)}$ in terms of the one-particle propagator $G^{\pm}$ one can show [10] that the dc-conductivity is determined by the transmission probability $T(r,r';E) = |\langle r|G^{+}|r'\rangle|^2$. If the average decay length λ defined consistently with (1) by

$$2/\lambda(E) = -\lim_{|r-r'|\to\infty} \overline{\log T(r,r';E)}/|r - r'| \tag{5}$$

is finite then the dc-conductivity will vanish in the limit of zero temperature. This will be the case if the states near E are exponentially localized in the sense of (1), as can be seen by using the spectral representation of G(E).

The inverse participation number can also be related to a diffusion process [11,12]. It can be shown that

$$\overline{P^{-1}} = \sum_{r\nu} \overline{|\psi_\nu(r)|^4\delta(E - E_\nu)}/\rho(E) = 2\pi \lim_{\varepsilon\to 0} \varepsilon\overline{A(r,E,\varepsilon)}/\rho(E) \tag{6}$$

where A denotes the average return probability $A(r,E,\varepsilon) = (\langle r|\delta_\varepsilon(E - H)|r\rangle)^2/4\pi^2$. $\rho(E)$ is the average density of states. δ_ε is the Lorentzian definition of the δ-function.

3. Numerical Methods

Among the numerical methods straightforward diagonalization of the secular matrix is the one which provides the most detailed information about the whole spectrum of eigenvalues and eigenstates. On the other hand, it requires the most extensive computational effort, the computational time being roughly proportional to the third power of the number of states in the system.

Often, the knowledge of the detailed information which is contained in the whole spectrum of eigenvalues and -states is not necessary. One is interested only in the states near a given energy, for instance. One of the most effective diagonalization methods for this purpose is the Lanczos method [13], by which the amount of computation can be drastically reduced if the matrices are sparse and if only certain eigenvalues and eigenfunctions are required. Numerical implementation of the renormalization group method [14] eliminates all unnecessary information keeping only the states near a given energy during the repeated diagonalization procedure. Thus, the computational time becomes essentially proportional to the size of the system. Such a procedure was used by LEE [15] and by SARKER and DOMANY [16] for the investigation of the scaling properties of the Anderson model.

Another method which was especially tailored for an economic use of the computer is the equation of motion method [17]. Here, the time-dependent Schrödinger equation is integrated numerically, and the resulting states are used to generate physical quantities as suitably defined time-averages. The method has been successfully applied for the calculation of the inverse participation number [17], the dc-conductivity [18], and to simulate quantum diffusion [19,20]. However, as in the case of the numerical diagonalization and the renormalization group method the results are somewhat uncontrolled with respect to the thermodynamic limit, especially for $d < 3$.

The first method which attempts to combine the scaling features near the Anderson transition with the recursive properties of the Schrödinger equation is the transfer matrix method [21], and/or the transmission method [22]. Both are based on the theorem that in 1d disordered systems all quantum states are exponentially localized, irrespective of their energy and the magnitude of the disorder [23]. Hence, the localization length is always a finite number in this case, the statistical error of which may be calculated as a function of the length of the system. The same holds true for systems of finite cross-section which are infinitely long in one direction. Then the investigation of the scaling properties of the localization length as the function of the cross-sectional diameter yields a controllable method of performing the thermodynamic limit, and of calculating the localization length (and the dc-conductivity) in the infinite system.

The method is especially tailored to provide a direct test of the one-parameter scaling theory of localization.

Specifically, we consider a quasi-1d system of a cross-sectional area M^{d-1} and length N ($\gg M$) described by the Anderson Hamiltonian. Disorder is introduced via randomly varying diagonal elements. The disorder parameter W is given by the width of their distribution.

The starting point is the definition of the localization length (5) together with the resolvent equation for the one-particle propagator.

Specializing (5) for the geometry considered here the localization length $\lambda_M(N)$ may be calculated from the transmission probability between the first and the Nth layer. It is finite.

In order to calculate the localization length of the d-dimensional system one has to consider both of the limits, $N \to \infty$ and $M \to \infty$. The essential feature of this method is that they can be performed in a numerically controllable manner: The statistical accuracy for any finite N can be determined, also recursively.

The essential idea necessary for the extrapolation $M \rightarrow \infty$ can be seen from Fig. 1.

The similar shape of the curves for different W suggests that an attempt to scale all of the data onto the same curve by changing the $1/M$-scale via multiplication with a scaling parameter $\xi(W)$ might be successful. Fig. 2 shows that this is indeed the case. The quantitative analysis shows that the deviations from the scaling function are less than the statistical accuracy of the raw data.

The scaling parameter has a physical meaning. Considering the lower branch of the scaling function in Fig. 2 we observe that for fixed W and $M \rightarrow \infty$ one moves downwards until the asymptotic region $\lambda_M/M \sim \xi(W)/M$ is reached. Thus, in this region

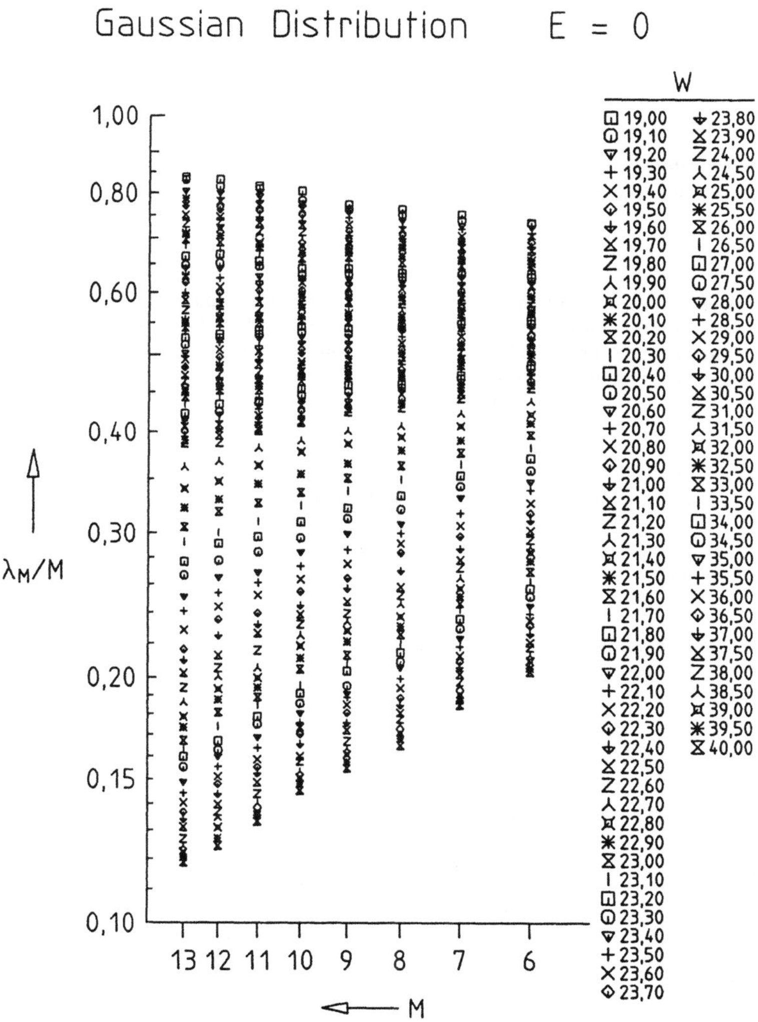

Fig. 1

Double logarithmic plot of λ_M/M as a function of $1/M$ for a 3d bar-shaped system (cross-sectional area MxM) described by the Anderson Hamiltonian on a simple cubic lattice with periodic boundary conditions for E=0, and Gaussian distribution of the site energies (width W). The relative statistical accuracy is better than 1%.

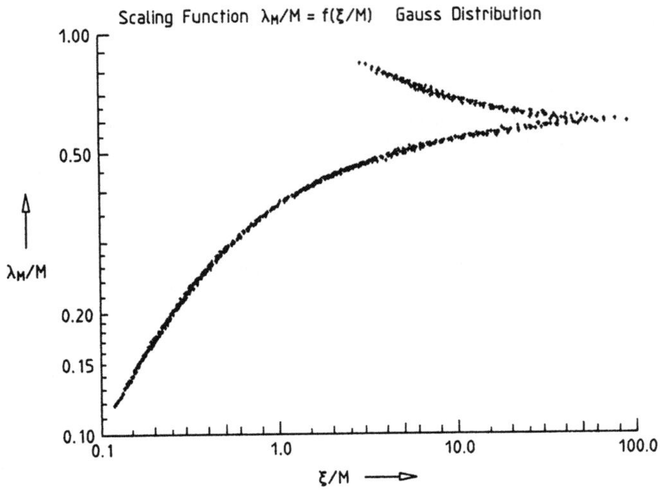

Fig. 2
The data of Fig. 1 scaled onto the same curve by multiplying the 1/M-scale with a disorder-dependent scaling parameter ξ [24]

$$\lim_{M \to \infty} \lambda_M = \xi \tag{7}$$

is the localization length in the infinite 3d system. For the upper branch the interpretation of ξ requires an additional assumption, namely that the conductance g is related to the transmission probability via Landauer's relation [25]. Then one obtains in the asymptotic region, where $\lambda_M/M \sim M/\xi(W)$, that [24]

$$\xi = \sigma_{dc}^{-1} \tag{8}$$

is the inverse of the dc-conductivity. From the behaviour of ξ as a function of W one may determine the critical exponents of the conductivity, s, and the localization length, ν.

In general, ξ will depend on W and E. $\xi(E,W)^{-1} = 0$ yields the phase diagram of localization.

Assuming a unique relation between the transmission probability and the conductance the scaling function may also be used to determine the ß-function for the conductance [22, 24, 26].

4. Results for the Anderson Model

Among the many numerical results from the "pre-scaling age" the following are especially worth mentioning.

(1) YOSHINO and OKAZAKI provided the first numerical evidence for localization in 2d disordered systems by diagonalizing secular matrices for systems containing up to 100x100 sites using the Lanczos method [27]. These calculations have been improved and extended to 3d systems by SCHREIBER who calculated the fractal dimensionality of the states from the inverse participation number, and determined the phase diagram of localization [28].

(2) WEAIRE et al. calculated the inverse participation ratio for 2d and 3d systems by using the equation-of-motion method [17].

(3) THOULESS and LICCIARDELLO estimated the conductivity by investigating the sensitivity of the eigenvalues to changes in the boundary conditions [29]. These calculations yielded the first hint that all the states might be localized in 2d systems.

(4) PRELOVSEK estimated the diffusion constant from a numerical simulation of quantum diffusion by using the equation-of-motion method [19]. These calculations have also been improved recently [20].
(5) An attempt was made in order to calculate the dc-conductivity from the Kubo formula by using the equation-of-motion method [18].

It must be noted that essentially all of these calculations suffered from the fact that in a numerical calculation one is always dealing with systems of finite sizes which are limited by the finite capacity of the computer, and that the extrapolations which are necessary in connection with the thermodynamic limit cannot be controlled.

To our best knowledge the first numerical investigation of localization with <u>each step being controllable</u> was the application of the transmission method and the transfer matrix method described in the last section. The main results of the work during the last years can be summarized as follows.

(1) It was demonstrated that in the band centre of the Anderson model a universal scaling function which depends only on the dimensionality does exist [24].
(2) d=2 was confirmed as the lower critical dimensionality. Within the parameter region investigated it was demonstrated that in 2d all of the states are exponentially localized. The existence of the AT in 3d was explicitly shown [24].
(3) The asymptotic behaviour for weak and strong disorder was recovered [24].
(4) Spin-orbit interaction, magnetic scattering as well as the influence of a homogeneous magnetic field have been investigated [30-32].
(5) The phase diagram of localization was determined for the box, the Gaussian, and the Lorentzian distribution of the site energies [33,34] (Fig. 3).
(6) An estimate for the critical exponents in the band centre was made yielding $s = \nu = 1.5 \pm 0.02$. $s = \nu$ is consistent with the one-parameter scaling theory. The numerical value is not in agreement with the results obtained from perturbation theory and the non-linear σ-model [24].
(7) The ß-function for the conductance was calculated from the numerically determined scaling function [24].

In recent calculations using larger systems and better accuracy the previous results except (6) have been confirmed. In addition, the calculation of the scaling function and the critical exponents have been extended to $E \neq 0$. In particular:
(1) Universality of the scaling function near the centre of the band was explicitly demonstrated [35,36].
(2) Detailed investigation of the critical behaviour near E = 0 showed that the numerical data are consistent with $s = \nu = 1$ [36].

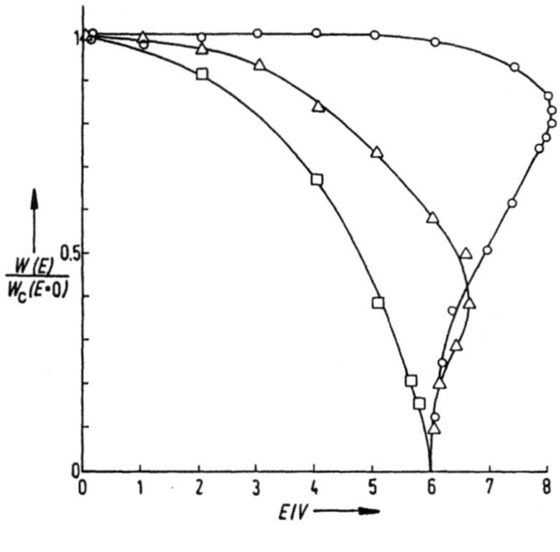

Fig. 3
Phase diagram of localization for the Anderson model with box ($W_c/V = 16.5 \pm 0.1$ (o)), Gaussian ($W_c/V = 20.9 \pm 0.1$ (△)), and Lorentzian distribution ($W_c/V = 3.8 \pm 0.5$ (□)) of the site energies

(3) Near the unperturbed band edge, and for small disorder the numerical scaling procedure does not work as well as in the band centre. However, the critical exponents turn out to be consistent with s = ν = 1, too, close to the transition. There is a crossover to ν = 1/2, away from the transition, on the isolating side. In the metallic region, the conductivity tends to a constant [36].

5. Discussion and Conclusion

The above described results are consistent with exponential localization only, in 2d and in the localized regime in 3d. This holds asymptotically and does not say anything about a power law decay of the states in an intermediate regime, as discussed in the literature [28].

The limits of weak and strong localization have both been detected numerically using the scaling method. The reason why one is able to calculate localization lengths which are of the order of 100 times larger than the maximum width of the systems considered (in 2d) lies in the fact that the scaling properties of the data provide additional information with respect to the asymptotic behaviour.

The results for the weak disorder regime are consistent with theory and experiments on weak localization [37]. Available experiments on the critical behaviour near the Anderson transition [7] yield both s = ν = 1/2 as well as s = 1. This leads to the conclusion that at least part of the experimentally investigated MITs is not solely due to disorder.

Acknowledgement

I gratefully acknowledge fruitful discussions with Michael Schreiber.

References

1. Localization, Interaction, and Transport Phenomena, ed. by B. Kramer, G. Bergmann and Y. Bruynseraede, Springer Ser. Solid-State Sci., Vol. 61 (Springer, Berlin, Heidelberg 1985)
2. P. A. Lee, T. V. Ramakrishnan: Rev. Mod. Phys. 57, 287 (1985)
3. F. J. Wegner: In Ref. 1, p. 99
4. W. Götze: In Ref. 1, p. 62
5. A. MacKinnon: J. Phys. C13, L1031 (1980)
6. T. Saso: In Proc. LITPIM Suppl. (PTB-PG-1), p. 30 (PTB Braunschweig 1984)
7. G. A. Thomas: In Localisation and Interaction in Disordered Materials and Doped Semiconductors ed. by D. M. Finlayson, p. 172 (SUSSP Publications, Edinburgh 1986)
8. I. M. Lifshitz: Sov. Phys. Usp. 7, 549 (1965)
9. N. F. Mott and E. A. Davis: Electronic Processes in Non-Crystalline Materials (Clarendon, Oxford 1971)
10. F. J. Wegner: Z. Phys. 25, 327 (1976)
11. D. J. Thouless: Phys. Reports C13, 93 (1974)
12. F. J. Wegner: Z. Phys. B36, 209 (1980)
13. J. H. Wilkinson: The Algebraic Eigenvalue Problem (Clarendon, Oxford 1965)
14. K. G. Wilson: Rev. Mod. Phys. 47, 773 (1975)
15. P. A. Lee: Phys. Rev. Letters 42, 1492 (1979)
16. S. Sarker, E. Domany: Phys. Rev. B23, 6018 (1981)
17. D. Weaire, A. R. Williams: J. Phys. C10, 1239 (1977); D. Weaire, V. Srivastava: J. Phys. C10, 4309 (1977)
18. B. Kramer, A. MacKinnon, D. Weaire, Phys. Rev. B23, 6357 (1981)
19. P. Prelovsek: Phys. Rev. B18, 3657 (1978)
20. G. M. Scher: J. Non-Cryst. Solids 59, 60, 33 (1983)
21. J. L. Pichard, G. Sarma: J. Phys. C14, L127, and L617 (1981)

22. A. MacKinnon, B. Kramer: Phys. Rev. Letters 47, 1546 (1981); A. MacKinnon: Z. Phys. B59, 385 (1985)
23. K. Ishii: Suppl. Progr. Theor. Phys. 53, 77 (1977); P. Erdös, R. C. Herndon: Adv. Phys., 31, 65 (1982)
24. A. MacKinnon, B. Kramer: Z. Phys. B53, 1 (1983)
25. R. Landauer: In Ref. 1, p. 38
26. E. Abrahams, P. W. Anderson, D. C. Licciardello, T. V. Ramakrishnan: Phys. Rev. Letters 42, 673 (1979)
27. S. Yoshino, M. Okazaki: J. Phys. Soc. Jpn 43, 415 (1977)
28. M. Schreiber: J. Phys. C18, 2493 (1985); Phys. Rev. B31, 6146 (1985)
29. D. C. Licciardello, D. J. Thouless: J. Phys. C8, 4157 (1975), and C11, 925 (1978)
30. A. MacKinnon: In Ref. 1, p. 90
31. A. MacKinnon, B. Kramer: In Application of High Magnetic Fields in Semiconductor Physics, Ed. By G. Landwehr, Lecture Notes in Physics Vol. 177, p. 74 (Springer, Berlin, Heidelberg 1983), and A. MacKinnon, B. Kramer, L. Schweitzer: Proc. LITPIM Suppl. (PTB-PG-1), p. 314 (PTB Braunschweig 1984)
32. H. Aoki, T. Ando: Phys. Rev. Letters 54, 831 (1985)
33. B. Bulka, M. Schreiber, B. Kramer: Z. Phys. B66, 21 (1987)
34. A.D. Zdetis, C.M. Soukoulis, E.N. Economou, G.S. Grest: Phys. Rev. B32, 8711 (1985); C.M. Soukoulis, A.D. Zdetis, E.N. Economou: Phys. Rev. B34, 2253 (1986)
35. B. Kramer, M. Schreiber: In Fluctuations and Stochastic Phenomena in Condensed Matter, ed. by L. Garrido, Lecture Notes in Physics, Vol. 268, p. 351 (Springer, Berlin, Heidelberg 1987)
36. M. Schreiber, B. Kramer, this volume
37. G. Bergmann: Phys. Reports 107, 1 (1984)

Critical Behaviour Near the Metal-Insulator Transition

M. Schreiber[1] *and B. Kramer*[2]

[1]Universität Dortmund, Institut für Physik, Postfach 500500, D-4600 Dortmund, Fed. Rep. of Germany
[2]Physikalisch-Technische Bundesanstalt Braunschweig, Bundesallee 100, D-3300 Braunschweig, Fed. Rep. of Germany

The scaling function for the 3d Anderson model with box and Gaussian distribution of the site energies is determined numerically. The critical behaviour near the disorder induced metal-insulator-transition is analyzed in the whole energy range, and the critical exponents for the localization length, ν, and the dc-conductivity, s, are estimated yielding $\nu = s = 1$.

1. Introduction

The behaviour of disordered systems near the metal-insulator-transition (MIT) has been controversially discussed [1], in particular the critical exponents s and ν describing the conductivity and the localization length close to the phase transition, respectively. While theoretical and numerical investigations yield conflicting values of 1.0 as well as 1.5, experimental evidence is also ambiguous between 0.55 and 1.07. Here we present results of an extensive numerical analysis of the Anderson Hamiltonian in the complete range of energies E. Evaluating considerably larger systems (up to 54 million sites) than previously with improved accuracy (relative error < 1%), the mentioned critical indices of 1.5 that had been derived with the same method reduce to 1.0 ± 0.1 in agreement with theory. Moreover, for weak disorder, a crossover to a critical index of 0.5 is observed which might be significant for the explanation of the low indices observed in the experiment.

2. Method

We consider the Anderson Hamiltonian for a disordered simple cubic system with constant transfer integral V(=1) between nearest neighbors only and random site energies governed by the box or the Gaussian distribution with width W^B or W^G. By means of the transfer matrix method a recursive computation of the quantum mechanical transmission probability through a very long bar with cross-section M^2 and length N ($\gg$M) is feasible [2]. It is an essential characteristic of our procedure that the accuracy can also be recursively determined, so that the statistical error can be controlled. The asymptotic behaviour of the transmission probability yields the localization length λ_M as the characteristic exponential decay length of the wave functions of this quasi-one-dimensional system.

Considering the dependence of λ_M on M for a given E and W, it is intuitively clear that the wave functions of the corresponding three-dimensional systems will be extended or localized, depending on whether or not λ_M increases more strongly than M. In this way one can determine the phase diagram of localization [3,4], i.e. the trajectory of the MIT in the E-W-diagram. An important feature, which is most pronounced for the box, but still significant for the Gaussian distribution, is the appearance of two MITs for certain energies, which have been attributed [3] to localization by quantum interference on the one hand (large W) and by potential fluctuations on the other hand (small W).

3. Scaling

For a more specific description of the phase transition we use a real space renormalization technique, attempting to scale the "raw data" λ_M/M versus $1/M$ for all E and W onto a single curve by choosing an appropriate scale factor $\xi(E,W)$. This curve as shown in Fig. 1 is composed out of rather short pieces for E = 0 and different W. However, due to the vast number of raw data (1328 values of λ_M/M) which were taken into account, multiple overlaps between these pieces occur. This allows an accurate extrapolation of λ_M with $M\to\infty$.

First of all, Fig. 1 convincingly demonstrates the universal validity of the scaling hypothesis at the band centre (E=0) independent of the employed distribution. The scaling function decomposes into two branches for the extended and the localized states yielding critical disorders $W_c^B = 16.5 \pm 0.1$ V for the box and $W_c^G = 20.9 \pm 0.1$ V for the Gaussian distribution, beyond which all states in the systems are localized. The formal scaling parameter ξ can be identified with the inverse dc-conductivity σ^{-1}, and with the localization length in three dimensions on the metallic and on the insulating side of the Anderson transition, respectively [2]. It is interesting to note that by this procedure we can reliably obtain localization lengths which are 100 times larger [2] than the cross-sectional diameter M of our samples, because the scaling properties of the raw data provide additional information for the extrapolation.

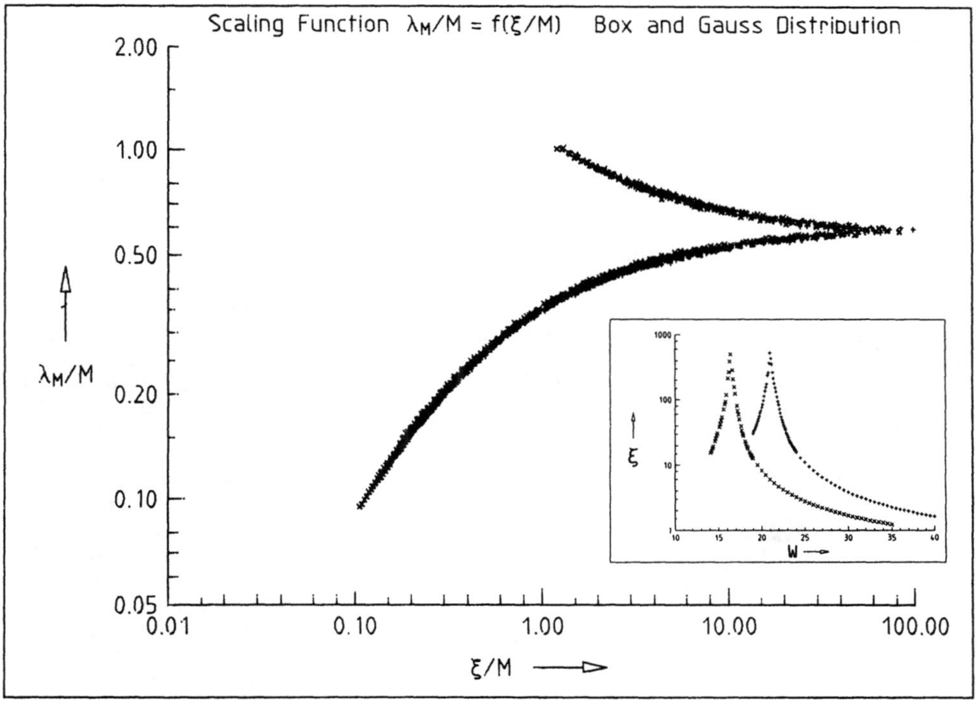

Fig. 1
Scaling function for the Anderson model at E = 0; M = 6, 7, ... 13;
W^B = 14.0, 14.1, ... 18.9, 19.0, 19.5, ... 34.5, 35.0
W^G = 19.0, 19.1, ... 23.9, 24.0, 24.1, ... 39.5, 40.0.
Insert shows the scaling parameter $\xi(W)$ for box (x) and Gaussian (+) distribution

4. Critical Behaviour

In analogy to the theory of second-order phase transitions we define critical exponents s and ν by $\xi \sim (W - W_c)^{-\nu}$ and $\sigma^{-1} \sim (W_c - W)^{-s}$ for constant E. From the scaling relations the equality $s = \nu$ can be derived. By fitting ln ξ as a function of ln $|W - W_c|$ to a straight line, we compute $s = \nu = 1.1$ for the box and $s = \nu = 0.9$ for the Gaussian distribution. The pure fitting error is only about 2%, but a much larger error stems from the fact that the scaling parameter ξ can be

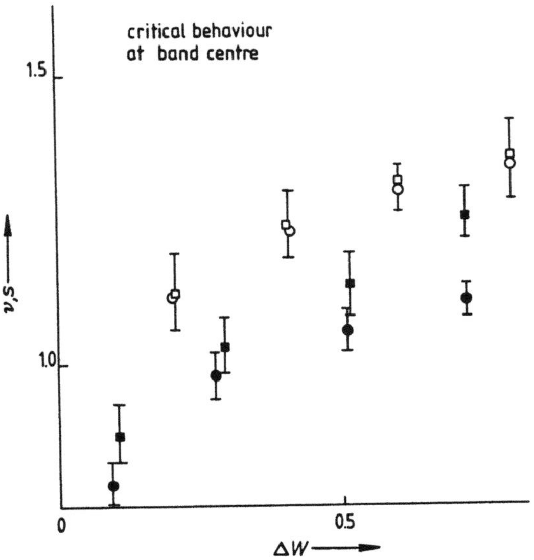

Fig. 2
Critical exponents s (squares) and ν (circles) for the Anderson model at E = 0 with box (open symbols) and Gaussian (full symbols) distribution of the site energies. Data for the scaling parameter ξ are omitted from of an interval ΔW around the critical disorder W_c (see text)

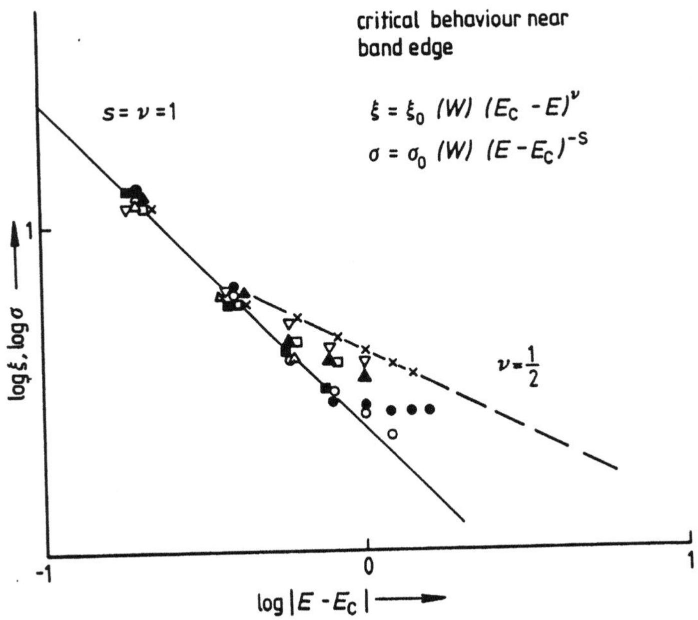

Fig. 3
Dependence of the scaling parameter on E for the Anderson model with box distribution on the isolating side W=2 (▽), 4 (□), 6 (x), 8 (▲), 10 (Δ) and on the metallic side, W=8 (■), 10 (O), 12 (●) of the phase transition. The critical energies were determined from the phase diagram of localization as E_c=6.0, 6.2, 6.6, 7.0, 7.4, and 7.8 for W=2, 4, ... 12, respectively [4]. The prefactors have been chosen so that the data for the smallest $|E - E_c|$ coincide. The slope of the full straight line is 1, the slope for the dashed line is ½

determined the less accurate the closer W is to the transition (i.e. in the flat part of the scaling curve in Fig. 1). This problem can be circumvented by omitting the data for ξ out of an interval ΔW around W_c, as displayed in Fig. 2. However, as criticality can only be expected to occur close to the transition, a large ΔW leads to a large error, too. We estimate an overall uncertainty of ±15% for the values given above, so that the results are consistent with previous theoretical work, i.e. $\nu = s = 1$ [5].

An inclusion of the localization lengths λ_M obtained for energies off the band centre does not significantly influence the scaling function, as long as the corresponding critical disorder remains large. Accordingly, Fig. 1 and the result $\nu = s = 1$ are not changed up to $E = 6V$, i.e. the band edge of the ideal crystal without disorder. For larger energies, however, the shape of the scaling function seems to change. Although the scaling parameters are not as well converged for this region as compared to $E = 0$ due to the smaller available set of raw data λ_M, we believe that this energy-dependent scaling function is not merely an artifact of the numerical calculation. This observation is, moreover, consistent with recent field theoretical arguments [6]. The physical reason for this behaviour which can be described by a "two-parameter scaling law" is the increasing importance of the localization due to potential fluctuations.

Remarkably, our analysis still yields a critical behaviour with $\nu = s = 1$ close to the phase trajectory in this regime. The same result follows from an investigation of the critical behaviour for constant W. Here the scaling parameter as a function of the energy yields critical indices $\nu = s = 1$ as shown in Fig. 3. However, the critical region shrinks. For small W, already close to the transition a crossover to $\nu = ½$ can be observed in Fig. 3. This is consistent with the penetration of the wave functions into the barriers between the potential wells, which confine these states.

5. Conclusion

In conclusion we state that the one-parameter theory of localization has been verified numerically for the centre of the band. The computed critical exponents are shown to be consistent with the results of perturbative treatments (ε-expansion) The observed deviations from one-parameter scaling beyond the unperturbed band edge remain to be investigated in more detail.

References

1. See the Conference Volumes Localization, Interaction, and Transport Phenomena, ed. by B. Kramer, G. Bergmann, and Y. Bruynseraede, Springer Ser. Solid-State Sci., Vol. 61 (Springer, Berlin, Heidelberg 1985); Localisation and Interaction in Disordered Materials and Doped Semiconductors, ed. by D. M. Finlayson (SUSSP Publications, Edinburgh 1986)
2. A. MacKinnon, B. Kramer: Z. Phys. B53, 1 (1983)
3. B. Bulka, M. Schreiber, B. Kramer: Z. Phys. B66, 21 (1987)
4. B. Kramer, this volume
5. F. J. Wegner: In Localization, Interaction, and Transport Phenomena, ed. by B. Kramer, G. Bergmann, and Y. Bruynseraede, Springer Ser. Solid-State Sci., Vol. 61 (Springer, Berlin, Heidelberg 1985), p. 99
6. K. B. Efetov: Sov. Phys. JETP 62, 605 (1985)

Renormalized Fermi Liquid Theory for Disordered Electron Systems and the Metal-Insulator Transition

C. Di Castro

Dipartimento di Fisica, Università di Roma "La Sapienza", I-00185 Roma, Italy

The basic concept of localization of electrons in a random potential was introduced by Anderson in 1958. Starting from the seventies a microscopic foundation to the theory of the underlying metal-insulator phase transition has been given via a renormalized perturbation theory /1/.

The electron-electron interaction modifies and completes /2/ the Anderson picture, thus providing a richer theoretical framework /3/ which, as it will be shown in the last section, is more suitable for the interpretation of the experiments.

The Anderson localization, being based on the one-electron picture, is specified once the amount of disorder present in the system is quantified by means of an effective coupling, which has been identified with the inverse conductance. The conductivity (or the diffusion constant) is renormalized and in strong disorder the system evolves towards an insulating phase.

When the mutual electron interaction is present several different scattering processes have to be considered. The related interaction amplitudes are strongly modified by the presence of disorder and additional effective couplings are required to specify the model /4/. The corresponding renormalization parameters have been identified in terms of physical quantities /5/ appropriate to the Landau Fermi liquid theory, i.e. the Landau parameters, related to the interaction amplitudes, are renormalized by the disorder and acquire a scale dependence. By studying their renormalization group equations various universality classes of the metal-insulator transition driven by the electron mutual interaction in the presence of disorder are obtained /3/. The resulting theoretical picture summarized according to the result of /4/ and /5/, is analyzed in comparison with the recent experiments.

1. ONE-ELECTRON PICTURE VERSUS EXPERIMENTS

In order to introduce some notation and to familiarize ourselves with the basic concepts of the theory we start by summarizing the one-electron picture /1/.

We start from the metallic phase and thus consider a free electron gas in the presence of a δ-correlated random potential U(x) defined via the averaging over the impurity distribution: $\overline{U(x)} = 0$, $\overline{U(x)U(x')} = \overline{u^2}\,\delta(x-x')$, $\overline{u^2}$ being proportional to the impurity concentration. In the Born approximation the scattering time τ_o is evaluated /6/ to be $\tau_o = \hbar(2\pi\,\overline{u^2}\,N_o)^{-1}$ where N_o is the density of states for the free electron gas in d-dimensions, $N_o \sim k_F^{d-2}/(\hbar^2/m)$.

Within the time τ_o between two collisions the electrons move along classical trajectories. No quantum interference effects are considered and the residual conductivity σ_o has the Drude form

$$\sigma_o = \frac{n e^2 \tau_o}{m} \, , \tag{1.1}$$

where n is the electron density.

The diffusive motion of the electrons implies that the density-density correlation function $\chi_{\rho\rho}$ has a diffusion pole in terms of the frequency Ω and the wave vector q

$$\chi_{\rho\rho} = -N_o \frac{D_o q^2}{-i\Omega + D_o q^2} \tag{1.2}$$

with the diffusion constant $D_o = (v_F^2 \tau_o/d)$.

In the static limit $\Omega = 0$, $q \to 0$ the density correlation function represents the change $\partial n/\partial\mu$ of the global particle density with respect to a statistic and homogeneous field, i.e. the chemical potential. In this simple case $\partial n/\partial\mu$ coincides with N_o. In the opposite limit ($q\to 0$, $\Omega\neq 0$), $\chi_{\rho\rho}$ vanishes, reflecting the conservation of the total number of particles. In the same way by the continuity equation one obtains the Einstein relation

$$\sigma_o = -\lim_{\Omega\to 0, q\to 0} e^2 \frac{\Omega}{q^2} \, \mathrm{Im}\, \chi_{\rho\rho} = e^2 N_o D_o \, .$$

When we increase the disorder present, corrections to the Drude theory should appear.To evaluate these corrections in perturbation theory, we have to identify a suitable expansion parameter. Since we start from the weak disorder region a good dimensionless effective coupling is given by the ratio between $N_o\overline{u^2}$ (which specifies the amount of disorder present in the system) and the Fermi energy ε_F (which for the free electron system is the other dimensional quantity specifying the model). This ratio $N_o\overline{u^2}/\varepsilon_F \sim \hbar/\varepsilon_F\tau_o \sim (k_F L)^{d-2}/g_o$ turns out to be inversely proportional to the conductance g_o in units of $e^2/\hbar$

$$g_o = \frac{\sigma_o L^{d-2}}{e^2/\hbar} \, , \tag{1.3}$$

where L is the characteristic size of the system. $t_o \sim 1/g_o$ is thus the effective coupling of the model, as has been envisaged in the phenomenological theory. From now on we put $\hbar=e=1$.

In the metallic phase the explicit dependence of t_o on L disappears in two dimensions, i.e. t_o is marginal according to the critical phenomena terminology at d=2 and acquires dimension ε=d-2 in terms of an inverse characteristic length at d=2+ε.

As usual, marginality manifests itself as logarithmic singularities in perturbation theory. These logarithmic corrections to the Drude term at first order in t_o physically correspond to the enhancement of the backscattering due to the quantum interference of the reverse patterns which are equivalent under time reversal.

The singularity appears as the logarithm(*) of the largest of the three ratios τ_L^{-1}/τ_o^{-1}, Ω/τ_o^{-1} and τ_T^{-1}/τ_o^{-1} where $\tau_L^{-1} \sim D_o L^{-2}$ and τ_T is the inelastic scattering time $\tau_T \sim T^{-p}$ (the power p depending on the choice of the possible scattering processes with p=2 for the electron-electron interaction).

At T=Ω=0 we have

* At d=3 the perturbative correction term has a square root rather than logarithmic behaviour and in the weak disorder region the conductivity should assume the form $\sigma=\sigma_o$-m $T^{p/2}$ with m<0 so that σ decreases with decreasing temperature.

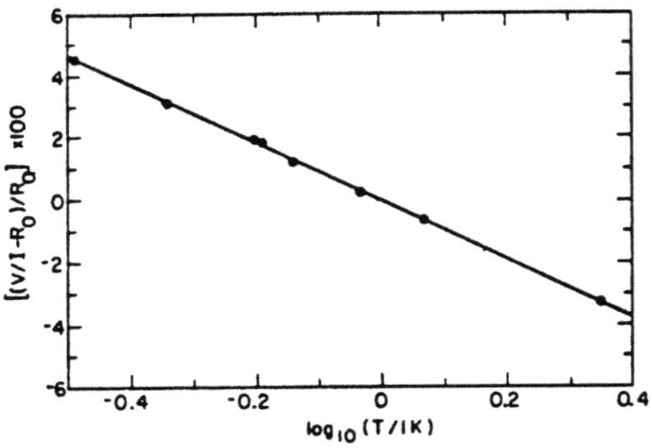

FIG.1 Resistivity rise plotted vs ln T for PdAu film. Dolan and Osheroff. Ref./7/

$$\sigma = \sigma_o \ (1+t_o \ln (D_o L^{-2} \tau_o)+...) \tag{1.4}$$

giving rise to a decrease of σ as L increases. At T=0 and for finite Ω, a $\ln \Omega\, \tau_o$ appears in eq.(1.4). At finite T, a ln T term in σ in the weak disorder regime is also reproduced in many experiments on amorphous metallic films /7/ (Fig.1).

The logarithmic singularities, which at d=2 dress the effective coupling t_o also, as is standard in critical phenomena, are summed up to a power law behaviour in d=2+ε. This is obtained by iterating the renormalization group equation up to the fixed point value $t^*\sim O(\varepsilon)$ of the renormalized effective coupling t via a continuous rescaling to infinity of the length ($L\rightarrow L'=sL$, $s\rightarrow\infty$). In d=2+ε the scaling theory of the Anderson transition is thus obtained. The behaviour of the system at the transition is governed by the divergence at T=0 of the characteristic coherence length ξ as a function of the disorder control parameter $\xi\sim(t_c-t_o)^{-\nu}$ where ν is the related critical index and t_c specifies the critical value of the disorder. In the localized regime ξ is the localization length. All the scaling quantities can be expressed in terms of the inverse of the coherence distance by means of their scaling dimension. The conductivity σ, for instance, scales as $\sigma\sim(t^*)^{-1}\,\xi^{-x_\sigma}$ with the scaling dimension x_σ=d-2=ε and therefore approaches zero at the transition with a power law behaviour in terms of the critical index μ satisfying the following scaling relation:

$$\sigma \sim (t^*)^{-1}\, \xi^{-x_\sigma} \sim (t_c-t_o)^{\mu} \quad ; \quad x_\sigma = \frac{\mu}{\nu} = \varepsilon . \tag{1.5}$$

The density correlation function $\chi_{\rho\rho}$ has many singular correction terms. By implementing global and local conservation laws it can be shown that $\chi_{\rho\rho}$ has the same diffusive form (1.2), where, however, D_o is replaced by the renormalized diffusion constant D, which scales as σ. According to scaling, the dimensions of T and Ω must be the same as the dimension of D ξ^{-2}, i.e. $x_T=x_\Omega=2+\varepsilon=d$. The critical index ν, at least for the first few orders in the ε-expansion, turns out to be $\nu=1/\varepsilon$ so that from eq.(1.5) we have μ=1 for the general case of non-magnetic impurities.

Breaking the time reversal invariance or the spin conservation, the interference mechanism between time reversed patterns is changed. In the presence of magnetic impurities the interference appears at the order t_o^2 among complicated diffusive processes. A scaling theory of the transition is still obtained with $\nu=1/2\varepsilon$ and μ=1/2 instead of unity. If spin-orbit coupling is present an antilocalization term is obtained at the order t_o and no phase transition appears in this context.

In all the cases discussed above, thermodynamic quantities like the single particle density of states, the spin susceptibility and the specific heat do not show any critical behaviour.

A comparison with the experiments /8,9/ shows that the theoretical investigation has to be extended to include the electron-electron interaction. We give here a very incomplete list of experimental results pointing in this direction:

1) Amorphous alloys like $Nb_x Si_{1-x}$ /8/ undergo a transition with a critical exponent $\mu=1$, however, the single-particle density of states N at the Fermi energy vanishes at the transition (Fig.2), contrary to theoretical expectations.

2) The interference term leading to the index $\mu=1$ in the Anderson localization is very sensitive to a time reversal invariance breaking term, due for example to an external magnetic field. In several systems as for instance in a-Si: Au /10/ the measured value $\mu=1$ is unchanged by the presence of the external magnetic field. The same result $\mu=1$ is obtained when the transition is driven by a large external magnetic field /11/, which result cannot be accounted for by the previous theoretical predictions.

3) Uncompensated doped semiconductors like Si:P have a continuous transition /9/ with an index $\mu=0.51\pm0.05$. A possible explanation for this value of the index could be that the correlation between the electrons induces a quasi-static spin motion so that the spin-flip universality class with $\mu=1/2$ should apply to this case. There are, however, many other experiments which imply the necessity of a full treatment of the interaction in the presence of disorder: a) The measured temperature-dependent term of the conductivity has a form

$$\sigma = \sigma_0 - m\ T^{1/2} \tag{1.6}$$

with a change of sign of m with varying the donor concentration, which cannot be taken into account by the disorder correction term in d=3. b) A strong temperature dependence of the spin susceptibility χ_s

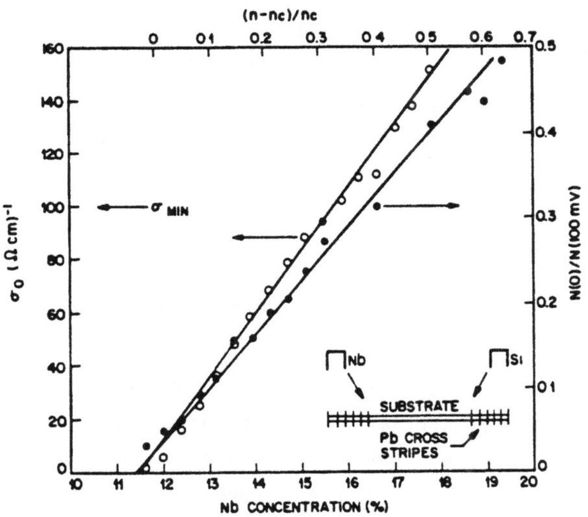

FIG.2 σ and N(0)/N(100 mV) vs Nb concentration. Inset shows method of sample preparation. Ref./8/

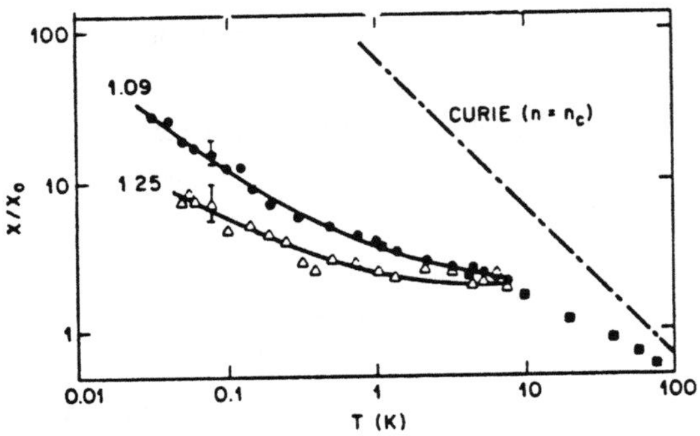

a

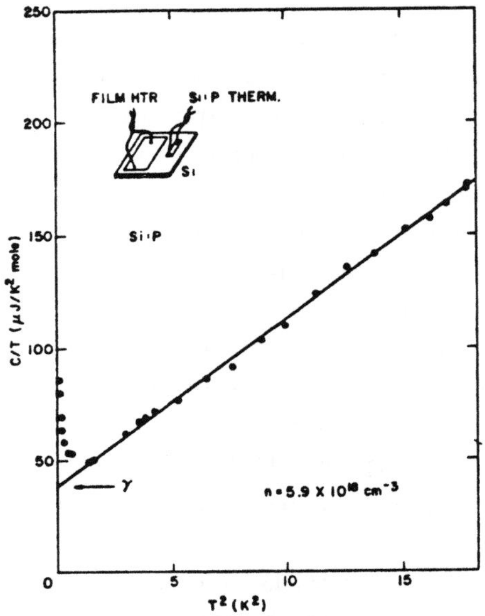

b

FIG.3a Spin susceptibility χ_S normalized to the Pauli value. M.A.Paalanen et al. Ref./12/

FIG.3b Specific-heat-to-temperature ratio, C/T, as a function of T^2, showing the low temperature deviation from the linear electronic specific heat. γ=C/T assumes a T-dependence and deviates from the linear zero-T intercept value indicated by the authors.

/12/, Fig.3a, and of the specific heat coefficient $\gamma=c_v/T$ /13/, Fig.3b, have been reported at low temperature.

4) When magnetic impurities are present a transition with an index $\mu=1$ rather than $\mu=1/2$ has been reported /14/. In the same way disordered Bi samples /15/ and Si-doped $Al_x\,Ga_{1-x}\,As$ mixed crystals /16/ show a metal-insulator transition with an index $\mu=1$ in the presence of a strong spin orbit coupling like in a-Si:Au /10/. Although in the weakly disordered films the antilocalization effect of the one-electron picture has been observed /15/, the transition cannot be driven by pure disorder and another mechanism must occur.

2. EFFECTIVE FERMI LIQUID THEORY

In dealing with more realistic systems the electron-electron interaction has to be considered together with a random potential.

Approaching the problem from the metallic side, in the absence of disorder the effects of the interaction are taken into account by the Landau Fermi liquid theory. In the low temperature region and for excitation energy near the Fermi surface where the inelastic scattering time becomes extremely long, the quasi-particles are treated as if they were free but for a few effective parameters entering various equilibrium physical quantities like the thermodynamic density of states, the spin susceptibility and the specific heat

$$\frac{\partial n^L}{\partial \mu} = \frac{\nu^L/N_o}{1+F_o^S} N_o\,, \quad \chi_S^L = \frac{\nu^L/N_o}{1+F_o^a} \chi_S^o\,, \quad \frac{C_v^L}{C_v^o} = \nu^L/N_o\,, \tag{2.1}$$

where $\nu^L/N_o = m^*/m$, m^* being the quasi-particle effective mass, and the superscripts L and 0 denote for each quantity the Landau form and the corresponding free-electron expression, respectively, i.e. $\chi^o{}_S = \mu_B{}^2 N_o$, $C^o{}_v = \gamma_o T = \pi^2/3\, N_o\, T$. The conventional Landau parameters $F_o{}^S$ and $F_o{}^a$ give a dimensionless measure of the interaction since they are related to the first coefficients of the expansion of the residual interaction among quasi-particles in terms of Legendre polynomials.

In the presence of disorder, repeated scattering processes with the impurities lead to a diffusive motion of the electrons and accordingly dress their mutual interaction into a frequency- and momentum-dependent form.

Singular correction terms, which are again logarithmic in d=2, appear in several physical quantities /2/. A renormalized perturbation theory is then built. Together with the renormalized inverse conductance t related to the disorder, the Landau parameters act as additional couplings, are renormalized and become scale dependent.

An effective Landau theory for disordered interacting electron systems is then obtained /3,4,5/:

$$\frac{\partial n}{\partial \mu} = Z_\rho \frac{\partial n^L}{\partial \mu}\,, \quad \chi_S = Z_S\, \chi_S^L\,, \quad C_v = Z\, C_v^L\,, \tag{2.2}$$

where Z_ρ, Z_S and Z are clearly identified via eq.(2.1) as the renormalization parameters of the interaction effective couplings and evolve according to a complicated set of renormalization group equations including t.

Historically this effective Landau picture has been obtained a posteriori. The interacting disordered electron system was mapped into a generalized non-linear σ-model in the first reference /4/. The renormalization parameters introduced to take care of the logarithmic singularities of this field theoretic model in 2+ε dimensions were afterward identified in terms of the modifications due to the combined effect of interaction and disorder on $\partial n/\partial \mu$, χ_S and C_v leading to the modified Landau expressions (2.2). This was achieved /5/ by implementing global and local conservation laws on the skeleton structure of the density, spin and energy correlation functions, which quite generally acquire a diffusive form

$$\chi_{\rho\rho} = -\frac{\partial n}{\partial \mu} \frac{D_\rho q^2}{-i\Omega + D_\rho q^2}\,, \qquad \chi_{SS} = \frac{\chi_S D_S q^2}{-i\Omega + D_S q^2}\,, \qquad \chi_{EE} = -TC_v \frac{D_E q^2}{-i\Omega + D_E q^2}\,, \tag{2.3}$$

with $\partial n/\partial \mu$, χ_S and C_v given by eqs.(2.2) and the particle, spin and heat transport coefficients renormalized as

$$D_\rho = \frac{D}{Z_\rho}(1+F_o^S)\,, \quad D_S = \frac{D}{Z_S}(1+F_o^a)\,, \quad D_E = \frac{D}{Z}\,, \tag{2.4}$$

where D is the diffusion constant appearing in the propagator of the diffusion mode of the theory.

Within this general framework the actual calculations have been carried out by considering that, as usual in the Landau scheme, the electron mutual interaction is effectively taken into account to all orders; the only expansion parameter of the theory is the coupling t_o specifying the disorder.

The first perturbative correction term for the conductivity σ in d=3 gives a $T^{1/2}$ temperature dependence in the same form (1.6) as obtained from the experiments. This gives a first hint that the theory is going in the right direction at least in the weak disorder region. Moreover the coefficient m depends on the various interaction effective couplings. Due to their evolution according to the group equations, the coefficient m may acquire an additional temperature dependence even in the weak disorder regime /17/, as has been experimentally observed in GeSb /18/, or change sign according to the experimental results of /9/.

The group equations /4,5/ for the effective couplings have been obtained by evaluating the singular correction terms at first order in t_o, in $d=2+\varepsilon$ dimensions and in the presence of a small time reversal symmetry breaking field to avoid additional complications due to the effect of the original quantum interference process on the electron mutual interaction.

The main consequences of the theory are now summarized and compared with the experiments.

The first general result of the perturbative analysis is that $\partial n/\partial\mu$ does not get renormalized and therefore $Z_\rho=1$. Z and Z_S besides acting as effective couplings of the theory enter multiplicatively some of the relevant physical quantities as indicated in eqs.(2.2) and (2.4). In spite of the fact that the thermodynamic density of states $\partial n/\partial\mu$ is not affected by the disorder in any relevant way, the density of states appearing in the renormalized specific heat coefficient γ is modified by Z and behaves accordingly. The single particle density of states has a third different behaviour and in all the cases where a localization transition driven by the interaction will appear, it goes to zero in various complicated forms and a pseudo-gap appears.

The specific heat renormalization parameter Z renormalizes also the temperature, whose scaling dimension is modified from the free electron value x^o_T,

$$x_T = x_T^o - x_Z , \qquad (2.5)$$

where x_Z is the scaling dimension of Z, and x^o_T, as we have seen assumes the value d for the Anderson transition and the value 2 for the free metallic case.

Depending on the asymptotic values assumed by the couplings as the length rescaling parameter goes to infinity under different symmetry conditions, very different physical scenarios are obtained.

In the general case of non-magnetic impurities at $d=2+\varepsilon$ and T=0, there exists a critical line $t_o=t_c\,(F_o^a,\varepsilon)$ (which goes to zero in two dimensions where all the states should be localized) such that in the region of low disorder ($t_o<t_c$) the system correctly scales to a conductor as the rescaling parameter goes to infinity, whereas in the region of strong disorder ($t_o>t_c$), where the system should evolve towards a localized regime, Z_S goes to infinity at a finite value of the rescaling parameter, signalling the presence of local magnetic moments. The group equations having been derived starting from the metallic regime cannot in this case be iterated to infinity.

At $t_o=t_c\,(F_o^a,\varepsilon)$ as the rescaling parameter goes to infinity, Z and Z_S diverge with a power law behaviour characterized by the scaling dimensions $x_Z=-3\varepsilon$ and $x_{Z_s}=-4\varepsilon$, respectively. At $t_o=t_c$ and finite temperature, χ_S and γ behave according to the following scaling forms:

$$\chi_S \sim T^{x_{Z_s}/x_T} \quad , \quad \gamma \sim T^{x_Z/x_T} \; . \qquad (2.6)$$

Considering the values of x_{Z_s} and x_Z given above and the relation (2.5), χ_S and γ diverge in this case as a function of the temperature

$$\chi_S \sim T^{-\frac{4\varepsilon}{2+3\varepsilon}} \underset{d=3}{\sim} T^{-4/5}, \quad \gamma \sim T^{-\frac{3\varepsilon}{2+3\varepsilon}} \underset{d=3}{\sim} T^{-3/5}. \tag{2.7}$$

We want to stress that χ_S and γ, being renormalized by two independent parameters, show different temperature behaviour.

In the group equation for the renormalized coupling t, the interaction introduces two competing terms. The divergence of Z_S prevents the localizing one from being effective. t approaches a fixed point value $t^*=0$ with a scaling dimension $x_t=\varepsilon$. Since $t\to 0$ the scaling low (1.5) is modified by the presence of the singular behaviour of $(t^*)^{-1}$ into $\mu=(\varepsilon-x_t)\nu$. At the order considered here $\mu=0$, i.e. σ remains constant at the transition, whereas according to eqs.(2.4) the spin and energy diffusion constants go to zero as Z and Z_S go to infinity. Higher order correction terms could modify this result leading to a vanishing conductivity in agreement with the experiments /9/ mentioned in the previous section.

Alternatively we can hypothesize that $t_0=t_c\,(F_0{}^a,\varepsilon)$ does not characterize the metal-insulator transition. The strong spin fluctuations, which are the relevant effect associated with $D_S\to 0$, lead instead to an instability line before the particle localization takes place. In this case the system, before reaching the instability, should make a cross-over to one of the universality classes with magnetic couplings which will be discussed later.

In any case the new relevant phenomenon characterized by the strong enhancement of χ_S and γ is indeed present in the experiments /12,13/ at temperatures below 1 K, Figs. 3a and 3b. The measured values of χ_S and γ also seem to suggest a stronger effect in χ_S than in γ /19/. When the temperature is sufficiently low that the spin scattering processes cut-off the effect of the Anderson term ($T<100$ mK according to the first reference of /12/) the agreement between theory (first formula (2.7)) and experiments appears to be more than qualitative.

Any mechanism which inhibits the spin fluctuation enhancement makes the localization term due to the interaction dominant in the equation for t. A finite fixed point value t^* is reached and a bona fide phase transition driven by the interaction is obtained. Whenever a magnetic coupling, due to magnetic impurities, spin-orbit or strong magnetic field ($\mu_B H >> kT$) is present, a finite $t^*\sim O(\varepsilon)$ value is indeed obtained at first order in t_0. At the transition the ordinary scaling law $\mu=\varepsilon\nu$ is valid with the critical index for the conductivity $\mu=1$.

In all the experiments /10,11,14,15,16/ designed to meet the previous physical situations an index $\mu=1$ is observed, thus solving the contradictions 2) and 4) of the list pointed out in the previous section while discussing the non-interacting theoretical framework in comparison with the experiments. With regard to point 2) the linear behaviour of σ as a function of the disorder control parameter discussed here for the interacting case is not affected by the presence of a magnetic field, as is now confirmed by many experiments in systems with spin-orbit coupling /10,15,16/. See Fig.4a. In the presence of spin-orbit coupling the localization term due to the interaction overcomes the antilocalization term due to the quantum interference, bringing back the localization transition with $\mu=1$. Of course the approach to criticality and the critical concentration itself is modified by the presence of the external field as verified by the experiments /16/, Fig.4a.

Although the critical behaviour of σ is characterized by the same value of the index $\mu=1$, the behaviour of Z distinguishes among the previous cases three different universality classes of the metal-insulator transition driven by the interaction in the presence of disorder.

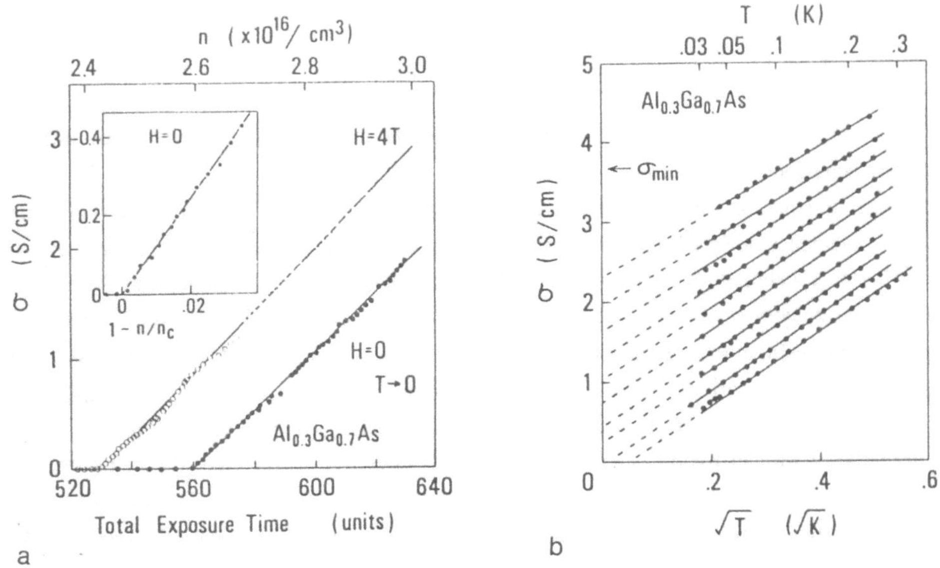

FIG.4a Extrapolated zero-temperature conductivity σ(0,n) as a function of the total exposure time (or the electron concentration) in the magnetic field of 0 T and 4 T. σ(0,n) near the metal-insulator critical point is plotted in the inset. Ref./16/

FIG.4b √T dependence of the conductivity below 300 mK in the vicinity of the critical point. Solid lines are fits to the data and dotted ones indicate the extrapolation to 0 K. Ref./16/

Z approaches a finite fixed point value Z^* in the presence of a strong magnetic field, i.e. its scaling dimension x_Z vanishes in this case. The specific heat remains linear in temperature and the scaling behaviour of σ as a function of temperature at criticality is $\sigma \sim T^{x_s/x_T} \sim T^{\varepsilon/d}$.

As the scaling parameter goes to infinity, Z approaches zero with a scaling dimension index $x_Z=\varepsilon/2$ and $x_Z=\varepsilon$ in the presence of magnetic impurities and spin-orbit coupling, respectively. Together with D_ρ, D_E also goes to zero in the first case. In the spin-orbit case instead with $x_Z=x_\sigma=\varepsilon$, Z and D vanish in the same way and D_E according to eq.(2.4) remains finite at the order considered.

γ and σ go to zero at criticality as a function of the temperature with a power law which is easily obtained by inserting the previous values of x_Z in their scaling forms. Experiments on the specific heat are not yet available to check these predictions. In the spin-orbit case the behaviour of σ at criticality as a function of the temperature is simply given by

$$\sigma \sim T^{x_\sigma/x_T} \sim T^{\varepsilon/2} \underset{d=3}{\sim} T^{1/2} , \tag{2.8}$$

where we have used the fact that in this case $x_T=d-x_Z=2$. The $T^{1/2}$ behaviour is here a result of scaling at criticality and must not be confused with the analogous perturbative term valid in the weak disorder regime. The experiments on $Al_{0.3}\,Ga_{0.7}\,As$ /16/ show at criticality no deviation from a $T^{1/2}$ behaviour: see Fig.4b.

We therefore conclude that whenever there is a magnetic coupling in the system, the predictions of the present theory for the disordered interacting electron systems fully agree with the experimental results. In the general case of non-magnetic impurities, although the theoretical predictions of a strong enhancement of the spin susceptibility and of the specific heat coefficient γ are indeed confirmed by the experiments, still more theoretical and experimental work is required to fully understand the transition and distinguish between the various hypotheses which have been suggested.

REFERENCES

1. P.W.Anderson, Phys.Rev. 109, 1492 (1958).
   For a review of Anderson localization see "Anderson Localization" Eds. Y.Nagaoka and H.Fukuyama, Springer-Verlag (1982) and Y.Nagaoka (ed.), Progr.Theor.Phys.Suppl. 84 (1985).
2. B.L.Altshuler, A.G.Aronov, Solid State Comm. 30, 115 (1979); Zh.Eksp.Teor.Fiz. 77, 2028 (1979);
   B.L.Altshuler, A.G.Aronov and P.A.Lee, Phys.Rev.Lett. 44, 1288 (1980);
   H.Fukuyama, J.Phys.Soc. Jpn. 48, 2169 (1980).
3. For a review see P.A.Lee and T.V.Ramakrishnan, Rev.Mod.Phys. 57, 287 (1985). C.Castellani and C.Di Castro in "Applications of Field Theory to Statistical Mechanics" Ed. L.Garrido. Springer-Verlag (1985). C.Castellani, C.Di Castro and G.G.Strinati in "Fluctuations and Stochastic Phenomena in Condensed Matter". Ed. L.Garrido. Springer-Verlag (1986). Localization and Interaction. Ed. D.M.Finlayson - Scottish University Summer School in Physics (1986).
4. A.M.Finkel'stein, Zh.Eksp.Teor.Fiz. 84, 168 (1983). Sov.Phys. JETP 57, 97 (1983);
   C.Castellani, C.Di Castro, P.A.Lee and M.Ma, Phys.Rev. B30, 527 (1984);
   B.L.Altshuler and A.G.Aronov, Solid State Comm. 46, 429 (1983).
5. A.M.Finkel'stein, Pis'ma Zh.Eksp.Teor.Fiz. 37, 436 (1983) and Z.Phys. B56, 189 (1984);
   C.Castellani, C.Di Castro, P.A.Lee, M.Ma, S.Sorella and E.Tabet, Phys.Rev. B30, 1596 (1984); B33, 6169 (1986);
   C.Castellani and C.Di Castro, Phys.Rev. B34, 5935 (1986);
   C.Castellani, G.Kotliar and P.A.Lee, Phys.Rev.Lett. 59, 923 (1987);
   C.Castellani, C.Di Castro, G.Kotliar, P.A.Lee and G.C.Strinati, Phys.Rev.Lett. 59, 477 (1987) and Phys.Rev. B in press.
6. See for instance A.A.Abrikorov, L.P.Gor'kov and I.Ye.Dzyaloshinskii, "Quantum Field Theoretical Method in Statistical Physics", Pergamon Press (1965).
7. G.J.Dolan and P.P.Osheroff, Phys.Rev.Lett. 43, 721 (1979);
   S.Kobayashi, F.Komori, Y.Ootuka and W.Sasaki, J.Phys.Soc. Jpn. 49, 1635 (1980).
8. G.Hertel, D.J.Bishop, E.G.Spencer, J.M.Rowell and R.C.Dynes, Phys.Rev.Lett. 50, 743 (1983).
9. G.A.Thomas, M.Paalanen and T.F.Rosenbaum, Phys.Rev. B27, 3897 (1983);
   T.F.Rosembaum, R.F.Milligan, M.Paalanen, G.A,Thomas and R.N.Bhatt, Phys.Rev. B27, 7509 (1983).
10. N.Nishida et al., Solid State Electronics Proceedings of the Santa Cruz Conference (1984).
11. S.Von Molnar, A.Briggs, J.Flouquet and G.Remenyi, Phys.Rev.Lett. 51, 706 (1983);
    R.M.Westervelt, M.J.Burns, P.F.Hopkins and A.J.Rimberg. This conference.

12. S.Ikeata and S.Kobayashi, Solid State Comm. 56, 607 (1985);
M.A.Paalanen, S.Sachdev and R.N.Bhatt, Phys.Rev.Lett. 57, 2061 (1986);
Y.Ootuka and N.Matsunaga , this conference.
13. G.A.Thomas, Y.Ootuka, S.Kobayashi and W.Sasaki, Phys.Rev. B24, 4886 (1981);
S.Ikheata, private communication.
14. S.Okuma, F.Komori, Y.Ootuka, S.Kobayashi, J.Phys.Soc. Jpn. 54, 2382 (1985).
15. F.Komori, S.Okuma and S.Kobayashi, J.Phys.Soc. Jpn. 56, 691 (1987) and this conference.
16. S.Katsumoto, F.Komori, N.Sano and S.Kobayashi, J.Phys.Soc. Jpn. 56, 2259 (1987) and this conference.
17. C.Castellani, C.Di Castro, H.Fukuyama, P.A.Lee and M.Ma, Phys.Rev. B33, 7277 (1986).
18. Y.Ootuko, S.Kobayashi, S.Ikehata, W.Sasaki and J.Kondo, Solid State Comm. 30, 169 (1979).
19. M.A.Paalanen, private communication.

Landau Fermi Liquid Theory for Disordered Systems and the Scaling Theory of the Metal-Insulator Transition

G. Kotliar

Department of Physics, Massachusetts Institute of Technology,
77 Massachusetts Avenue, Cambridge, MA 02139, USA

We pedagogically outline a Fermi liquid framework to describe interacting electrons in a random potential in complete analogy with Landau's theory for translationally invariant system and use this language to interpret the different physical scenarios for the metal insulator transition.

1. LANDAU THEORY: CLEAN CASE

Landau Fermi liquid theory provides a rigorous description of the low energy long wavelength properties of strongly interacting Fermi systems. The description of strongly interacting Fermions in terms of weakly interacting quasiparticles (QP), originally formulated by Landau[1] for translationally invariant systems has also been used to describe finite systems like atomic nuclei[2]. In this talk we will outline a Fermi liquid framework to describe interacting electrons in a random potential which closely parallels Landau's construction. This framework is useful to parametrize all the response functions of a strongly disordered metal and naturally leads to the scaling variables relevant to the metal insulator transition originally proposed by Finkelstein[3] and developed by Castellani, et al.[11,12] The purpose of these notes is pedagogical. We shall first review the key concepts of the Landau approach in the translational invariant case, and then we will indicate how a very similar construction can be carried out for the disordered electron gas. We conclude by interpreting the different scenarios for the metal insulator transition (MIT) in terms of the QP parameters. This exposition is based on refs. 4 and 5. Refs. 6 and 7 and in particular ref. 8, contain earlier references, and stress the relevance of Fermi liquid ideas to disordered Fermi systems. Ref. 9 contains excellent reviews of Landau theory.

Basic to Landau's theory is the idea of quasiparticles which are well defined low lying excitations of a many body system. In a translational invariant system the QP states are labelled by their momenta k, their energy is given by $\varepsilon_k=k^2/2m^*$. The interactions are taken to invariant under spin rotations and σ is a spin index. The energy of a weakly excited state characterized by a definite number of QP excitations $n_{p\sigma}$ in a unit volume, is written as

$$F=\sum_{p\sigma}(\varepsilon_p-\mu)\,\delta n_{p\sigma}+1/2\sum_{pp'\sigma\sigma'} f_{p\sigma,p'\sigma'}\delta n_{p\sigma}\delta n_{p\sigma'} \tag{1}$$

$f_{p\sigma,p'\sigma'}$ describes the interactions between the QP while $\delta n_{p\sigma}=n_{p\sigma}-n_{p\sigma}^{\circ}$ is the deviation of the QP occupation number from the equilibrium value $n_{p\sigma}^{\circ}$ which is given by the Fermi distribution. In the presence of an external field V_σ it obeys a linearized Boltzman equation which in the collisionless regime becomes

$$\frac{\partial\delta n_{p\sigma}}{\partial t}+v_p\,\frac{\partial\delta n_{p\sigma}}{\partial r}+F_{p\sigma}\,\frac{\partial n^{\circ}_{p\sigma}}{\partial p}=0 \tag{2}$$

which in turn can be derived from the equations of motion $p/m^*=v_p=\partial H/\partial p$, $F_{\sigma p}=-\partial H_\sigma/\partial r$ for the semiclassical Hamiltonian

$$H_\sigma = \frac{p^2}{2m^*} + \sum_{p'\sigma'} \delta n_{p'\sigma'} f_{p\sigma,p'\sigma'} + V_\sigma(r). \tag{3}$$

The kinetic equation can be used to derive the spin-spin, density-density and heat-heat correlation functions. Defining the Landau paramters F_o^s, F_o^a by

$$2\rho_Q f_{p\sigma p'\sigma'} = \sum_\ell P_\ell(\cos\theta)(F_\ell^s+\sigma\sigma' F_\ell^a) \tag{4}$$

with $\rho_Q = m^*k_F/2\pi^2$ the QP density of states per unit spin in 3-dimensions. Solving (2) for the response functions one gets in the static the well known results for the compressibility, susceptibility and specific heat.

$$\frac{dn}{d\mu} = \frac{2\rho_Q}{1+F_o^s} \qquad \chi = \frac{2\rho_Q \mu_B^2}{1+F_o^a} \qquad C_V = \frac{2\pi^2\rho_Q T}{3}. \tag{5}$$

The microscopic justification of Landau theory is based on two key insights. The first focuses on the pole structure of the one particle Green's function for small ω and $\varepsilon_k-\mu$:

$$G(k,\omega) \approx \frac{a}{\omega-(\varepsilon_k-\mu)+i\,\mathrm{sign}\,\omega/2\tau_{in}}. \tag{6}$$

The location of the pole ε_k defines the quasiparticle energy, the residue a is identified as the quasiparticle weight while τ_{in} is the inverse width of the level and diverges as $\varepsilon_k-\mu$, ω and the temperature goes to zero. The second insight is that the pole structure in G induces a singular dependence of the response functions and four point functions in the ratio $r=\Omega/qv_F$. This arises because the skeleton structure of these quantities contain a singularity from the free propagation of a particle hole pair represented by the product of two Green functions. For small Ω and q^9.

$$G(k+q/2,\ \omega+\Omega/2)\ G(k-q/2,\ \omega-\Omega/2) = \ln_{r\to 0}\,[G^2] + \tilde{R}(k\omega,q\Omega) \tag{7}$$

with all the Ω and q dependence contained in the singular function:

$$\tilde{R} = \frac{2\pi i a^2\ \delta(\varepsilon_k-\mu)\ \delta(\omega)\ \Omega}{\Omega - q.v_k + i\ \mathrm{sig}\ \Omega/\tau_{in}}. \tag{8}$$

The singular factor $\tilde{R}$ enters as an intermediate state in the skeleton decomposition of the scattering amplitude (four-point function with amputated external legs) with small momentum transfer in the particle hole channel $\Gamma\ \gamma\delta\,\alpha\beta\ (P_1\varepsilon_1,\ P_2\varepsilon_2,\ q\Omega)$ (see Figs.1,2).

As a result $\bar{\Gamma}$ has two different values $\bar{\Gamma}^\omega$ and $\bar{\Gamma}^q$ in the limit $\omega\to 0$ $q=0$ and $q\to 0$ $\omega=0$ respectively. They identify the Landau parameters F_ℓ^s, F_ℓ^a, A_ℓ^s, A_ℓ^a microscopically via the relations

$$2\rho_Q a^2 \tilde{\Gamma}^\omega_{\alpha\beta\gamma\delta}(p_1p_2) = \sum_{\ell=o}^{\infty} P_\ell(\cos\theta)\,\{F_\ell^s\delta_{\alpha\gamma}\delta_{\beta\delta}+F_\ell^a\sigma^i_{\alpha\gamma}\sigma^i_{\beta\delta}\}$$

$$2\rho_Q a^2 \tilde{\Gamma}^q_{\alpha\beta\gamma\delta}(p_1p_2) = \sum_{\ell=o}^{\infty} P_\ell(\cos\theta)\,\{A_\ell^s\delta_{\alpha\gamma}\delta_{\beta\delta}+A_\ell^a\sigma^i_{\alpha\gamma}\sigma^i_{\delta\beta}\} \tag{9}$$

with p_1p_2 on the Fermi surface and $P_F^2\cos\theta = p_1\cdot p_2$. All the Ω and q dependence of the response functions which enters via the singular factor $\bar{R}$ is extracted

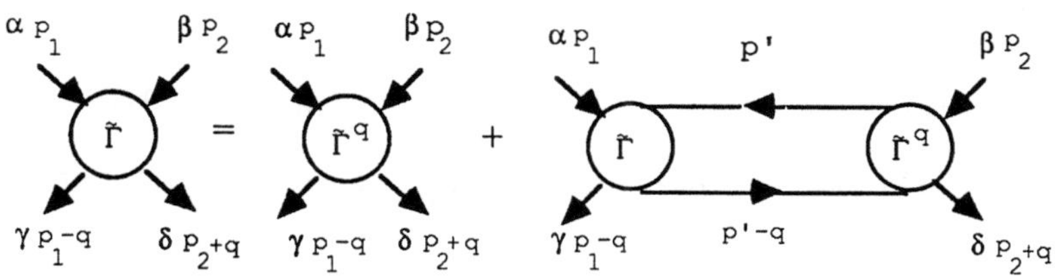

FIG. 1 The full scattering amplitude $\tilde{\Gamma}$ is expressed in terms of the static amplitude $\tilde{\Gamma}^q$ and the singular function $\bar{R}$. (Product of two QP Green's functions). In the disordered case R is replaced by L

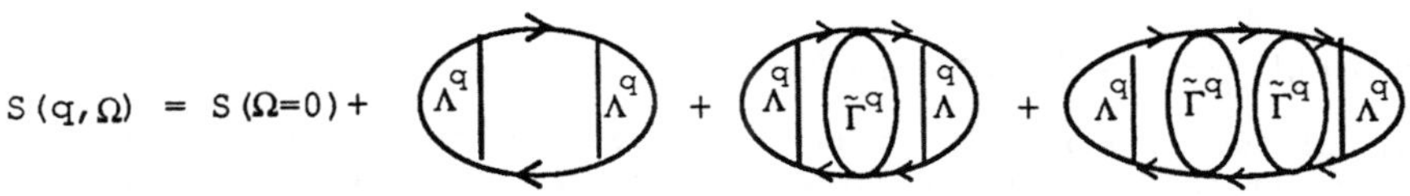

FIG. 2 Skeleton expansion of a response (spin-spin, density-density, heat-heat) function $S(q,\Omega)$. The Ω,q dependence is contained in the product of two intermediate Green's function lines, i.e., the singular function R in the clean case or L in the disordered case. The first blob is the static (singlet, triplet, or energy) vertex Λ^q. The blob $\tilde{\Gamma}^q$ denotes the static (singlet for the density or heat for the spin triplet) amplitude

using skeleton decompositions in vertex parts irreducible with respect to $\bar{R}$. As shown in Fig.2.

The physical fact that QP carries unit charge, spin and heat is reflected in the Ward identities for the charge (singlet) Λ_s^ω, spin (triplet) Λ_t^ω, and heat Λ_H^ω vertices:

$$a\Lambda_s^\omega = 1, \quad a\Lambda_t^\omega = 1, \quad a\Lambda_H^\omega = 1. \tag{10}$$

The previous discussion applies to short range interactions. In the presence of long range forces one separates the long range part of the interaction which is treated as a self consistent electric potential from the short range part which is treated as above. The short range part of the interaction gives rise to $f_{p\sigma,p'\sigma'}$ and to the Landau parameters while the self-consistent potential which arises from the long range part of the interaction is added to V_σ in the kinetic equation. Diagramatically the short range part of the interaction $\Gamma_{S.R}$ is identified as the sum of all diagrams contributing to the scattering amplitude which are irreducible with respect to a coulomb line . They sum up to $(\Lambda^q{}_s)^2/(dn/d\mu)$. Adding the long range and the short range part to construct the total singlet static amplitude one finds the general result $2a^2\,\rho_Q\,\Gamma_s{}^{tot} = 1$.

2. LANDAU THEORY: DISORDERED CASE

Fermi liquid theory survives the introduction of disorder and is valid up to the metal insulator transition.

As a result, the QP residue a disappears from all the response functions[9] and one obtains (5). In the absence of translation invariance k is not conserved and cannot characterize quasiparticle eigenstates. We will use ε_n the energy of a quasiparticle in isolation (analogous to $\varepsilon_k = k^2/2m^*$) to label QP states. The free energy as a functional of the QP distribution function $n_\sigma(\varepsilon_n, r)$ is written by analogy to (1)

$$F(n_\sigma)=\sum_\sigma \int dr\, n_\sigma(\varepsilon_n r)(\varepsilon_n-\mu) + 1/2 \int d^d r \qquad \delta N_\sigma(r)\, \delta N_{\sigma'}(r^*)\, f_{\sigma\sigma'} . \tag{11}$$

$N_\sigma(r)=\sum n_\sigma(\varepsilon_n r)$ is the total density per spin and $f_{\sigma\sigma'}=f_\sigma+\sigma\sigma' f^a$ is the QP interaction function.

Comparing (2) and (11) one sees that in the dirty case the angular dependence of f is neglected. This is because $n_\sigma(\varepsilon_n r)$ is intended to provide a course grained description of the system over distances longer than the mean-free path where only s-wave scattering is important. In an external spin dependent field v_σ the QP distribution function obeys a kinetic equation

$$0= \frac{\partial}{\partial t} n_\sigma(-D_Q\Delta^2)n_\sigma+ \frac{\partial n_\sigma}{\partial \varepsilon} (-D_Q\nabla^2)\,[V_\sigma+\sum_\sigma f_{\sigma\sigma'}N_{\sigma'}]=0 . \tag{12}$$

D_Q is the QP diffusion constant to be identified microscopically below. Equation (12) is obtained from the kinetic equation of the clean system (2) by replacing $v_p\partial/\partial r$ by $-D_Q\, \partial^2/\partial r^2$. Physically, this substitution reflects the crossover from ballistic to diffusive transport as the mean free path is decreased. It can be derived from a semiclassical stochastic dynamics

$$\frac{dr}{dt} = \xi \qquad \frac{d}{dt}\,[\varepsilon_n+ \sum_{\sigma'} f_{\sigma\sigma'}\, \delta N_{\sigma'}(r)+V_\sigma]=0 \tag{13}$$

for the "coordinates" r and ε_n. ξ is a Brownian motion with covariance $<\xi(t)\xi(0)>=D_Q\delta(t)$ while the second equation is the statement of conservation of energy.

Solving (12) for the density-density, spin-spin and heat-heat correlation function π, χ, χ_H, one find the standard diffusive form[11,12,5]

$$\pi(q,\omega) = \left(\frac{dn}{d\mu}\right) \frac{D_\rho q^2}{D_\rho q^2-i\omega} \qquad \chi(q,\omega) = \frac{\chi D_s q^2}{D_s q^2-i\omega} \qquad \chi_H(q,\omega) = \frac{TC_V D_H q^2}{D_H q^2-i\omega} \tag{14}$$

with $dn/d\mu=2\rho_Q/1+F_o{}^s$, $\chi=2\rho_Q\mu_\beta{}^2 / 1+ F_o{}^a$ and $C_V=2\pi^2\rho_Q T/3$, $2\rho_Q f^s=F_o{}^s$, $2\rho_Q f^2=F_o{}^{a\ a}$ as in the clean case and

$$D_\rho= D_Q\ (1 + F_o{}^s),\ D_\rho = D_Q\ (1 + F_o{}^a) \qquad D_H = D_Q \tag{15}$$

This structure has many interesting implications. As we approach the MIT the transport coefficients $\sigma=(dn/d\mu)D\rho$, $\sigma_s=\chi D_\rho$ and $\kappa=C_V D_H$ vanish but their ratio remains constant. In particular the Wiedeman Franz law holds up to the MIT[5]. As one approaches the MIT, χ is expected to diverge since it is infinite on the insulating side, while $dn/d\mu$ remains finite. Therefore, $D_s/D_\rho = (dn/d\mu)/\chi$ goes to zero, i.e., spin diffusion is much slower than charge diffusion close to the MIT.

To justify the phenomenological picture we focus on the one and two particle Green's functions. In a disordered medium one lacks translational invariance and there is no basis (like plane waves in the ordered case) which a priori diagonalizes the one-particle Green's function $G(x,x',\omega)$. Nevertheless, as shown in ref.(6) at zero frequency $G(x,x',\omega=0)$ is an Hermitian matrix with eigenvectors $\varphi_n(x)$. In this basis the QP assumption states that for small frequency G as a pole structure

$$G(x,x',\omega) \approx \sum_n \frac{a_n\ \varphi^*{}_n(x)\ \varphi_n(x')}{[\omega-(E_n-\mu)+i/2\tau_{in}\mathrm{sig}\ \omega]} \tag{16}$$

which closely parallels (6). Equation (16) holds for any realization of the random potential. E_n and φ_n are the exact poles and eigenvectors of $G(\omega=0)$ and are very different from the exact eigenvalues and eigenvectors of the reference Anderson Hamiltonian obtained by turning off the interactions. In particular we will see that they lead to diffusion constants and density of states which behave very differently than that of the non-interacting case. To make further progress we consider disorder averaged, denoted by <>, quantities. We identify the QP density of states per unit spin ρ_Q as

$$\rho_Q = < \sum_n \delta(E_n-\mu) > . \tag{17}$$

The ratio of ρ_Q to ρ_o, the density of states of the non-interacting system plays the role of m^*/m in the translationally invariant case. The quasiparticle diffusion constant D_Q is identified by constructing a wave packet out of QP wave functions $\varphi_n(x)$ and finding its long time behavior. Assuming[10]

$$<|\int\varphi_m^*(x)\, e^{iqx}\varphi_n(x)\,|^2> = \frac{\rho_Q}{\pi}\;\frac{D_Q q^2}{D_Q q^2 + |\varepsilon_m-\varepsilon_n|} \tag{18}$$

one finds that the wave packet diffuses with diffusion constant D_Q.

As in the clean case the pole structure in the single particle Green's function gives rise to a singular dependence in the ratio $r=\omega/D_Q q^2$ (which is analogous to the ratio $\omega/v_F q$ when one considers the propagation of a particle hole pair. This enters into intermediate states in the calculation of the scattering amplitude and the correlation factors. From (16) and (18) it follows

$$\int d(x-x')<G(x,x',\varepsilon+\omega)\,G(x,x'\varepsilon)>e^{iq(x-x')}=L(q,\omega,\varepsilon)+(\text{non-singular term}) \tag{19}$$

with all the singular dependence in q and Ω contained in

$$L(q,\Omega,\varepsilon) = \frac{(2\pi)\; a^2\rho_Q\delta(\varepsilon)\,\Omega}{[D_Q q^2 - i\Omega+\tau_Q^{-1}]} . \tag{20}$$

The analogy with the clean case becomes obvious once we integrate (7) and (8) over the fast momenta k which has no significance in the disordered case. The factor $\delta(\varepsilon)$ expresses the fact that $L=0$ unless $(\varepsilon+\omega)\varepsilon<0$. Once more the ballistic transport term $v_k.q$ is replaced by $D_Q q^2$ as in the kinetic equation. At this point is useful to indicate the expression for Finkelstein scaling variables in the perturbation theory framework of Castellani et al.[11,12] in terms of QP parameters: $D_Q=D/Z$, $a=\xi/Z$, $\rho_Q=\rho_o Z$, $\tau_Q=\tau_{ph}/Z$ with τ_{ph} the particle hole dephasing time introduced in ref.15.

The scattering amplitude (see Fig.1) has now a singular dependence in $\omega/D_Q q^2$. At this point the theory keeps only $\ell=0$ spherical harmonic

$$\tilde\Gamma_{\alpha\beta\gamma\delta} = \tilde\Gamma^s\delta_{\alpha\gamma}\delta_{\beta\delta} + \tilde\Gamma^t\sigma^i{}_{\alpha\gamma}\sigma^i{}_{\beta\delta} . \tag{21}$$

In fact in perturbation theory in $1/k_F\ell$ the higher ℓ terms are less divergent than the $\ell=0$, and one assumes they are unimportant for the long distance behavior. The scattering amplitudes are related to the static amplitudes by the usual Bethe Salpeter equation, which in the disorder case, reduces to algebraic equations[11,12]. Fig.1 with the intermediate Green's function lines represented now by L, then gives

$$\tilde\Gamma^s(q,\omega)= \tilde\Gamma^\omega\,\frac{D_Q q^2 -i\omega}{[\,D_Q q^2 -(1-A^\circ_s)\,\omega]}\qquad \tilde\Gamma^t(q,\omega) = \tilde\Gamma_t^\omega\,\frac{D_Q q^2 -i\omega}{[D_Q q^2 -(1-A^\circ_t)\,\omega]} . \tag{22}$$

As in the clean case one considers the static and dynamic limits. The $\omega=0$ $q\to0$ gives the static amplitude from which we identify the Landau parameters. $A_o^s = 2\rho_Q a^2 \Gamma_s$, $A_o^t = 2\rho_Q a^2 \Gamma_t$. The $q=0$ $\omega\to0$ limit of Γ defines the Fermi liquid parameters F as in (4). Equation (22) then gives the relation

$$F_o^s=A_o^s/1-A_o^s \qquad F_o^a=A_o^a/1-A_o^a. \tag{23}$$

The ward identities (10) which only rely on conservation laws remain valid in the dirty limit. Expressing them in terms of the static vertices via the relations $\Lambda^{\omega}_s = \Lambda^{k}_s{}^0\,(1+F_o^s)$ $\Lambda^{\omega}_t = \Lambda^{k}_t(1+F_o^a)$ obtained from a Bethe Salpeter equation similar to (22) for the vertex functions, one finds the results[12,3]

$$a\Lambda_s^k = \frac{1}{(1+F_o^s)}, \qquad a\Lambda_t^k = \frac{1}{(1+F_o^a)}, \qquad a\Lambda_H^k = 1 , \tag{24}$$

As in the clean case one studies the ω and q dependence of the response functions by isolating the singular factor L in the skeleton expansion of Fig.2. The density-density correlation function is expressed in terms of its static limit $dn/d\mu$, L, the singlet static vertex Λ^q_s and the singlet state scattering amplitude Λ^q_H. The spin-spin correlation function is expressed in terms of its static limit χ, and triplet static vertex Λ^q_t and the triplet scattering amplitude Λ^q_t. The heat-heat correlation function involves an energy vertex Λ^q_H and singlet quantities, but only the first two terms in the skeleton expansion are relevant, therefore, identifying the heat-diffusion constant with the QP diffusion constant without Fermi liquid corrections[5]. Using Fig.2 and (20)-(24) one justifies microscopically the fundamental relations (14)-(15).

The QP residue cancels against the vertex and disappear from all two particle observables. It appears only in the one paricle density of states ρ_1 which is proportional to the imaginary part of the one-particle Green's function. From (6) it follows that $\rho_1=a\rho_Q$.

3. THE MIT AS A BREAKDOWN OF FERMI LIQUID THEORY

The metal insulator transition can be viewed as a gradual breakdown of the Fermi liquid state. As one varies a control parameter n, i.e. the strength of the disorder, the donor concentration of uniaxial stress, the Fermi liquid parameters D_Q, ρ_Q, a, A^a_o, which describe the low energy physics, change continuously. As one approaches a critical value n_c some of these parameters go to zero or infinity. This behaviour is related to the divergence charactertized by a critical exponent ν of a length $\xi = 1/|n-n_c|^{\nu}$ which can be interpreted in the insulating side of the transition as the scale inside which a Fermi liquid description of the system holds.

It is a non trivial assumption, justified in the framework of Finkelstein's non linear σ model[3] and by lowest order perturbation theory[11] that $1+F_o^s$ scales with ρ_Q leaving $dn/d\mu$ unrenormalized. We are therefore left with two dimensionless scaling variables describing the physics at scale L close to the MIT: the dimensionless inverse conductance familiar from the Anderson localization problem $t(L)= (e^2/\hbar)/\rho_Q D_Q L^{(d-2)}$ and the triplet Fermi liquid scattering amplitude $A_o^a = 2\rho_Q a^2\tilde{\Gamma}_t^q$. When the Cooper channel is present one also includes the corresponding scattering amplitude in this Channel $\gamma_c=2\rho_Q a^2\tilde{\Gamma}_c$. The renormalization-group equations for the scaling variables were derived and the critical exponents to lowest order in ε in $2+\varepsilon$ dimensions were calculated by Finkelstein and Castellani, et al., in refs.11-13-14 and several universality classes were found. It is illuminating to interpret them in terms of the behaviour of the QP paramenters in the different cases.

From a scaling analysis point of view there are two very different situations.

a) The transition takes place at t=0, $A^a_0=\infty$. The critical point occurs at a finite value of the product $tA_0{}^a$ which can be interpreted as a ratio of the on-site repulsion to the bandwidth of a Hubbard model. Hence this transition has the character of a Mott transition in a weakly disordered enviroment. As one approaches the transition the QP density of states diverges with an exponent x, $\rho_Q \sim \xi^x$ while $A_0{}^a$ scales with an exponent $A_0{}^a \sim \xi^\theta$ which also gives the rate at which the dangerously irrelevant variable t approaches 0. At the transition the conductivity scales as $\sigma \sim 1/\xi^{d-2-\theta}$, this modifies the usual relation between the conductivity and the localization length which now becomes

$$\sigma \sim 1/\xi^{d-2-\theta} \sim (n-n_c)^{\nu(d-2-\theta)}.$$

In this framework it is still possible to have minimum metallic conductivity and a continuous second order MIT provided that $\theta=d-2$. This seems to be the case, at least to lowest order in the $2+\varepsilon$ expansion[13]. This strange behavior can be understood as a cancellation between the vanishing of the QP diffusion constant $D_Q \sim 1/\xi^{d-2+x-\theta}$ $(d-2+x-\theta=3\varepsilon+O(\varepsilon^2))$ and the divergence of the QP density of states $x\approx3\varepsilon+O(\varepsilon^2)$ in the product $\sigma=\rho_Q D_\theta$. The QP weight a scales to zero as $a\sim1/\xi^\beta$, $\beta=3\varepsilon+O(\varepsilon^2)$ this fact together with the divergence of $\chi\sim\xi^{x+\theta}$ and $C_V/T\sim\xi^x$ suggest a physical picture of the MIT which takes place as a gradual conversion of QP degrees of freedom into spins.

It is interesting to contrast the different behavior of the 3 density of states C_V/T, $dn/d\mu$, and ρ_1. The linear coefficient of the specific heat is enhanced due to the divergence of the QP density of state. $\rho_1=a\rho_Q$ is driven to zero because a vanishes faster than ρ_Q diverges while $dn/d\mu$ stays finite. Notice that since $A^s{}_0\approx1$ while $A^a{}_0$ diverges the Landau sum rule

$$\sum_{\ell=0}^{\infty}(2\ell+1)\;(A^s_\ell+A^a_\ell) = 0$$

cannot be satisfied with just two amplitudes. Presumably, a large number (of the order of some power of ξ) of amplitudes $A^s{}_\ell$, $A^a{}_\ell$ become large as one approaches the transition.

b) When spin is not conserved the MIT takes place at a finite value of the disorder parameter t. The usual scaling relation $\sigma\sim1/\xi^{d-2}$ holds. This scenario is realized in the presence of magnetic impurities, strong magnetic fields, and spin orbit scattering occur at a finite value of the disorder parameter t. Since $\sigma=\rho_Q D_Q$, one could imagine a MIT situation driven by a vanishing ρ_Q, a vanishing D_Q or both. In $2+\varepsilon$ dimensions, in all cases the QP lifetime $\tau_Q{}^{-1}$ vanishes faster than the temperature and the QP are well defined.

To lowest order in ε, in the $2+\varepsilon$ expansion each of these possibilities is realized in some universality class. In the spin orbit case ρ_Q vanishes while D_Q stays finite. So to this order, this is not a localization transition at all. In the strong magnetic field case, D_Q vanishes and ρ_Q stays finite while the MIT in the presence of magnetic impurities is driven by the simultaneous vanishing of ρ_Q and D_Q.

Experimentally there is growing evidence that uncompensated semiconductors, having a conductivity exponent $s\approx1/2$, are in a different universality class than compensated semiconductors, amorphous alloys, and semiconductors in a magnetic field, which have a conductivity exponent $s\approx1$. Mott's bound $\nu\geq2/d$ has been put on rigorous mathematical grounds[16]. If $\theta=0$, this gives a bound on the conductivity exponent $s>2/3$ in three dimensions suggesting that uncompensated semiconductors belong to case a and have a non-zero θ. A crucial test of the theory would be to check whether systems having $s\approx1$ have a spin lifetime which is short compared to the temperature.

References

1. Sov. Phys. JETP 3, 920 (1956), 5, 101 (1957), 7, 182 (1958), 8, 70 (1959).
2. A. Migdal, Nuclear Theory: The Quasiparticle Method, Benjamin (1968).
3. A.M. Finkelstein, Sov. Phys. JETP 57, 97 (1985), 59 212 (1984).
4. C. Castellani, G. Kotliar, and P.A. Lee, Phys. Rev. Lett. 59, 323 (1987).
5. C. Castellani, C. DiCastro, G. Kotliar, P.A. Lee and G. Strinati, Phys. Rev. Lett. 59, 477 (1987).
6. L. Fleishman and P.W. Anderson, Phys. Rev. B 21, 2366 (1980).
7. W. McMillan, Phys. Rev. B 31, 2750 (1985).
8. C. Castellani and C. diCastro in Localization and Metal Insulator Transition, H. Fritzche and D. Adler, eds., Plenum (1985).
9. D. Pines and P. Noziers, The Theory of Quantum Liquids, Benjamin (1986); Landau and Lifshitz, Course of Theoretical Physics, Vol.9, Pergamon (1980); P. Noziers, Theory of Interacting Fermi Systems, Benjamin (1964).
10. McMillan, E. Abrahams, P.W. Anderson, P.A. Lee and T.V. Ramakrishnan, Phys. Rev. B 24, 6783 (1982).
11. C. Castellani, C. DiCastro, P.A. Lee, and M. Ma, Phys. Rev. B 30, 527 (1984).
12. C. Castellani, C. DiCastro, P.A. Lee, M. Ma, S. Sorella and E. Tabet, Phys. Rev. B 33, 9 (1986).
13. A.M. Finkelstein, Z. Phys. B 56, 189-196 (1984).
14. C. Castellani, C. DiCastro, G. Forgacs, and S. Sorella, Solid State Comm. 52, 261 (1984).
15. C. Castellani, C. DiCastro, G. Kotliar, and P.A. Lee, Phys. Rev. Lett. 56, 1179 (1986).
16. J. Chayes, L. Chayes, D. Fisher, and T. Spencer, (to appear in Phys. Rev. Lett. 1987).

Thermal Properties of Disordered Interacting Electronic Systems Near the Metal-Insulator Transition

*C. Castellani, C. Di Castro, and G. Strinati**

Dipartimento di Fisica, Università di Roma "La Sapienza",
I-00185 Roma, Italy
*Scuola Normale Superiore, I-56100 Pisa, Italy

The thermal properties of disordered electronic systems at low temperatures (such as the specific heat, the thermal conductivity, and the coefficient of thermoelectric power) can be inferred from the knowledge of the Fourier transform of the heat-heat (χ_{QQ}) and heat-number (χ_{Qn}) density correlation functions in the hydrodynamic regime:

$$\chi_{\alpha\beta}(\vec{q};\omega) = \vec{q}^2 \frac{(D_n D_Q \vec{q}^2 \chi^{st}_{\alpha\beta} + iL_{\alpha\beta}\omega)}{(D_n \vec{q}^2 - i\omega)(D_Q \vec{q}^2 - i\omega)} . \tag{1}$$

Here, α and β stand for either n or Q, D_n and D_Q are the number and heat diffusion coefficients, respectively, and $\chi^{st}_{\alpha\beta}$ and $L_{\alpha\beta}$ are the matrices of thermodynamic derivatives and transport coefficients.

Since in the hydrodynamic regime the constraints originating from number and energy conservation play a crucial role, we have determined a Ward identity for the heat vertex in the long wavelength limit with inclusion of interaction effects. This identity has enabled us to complete the identification of the parameters entering the skeleton structure of the renormalized perturbation theory, in terms of physical quantities /1/. Specifically, the parameter z which was introduced by Finkel'shtein to renormalize the frequency of the dressed diffusion propagator /2/, is shown to coincide with the ratio between the specific heats in the presence and in the absence of disorder /3/. This result and the observation that the parameter z also affects the heat diffusion coefficient, yield then a generalization of the Wiedemann-Franz law near the localization transition with inclusion of interaction effects /1/.

For the interacting disordered systems we are interested in, we consider the heat-heat correlation function

$$\chi_{QQ}(\vec{r}-\vec{r}';\, t-t') = -i\theta(t-t')\langle[K(\vec{r},t),\, K(\vec{r}',t')]\rangle , \tag{2}$$

where $K(\vec{r},t)$ is the grand-canonical Hamiltonian density

$$K(\vec{r},t) = \tfrac{1}{2} \Sigma_\xi \Psi^\dagger_\xi(\vec{r},t)\left[i\partial_t + h_0(\vec{r})\right]\Psi_\xi(\vec{r},t) \tag{3}$$

and $h_0(\vec{r}) = -\vec{\nabla}^2/2m+u(\vec{r})$ is the single-particle term in the presence of a random impurity potential $u(\vec{r})$ (ξ being a spin label). The Fourier transform of (2) is then naturally partitioned into a <u>static</u> part, that provides the specific heat at low temperature T

$$c_V = -\frac{1}{T} \lim_{\vec{q}\to 0} \lim_{\omega\to 0} \chi_{QQ}(\vec{q};\omega) \equiv -\frac{1}{T}\chi^{st}_{QQ} , \tag{4}$$

and a dynamic part according to the symbolic representation of Fig.1. Here $\bar{\Gamma} = \Gamma_s - \Gamma_q$ is the proper singlet interaction amplitude that excludes insertion of a Coulomb line, Λ_Q is the heat vertex, and L stands for the dressed diffusion propagator between electron and hole lines with positive (+) and negative (-) frequencies

$$L(\vec{q};\omega) = \frac{\zeta^2}{-iz\omega + D\vec{q}^2 + 1/\tau_{ph}} . \tag{5}$$

In (5), the wavefunction renormalization ζ can be identified with N/ν where N and ν are the single-particle densities of states in the presence and in the absence of disorder, respectively, /2,4/ D is a diffusion coefficient, and the dephasing time τ_{ph} eventually disappears from the heat /1/ as well as from the number /5/ response functions. The dynamic part of the heat response function can then be cast in the following form

$$\chi^{dyn}_{QQ}(\vec{q};\omega) = -c_V^{(L)} T \frac{\Lambda_Q i\omega\zeta^2\Lambda_Q}{-i\omega z + D\vec{q}^2} , \tag{6}$$

where $c_V^{(L)}$ is the Landau electronic specific heat in the absence of disorder. Notice that only z enters the denominator of (6) since the presence of a frequency at each vertex makes the amplitude-ladder resummation negligible.

Global number and energy conservation, in the form $\chi_{QQ}(\vec{q}=0;\omega)=0$, can be combined with (4) and (6) to yield the diffusive behavior

$$\chi_{QQ}(\vec{q};\omega) = -Tc_V \frac{D_Q\vec{q}^2}{-i\omega + D_Q\vec{q}^2} \tag{7}$$

with D_Q given by

$$D_Q = D/z \tag{8}$$

and with the condition

$$\zeta^2 \Lambda_Q^{\,2} = z\, c_V/c_V^{(L)} . \tag{9}$$

An additional condition is required to identify both Λ_Q and z in terms of physical quantities. To this end, we consider the temperature function corresponding to (2) and its Fourier transform, and we introduce the vertex

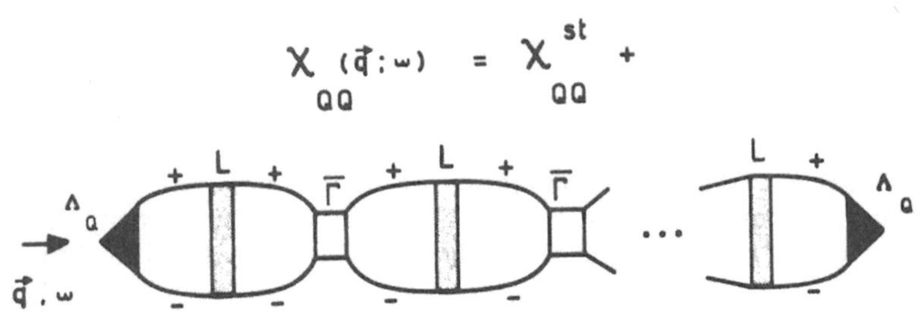

Fig.1: Skeleton structure for the dynamic part of the heat response function

$$\Lambda(1;2,2') = \langle T_\tau \left[K'(1)\Psi^\dagger(2')\Psi(2)\right]\rangle \tag{10}$$

where T_τ is the imaginary-time ordering and $K'(1) = K(1) - \langle K(1)\rangle$. Making use of the continuity equation for $K'(1)$ we then obtain the following Ward identity

$$\int d\vec{r}_1 \, \Sigma_{\xi_1} \partial_{\tau_1} \Lambda(1;2,2')$$

$$= \int d\vec{r}_1 \, \Sigma_{\xi_1} \left[\delta(1,2')\partial_{\tau_1} G(2,1) + \delta(1,2)\partial_{\tau_1} G(1,2') \right] \tag{11}$$

where G is a temperature single-particle Green's function. The identity (11) reduces the long wavelength limit of the dynamic part of the heat response function to the form

$$\chi^{dyn}_{QQ}(\vec{q}=0;\omega) = \Lambda_Q \, c_V^{(L)} \, TN/\nu \qquad (\omega \to 0), \tag{12}$$

and a combination of (12) with (6) relates Λ_Q and z to the wavefunction renormalization ζ, giving the required condition

$$\zeta \, \Lambda_Q = z. \tag{13}$$

From (9) and (13) we identify eventually z with the specific heat renormalization according to

$$z = c_V / c_V^{(L)}. \tag{14}$$

A generalization of the Wiedemann-Franz law in the presence of interaction results by combining (14) with (8) and with the Einstein relation for heat diffusion, namely, $\kappa = C_V D_Q = C_V^{(L)} D$ where κ is the thermal conductivity. We obtain:

$$\kappa / T\sigma = c_V^{(L)} / T\nu, \tag{15}$$

where σ is the electrical conductivity. The validity of the Wiedemann-Franz law (15) (in the presence of strong disorder and including the electron-electron interaction) finds here a simple physical explanation since we have been assuming a collisionless regime of the effective Landau quasi-particle picture /1/.

We have further determined the coefficient Q of thermoelectric power for the non-interacting case and shown that the ratio Q/T diverges at the localization transition /6/. Here Q can be obtained from the hydrodynamic form (1) of χ_{Qn}, where for non-interacting electrons $D_Q = D_n$ /7/ and

$$L_{12} = (\pi^2/3) \, k_B^2 \, T^2 \left[dS(E)/dE \right]_{E=\mu} \tag{16}$$

with

$$S(E) = \sigma(E)/e^2 \tag{17}$$

being proportional to the conductivity evaluated at chemical potential E. From the knowledge of the power-law behavior of the electrical conductivity near the Anderson transition, namely,

$$\sigma(\mu) \sim (\mu - E_c)^s, \tag{18}$$

where E_c is the mobility edge and s is the conductivity exponent, we can thus obtain the asymptotic behavior of L_{12}

$$L_{12} \sim T^2(\mu - E_c)^{s-1} . \qquad (19)$$

For the coefficient of thermoelectric power we then find

$$Q \sim T\left[dS(E)/dE\right]_{E=\mu} S(\mu)^{-1} \sim T(\mu - E_c)^{-1} , \qquad (20)$$

that is, Q/T diverges at the localization threshold.

TING, HOUGHTON, and SENNA /8/ by taking into account the quantum interference in two dimensions due to the effect of non-magnetic impurities at the one-loop approximation, found the same logarithmically singular contribution to σ and L_{12}, thereby suggesting the same exponentiation for these two quantities and Q/T = constant at the transition. The results (19) and (20) are instead correct since, in the presence of non-magnetic impurities, we have verified that the one-loop contribution to S(E) is actually independent of E and thus does not contribute to L_{12} /6/. This result is in turn consistent, at the order here considered, with the scaling behavior (19) with s=1. In the presence of magnetic impurities, on the other hand, the first correction in the $2+\varepsilon$ expansion yields /6/

$$L_{12} = L_{12}^{0}\left[1 - e^4(2\pi^2\sigma^0(\mu))^{-2}\ln(\omega\tau)\right] , \qquad (21)$$

where the suffix 0 refers to the Drude values and τ is the impurity scattering lifetime. At the fixed-point value of the dimensionless conductance, upon exponentiation of the expression (21) for L_{12} we obtain the scaling behavior

$$L_{12}T^{-2} \sim (\mu - E_c)^{-\nu\varepsilon} , \qquad (22)$$

where we have introduced the localization length $\xi \sim (\mu-E_c)^{-\nu}$ with exponent $\nu=(2\varepsilon)^{-1}$ in the present case. Agreement with (19) is thus recovered from the diagrammatic analysis in both cases /6/.

1. C. Castellani, C. Di Castro, G. Kotliar, P.A. Lee, G. Strinati: Phys. Rev. Lett. (in press), and unpublished
2. A.M. Finkel'shtein: Sov. Phys. JEPT 57, 97 (1983)
3. C. Castellani, C. Di Castro: Phys. Rev. B34, 5935 (1986)
4. C. Castellani, C. Di Castro, P.A. Lee, M. Ma: Phys. Rev. B30, 527 (1984); C. Castellani, C. Di Castro, P.A. Lee, M. Ma, S. Sorella, E. Tabet: Phys. Rev. B30, 1596 (1984); ibid. B33, 6169 (1986)
5. C. Castellani, C. Di Castro, G. Kotliar, P.A. Lee: Phys. Rev. Lett. 56, 1179 (1986)
6. C. Castellani, C. Di Castro, M. Grilli, G. Strinati: unpublished
7. C. Castellani, C. Di Castro, G. Strinati: Europhys. Lett. (in press)
8. C.S. Ting, A. Houghton, J.R. Senna: Phys. Rev. B25, 1439 (1982)

On the Universality of the Metal-Insulator Transition of Interacting Electrons

A. Kawabata

Department of Physics, Gakushuin University, 1-5-1 Mejiro, Toshima-ku, Tokyo, Japan

Abstract

Metal-insulator transition of disordered interacting electron system is investigated from the point of view of the universality. A formulation is proposed to deal with the insulator phase as well as the metal phase quantitatively within the framework of renormalization group theory. The case of strong spin-orbit coupling is treated, and it is shown that there is a universal quantity associated with the transition besides the critical exponents.

1. Introduction

In recent years theory of metal-insulator transition has extensively been studied [1-3]. The scaling theory [1,2](or renormalization group theory) predicts that in the vicinity of the transition the conductivity σ and the dielectric constant ϵ behave like:

$$\sigma = \sigma_0 (x/x_c - 1)^s , \qquad (1.1)$$

$$\epsilon = 1 + \chi_0 (1 - x/x_c)^{-s'} , \qquad (1.2)$$

where x is a parameter which controls the transition and x_c is its critical value. The present author has discussed the universality of metal-insulator transition [4,5]: Within one electron theory the exponents s and s' are related with each other like s' = 2s , and the quantity $\sigma_0(\chi_0/N_0)^{1/2}$ takes a universal value which is dependent only on the universality class [6-9] to which the system belongs, where N_0 is the density of states at the Fermi level (the relation s' = 2s was first derived phenomenologically by IMRY, GEFEN and BERGMAN [10]).

These relations have to be modified in the case of interacting electrons, and the purpose of this paper is to discuss this problem. Among various interaction theories [11-19], FINKEL'STEIN's theory [14-16], which was rederived by CASTELLANI et al. [18,19], seems to give the most reasonable and consistent description of the metal-insulator transition, although the interpretation of the outcome of this theory is not yet established. The complexity of the theory is brought about by the appearance of the low-frequency spin fluctuations near the transition. Therefore, the theory is much simplified if this effect is suppressed by rapid spin-flip scattering due to spin-orbit interaction or magnetic impurities, as was pointed out by ALTSHULER and ARONOV [17]. So far only the critical exponents have been discussed on these cases [15,17,18,20], and it is worth while to investigate the simple case in detail in order to interpret the experimental results. Therefore, in the following we will treat only the case of strong spin-orbit interaction.

2. Renormalization Group Equation

We start with the scaling assumption on the size and frequency dependence of the conductivity σ at 0K [4]: If the static conductivity σ' at system size L' is given, the frequency-dependent conductivity σ of the system of size L with the same microscopic structure is uniquely determined. This is the extension of the scaling assumption on static conductivity by ABRAHAMS et al. [1].

We define the dimensionless resistance of the system of size L' by

$$t = \alpha_d e^2 \mu^\varepsilon / \sigma' \hbar, \tag{2.1}$$

where $\varepsilon = d - 2$, d being the dimension of the system, and $\alpha_d = S_d/\pi(2\pi)^d$, S_d being the surface area of the d-dimensional unit sphere. The frequency ω is scaled by D'/L'^2, D' being the diffusion coefficient at system size L', for in the scaling region the motion of an electron is diffusive. Therefore we define the scaled frequency

$$h = 2z\omega/D'\mu^2 = 2\hbar z\omega N_0 t/\alpha_d \mu^d \quad , \tag{2.2}$$

where $\mu = 1/L'$ and z is the frequency renormalization factor [7,14] (note that $\sigma' = 2e^2 D' N_0$, N_0 being the density of states without interaction correction). We assume that ω enter the expression for σ always in the form $z\omega$ (Here ω is the Matsubara frequency and should be replaced with $-i\omega$ in the final results).

Then, σ, the conductivity at system size L and frequency ω, is a function of L, μ, t and h;

$$\sigma = \sigma(L, \mu, t, h). \tag{2.3}$$

On the other hand, μ is arbitrary and σ is independent of it. Therefore, we obtain

$$\frac{d\sigma}{d\mu} = \{ \mu\frac{\partial}{\partial\mu} + \beta(t)\frac{\partial}{\partial t} + \theta(t)h\frac{\partial}{\partial h} \} \sigma(L, \mu, t, h) = 0 , \tag{2.4}$$

where

$$\beta(t) = \mu\frac{dt}{d\mu} = \varepsilon t - t^2 \tag{2.5}$$

and

$$\theta(t) = \mu\frac{d}{d\mu} \log h = \zeta(t) + \beta(t)/t - d \tag{2.6}$$

with

$$\zeta(t) = \mu\frac{d}{d\mu} \log z = t , \tag{2.7}$$

where the second equation of (2.5) and (2.7) are the results of the perturbation calculation [17,18,20]in the lowest order in t in which the localization effect [6,7] is included.

Let $\sigma(\omega)$ be the conductivity of the infinitely large system. Then from the dimsnsional analysis we can put

$$\sigma(\omega) = \alpha_d \frac{e^2}{\hbar} \mu^{\varepsilon} G(t, h) \tag{2.8}$$

and we easily find that (2.4) has a solution of the form

$$G(t, h) = C(t)^{\varepsilon} F(hB(t)) , \tag{2.9}$$

where $F(y)$ is an arbitrary function and

$$C(t) = \exp\{ - \int_{t_0}^{t} \frac{dt'}{\beta(t')} \} \tag{2.10}$$

and

$$B(t) = \exp\{ - \int_{t_0}^{t} \frac{\theta(t')}{\beta(t')} dt' \} , \tag{2.11}$$

where the value of t_0 will be determined later.

Let t^* be the fixed point value of t, i.e.,

$$\beta(t^*) = 0 . \tag{2.12}$$

Then it is easy to see that when $t \simeq t^*$ and $t < t^*$ $C(t)$ behaves like (we take $0 < t_0 < t^*$ in (2.10) and (2.11)).

$$C(t) = c_0(t^* - t)^{\nu} , \tag{2.13}$$

where c_0 is a positive constant and

$$\nu = - 1/\beta'(t^*) , \tag{2.14}$$

$\beta'(t)$ being the derivative of $\beta(t)$. From (2.6) and (2.11), we find that when $t \simeq t^*$ and $t < t^*$

$$B(t) \propto (t^* - t)^{\nu(\zeta^* - d)} , \tag{2.15}$$

where

$$\zeta^* = \zeta(t^*) . \tag{2.16}$$

As is seen from (2.13) and (2.15), $C(t)$ and $B(t)$ are singular at $t = t^*$ except when their exponents are integers. We rewrite (2.9) in the form in which these singularities do not appear explicitly [4]:

$$h = \frac{C(t)^{\zeta^*-d}}{B(t)} G^{(d-\zeta^*)/\varepsilon} f[- C(t)^{1/\nu} G^{-1/\varepsilon\nu}] , \tag{2.17}$$

where $f(y)$ is a function related with the inverse function of $F(y)$ in (2.9) and is assumed to be analytic at $y = 0$. Then we may extend this expression to the region $t < t^*$, i.e., to the insulator phase. The functional form of $f(y)$ will be determined from the requirements for it in various limits.

3. Determination of f(y) and the Universal Quantity

In the metal phase, solving (2.5) with a boundary condition $t = t_L$ for $\mu = 1/L$, we obtain

$$\mu L = \exp\left[\int_{t_L}^{t} \frac{dt'}{\beta(t')} \right] \tag{3.1}$$

and hence from (2.1) and (2.8)

$$G(t, 0) = \lim_{L\to\infty} 1/t_L (L\mu)^{\varepsilon} = \frac{1}{t} \exp\left(\int_0^t \left\{ \frac{1}{t'} - \frac{\varepsilon}{\beta(t')} \right\} dt' \right), \tag{3.2}$$

where it should be noted that $t_L \to 0$ when $L \to \infty$ in metal phase, and that the integral is convergent at the lower limit because of (2.12). Then, if we choose t_0 in (2.10) and (2.11) so that

$$\exp\left(\int_0^{t_0} \left\{ \frac{1}{t'} - \frac{\varepsilon}{\beta(t')} \right\} dt' \right) = t_0 \tag{3.3}$$

(3.2) becomes

$$G(t, 0) = \exp\left(\int_{t_0}^{t} \left\{ - \frac{\varepsilon}{\beta(t')} \right\} dt' \right) = C(t)^{\varepsilon} . \tag{3.4}$$

For small but finite ω, the perturbation theory gives

$$G(t, h) = C(t)^{\varepsilon} + h^{\varepsilon/2}(4/d - 1)/\varepsilon . \tag{3.5}$$

From (2.5), (2.7), (2.12), (2.14) and (2.16), we find that

$$t^* = \zeta^* = 1/\nu = \varepsilon , \tag{3.6}$$

and hence that (2.17) is satisfied by (3.5) if

$$f(y) = A(1 + y)^{2/\varepsilon} \tag{3.7}$$

with

$$A = \{(4/d - 1)/\varepsilon\}^{-2/\varepsilon} . \tag{3.8}$$

In the insulator phase, the dielectric constant is given by

$$\epsilon = 1 + \lim_{\omega\to 0} 4\pi\sigma(\omega)/\omega . \tag{3.9}$$

Therefore, in order that the dielectric constant be finite, we must have $G \propto h$, when $h \to 0$. Then, as $t > t^*$, from (2.13) and (2.17) it follows that when $y \gg 1$ the function f(y) must behave like

$$f(y) = A'y^{\alpha} \tag{3.10}$$

with

$$\alpha = \nu(2 - \zeta^*) \quad , \tag{3.11}$$

A' being a constant. Because of (3.6) we have $\alpha = (2/\varepsilon)(1 - \varepsilon/2)$, and if we write $y^{\alpha} = y^{\varepsilon/2}\{1 - (\varepsilon/2) \ln y\}$, we can match (3.7) with (3.10) by putting $A' = A$ in the lowest order in ε (we can show that this relation holds up to the higher order).

From (2.13) and (2.15) we find that in (2.17)

$$R \equiv \lim_{t \to t^*} \frac{C(t)^{\zeta^* - d}}{B(t)} \tag{3.12}$$

is finite. Therefore when $t \simeq t^*$ and $t > t^*$, from (2.17) and (3.10) we obtain

$$h = RGA\{c_0(t - t^*)\}^{\alpha} \tag{3.13}$$

and hence the dielectric constant is given by

$$\epsilon = 1 + 8\pi e^2 z N_0 t^*/\{ RAc_0{}^{\alpha}\mu^2 (t - t^*)^{\alpha} \} . \tag{3.14}$$

The expression for the static conductivity is given from (2.8) and (3.5);

$$\sigma(0) = \alpha_d \frac{e^2}{\hbar} \mu^{\varepsilon}\{ c_0(t^* - t) \}^{\nu\varepsilon} . \tag{3.15}$$

We assume that t is a smooth function of x in (1.1) and (1.2) at $x = x_c$ ($t = t^*$), then we can write

$$C(t)^{1/\nu} \simeq c_1(x/x_c - 1) \quad , \tag{3.16}$$

and we can write (3.14) and (3.15) in the form of (1.1) and (1.2) with [15,18]

$$s = \nu\varepsilon \; , \qquad s' = \alpha = \nu(2 - \zeta^*) \quad , \tag{3.17}$$

$$\chi_0 = 8\pi e^2 z N_0 t^*/RA\mu^2 c_1{}^{s'} \tag{3.18}$$

and

$$\sigma_0 = \alpha_d \frac{e^2}{\hbar} \mu^{\varepsilon} c_1{}^{s} . \tag{3.19}$$

In the above μ is arbitrary and we take $\mu = k_F$, the Fermi wave number, and we let z_0 be z for $\mu = k_F$. The renormalization group treatment is valid when $\mu \lessapprox 1/\lambda$, λ being the mean free path, and $z \sim 1$ when $\mu \sim 1$. On the other hand the transition occurs when $k_F\lambda \sim 1$, and hence we have $z_0 \sim 1$.

Thus we find that instead of (1.3)

$$Q \equiv \left(\frac{\sigma_0}{e^2 k_F{}^{\varepsilon}/\hbar} \right)^{1/s} \left(\frac{\chi_0}{e^2 N_0 \;/k_F{}^2} \right)^{1/s'} = \alpha_d{}^{1/s} \left(\frac{8\pi z_0 t^*}{RA} \right)^{1/s'} . \tag{3.20}$$

The quantities appearing in the right-hand side of (3.20) are universal except for z_0. In fact, in three dimension ($d = 3$), from (3.6), (3.8), (3.9) and (3.15) we obtain [20]

$$s = s' = t^* = R = 1 \quad , \tag{3.21}$$

$$A = 9 \quad , \tag{3.22}$$

and hence

$$Q = 2\pi^2\hbar^3\sigma_0\chi_0/m^*e^4 = 4z_0/9\pi^2 . \tag{3.23}$$

As was discussed above, z_0 is not universal but is of order unity. Therefore, (3.23) will be used to test the validity of the renormalization group theory. In order to eliminate z_0, Q must be combined with some other quantity.

Just on the metal-insulator transition, the argument of f(y) in (2.22) is 0 and hence we obtain

$$h = RG^{(d-\zeta^*)/\varepsilon}f(0) \quad , \tag{3.24}$$

where R is defined by (3.12). Therefore, from (2.3) and (2.8), it follows that

$$\sigma(\omega) = P\frac{e^2}{\hbar}\{ k_F{}^{-\zeta^*}\hbar\omega N_0 \}^{1/\delta} \quad , \tag{3.25}$$

where

$$\delta = (d - \zeta^*)/\varepsilon \quad , \tag{3.26}$$

and

$$P \equiv \alpha_d \{ 2z_0t^*/\alpha_d Rf(0) \}^{1/\delta} . \tag{3.27}$$

Since $\delta = 2$ from (3.6) for $d = 3$, we find that

$$P^2/Q = A/(4\pi f(0)) \quad , \tag{3.28}$$

is a universal quantity, for f(y) is a universal function. A more detailed investigation of f(y) leads to $f(0) = A/2$, and hence

$$P^2/Q = 1/2\pi \quad . \tag{3.29}$$

4. Summary

In contrast to the non-interacting electron case, in the case of interacting electrons we can not construct a universal quantity only from the amplitudes σ_0 and χ_0 of the conductivity and the dielectric constant. A universal quantity is obtained by combining them with the coefficient of the frequency-dependent conductivity at the transition point. We can discuss the temperature dependence of the conductivity in the same way, and we can show that a similar relation holds also for the coefficient of the temperature-dependent conductivity.

Acknowledgments

The author is grateful to Prof. C. DI CASTRO for valuable discussions. This study is partly supported by ABE Nosei Memorial Foundation.

References

1. E. Abrahams, P.W. Anderson, D.C. Licciardello and T.V. Ramakrishnan: Phys. Rev. Lett. 42, 673 (1979)
2. F.J. Wegner: Anderson Localization (Proc. 4th Taniguchi Int. Symp., Sanda-shi, 1981), ed. by Y. Nagaoka and H. Fukuyama, Springer Ser. Solid-State Sci., Vol 39 (Springer Berlin, Heidelberg, 1982) p.8
3. N.F. Mott: Phil. Mag. B44, 265 (1981)
4. A. Kawabata: J. Phys. Soc. Jpn. 53, 1429 (1984), and Prog. Theor. Phys. Supple. No. 84, 16 (1985)
5. A. Kawabata: Proc. Int. Conf. Heavy Doping and the Metal-Insulator Transition in Semiconductors, ed. by P.T. Landsberg, Solid State Electronics 28, 131 (1985)
6. S. Hikami, A.I. Larkin and Y. Nagaoka: Prog. Theor. Phys. 63, 707 (1980)
7. S. Hikami: Anderson Localization (cf. ref. 2) p.15
8. F.J. Wegner: Z. Phys. B35, 207 (1979)
9. K. Jungling and R. Oppermann: Z. Phys. B38, 93 (1980)
10. Y. Imry, Y. Gefen and D.J. Bergman : Anderson Localization, ref. 2, p. 138
11. W.L. McMillan: Phys. Rev. B24, 2739 (1981)
12. G.S. Grest and P.A. Lee: Phys. Rev. Lett. 50, 693 (1983)
13. C. Castellani, C. Di Castro, G. Forgacs and E. Tabet: Nucl. Phys. B[FS3]225, 441 (1983)
14. A.M. Finkel'stein: Zh. Eksp. Teor. Fiz. 84, 168 (1983) (Sov. Phys. JETP 57, 97 (1983))
14. A.M. Finkel'stein: Zh. Eksp. Teor. Fiz. 86, 367 (1984) (Sov. Phys. JETP 59, 212 (1986))
16. A.M. Finkel'stein: Z. Phys. B56, 189 (1984)
17. B.L. Altshuler and A.G. Aronov: Solid State Commun. 46, 429 (1983)
18. C. Castellani, C. Di Castro, P.A. Lee and M. Ma: Phys. Rev. B30, 527 (1984)
19. C. Castellani, C. Di Castro, P.A. Lee, M. Ma, S. Sorella and E. Tabet: Phys. Rev. B30, 1596 (1984)
20. C. Castellani, C. Di Castro, G. Forgacs and S. Sorella: Solid State Commun. 52, 261 (1984)

Localization Properties and Parameters of a-SiSn: H and a-GeSn: H Alloys

A.D. Zdetsis

Department of Physics, University of Crete and Research Center of Crete, P.O. Box 1527, GR-711 10 Heraklio, Crete, Greece

By extending and generalizing, in order to handle alloys, an earlier model developed for a-Si:H, the electronic and transport properties of a-SiSn:H and a-GeSn:H alloys have been studied. The model, based on the coherent potential approximation and the potential well analogy, can give explicit results for localization parameters such as the mobility edge, the density of states at the mobility edge, the mean free path, the localization length, etc. The theoretical results obtained here for the variation of the optical gap with tin-concentration are in good agreement with existing optical absorption experimental data.

1. INTRODUCTION

The present model, discussed more thoroughly earlier/1-3/ for a-Si:H, is based on a combination of the coherent potential approximation (CPA) method and the potential well analogy (PWA). The CPA method starting from a zeroth order crystalline Hamiltonian treats the disorder by introducing an effective medium in which, at the first level of approximation, the on-site matrix elements of the underlying tight-binding Hamiltonian are replaced by a common complex self-energy Σ. The simplest deviation from the CPA uniform medium, a simple potential well, has been shown/1-3/ to be enough to simulate and describe the localization properties and parameters of a disordered medium such as an amorphous semiconductor. This method has been very successful for elemental hydrogenated amorphous semiconductors such as a-SiH_x/3/. In the present work this method is applied to a-SiSn:H which has attracted recently attention for potential applications as a variable gap semiconductor. For a comparison a-GeSn:H is also studied.

2. THE METHOD AND RESULTS

The Green's function $\tilde{G}_e$ of the CPA effective medium is the key quantity which must be determined iteratively through the CPA condition of zero scattering on the average. To evaluate $\tilde{G}_e$, the crystalline semiconductor Green's function and Hamiltonian is needed which is provided by the Slater-Koster (SK) tight-binding interpolation scheme/4/.The matrix elements of the SK Hamiltonian in the sp^3 basis are obtained by fitting first principles band structure calculations and are available in the literature/5/.The suitability of these matrix elements for tin is questionable in view of the lack of d-orbitals in the basis set. However, the results obtained here seem to indicate that this problem might not be so serious, after all. Matrix elements involving Si-Hi and H-H sites have been obtained from molecular calculations/4/.Other semiconductor-hydrogen matrix elements have been obtained by some simple but arbitrary scaling arguments/6/. .The CPA programme outlined above yields the self energies Σ and the Green's function $\tilde{G}_e$ from which the density of states (DOS) and related quantities are obtained as functions of energy E. The DOS for a-SiSn: is shown in Fig.1 for 75% Sn, 20% Si and 5% hydrogenated vacancies, with an equivalent total amount of hydrogen of 20%. The DOS of a-GeSn:H for two different concentrations of tin and for 20% hydrogen is also shown in Fig.2 for a comparison. As can be seen, the DOS are dominated by tin, due to its large percentage. The CPA energy gap, strangely enough, persists even at these high tin concentrations.

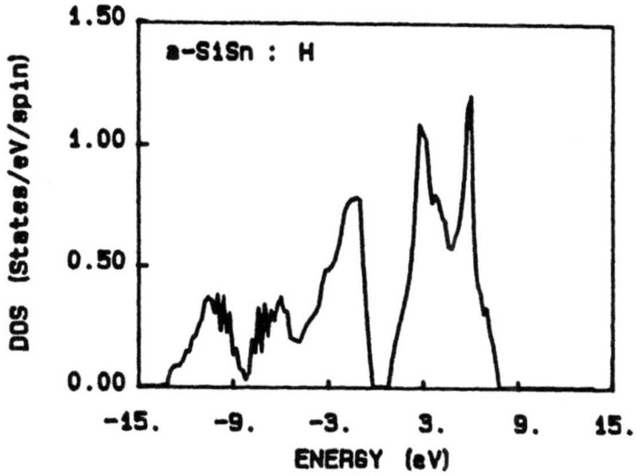

Figure 1
DOS for a-SiSn:H for a large percentage (75%) of Sn. The Si concentration is 20% and the concentration of hydrogenated vacancies is 5%.

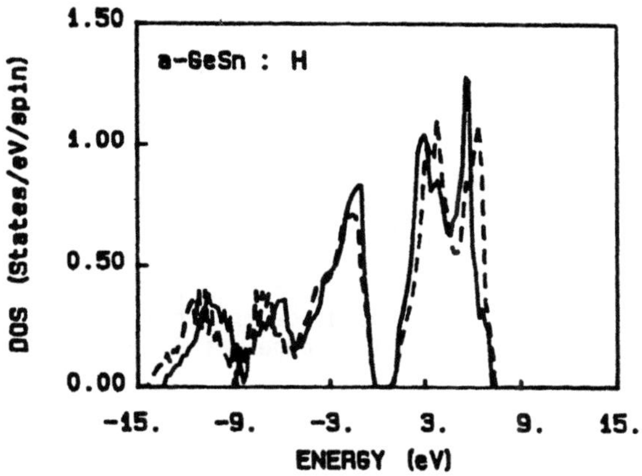

Figure 2
DOS for a-GeSn:H for a small percentage 20% of Sn (solid line) and for a large percentage 75% of Sn (broken line). The concentration of hydrogenated vacancies is 5% for both curves.

Within the CPA regime and through the Kubo-Greenwood formula the DC-conductivity $\sigma(E)$ and the AC-conductivity $\sigma_{AC}(\omega)$ can be calculated. From the AC-conductivity the optical absorption and the optical gaps E_g and E_{04} can be calculated, concluding the results obtained by the CPA part of the model. Fig.3 shows the optical gap E_g of a-SiSn:H as a function of tin concentration for two different values of hydrogen concentration. The comparison with the displayed experimental data/7/ shows a good agreement, especially in view of the quite large experimental uncertainties in the amount of hydrogen. and tin.

The second part of the model involving the PWA method, utilizes the CPA results and in particular the CPA conductivity $\sigma(E)$ to locate E_c and to implement localization corrections on $\sigma(E)$. The PWA method maps the localization criterion to the criterion of existence of a bound state in an effective potential well. This effective well is constructed for each energy E from the constant energy surface S(E), which for a disordered solid is the same as in the crystalline case but with E measured from the CPA band edge/1-3/,and the mean free path, $\ell(E)$. This last quantity and in turn E_c are obtained from $\sigma(E)$, using the relations:

$$\sigma(E) = (e^2/12\pi^3\hbar)\, \ell(E)\, S(E)\ , \qquad S(E_c)\, \ell^2\, (E) = 8.96 \qquad (1)$$

The second part of (1) describes the localization of criterion from which E_c is obtained. The energy surface S(E) and the mean free path $\ell(E)$ can be further used to implement localization corrections/2-3/ to the CPA conductivity $\sigma(E)$. The conduct-

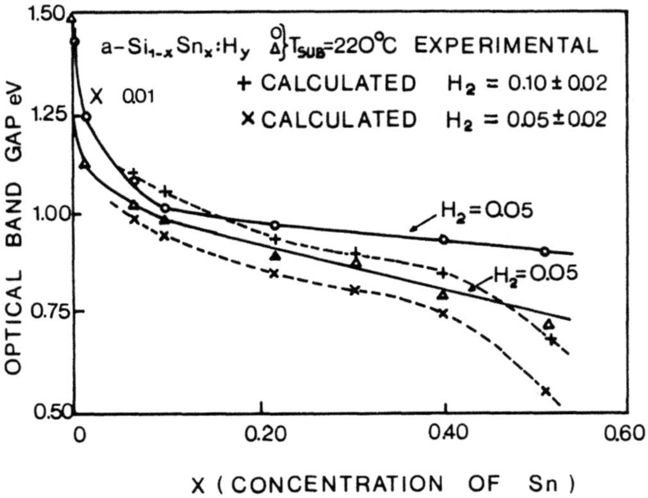

Figure 3
Variation of the optical gap E_g with tin concentration. Comparison of calculated and measured results. (Experimental data are from D. Girginoudi et al /7/).

ivity $\sigma(E)$ around the band edges is behaving like a straight line, the slope of which is sufficient to determine most of the parameters calculated here. In view of the expected inadequacies of this model for relatively high tin concentrations, (due mainly to the omission of d-orbitals from the SK basis) we have considered here only a-SiSn:H with a high Si concentration of 92% with 5% Sn and 3% hydrogenated vacancies. The slope of the conductivity curve is 4300 $(\Omega cm\ eV)^{-1}$ giving a mean free path $\ell \sim 13$ Å. The position of E_c relative to the valence band edge is found to be $E_c \sim 0.01$ eV. The DOS at E_c, which has square root dependence is estimated to be $\sim 2x10^{19}\ eV^{-1}\ cm^{-3}$. Finally the mobility, μ, at the mobility edge was found to be $\mu(E_c) \sim 19\ cm^2/V$ sec.

3. CONCLUSIONS

The values obtained for σ, E_c, compared to the small percentage of Sn do show some rather large differences from the a-SiH$_x$ case/3/ with the same amount of hydrogen. This is also reflected in the gap variation of Fig.3 which is rather large for relatively small tin concentrations. In Fig.3, showing the gap variation with concentration of tin, a rather abrupt variation of E_g is observed around the site percolation concentration of 0.43. This could be an artifact of the calculation related with the virtual crystal approximation (VCA) employed to estimate the off-diagonal matrix elements of the SK Hamiltonian or with the omission of d-orbitals in the basis set. This will be settled in a forthcoming calculation incorporating off-diagonal disorder and d-orbitals in the basis set. If, however, this is not a spurius effect of the calculational scheme, it could be related with a strong Sn-H bond at about this concentration, as was indicated by recent experimental data/7/, or it could have far reaching relationships and connections with the general quantum mechanical percolation problem.

As was illustrated, this model provides a reasonable description of a-SiSn:H and other amorphous hydrogenated alloys in the same way as in a-SiH$_x$. The reason for this success must be related to the fact that only certain gross features of the system are truly significant for a good physical description near the band edges/2-3/ due to an underlying universality.

The present model although omits reconstruction, topological and off-diagonal disorder, and although it possesses an underlying crystallinity, seems to be able to provide a consistent and realistic picture of a-SiSn:H and a-GeSn:H.

Similar results for other hydrogenated amorphous semiconductor alloys/6-9/ verify this, as well.

REFERENCES

1. D.A. Papaconstantopoulos and E.N. Economou, Phys. Rev. B22, 2903 (1980); Phys. Rev. B24, 7233 (1981); W. Pickett, D.A Papaconstantopoulos, E.N. Economou, Phys. Rev. B28, 2232 (1983)
2. E.N. Economou, C.M. Soukoulis and A.D. Zdetsis, Phys. Rev. B30, 1686 (1984); B31, 6172 (1985)
3. E.N. Economou, A.D. Zdetsis and D.A. Papaconstantopoulos, J. of Non Cryst. Sol. 77-78 (1985)
4. J.C. Slater and G.F. Koster, Phys. Rev. 94, 1498 (1954)
5. D.A. Papaconstantopoulos, Handbook of the band structure of elemental solids, (Plenum, New York, 1986)
6. A.D. Zdetsis, Journal of non crystalline solids, Proceedings of the 12th ICALS, Praha (North-Holland, 1987)
7. D. Girginoudi, A. Thanailakis and A. Christou, Proc. Mat. Res. Sci., Dec. 7-12, Boston, MA (1986)
8. D.A. Papaconstantopoulos, E.N. Economou and A.D. Zdetsis, "Proceedings of the 17th International Conference on the physics of semiconductors", edited by J.D. Chadi and W.A. Harisson, 795 (Springer-Verlag, New York 1985)
9. A.D. Zdetsis, D. Girginoudi, G. Kiriakidis, Z. Hatzopoulos, A. Thanailakis and A. Christou, "Proceedings of the 12th ICALS, Praha (North-Holland, 1987)

Quantum Disordered Spin Models and Bose Condensation

J.M.F. Gunn[1] *and M.A. Brackstone*[2]

[1]Rutherford Appleton Laboratory, Chilton, Didcot, Oxfordshire OX11 0QX, UK

[2]Department of Physics, University of Southampton, Highfield, Southampton SO9 5NH, UK

In recent years there has been a sustained effort to understand the effect of interactions on the behaviour of electrons in random potentials (e.g. the review by LEE and RAMAKRISHNAN /1/). There has been less work on the problem of interacting bosons in random potentials, but now under experimental stimulation this is increasing. The main experimental system is ^{4}He absorbed on random substrates, such as mylar glass (CROOKER et al /2/). Another possible "Bose Glass" is granular superconductivity.

On the theoretical side the pioneering work of HERTZ et al. /3/ predicted a transition at zero temperature, from a localised "Bose Glass" to a Bose condensed system, as the chemical potential of the interacting bosons was raised. They also pointed out that the non-interacting limit was uninteresting - all bosons would condense (unrealistically) into the lowest, in energy, localised state.. Subsequent work has found transitions between Bose condensed and Anderson localised (ANDERSON /4/) phases by a variety of techniques e.g. GOLD /5,6/, MA et al /7/.

In the absence of any soluble, simple, limit of the Bose problem, we turn to an alternative formulation of the problem in terms of quantum spins: For ordered problems this was first represented by MATSUBARA and MATSUDA /8/ and for the disordered problem by MA et al /7/. We will show that there is a non-trivial limit of boson problem which is natural in the spin treatment, which allows an "RPA" discussion of the localisation/condensation transition and shows how interacting bosons may screen out a random potential. The mathematical details will be published elsewhere (BRACKSTONE and GUNN /9/).

The principle is to note the equivalence for "hard-core" bose operators (i.e ones obeying $(b^+_i)^2=0$) and the Pauli spin-matrices.

$$b^+_i \Leftrightarrow \sigma^+_i \qquad b_i \Leftrightarrow \sigma^-_i \qquad n_i \Leftrightarrow \sigma_i^z + 1 \tag{1}$$

Thus the Hamiltonian for hard-core bosons (e.g. ^{4}He atoms) in a disordered lattice may be represented as:

$$H = -J \sum_{\langle ij\rangle} (S^x_i S^x_j + S^y_i S^y_j) - \sum_i (h_0 + h_i) S^z_i \tag{2}$$

where the "hopping" term in, $s^+_i s^-_j$, has been rearranged to yield a ferromagnetic X-Y coupling, and the chemical (random) potential becomes a (random) magnetic field in the Z-direction. The modulus (angle) of the magnetisation in the X-Y plane corresponds to the modulus (phase) of the order parameter for bose condensation. MA et al /7/ studied the Hamiltonian, for spin 1/2 operators, and argued that there was a condensation/localisation transition as a function of the ratio of the variance of the h_i to the ferromagnetic coupling, J.

However spin-1/2 operators are difficult to deal with (Ma et al used Renormalisation Group techniques) so it is natural to ask about the properties of the Hamiltonian in the classical limit ($S\to\infty$). We will show that this limit is soluble (at least in the "strong disorder limit", where the r.m.s value of h_i, Δ, is large compared to J.) The significance of the $S\to\infty$ limit in Bose language may be seen by noting that the two spin states for the S=1/2 case correspond to zero bosons and one boson on a site. Thus the (2S+1) spin states (for the spin-S case) to correspond to 0,1,..2S bosons on a site. Thus $S\to\infty$ corresponds to a weak-interaction limit (as increasingly large numbers of bosons are allowed to inhabit a site). We will see it is a non-trivial limit as the cases of interest correspond to $\langle S^z\rangle/S=\alpha$, where $-1<\alpha<1$, and $\langle S^z\rangle$ is the average of S^z over the ground state and sites. This implies that the average number of "bosons" on a site will diverge as $S\to\infty$, for fixed α. Thus the $S\to\infty$ limit corresponds to the interaction strength of the bosons tending to zero and the density tending to infinity, so that the product remains finite. It is this, non-trivial, limit that we will find so useful.

We now examine the classical ground state and then discuss the quantised version. We will work in the strong disorder limit, $JS/\Delta \ll 1$, Δ being the r.m.s value of h_i (i.e. the random potential). We note that for a single-particle problem all states would be localised in that limit. Let us construct the classical ground state in three stages of increasing accuracy (for the case of $h_0=0$):

(i) since $\Delta \gg JS$, assume that the $h_i S^z_i$ term is dominant. Then every spin is parallel or anti-parallel to the z-axis - depending on the sign of h_i.

(ii) there will be of order $(JS/\Delta)^2$ sets of nearest neighbour pairs which will have sufficiently weak z-fields that they prefer to cant into the x-y plane to gain energy from the x-y exchange.

(iii) these pairs will interact by polarising the intervening set of spins away from the z-axis. (To see this consider the net field on a spin which is a neighbour of a pair which has canted over into the x-y plane.)

The conclusion of this construction is that there is always a magnetised (condensed) ground state in the classical limit, albeit of a peculiar, inhomogeneous, form - with the magnetisation residing in dilute, weakly interacting, pairs of spins. Since the interaction is of a Yukawa form (BRACKSTONE and GUNN /9/), the mean field estimate of the transition temperature will vary exponentially with the density (and hence JS/Δ) of the pairs.

So far we have assumed that the average external field, h_0, was zero (i.e. $\mu=0$ in Bose language). Thus S^z and $-S^z$ are identical (i.e. a particle-hole symmetry for the bosons). To consider the case where either the particles or holes are predominant, a non-zero chemical potential must be added (i.e. a net magnetic field in the spin problem). So to generate a dilute set of particles, we add a large negative h_0. Almost all sites will have their spin parallel with the negative z-axis (i.e. unpopulated by bosons). Only those with a random field which is larger than $-h_0$ will be populated. To see the structure of the condensate (or magnetisation) in this limit, note that only those pairs with z-fields between $-h_0-JS$ and $-h_0+JS$ will be canted into the x-y plane. This number is roughly $(JS/\Delta)\exp(-h_0^2/\Delta^2)$. Thus if $h_0 \gg \Delta$ the number of pairs is very small. In localisation language the condensation is occurring in the "Lifshitz tail" of the single particle density of states. This rather remarkable result - condensation occurring in what is an extreme localisation limit in the single particle problem - will be discussed later.

Since we have isolated the important degrees of freedom as being the canted pairs of spins, let us write down an effective Hamiltonian which only involves those variables. It is plausible (and is proved in /9/) that the lowest energy excitations are those where the spins within each pair remain parallel, but the pairs are not all exactly parallel. These "acoustic" spin waves should have an energy that is lower than the "optic" ones by the ratio of the effective Yukawa coupling to the ordinary inter spin coupling, J. We are thus led to an effective Hamiltonian of the form:

$$H^{eff} = -\sum_{ij} J_{ij} \underline{\Sigma}_i \cdot \underline{\Sigma}_j + \sum_i U_i (\delta\Sigma^z{}_i)^2 \tag{3}$$

Here $\underline{\Sigma}_i$ is the total spin of the ith pair, $\underline{\Sigma}_i$ is the projection of $\underline{\Sigma}_i$ onto the x-y plane and J_{ij} is the effective Yukawa coupling. U_i represents the energy to twist the total spin away from the equilibrium value for its z-component. The origin of this term in the case of $h_i=0$ is clear: twisting the total spin out of the x-y plane causes the x-y coupling (between spins in the pair) to diminish. For $h_i \neq 0$ both the expression for U_i and its interpretation is complicated (see /9/).

The most important consequence of the creation of the effective Hamiltonian is that it can be averaged. (It is clear that if the original Hamiltonian were averaged then the random potential disappears from the problem.) This is because the effects of the random potential are contained in the selection of the important degrees of freedom and in the average values of J_{ij} and U_i. We will now mention the important properties of the averaged model:

$$H^{av} = -J \sum_{\langle ij \rangle} \underline{S}_i \cdot \underline{S}_j + U \sum_i (S^z{}_i)^2 \tag{4}$$

Classically the two terms are not in conflict: the second term confines the spins to the x-y plane and the first term aligns them in that plane. However quantum mechanically there is a conflict due to the non-commutativity of the spin components. We may check that increasing U, and hence forcing each spin into the $S^z=0$ eigenstate, the magnetic order is destroyed by calculating $\langle S^x{}_i S^x{}_j \rangle$ for $|\underline{R}_i - \underline{R}_j| \to \infty$ (/9/). The long-distance limit of the correlation function is driven to zero as U is increased. It reaches zero at

$$(2U/ZJ) = (2S/A)^2 \tag{5}$$

where

$$A = 1/N \sum_{\underline{k}} \left\{ \sum_{i=1}^{d} \sin^2 k_i/2 \right\}^{-1} \tag{6}$$

It is natural to interpret this as a localisation transition, although the localised phase is not accessible in the semi-classical spin model. Inserting the averaged values of J and U yields (for $h_0=0$):

$$\gamma \exp\{[\ln(\Delta/JS)(\Delta/JS)^{2/d}]/d\} \leq S^2 \tag{7}$$

for order to occur. γ is a complicated prefactor (ref. /9/). The results for $h_0 \neq 0$ are immediately deduced by replacing Δ by $\Delta\exp[(h_0/\Delta)^2]$. As $h_0 \to \infty$ the spins must be very classical to yield a magnetised ground state.

Let us now interpret the above results in boson and Anderson Localisation language. The result, stated in boson terms, is: the interaction strength ($\propto 1/S$) must be sufficiently small, for given disorder, to yield condensation. The inclusion of a non-zero chemical potential places an extremely stringent restriction on the strength of the interactions. To understand the result let us turn to a localisation viewpoint. The exponential on the left hand side is the square of the "hopping integral" between the sites on which the number of particles is allowed to fluctuate. The right hand side is the square of the inverse of the interaction strength. Thus we have something like an Anderson criterion, where the matrix element is the hopping term, but the typical energy denominator for shifting a particle from one site to another is the <u>interaction strength</u> and not the characteristic energy of the site energy distribution, Δ. It is this replacement of Δ, which may be very large compared to the effective hopping, by S^{-1} which allows condensation even in the Lifshitz tails to occur in the classical, non-interacting limit. To see why the bosons "screen out" the random potential let us consider an interacting Anderson model, with an on-site interaction strength U:

$$H_0 = \sum_{\underline{m}} \varepsilon_{\underline{m}} a^+_{\underline{m}} a_{\underline{m}} + U/2 \sum_{\underline{m}} n_{\underline{m}}(n_{\underline{m}}-1) - \mu \sum_{\underline{m}} n_{\underline{m}}$$

$$H_1 = -t \sum_{\underline{<nm>}} a^+_{\underline{m}} a_{\underline{n}} \tag{8}$$

H_0 may be diagonalised exactly - each site is "filled-up" with particles to within U of μ, so that the random site energy, ε_n, balances the Hartree energy to yield μ to the accuracy that the quantisation of the Hartree potential, in units of U, allows. This means that the distribution of "effective" (i.e. including the Hartree energy) site energies is changed. The occupied sites now form a bump at μ of width U, which is pinned to μ and grows in strength as μ is raised. As the number of occupied sites increases, as μ is raised, the effective hopping term between those sites increases as well. The combination of these two effects will imply an Anderson Transition to a condensed phase as μ increases, as long as $U<Zt$ - which is the limiting Anderson criterion when all sites are occupied and the motion of one particle is considered in the frozen Hartree potentials of the others. It should be noted that off-diagonal disorder may also be screened in a similar way, by considering pairs of sites.

References

1. P.A. Lee and T.V. Ramakrishnan: Rev. Mod. Phys. 57 287 (1985)

2. B.C. Crooker, E. Hebard, E.N. Smith, Y. Takano and J.D. Reppy: Phys. Rev. Lett. 51 666 (1983)

3. J.A. Hertz, L. Fleishman and P.W. Anderson: Phys. Rev. Lett. 43 942 (1979)

4. P.W. Anderson: Phys. Rev. 109 1492 (1958)

5. A. Gold: Z Phys B52 1 (1983)

6. A. Gold: Phys. Rev A33 652 (1986)

7. M. Ma, B.I. Halperin and P.A. Lee: Phys. Rev. B34 3136 (1986)

8. T. Matsubara and H. Matsuda: Prog. Theo. Phys. 16 569-582 (1956)

9. M.A. Brackstone and J.M.F. Gunn: to appear in J. Phys. C (1987)

Part IV

Weak Localization Effects in Metals

The Physical Interpretation of the Interaction Effect

G. Bergmann

Physics Department, University of Southern California,
Los Angeles, CA 90089-0484, USA

Two-dimensional disordered conductors show a resistance anomaly at low temperatures. In addition to weak localization there is also a contribution due to the Coulomb interaction which was discovered by Altshuler, et al. In this article the Hartree contribution is translated into transparent physics. The underlying physics is rather beautiful and has some similarity with optical holography. If one considers the propagation of an electron on a closed loop (after many scattering events) then it interferes with itself and forms a charge pattern which contains all information about the impurities (and corresponds to the hologram). A conduction electron which is scattered by the same impurities generally experiences the same phase shifts. However, if it is scattered by the "charge hologram" the dephasing is readjusted, and one obtains a constructive interference.

1. Introduction

At low temperature the resistance of two dimensional disordered electronic systems such as thin films show interesting deviations from the Boltzmann theory, in particular a logarithmic temperature dependence. These anomalies are due to electron interference. For normal metals one distinguishes essentially two contributions: quantum interference - generally called weak localization - and the Coulomb anomaly. The Coulomb anomaly was discovered by ALTSHULER and ARONOV [1] in three dimensional disordered conductors and extended to two dimensions by ALTSHULER, et al., [2] and FUKUYAMA [3]. During the meantime the effect of Coulomb interaction in disordered electron systems has been intensively studied and reviewed by several authors [4-6]. Although the evaluation of the Kubo diagrams is straight forward within the formalism of the Green functions it has only recently been discussed in terms of graphical physics [7]. (By graphical I mean "physically transparent" corresponding to the German word "anschaulich").

The diagrams in Fig.1 describe the Hartree correction to the conductance. In section II we will discuss its physical meaning. For its physical understanding it is quite helpful to consider first the so-called self-energy of a conduction electron, i.e. the effect the Coulomb interaction has on the frequency and damping of a conduction electron. The corresponding diagrams are shown in Fig. 2. The physics behind the different diagrams is quite similar although they partially compensate each other. We will discuss only

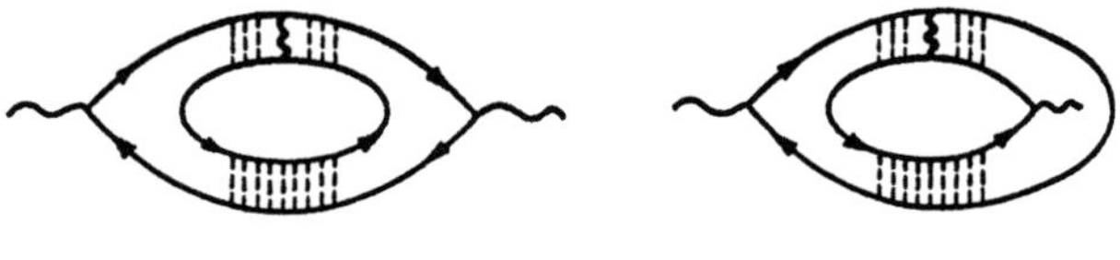

Fig. 1: Kubo diagrams which yield the conductance correction for the Hartree term in the presence of the Coulomb interaction

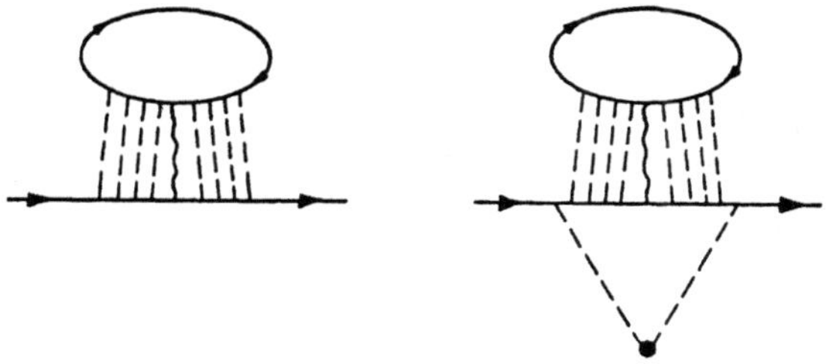

Fig. 2: The Hartree diagrams of self-energy

the first one. Essentially we have to unfold this diagram and for our understanding the description in real time is most helpful. Such a diagram contains the recipe for the electron propagation and the information about the development of the electron phase. This will be considered later in detail.

The propagation of electrons may be evaluated in k-space as well as in r-space. Both pictures give (different) physical insight into the problem. We will concentrate our discussion on the electron propagation in real space. In real space it turns out that we have to consider the motion of an electron on a closed loop. By interference with the original (self-reconstructing) wave function it builds up a charge pattern. At finite temperature such a charge pattern is only possible if the thermal coherence length $\ell_T = \hbar v_F/(2\pi k_B T)$ is larger than the diffusion path on the closed loop. This brings the temperature into play. The propagation in real space yields an understanding of the influence of a magnetic field and the role of inelastic scattering processes.

In this article we concentrate on the Hartree term which is shown in Fig.1. The Fock term has a similar meaning because it follows the same physics. On the other hand it has an additional inner dynamics because the effective interaction depends on the energy difference of the two electrons involved.

2. Diffusion in Real Space

Let us consider an electron gas in a box with impurities as scattering centers. All states up to the Fermi frequency are occupied. In the following we use Huygens principle for the electronic wave function. We start with the assumption that the electron wave functions in the space between the impurities are free electron wave functions ψ''. Then, according to Huygens principle, every point r' acts as a continuous source of an electron wave which propagates radially in all directions. In the presence of impurities the amplitude is damped since every impurity generates a new spherical wave (we assume s-scattering in the following consideration). In d=2 or 3 dimensions this damped spherical electron wave has an amplitude at the position r

$$(1/2i)\ \psi''(r')/|r-r'|^{(d-1)/2}\ e^{i|k||r-r'|}\ e^{-|r-r'|/2\ell}\ . \qquad (1)$$

Let us consider an electron source at the position C in the lower part of Fig. 3. We follow the trajectory along the closed loop C → 1" → 2" → . . → 9" → 10" → C. The electron wave returns to C with amplitude A. The frequency of the electron source may be the Fermi frequency ε_F. On the closed loop the electron wave experiences a phase shift δ which is included in the amplitude A. If L is the total length of the closed loop ($L=v_F t$, where t = diffusion time) then the main phase shift between the source and the returning amplitude A is given by Lk_F. The scattering by the impurities

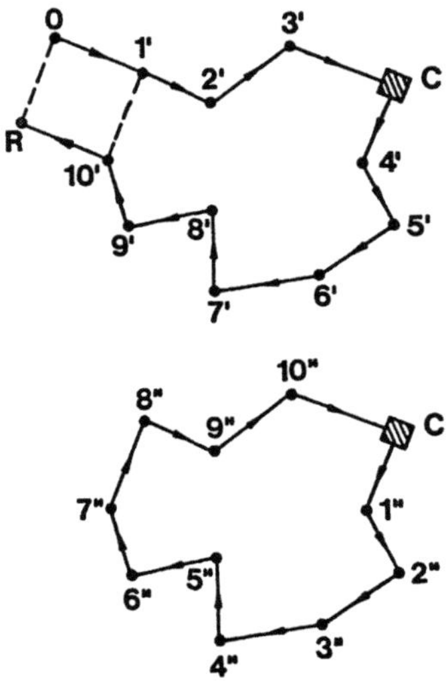

Fig. 3.: The scattering sequence of an electron (") (lower part) which propagates on a closed loop from C → 1" → 2" → .. → C. At C it interferes with itself and forms a charge hologram by which another electron (') (upper part) can be scattered due to the Coulomb interaction

causes additional phase shifts. Electrons which differ in their frequency by ε'' from the Fermi frequency experience an additional phase shift $\varepsilon''t$ along the closed loop. The interference between the electron source and the returning wave modifies the (homogeneous) local electron charge by a factor $[1+A\exp(+\varepsilon''t)+A^*\exp(-i\varepsilon''t)]$.

The amplitude of the charge pattern can be estimated. We concentrate on the charge pattern $A^*\exp(-i\varepsilon''t)$ resulting from the interference after encircling the closed loop. All occupied states $\psi''(r)$ contribute to the charge pattern according to the thermal occupation number $f(\varepsilon'')$. The summation over all occupied states yields for the charge density

$$\rho(C) = A^*N_0\hbar \int d\varepsilon''\, f(\varepsilon'')\exp(-i\varepsilon''t) = i\varepsilon_T A^*N_0\hbar \sum_{\ell\geq 0}\exp(-\varepsilon_\ell t) \tag{2}$$

where N_0 ($=m/(\pi\hbar^2)$ in two dimensions) is the density of states ($N_0\hbar$ is the density of frequencies), ε_T is the characteristic frequency for the temperature T and τ_T the corresponding time

$$\varepsilon_T = 1/\tau_T = 2\pi k_B T/\hbar \ . \tag{3}$$

The Matsubara frequencies are defined as

$$\varepsilon_\ell = (\ell+1/2)\varepsilon_T \ . \tag{4}$$

For short times the sum at the right side of (2) can be replaced by an integral and yields

$$\rho(C) = iA^*N_0\hbar/t \ ; \ \text{for } t < \tau_T \ . \tag{2A}$$

For $t>\tau_T$ the charge pattern reduces proportionally to $\exp(-\varepsilon_T t/2)$, which describes the fact that at finite temperature the electron wave packets have a coherence time of τ_T. For simplicity we replace the time dependence by a $1/t$-law with a cut-off at $\tau_T = 1/\varepsilon_T$. Therefore one obtains the important result that the charge pattern is temperature dependent. Of course, every other closed loop yields its own pattern which characterizes the phase shift on this loop.

In the next step we consider the propagation of a probe electron (') from the position point 0 (at the origin) to R as is shown in the upper part of Fig. 3. Its amplitude decays as given by equation (1). On the other hand the electron wave may also propagate on the same closed loop from 0 to point R. The electron propagates from point 0 via 1' → 2' → 3' to point C. Here it is scattered by the charge distribution ρ(C) of the other electrons which we discussed above and continues its closed loop via 4'... to R, where it interferes with its original wave function. For simplicity we assume that the distances 0-1' and 10'-1' are the same as are the distances 0-R and 10'-R. Therefore the amplitudes at point R for the Coulomb scattered wave (along the closed loop) and for the direct path differ only by the propagator along the closed loop. Along the closed loop the probe electron experiences the same phase shift as if it had started at position C, propagated on the same closed loop to C and performed the Coulomb scattering at C. The amplitude after the propagation along the closed loop from C to C is the same as for the electron (") i.e. $Ae^{i\varepsilon' t}$. The Coulomb potential for the screened charges can be expressed by a screened potential $F\rho(r)/N_0$. The Coulomb scattering contributes to the final amplitude of the returning electron wave (') with the factor

$$-i\langle U\rangle\tau_0/\hbar = -(i/\hbar)\tau_0\rho(r)\, F/N_0 = F\tau_0 A^*/t \; . \tag{5}$$

The total amplitude of the returning electron wave (') on the closed loop including one Coulomb scattering is just $F|A|^2$ as long as the diffusion time is less than τ_T. For times larger that τ_T there is an exponential cut-off.

The value of $|A|^2$ depends on the impurity distribution. However, if we sum over all loops and average over all impurity distributions, then we expect the probability for back scattering on a closed loop to be described by the diffusion probability. In two dimensions the total probability to return to the origin is $\Sigma|A|^2 \propto 1/(Dt)$ where Σ represents the sum over all closed loops.

The relative "amplitude" of the Coulomb scattered electron wave at the point R for times $t<\tau_T$ is proportional to F/(Dt). In other words, besides the direct electron wave amplitude (for propagation from 0 to R) we also find a coherent amplitude at the position R which comes from the partial waves which started before the time t and propagated along all possible closed loops and reached finally the point R. Since all partial waves which started in the time interval $-\tau_T<t'<0$ contribute coherently at the time t=0, their amplitudes add up to $(F/D)\, \ln(\tau_0/\tau_T)$.

The effect of the hologram scattering on the electron wave which passes through the closed loop is a partial coherent repair of the electron wave function propagating from 0 to R. The "direct" propagator fades with increasing distance from 0 because of the impurities which scatter this wave and cause an overall decay as $\exp[-r/2\ell]$. The partial waves scattered by the hologram replace this loss of "coherent" amplitude partially and reduce the decay or increase the effective decaying length to $2\ell\, [1+F/(2\pi k_F\ell)\ln(\tau_T/\tau_0)]$. The relative difference between this and the old amplitude as a function of the distance r between 0 and R is $(r/2\ell)(F/2\pi k_F\ell)\, \ln(\tau_T/\tau_0)$. The effect of the hologram scattering is just that the metal appears to have a larger mean free path $\ell' = \ell[1+F/(2\pi k_F\ell)\, \ln(\tau_T/\tau_0)]$.

3. The Correction to the Diffusion Constant

In ref. [7] the contribution of the diagram in Fig. 1 is discussed in some detail in k-space. Since such a discussion is rather complex I prefer at this place to discuss qualitatively the contribution of the Hartree diagram to the diffusion in real space. We are interested in the additional

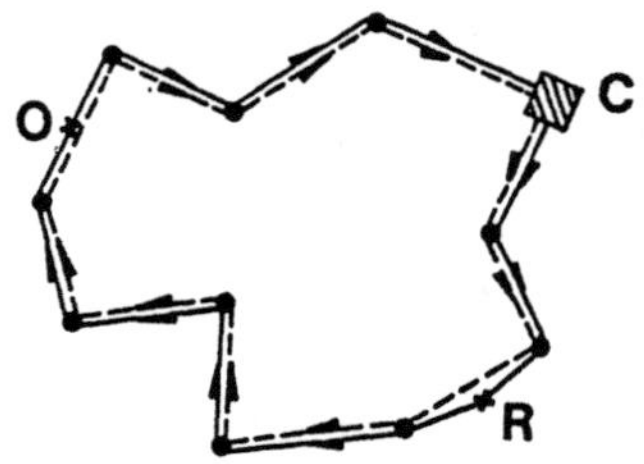

Fig. 4: The two contributions of the Coulomb satellite to the diffusion in real space

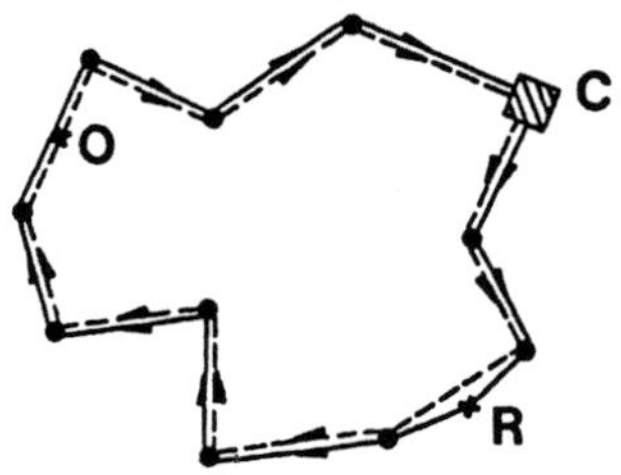

diffusion from point O to R due to the Coulomb interaction. Generally one is not interested in the history of the electron at the position O but there is a well defined probability that the electron has visted the point R before. In this case it propagated on a closed loop and can be coherently scattered by the appropriate charge hologram at C yielding an additional diffusion channel. This process is drawn in Fig. 4a. The electron (") starts as usual at the position C, performs a closed diffusion loop and interferes with itself at C generating the charge hologram. The probe electron (') may diffuse from the position R along the closed loop back to R being Coulomb scattered by the charge hologram at C. At the time t=0 it may be located at the position O. Therefore this propagation yields an additional diffusion channel from O to R.

The second contribution is shown in Fig. 4b. Here the electron (") propagates from C to C along the closed loop and forms the charge hologram at C. At the time t=0 we find it at the position O. The other electron (') starts at R, propagates along the closed loop including the Coulomb scattering at C and returns to R. This is an indirect diffusion from O to R because the electron (') reaches only the point R when the electron (") is at the time t=0 at the position O. If we would remove the electron (") at t=0 from the position O the electron (') would not reach the point R (by this process). Therefore both processes modify the diffusion in real space.

4. Conclusions

In a disordered metal the electronic wave function has many possibilities to interfere with itself. As a consequence one finds an inhomogeneous electron density which stores the distribution of the impurity positions and which represents an electron hologram. This hologram contains the charge pattern for every closed scattering loop and acts as a (Coulomb) scattering potential itself. A second conduction electron which propagates on the same closed loop can readjust its phase, interfere coherently at its starting point with its original wave function and change the mean free path.

The coherence of the electronic wave functions is an essential requirement for the mechanism of the Hartree (and Fock) diagram. This coherence is limited by the thermal coherence time τ_T.

In addition the Coulomb anomaly must be sensitive to inelastic scattering which destroys the coherence of the wave function. However, the effect of inelastic scattering is rather sophisticated. If we consider for example the electron paths in Fig.3 there the electrons (') and (") propagate from 1'=8" to C at the same time along the same impurities. Any inelastic scattering on this part of the closed loop has little effect on the interference because the inelastic process occurs for both electrons with essentially the same probability and phase. However, along the path from C to 10'=7" the two electrons travel at quite different times and inelastic processes are destructive for the phase coherence. This demonstrates that the understanding of the physical processes in real time are important for the evaluation of the phase coherence.

It is important that an electric field does not influence the coherence between $\psi'(r)$ and the Coulomb satellite. Of course, the electric field changes (in real space) the phase of the amplitude A" and therefore the charge pattern. However, the phase shift is the same for both electrons on the same path and therefore cancels in the amplitude of the Coulomb satellite.

The same applies for a magnetic field. For the paths in Fig. 4 as well as for Fig. 3 we recognize easily that a magnetic field changes the electron hologram because the flux which is encircled by the closed loop changes the phases of the two partial waves and therefore the interference. As a consequence the phase shift caused by the magnetic field is also stored in the electron hologram. This (magnetic) phase shift compensates the phase shift which the conduction electron experiences on the same closed loop because the coherent Coulomb scattering selects just the complex conjugated component of the charge wave.

The physical interpretation of the Hartree diagram gives a simple understanding of the magneto-resistance in a magnetic field derived by LEE and RAMAKRISHNAN [8]. In a magnetic field the spin up and spin down electrons have different Fermi-wave lengths and therefore the hologram of spin up electrons (at the Fermi frequency) does not perfectly compensate the phase shifts of a spin down conduction electron and vice versa. For $\mu_B H \approx k_B T$ half of the electron hologram becomes ineffective in the phase adjustment and no longer contributes to the Coulomb-anomaly. This yields the magneto-resistance with the weight F/2.

Besides the Hartree diagram in the particle-hole channel the theory yields also a Hartree diagram in the particle-particle channel. For a single Coulomb scattering, as has first been discussed by FUKUYAMA [3], the physical interpretation is essentially given by Fig. 3. The only difference is that the electron (") propagates on the closed loop in opposite direction. This does not change the charge hologram. However, the effect of a magnetic field is no longer compensated on the two paths but adds up and yields a magnetic phase shift which is $2e/\hbar$ times the enclosed magnetic flux. This contribution will therefore be destroyed by a magnetic field as in weak localization.

We want to comment shortly the Fock diagram. Since the Fock diagram is caused by the same Coulomb interaction as is the Hartree diagram, it describes essentially the same physics. One has to treat the Coulomb interaction in the correct two-particle wave function. In our description this means that we have to include in particular the "Pauli interference", i.e. the fact that the two different electron states can interfere because the two electrons are in both states at the same time. In addition the charge hologram oscillates with time and requires that the screening is calculated dynamically. In the Fock diagram the two electron generally diffuse on a string, not on a closed loop and the Coulomb interaction acts over large distances.

5. References

1. B.L. Altshuler and A.G. Aronov, Solid State Commun. 36, 115 (1979)
2. B. L. Altshuler, A.G. Aronov and P.A. Lee, Phys. Rev. Lett. 44, 1288 (1980)
3. H. Fukuyama, J. Phys. Soc. Jpn. 48, 2169 (1980)
4. P.A. Lee and T.V. Ramakrishnan, Rev. Mod. Phys. 57, 287 (1985)
5. B.L. Altshuler and A.G. Aronov, in Modern Problems in Condensed Sciences, edited by A.L. Efros and M. Pollak North-Holland, Amsterdam, 1985), p. 1
6. H. Fukuyama, in Modern Problems in Condensed Sciences, edited by A.L. Efros and M. Pollak, (North-Holland, Amsterdam, 1985), p. 155
7. G. Bergmann, Phys. Rev. B35, 4205 (1987)
8. P.A. Lee and T.V. Ramakrishnan, Phys. Rev. B26, 4009 (1982)

Nonuniversality of the Mooij Correlation

C.C. Tsuei

IBM Thomas J. Watson Research Center, P.O. Box 218,
Yorktown Heights, NY 10598, USA

Recently compiled and updated experimental data have demonstrated that the correlation between the temperature coefficient of electrical resistivity $\alpha(\equiv\rho^{-1}\partial\rho/\partial T)$ and resistivity ρ for disordered metals at room temperature is not universal. The origin of the non-universality of the Mooij correlation stems mostly from a competition between the quantum mechanical effects of incipient localization and the classical Boltzmann electron transport. Numerical analysis based on the theory of weak localization accounts well for the observed non-universality in the high resistivity regime. For a specific disordered metallic system, this study shows that there is a unique and monotonic decrease of α with increasing ρ, in agreement with the new Mooij α vs ρ plot. Experimental data also suggest that thermopower, superconductivity transition temperature and upper critical field correlate with ρ in various disordered metals. These generalized Mooij correlations can also be understood in terms of weak localization.

1. Introduction

The temperature dependence of the electrical resistivity, $\rho(T)$, of relatively clean metals can be generally described by the following expression:

$$\rho(T) = \rho(0) + CT^p \qquad (1)$$

where $\rho(0)$ is the residual resistivity due to the elastic scattering of conduction electrons [1]. The effect of inelastic scattering as a function of temperature is represented by the term CT^p and the condition of $C > 0$ is always observed experimentally. The exponent p depends on specific inelastic scattering mechanism. For example, p is 2 for classical electron-electron scattering, and p ~ 2-5 for electron-phonon interaction at relatively low temperatures ($T \lesssim \theta_D$, the Debye temperature) and p=1 for higher temperatures ($T \gtrsim \theta_D$). Furthermore, many high purity metals are characterized by a very large ratio of room temperature resistance to its low temperature value. For example, the value of $\rho(T = 300K)/\rho(T = 4.2)$ for a Cu single crystal was found to be about 7000. When a minute amount of disorder (impurity, and/or structural defects) is introduced into a clean metallic system, the residual resistivity increases and the temperature dependence of the resistivity at higher temperature is essentially not affected (Matthiessen's rule). When more disorder is added, however, all the characteristic features of $\rho(T)$ just described are significantly altered. In addition to a very large value of $\rho(0)$, the parameter C becomes very small and can even change its sign (i.e., $C \lesssim 0$). As a result, the ratio $\rho(300)/\rho(4.2)$ is often close to or less than one. The exponent p is usually smaller than two. A strong deviation from the Matthiessen's rule is the rule rather than the exception.

In 1973, J. H. MOOIJ [2] discovered a remarkable correlation between the magnitude of the room-temperature resistivity, ρ and its temperature coefficient $\alpha(\rho)$ in many disordered metallic systems. The Mooij correlation is important since, in a surprisingly simple form, it summarizes the complex resistivity behavior as a function of disorder in many metals which encompass amorphous and crystalline metallic conductors, in thin films as well as in bulk. It strongly suggests that resistivity is a viable measure of disorder in metals. Mooij also noted that, "No negative α is observed for resistivities below 100 $\mu\Omega$ cm while above 150 $\mu\Omega$ cm, hardly any positive α is found", suggesting that α changes sign only in a relatively narrow range of resistivity, (i.e., the critical resistivity for which α = o, $\rho_c \approx 100 - 150$ $\mu\Omega$ cm). In the literature, a fundamental significance has often been attached to a ρ_c of 150 $\mu\Omega$ cm in the sense that it is considered as a universal boundary between the positive and negative α's. Recently, a larger set of experimental data [3] has revealed that the correlation between α and ρ is not universal and is more complex than the original Mooij correlation. It is the purpose of this review article to discuss the origin(s) of the nonuniversality of the Mooij correlation in terms of existing theoretical models with emphasis on the effects of weak localization and electron-electron interaction. Related correlations in thermopower, superconducting transition temperature and upper critical field as a function of disorder will be briefly described.

2. Nonuniversality of the Mooij Correlation - Experimental Evidence

In the original Mooij α vs ρ plot, there were totally 133 data points. Of these, 122 points were from crystalline disordered metallic systems and only 11 points representing the amorphous alloys. As shown in Fig. 1, the solid lines outline the region where the original Mooij data points were distributed. A larger set (~ 500 points) of recently compiled experimental data [3 - 6] of both amorphous and crystalline metals, has demonstrated that the original, approximately linear correlation between α and ρ is not an accurate representation of the experimental data (see Fig. 1). Many data points lie well outside of the Mooij region. It is interesting to note that the critical resistivity, ρ_c, is strongly material-dependent and can vary from about 30 $\mu\Omega$ cm to as high as 400 $\mu\Omega$ cm Clearly, the perception of 150 $\mu\Omega$ cm as a universal value for ρ_c is not substantiated by the experimental data. In fact, as the results in Fig. 1 indicate, for materials with room-temperature resistivity $\rho \lesssim 200$ $\mu\Omega$ cm, there appears no correlation between α and ρ. For $\rho > 200$ $\mu\Omega$ cm, the occurrence of negative α is more favored with increasing ρ.

The data shown in Fig. 1 are useful in demonstrating that ρ_c is not universal and in showing a general trend of α as a function of ρ at a given temperature. Several important features about the α–ρ correlation can not be seen from Fig. 1 and are worth mentioning here: (1) There is a well-defined, monotonic de-

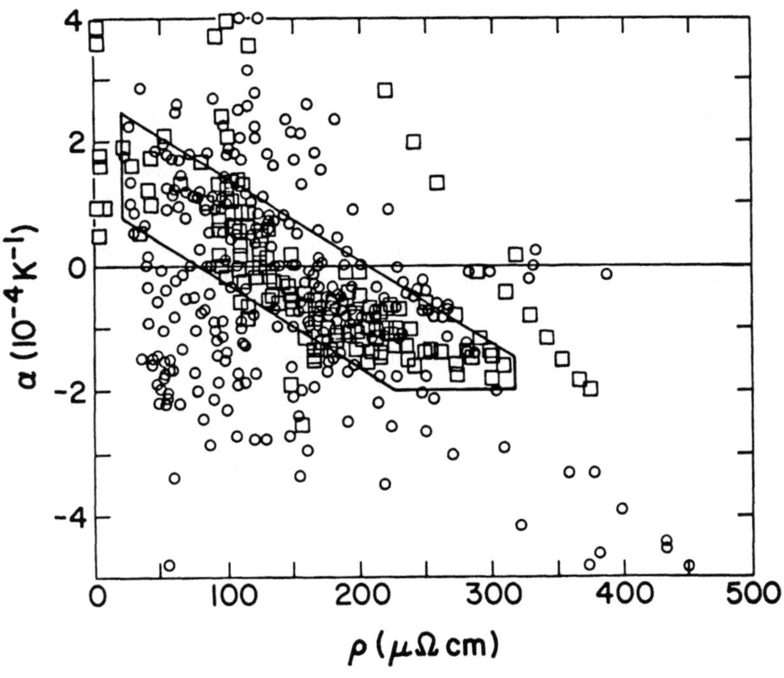

Fig. 1. The temperature coefficient of electrical resistivity (α) vs resistivity (ρ) for various crystalline (open boxes) and amorphous (open circles) metallic conductors at room temperature [3]. The solid lines are for outlining the region where the original Mooij data points were distributed.

creasing of α with increasing ρ for a specific disordered metal, even if ρ is less than 200 $\mu\Omega$ cm. The exact value of ρ at which α changes its sign (i.e., ρ_c) depends sensitively on the individual material. In this sense, the magnitude of resistivity ρ is still a good measure of the amount of disorder. As an example, α as a function of room-temperature resistivity ρ is plotted in Fig. 2 for the α-particle irradiated $LuRh_4B_4$ films. The experimental results (dots) are taken from the work of Dynes, et al., [7] on ρ of $LuRh_4B_4$ films (3000 Å) as a function of temperature and disorder induced by 1.8 – MeV α-particle irradiation. As shown in Fig. 2, in the $\alpha > o$ regime, α decreases rapidly and non-linearly with increasing ρ which is proportional to the α-particle dose. For $\alpha < o$, α decreases linearly with ρ at a much reduced rate; (2) It appears that the essential features of the correlation between α and ρ are not sensitive to what type of disorder is responsible for the increase of resistivity; (3) The Mooij correlation can be generalized to temperatures other than the room-temperature and to physical properties other than the temperature coefficient of resistivity. For example, the Mooij correlation for disordered Nb_3Ge holds at room-temperature and at a lower temperature.[8] There are experimental data to indicate that there is a correlation between thermoelectric power and resistivity in amorphous metals. For superconductors, a large amount of data suggest a correlation between the superconductivity transition temperature T_c and the normal state resistivity near T_c.

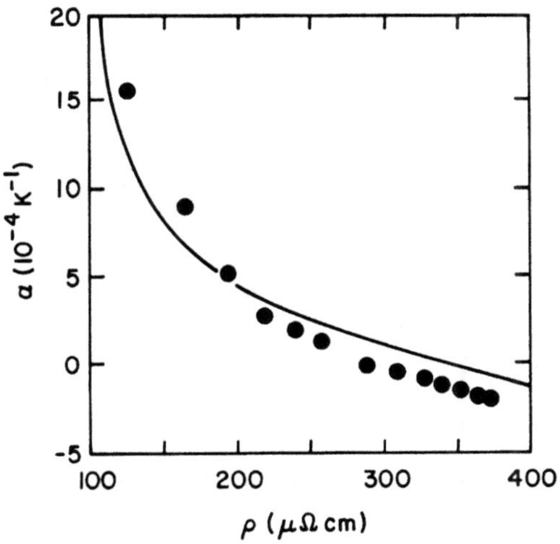

Fig. 2. The temperature coefficient of resistivity α as a function of resistivity ρ at room temperature for α-particle irradiated $LuRh_4B_4$ films (3000Å). The dots are the digitized values for α from the experimental ρ vs T curves of Fig. 5 in Ref. [7]. The theoretical curve is from a numerical calculation based on Eqs. [4] and [5].

3. Theoretical Models

There have been numerous theoretical studies [9] made to understand the Mooij correlation in its original form. A theoretical consideration based on the Kaveh-Mott weak localization model [10] will be presented to explore the origin of the Mooij correlation and its nonuniversality. Other models will also be briefly discussed at the end of this section.

3.1 Weak Localization Model

In terms of the weak localization theory, the effect of disorder on the conductivity of metals can be described by a quantum correction to the Boltzmann conductivity $\sigma_B(T)$:

$$\delta\sigma = \sigma(T) - \sigma_B(T) \cong \frac{e^2}{\pi^2\hbar}\left(\frac{1}{L_i} - \frac{1}{\ell_e}\right) \tag{2}$$

where $\sigma_B(T)$ is related to the Boltzmann resistivity $\rho_B(T)$ by $\sigma_B^{-1}(T) = \rho_B(T) = 3\pi^2\hbar/e^2k_F^2\ell$, and the electron mean free path ℓ is expressed in its elastic and inelastic components by $\frac{1}{\ell} = \frac{1}{\ell_e} + \frac{1}{\ell_i}$. The inelastic diffusion length L_i, is defined as $L_i = \sqrt{\ell_e\ell_i/3}$ and k_F is the Fermi wavevector.

The weak localization assumption inherent in Eq. (2) is consistent with the fact that the resistivity of most highly disordered conductors is only a fraction of the Mott resistivity $\rho_M = 3\pi^2\hbar/e^2k_F$ and that $k_F\ell_e$ for many amorphous metals is about 4 — 10 significantly above the Ioffe-Regel criterion for a metal-insulator transition. Instead of claiming that disorder leads to a total breakdown of the standard Boltzmann theory, the validity of combining the classical Boltzmann transport with the quantum mechanical phase-coherent effects (as done in Eq. (2) in the weakly localized regime) is justified. Recently, HERSHFIELD and AMBEGAOKAR [11] have generalized the Boltzmann equation to include the conduction electron coherent backscattering which results in weak localization.

In disordered metals, the effect of electron-electron interaction [9] can significantly modify the conventional Boltzmann electron transport, especially for the case of low dimensional metallic systems at low temperatures. For a three dimensional disordered metal at room temperature, the effect of interaction is, in general, small as compared with that of localization. This can be seen from the following:

$$\frac{\delta\sigma_I^T}{\delta\sigma_L^T} \cong \frac{0.23}{\hbar}\sqrt{\ell_i(T)mk_BT/k_F} \tag{3}$$

where $\delta\sigma_I^T$ and $\delta\sigma_L^T$ are the temperature-dependent corrections on $\sigma_B(T)$ due to interaction and localization respectively. With reasonable values for ℓ_i, and k_F and $T = 300K$, one finds that $\delta\sigma_I^T/\delta\sigma_L^T$ is about 0.02 to 0.05.

In short, for the sake of discussing the Mooij correlation, it is reasonable to approximate the total conductivity for a three dimensional disordered metal by the following:

$$\sigma(T) = \sigma_B(0)\left[\frac{\ell_i}{\ell_e+\ell_i} + \frac{3\ell_e}{(k_F\ell_e)^2}\left\{\frac{1}{L_i} - \frac{1}{\ell_e}\right\}\right] \tag{4}$$

From Eq.(4), one can readily calculate $\alpha(T) = \rho^{-1}(T)\,\partial\rho(T)/\partial T = -\sigma^{-1}(T)\,\partial\sigma(T)/\partial T$ and obtain:

$$\alpha(T) = \frac{P}{T}\left[\frac{\ell_e}{(\ell_e+\ell_i)^2} - \frac{3\sqrt{3}\,\ell_e^{1/2}}{2(k_F\ell_e)^2\,\ell_i^{3/2}}\right]\left[\frac{\ell_i}{\ell_e+\ell_i} + \frac{3\ell_e}{(k_F\ell_e)^2}\left\{\frac{1}{L_i} - \frac{1}{\ell_e}\right\}\right]^{-1} \tag{5}$$

where the temperature dependence of ℓ_i is assumed to be of the form $\ell i(T) = AT^{-P}, P > 0$. The condition for ρ_c (i.e. $\alpha = 0$) leads to:

$$\frac{\ell_i}{\ell_e} = \frac{4}{27}\frac{(k_F\,\ell_e)^4}{(1+\frac{\ell_e}{\ell_i})^4} \tag{6}$$

and the critical resistivity ρ_c can be expressed as:

$$\rho_c^{3D} = \rho_B(0)\left[\frac{1}{1+\ell_e/\ell_i} - \frac{3}{(k_F\,\ell_e)^2} + \frac{6.75(1+\ell_e/\ell_i)^4}{(k_F\,\ell_e)^4}\right]^{-1} \tag{7}$$

The results as written in Eqs. (6) and (7) are the same as those derived in Ref. [3] if the condition of $\ell_e << \ell_i$ is assumed. The expression for ρ_c is then :

$$\rho_c^{3D} \cong \frac{1213.5}{k_F^2 \ell_e} \left[1 - \frac{3}{(k_F \ell_e)^2} + \frac{6.75}{(k_F \ell_e)^4} \right]^{-1} \tag{8}$$

in $\mu\Omega$ cm., if k_F and ℓ_e are in $\AA^{-1}$ and Å respectively. From Eq. (8), it is clear that ρ_c is not universal and depends on two parameters: k_F and ℓ_e . The former is a material-dependent property; the latter, a measure of disorder. As pointed out in [3], Eq. (8) is capable of yielding a range of values for ρ_c i.e., $\rho_c \approx 30 - 500\ \mu\Omega$ cm . This is due to the fact that k_F for most metallic systems lies in the range of $1\AA^{-1} < k_F < 2\AA^{-1}$ and ℓ_e is about 2 to 3 Å in the dirty limit. Equation (8) also suggests that the preponderance of ρ_c data around 150 $\mu\Omega$ cm (as depicted in Fig. 1) is only a manifestation of the fact that many, if not most, disordered metals are characterized by a ℓ_c of $5 \pm 2\AA$ and a k_F of $1.5 \pm 0.3\AA^{-1}$. For $\rho_c < 100\mu\Omega$cm, Eqs. (6) and (8) imply an unrealistic value of ℓ_i at room temperature[12] and the condition $\ell_e << \ell_i$ on which Eq. (2) and the essence of electron localization are based can not be satisfied. For low resistivity disordered metals ($\rho < 100\mu\Omega$ cm), mechanisms [13, 14, 15] other than electron localization have been suggested for explaining the relatively low ρ_c's and large negative α's as shown in Fig. 1. It should be emphasized that, however, sample inhomogeneity and granularity can shift ρ_c to a value higher or lower than those predicted by Eq. (8), depending on the details of the minority phase and/or the intergranular material [16].

Based on Eqs. (4) and (5), one can numerically plot α as a function of ρ for a given temperature. If k_F and T are kept constant, the results of such a numerical analysis always show that α decreases with increasing ρ, rapidly and nonlinearly for $\rho < \rho_c$ and linearly and slowly for $\rho > \rho_c$. As an example, a comparison between the theory and the experiment for the α-particle irradiated films of $LuRh_4B_4$ is presented in Fig. 2. The solid curve is based on the assumptions of $k_F = 1.7\AA^{-1}$ ($E_F = 11$eV) and $A = 5000\AA K$, and $T = 300K$. A Fermi level of 11eV is suggested by a self-consistent energy-band calculation [17] on a similar ternary boride $ErRh_4B_4$. The only fitting parameter $A=5000\ \AA K$ leads to an inelastic mean free path ℓ_i of about 15Å at T = 300K. As seen from Fig. 2, the theoretical curve can account for the essential features of the experimental data. The agreement between theory and experiment probably could be improved upon if spin-orbit scattering, electron-electron interaction and the effect of disorder on k_F and ℓ_i are included in the numerical analysis. It appears, to a large extent, the weak localization model can account for the α data as a function of ρ and offers a simple explanation for the nonuniversality of the Mooij correlation. In addition to the resistivity dependence of α, there are other properties such as low-temperature magneto-conductivity, thermopower, superconducting transition temperature and upper critical field in the case of disordered superconductors, lending supporting evidence for localization in three dimensional disordered metallic systems.

3.2 Other Models

In view of the diversity of the data shown in Fig. 1, it is not surprising that there are other models that can explain the Mooij correlation, at least, in certain special cases. For example, the Ziman diffraction model [13] and the grain boundary reflection model [16] are probably valid in the low resistivity regime. The inter-band models [15] are relevant to transition-metal alloys. Other alternative approaches include models based on phonon-assisted tunneling [14], and inter-granular tunneling [16].

4. Generalized Correlations

The well-defined correlation between α and ρ for a specific disordered metal strongly suggests that resistivity value is a viable measure of disorder. If this conclusion is true, Mooij-type correlations should be observed in the resistivity dependence of physical properties other than α. This is indeed the case. A few examples of generalized Mooij correlations will be discussed as follows:

4.1 Thermopower

A correlation between thermopower and resistivity in amorphous metals was first suggested by Nagel, [18] who found that thermopower S tends to be positive and α is likely negative for metallic glasses with

$\rho > 150\mu\Omega$cm. This experimental finding was interpreted by Ziman-type theory. There is also a correlation between S and α for amorphous metallic alloys [4]. More recently, Kaiser [19] found a thermopower analog of the Mooij correlation (i.e., the magnitude of thermopower $|S|$ is proportional to ρ^2). This correlation can be understood in terms of incipient localization. If the quantum corrections on the Boltzmann conductivity σ_B, as described in Eq. (2), are included in the formula for thermopower. The dominant correction terms in S due to incipient localization is found to increase with ρ^2. Sivan and Imry calculated the thermopower for strongly disordered metallic systems and concluded that S increases linearly with ρ as the metal-insulator transition is approached [20].

4.2 Superconducting Transition Temperature

There have been numerous studies, experimental [21] and theoretical, [22, 23, 24] on the effect of disorder on superconductivity. Except in the very weak disorder regime where the Anderson theorem applies, the superconducting transition temperature T_c and other superconducting properties such as upper critical field H_{c2} are sensitively affected by disorder. In particular, it is well established experimentally that T_c decreases with increasing normal state resistivity near T_c, $\rho(\sim T_c)$, independent of whatever type of disorder causing the resistivity increase. In fact, the striking similarity among the T_c vs $\rho(\sim T_c)$ curves for various disordered superconducting systems has lead to the term "universal degradation" of T_c in disordered superconductors.

Considerable amount of theoretical effort has been devoted to the understanding of the interplay between superconductivity and disorder. It is recognized only recently that disorder-induced degradation effects on superconductivity can not be adequately interpreted in terms of a simple enhancement in Coulomb-interaction, or a reduction in electron-phonon interaction and/or the density of states. Many inter-related effects due to disorder have to be considered on an equal footing and at microscopic level. In this regard, significant progress has been made steadily throughout recent years. [22 - 25] In particular, a recent theory by Belitz [25] has taken into account various electron self-energy and vertex corrections. The Belitz theory extends the validity regime of standard dirty-limit theory substantially toward strong disorder. This theory has been applied in studying the so-called universal degradation of T_c in various superconductors such as the A15 compounds and rare earth borides [26]. The theory is, in general, in good agreement with the experimental data, T_c as a function of $\rho(\sim T_c)$, for a specific alloy system. As reported in the literature, [21] when all the data of T_c/T_{co} (T_{co} is the superconducting transition temperature in the clean limit) are plotted against $\rho(T = 300K)/\rho(\sim T_c)$, the data points are found to distribute over a characteristic band instead of a universal curve. The nonuniversality of the T_c degradation as a function $\rho(300K)/\rho(\sim T_c)$ can be understood only if the localization effect on $\sigma(T)$ is taken into account and material-dependent parameters such as k_F, θ_D in the Belitz theory are allowed to vary. [26] In particular, the T_c formula by Belitz has been used to compare the T_c data of Dynes, et al., as a function of disorder for the α-particle irradiated films of $LuRh_4B_4$. In the Belitz formula, as shown below, various renormalizations due to disorder are expressed in the f parameters (i.e., $f_1, f_2 \ldots$) which are functions of the resistivity ratio of the normal state resistivity to the Mott resistivity, $\hat{\rho} = \rho(\sim T_c)/\rho_M$:

$$T_c = \frac{\theta_D}{1.45} \exp\left[- \frac{1.04[1 + \lambda f_1(\hat{\rho}) + f_2(\hat{\rho})]}{\lambda f_1(\hat{\rho}) - \mu^*(\hat{\rho})\{1 + 0.62\lambda f_1(\hat{\rho})/[1 + f_2(\hat{\rho})]\}} \right] \tag{9}$$

where the parameters such as $\hat{\rho}$, $f_1(\hat{\rho})$ ... are defined in Ref. [25]. The ratio of $\rho(T)/\rho(\sim T_c)$ can be obtained from Eq. (2), as follows:

$$\frac{\rho(T)}{\rho(\sim T_c)} = \frac{1}{\hat{\rho}\rho_M\sigma(T)} \approx \frac{\sigma(0)}{\sigma(T)} = \frac{\left[1 - \dfrac{3}{(k_F \ell_e)^2}\right]}{\left[\dfrac{\ell_i}{\ell_e + \ell_i} + \dfrac{3\ell e}{(k_F \ell_e)^2}\left(\dfrac{1}{L_i} - \dfrac{1}{\ell_e}\right)\right]} \tag{10}$$

The results of a numerical analysis based on Eqs. (9) and (10) are compared with the experimental data for the α-particle irradiated $LuRh_4B_4$ films taken from Ref. [7]. The agreement between the theory and experiment is quite remarkable [26]. The parameters used in calculating $\rho(300K)/\rho(\sim T_c)$ are the same as those [3] in determining α as a function of $\rho(300K)$, suggesting the internal consistency of the localization model.

4.3 Upper Critical Field

Disordered superconductors are characterized by a large normal state resistivity, ρ, and an upper critical field, H_{c2} , which is significantly larger than the values predicted by the conventional dirty-limit theory for H_{c2} . The size of the deviation from theory correlates with ρ [27]. The H_{c2} enhancement as a function of ρ can be attributed to the de-localization effect by the application of a magnetic field [23, 28].

5. Concluding Remarks

Newly compiled and updated experimental data have demonstrated that the correlation between α and ρ for various disordered metals at room temperature is not universal, contrary to the Mooij correlation in its original form. The origin of the non-universality of the Mooij correlation stems mostly from a competition between the quantum mechanical effects of incipient localization and the classical Boltzmann electron transport. A numerical analysis based on the theory of weak localization accounts for the observed nonuniversality in the high resistivity regime. The critical resistivity ρ_c (for $\alpha = 0$) is found to depend on Fermi level (k_F) and disorder (ℓ_e) . For a specific disordered metallic system, this study shows that there is unique and monotonic decrease of α with increasing ρ, in accordance with the essential features of the new Mooij α vs ρ plot.

The concept of Mooij correlation between α and ρ has been generalized to physical properties other than α. For instance, experimental data suggest that thermopower S, superconducting transition temperature T_c and upper critical field H_{c2} correlate with ρ in various disordered metals, thus underscoring that ρ is a valid measure of disorder. These generalized correlations and their nonuniversality can be understood in terms of weak localization.

The author wishes to thank C. C. Chi, D. H. Lee, D. Mukamel, S. I. Park and P. Santhanam for useful discussions.

References

1. J. M. Ziman: Electrons and Phonons (Oxford, 1963)
2. J. H. Mooij: Phys. Status Solidi A**17**, 521 (1973)
3. C. C. Tsuei: Phys. Rev. Lett. **57**, 1943 (1986)
4. U. Mizutani: Prog. in Mater. Sci. **28**, 97 (1983)
5. S. R. Nagel: Adv. in Chem. Phys. **51**, 227 (1982)
6. D. G. Naugle: J. Phys. Chem. Solids **45**, 367 (1984)
7. R. C. Dynes: J. M. Rowell, and P. H. Schmidt: In Ternary Superconductors, ed. by G. K. Shenoy, B. D. Dunlap, and F. Y. Fradin (North-Holland,Amsterdam, 1981), p. 169
8. F. Rullier-Albenque, L. Zuppiroli and F. Weiss: J. Physique **45**, 1689 (1984)
9. P. A. Lee and T. V. Ramakrishnan: Rev. Mod. Phys. **57**, 287 (1985)
10. M. Kavah and N. F. Mott: J. Phys. C **15**, L707 (1982); M. A. Howson: J. Phys. F, **14**, L25 (1984); Y. Imry: Phys. Rev. Lett. **44**, 469 (1980); M. Jonson and S. M. Girvin Phy. Rev. Lett. **43**, 1447 (1979)
11. S. Hershfield and V. Ambegaokar: Phys. Rev. B**34**, 2147 (1986)

12. A. B. Kaiser: Phys. Rev. Lett. **58**, 1384 (1987)
13. J. M. Ziman: Philos. Mag. **6**, 1013 (1961); K. F. Kelton and F. Spaepen: Phys. Rev. **B30**, 5516 (1984)
14. D. Belitz and W. Schirmacher: J. Phys. C, **16**, 913 (1983)
15. P. B. Allen and B. Chakraborty: Phys. Rev. **B23**, 4815 (1981); N. F. Mott: Philo. Mag. **26**, 1249 (1972)
16. G. Reiss, J. Vancea, and H. Hoffmann: Phys. Rev. Lett. **56**, 2100 (1986); J. Vancea, S. Pukowietz, G. Reiss, and H. Hoffmann: Phys. Rev. **B35**, 9067 (1987); O. Entin-Wohlman and Y. Shapira: J. Phys. C, **15**, 975 (1982)
17. T. Jarlborg, A. J. Freeman, and T. J. Watson-Yang, Phys. Rev. Lett. **39**, 1032 (1977)
18. S. R. Nagel: Phys. Rev. Lett. **41**, 990 (1978)
19. A. B. Kaiser: Phys. Rev. **B35**, 2480 (1986)
20. U. Sivan and Y. Imry: Phys. Rev. **B33**, 551 (1986)
21. L. R. Testardi, J. M. Poate, and H. J. Levinstein: Phys. Rev. **B15**, 2570 (1977); A. R. Sweedler, C. L. Snead, L. Newkirk, F. Valencia, T. H. Geballe, R. H. Schwall, B. T. Matthias and E. Corenzwit: Proc. Intl. Conf. on Radiation Effects and Tritium Technology (Gatlinburg, TN), (1976) Vol. III, p. 422; A K. Ghosh, H. Weissman, M. Gurevitch, H. Lutz, O. F. Kammerer, C. L. Shead, A. Goland and M. Strongin: J. Nucl. Mater. **72**, 70 (1978); A. R. Sweedler D. E. Cox and S. Moehlecke: J. Nucl. Mater. **72**, 50 (1978)
22. P. W. Anderson, K. A. Muttalib and T. V. Ramakrishnan: Phys.Rev. **B28**, 117 (1983)
23. S. Maekawa, H. Ebisawa and H. Fukuyama: Prog. theor. Phys. Suppl. No. 84, 154 (1984); H. Fukuyama: Physica, **126B**, 306 (1984); H. Fukuyama, H. Ebisawa, and S. Maekawa: J. Phys. Soc. Japan **53**, 3560 (1984)
24. D. Belitz: J. Phys. F, **15**, 2315 (1985)
25. D. Belitz: Phys. Rev. **B35**, 1636 (1987); ibid **35**, 1651 (1987)
26. C. C. Tsuei, R. Wordenweber, P. H. Kes (to be published)
27. M. Tenhover, W. L. Johnson, and C. C. Tsuei: Solid State Commun. **38**, 53 (1981); M. Ikebe, Y. Muto, S. Ikeda, H. Fujimori, and K. Suzuki, Physica **107B**, 387 (1981)
28. L. Coffey, K. Levin and K.A. Muttalib: Phys. Rev. **B32**, 4382 (1985)

Spin-Glass Transition and Resonant Kondo Scattering in Weakly Localized Metal Films

C. Van Haesendonck, J. Vranken, and Y. Bruynseraede

Laboratorium voor Vaste Stof-Fysika en Magnetisme,
Katholieke Universiteit Leuven, B-3030 Leuven, Belgium

We report detailed measurements of the resistance as a function of temperature and magnetic field in thin films of magnetic alloys. In CuCr films with a thickness $t \simeq 4.0$nm and a Cr concentration $c < 100$ ppm, the temperature dependence of the spin-flip scattering in the vicinity of the Kondo temperature can be obtained from the low field magnetoresistance caused by the weak electron localization. In CuCr and AgMn films with a thickness $t > 10$ nm and $c \sim 1$at%, the bulk transport properties are recovered. While for AgMn the spin-glass resistivity remains unaffected when the film thickness is decreased, the more concentrated CuCr films show a spin-glass to Kondo alloy transition. The different behavior of AgMn and CuCr films can be explained when the damping of the RKKY interaction by elastic defect scattering and the strength of the Kondo effect are taken into account.

1. Introduction

The presence of magnetic impurities (Cr, Fe, Mn) in noble metals, strongly influences the low temperature metallic properties. For very low impurity concentration c (< 100 ppm), the magnetic scattering is governed by the single impurity Kondo effect [1]. Below the Kondo temperature T_K, the impurities lose their magnetic moment due to the anti-parallel screening by the conduction electron spins. For $T > T_K$, the screening is heralded by a logarithmic divergence of the magnetic scattering, causing the well-known Kondo minimum in the low temperature resistivity. Within the Suhl-Nagaoka approximation the resonant spin-flip scattering rate in the vicinity of T_K is given by:

$$\tau_{sf}^{-1} = \frac{c}{2\pi\hbar N(E_F)} \frac{\pi^2 S(S+1)}{\pi^2 S(S+1) + \ln^2(T/T_K)} . \tag{1}$$

τ_{sf} is the finite life-time of the conduction electrons due to spin-flip scattering. In (1), $N(E_F)$ is the conduction electron density of states near the Fermi level (for one spin direction), S represents the impurity spin and c is the number of magnetic impurities per unit volume. Equation (1) reproduces the essential features of the Kondo effect: (i) the scattering rate is proportional to the concentration c, (ii) there occurs a logarithmic divergence of the magnetic scattering for $T \gg T_K$ and (iii) the spin-flip scattering vanishes for $T \ll T_K$.

For more concentrated alloys ($c \sim 1$ at%), interaction effects between magnetic impurities become important. Well below the percolation limit, direct exchange between magnetic moments can be neglected. In this dilute spin-glass [2] regime, the interaction is dominated by the indirect RKKY process. The interaction between two magnetic moments S_i and S_j is given by

$$\mathcal{H}_{RKKY} = V_o r_{ij}^{-3} \cos(2k_F r_{ij}) \vec{S}_i . \vec{S}_j . \tag{2}$$

V_o is the coupling strength and r_{ij} is the distance between the two spins. Due to the oscillation of $\mathcal{H}_{RKKY}$ on an atomic length scale $\lambda_F = 2\pi/k_F$, the interaction changes between ferromagnetic and anti-ferromagnetic for different impurity pairs. The interaction mechanism (2) causes a freezing of the magnetic moments in random directions for $T \to 0$. The onset of the freezing process results in a cusp in the magnetic susceptibility at the spin-glass freezing temperature T_f. For dilute spin-glasses, $k_B T_f$ is comparable to the average RKKY interaction strength Δ_c , provided the Kondo effect is weak near T_f. For an homogeneous impurity distribution (no clustering), the average distance between impurities is $c^{-1/3}$. Equation (2) therefore implies that T_f varies proportional to c . While the magnetic susceptibility shows an anomaly near T_f, the temperature dependence of the resistivity only shows a broad maximum [3] at $T_{max} \neq T_f$. The maximum is a direct consequence of the transition between the Kondo scattering at higher temperatures and the spin-glass freezing which inhibits the spin-flip scattering when the temperature is lowered. In contrast to the single impurity Kondo effect, the impurities conserve their magnetic moment for $T \to 0$.

The presence of a phase transition at T_f is still controversial. If the critical dimensionality for the transition is close to 3, one could expect an important modification of the spin-glass order in thin film alloys. Unfortunately, thin films always contain many defects. The signature of a possible dimensional crossover will be considerably disturbed by the damping of the RKKY interaction due to the elastic defect scattering. A first order calculation [4, 5] indicates that (2) should be multiplied by a damping factor $\exp(-r_{ij}/\ell_{el})$, where ℓ_{el} is the elastic mean free path. The spin-glass temperature T_f will therefore drop rapidly with decreasing mean free path. At the same time, the maximum in the temperature dependence of the resistivity will also shift towards lower temperatures and the Kondo effect is recovered for $T > T_{max}$ [3].

In this paper, we will present electrical transport measurements, which confirm the presence of Kondo scattering and spin-glass order in very thin CuCr and AgMn films with $0 < c \lesssim 1$ at% and thickness t ranging between 4.0 nm and 40.0 nm. In the thinnest films the weak electron localization causes an anomalous low field magnetoresistance, from which the temperature dependence of the magnetic scattering rate near T_K and T_{max} can be calculated (Section 3). In the thickest films, the temperature and magnetic field dependence of the resistivity are dominated by the magnetic scattering. The bulk properties are recovered when the elastic scattering is taken into account (Section 4).

2. Experimental Details

We would like the transition metal impurities (Cr, Mn) to be distributed homogeneously and randomly within the noble metal films (Cu, Ag). Because of distillation effects, slow thermal evaporation of pieces of a bulk alloy affects the final impurity concentration. It is therefore necessary to flash evaporate very small pieces of the bulk alloy. In order to prevent clustering of the magnetic impurities, we reduced their mobility by quench condensation onto a liquid nitrogen cooled glass substrate. Since residual oxygen in the vacuum chamber can neutralize the magnetic moments, we reduced the partial oxygen pressure using a Ti sublimator. In order to avoid surface oxidation, the films are covered in situ with a 100 nm layer of SiO.

For the weakly doped CuCr films ($c < 100$ ppm), the impurity concentration c can be obtained by measuring the Kondo anomaly in the temperature dependence of the resistance. The disorder enhanced Coulomb repulsion between the electrons causes an additional logarithmic divergence of the resistance [6]. For a change in temperature of one decade, the variation of the sheet resistance $R_\square$ is given by:

$$\Delta R_\square = \Delta R_\square^C + \Delta R_\square^K = \alpha \frac{e^2}{2\pi^2\hbar} \ln 10 \, R_\square^2 + \frac{\Delta\rho_K}{t} \, . \tag{3}$$

The value of the constant $\alpha \simeq 1$ is determined by measuring the Coulomb effect in a pure Cu film. The resistivity change $\Delta\rho_K$ has been measured previously for bulk CuCr [7]: $\Delta\rho_K = 0.55 \pm 0.05$ nΩcm/ppm. For the more concentrated CuCr and AgMn alloys (c ~ 1 at%), we have to rely upon neutron activation measurements and chemical absorption spectroscopy to estimate c. Although these methods provide an accurate value for the Cr or Mn content within a 10 % error margin, they do not allow to check whether these impurities are still magnetic (not oxidized) and are not clustering. For the CuCr Kondo alloys, the Cr concentration obtained from (3) is always comparable to concentration in the bulk alloy. We may therefore assume that the concentration of active magnetic impurities in the spin-glass alloys is within 10 % the same as the one obtained from the chemical analysis.

3. Weak Electron Localization in Magnetic Alloys

Weak electron localization (WEL) in a thin film results from an enhanced back-scattering probability due to the interference between electron waves traveling along time-reversed paths [8]. A magnetic field perpendicular to the plane of the film, modulates the phase of the interfering electron waves and causes an anomalous low field magnetoresistance. The detailed shape of the magnetoresistance curve strongly depends upon the characteristic scattering times of the conduction electrons [9]. Several experiments have shown that an analysis of the experimental magnetoresistance is a unique tool to determine the temperature dependence of the inelastic scattering time τ_{in} (due to electron-phonon or electron-electron scattering) and the spin-orbit scattering rate τ_{so} [10]. Recently it has become clear that WEL can also be used to study the temperature dependence of the magnetic scattering rate τ_m^{-1}, caused by the presence of localized magnetic moments [11, 12]. WEL allows to determine the phase breaking rate:

$$\tau_\varphi^{-1} = \tau_{in}^{-1} + 2\tau_m^{-1} . \tag{4}$$

τ_φ is the time during which the electrons conserve their phase memory. For a clean metal film, $\tau_\varphi \simeq \tau_{in} \propto T^{-p}$ (p depends on the dominant inelastic scattering process). For sufficiently strong magnetic scattering, the power law dependence of τ_φ will be disturbed.

Fig. 1 shows the magnetoresistance measured in a perpendicular field at different temperatures for a CuCr film with $c_{Cr} = 40$ ppm and $R_\square = 24.9\ \Omega/\square$. The positive magnetoresistance indicates the presence of a strong spin-orbit scattering: $\tau_{so} << \tau_{in}$ [9]. Theory predicts that the magnetoresistance will be a universal function of the reduced field $B/B_\varphi(T)$. The characteristic field $B_\varphi(T)$ is linked to the scattering time $\tau_\varphi(T)$ via:

$$B_\varphi(T) = \frac{\hbar}{4eD\tau_\varphi(T)} , \tag{5}$$

where $D = v_F\ell_{el}/3$ is the diffusion constant for the disordered CuCr film. From Fig. 1 it is clear that the phase breaking rate $\tau_\varphi^{-1}(T)$ saturates between 4.2 K and 1.3 K (magnetoresistance is nearly temperature independent). Since B_φ decreases monotonically with decreasing temperature for a pure Cu film [10], we are tempted to identify this saturation with the resonance in the spin-flip scattering near the Kondo temperature $T_K \simeq 2$ K. When we substract from the $B_\varphi(T)$ values for the CuCr film the $B_\varphi(T)$ values for a pure Cu film prepared under similar conditions, we obtain the characteristic field $B_m(T) \propto \tau_m^{-1}(T)$ (see (4)). From (5) we calculate the temperature dependence of the magnetic scattering rate τ_m^{-1} shown in Fig. 2 . There we also show the result obtained for a CuCr film with $c_{Cr} = 13$ ppm. The full curves were calculated using (1) for the spin-flip scattering rate. If we also take into account the spin-nonflip scattering rate to calculate τ_m^{-1}, we obtain the dashed curve. From Fig. 2 it is obvious that the spin-flip scattering causes the destruction of the WEL near T_K.

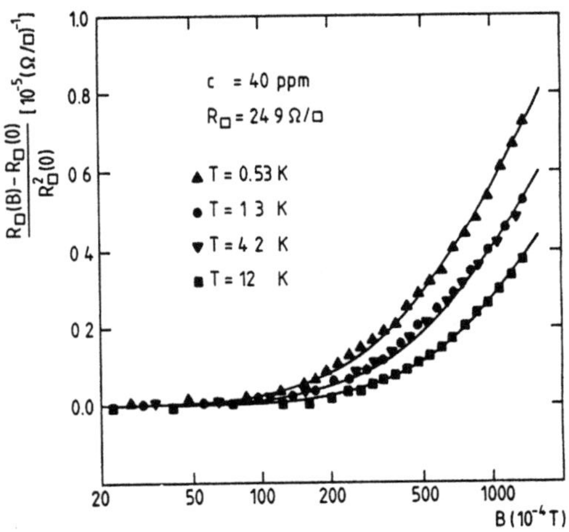

Fig. 1: Normalized low-field magnetoresistance for a CuCr film in the Kondo regime [12]

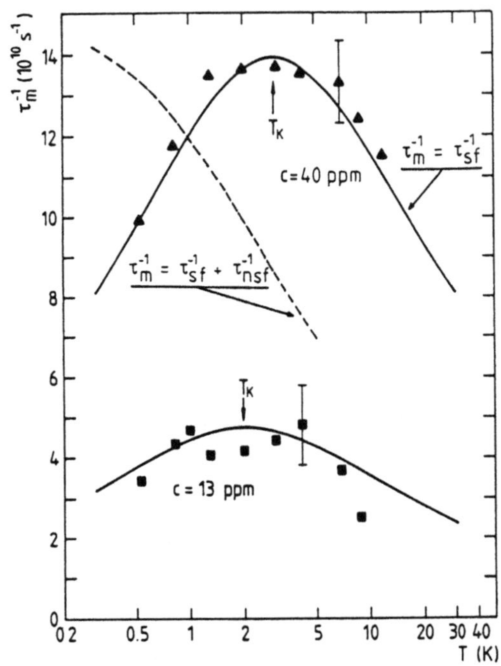

Fig. 2: Temperature dependence of the magnetic scattering rate in weakly doped CuCr films [12]

For the CuCr films shown in Fig. 2, the destruction of the WEL is caused by a single impurity Kondo effect ($\tau_m^{-1} \propto c$). Even when the RKKY interactions can no longer be neglected ($c > 100$ ppm), the WEL method can still be used to probe the $\tau_m(T)$ variation. The only condition for this method to be reliable, is a WEL magnetoresistance much larger than the resistance changes due to the $\tau_m(T,B)$ variation ($R_\square > 100\ \Omega/\square$). The WEL method can be applied as long as the thin films are homogeneous and continuous. Up to now, our studies of the WEL for $c > 100$ ppm have only been qualitative.

4. Spin-glass Order in Thin Films

We have studied the spin-glass order in CuCr ($c_{Cr} = 1.0 \pm 0.1$ at%) and AgMn ($c_{Mn} = 0.75 \pm 0.10$ at%) films with a thickness t varying between 4.0 and 40.0 nm. For the CuCr and AgMn films, the average separation between the impurities is $c^{-1/3} \simeq 1.0$ nm. On the other hand, ℓ_{el} typically ranges between 10.0 nm in the thickest films and 2.0 nm in the thinnest films. The damping of the RKKY interaction by elastic scattering will therefore mainly affect the interaction between impurities that are not the nearest neighbors.

LARSEN [13,14] showed that the resistivity $\rho(T/T_K)$ only depends upon the ratio between the average RKKY interaction strength Δ_c and the Kondo temperature T_K. When the elastic mean free path is reduced, Δ_c will be depressed and the ratio Δ_c/T_K will decrease. AgMn is known to be a archetypical spin-glass with $T_f \simeq 4$ K for $c_{Mn} = 0.75$ at% [2] and a very weak Kondo effect: $T_K < 0.001$K and $\Delta\rho_K \simeq 0.02$ nΩcm/ppm [15] The elastic scattering is unable to destroy the low temperature spin-glass resistivity when $\ell_{el} > c^{-1/3}$. For CuCr, $k_B T_K$ is comparable to the average RKKY interaction energy Δ_c for $c_{Cr} \simeq$ 1at% [3]. Even a moderate damping of the RKKY interaction can therefore induce a spin-glass to Kondo alloy transition.

The influence of the film thickness on the temperature dependence of the resistivity in CuCr films, is illustrated in Fig. 3 . When the thickness is reduced from 40.0 nm to 4.0 nm, we observe a gradual transition from a spin-glass towards a Kondo behavior. As already indicated in Section 2, the disorder enhanced Coulomb repulsion contributes to the temperature dependence of the resistance. When we substract this logarithmic divergence from the experimental data, we obtain the $\rho(T)$ variation caused by the magnetic scattering. The result is shown in Fig. 4 for films (a) and (c), having a thickness of 36.5 nm and 4.8 nm respectively. While T_{max} has shifted towards lower temperatures for film (c), we also note that the resistivity does not reach its minimum value for $T \to 0$. According to Larsen's theory [14], this indicates that $\Delta_c/T_K \simeq 1$ for a thickness 4.0 nm < t < 40.0 nm. For film (d), the maximum has completely disappeared and the single impurity Kondo effect has largely been restored. Fig. 5 shows the isotropic (independent of magnetic field direction) magnetoresistivity for three CuCr films with different thickness at T = 1.3 K. The alignment of the impurity spins in the magnetic field gives rise to a negative magnetoresistivity. The amplitude of the effect is much smaller in the spin-glass regime. This may be explained by the fact that the alignment of the spins is less important when they are already frozen into random directions.

The temperature dependence of the resistivity for four AgMn films with different thickness is shown in Fig. 6 . For film (d), the logarithmic divergence of the resistance is in good agreement with the $\Delta R_\square/R_\square^2$ value for clean Ag films. After substracting this contribution caused by the Coulomb repulsion, we obtain a similar $\rho(T)$ variation for films (a) and (c) as shown in Fig. 7 . The theoretical full curve has been calculated assuming bulk properties for the AgMn. Fig. 8 shows the magnetoresistivity in a perpendicular field for three AgMn films with different thickness at T = 0.5 K. Since the Kondo effect is much smaller for AgMn than for CuCr, the ampli-

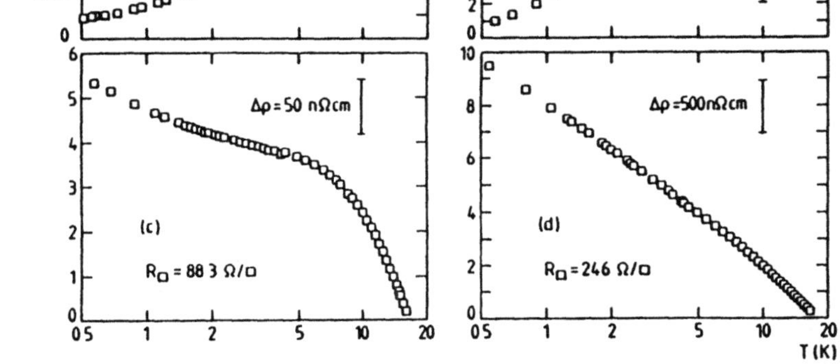

Fig. 3: Temperature dependence of the resistivity in CuCr films ($c_{Cr} \simeq$ 1.0 at%) with different thickness

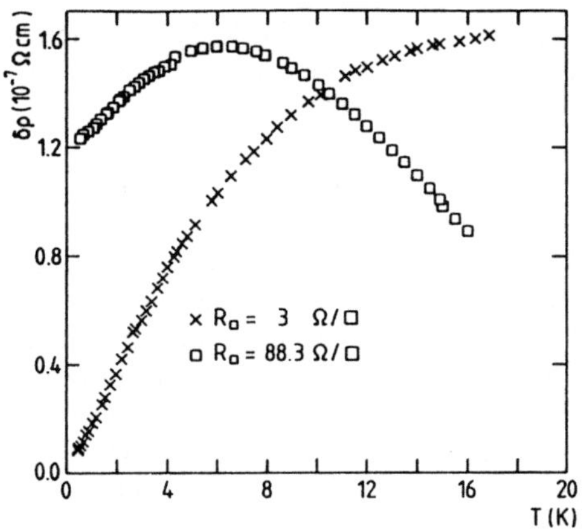

Fig. 4: Temperature dependence of the resistivity for the films (a) and (c) shown in Fig. 3, after subtraction of the Coulomb interaction effects

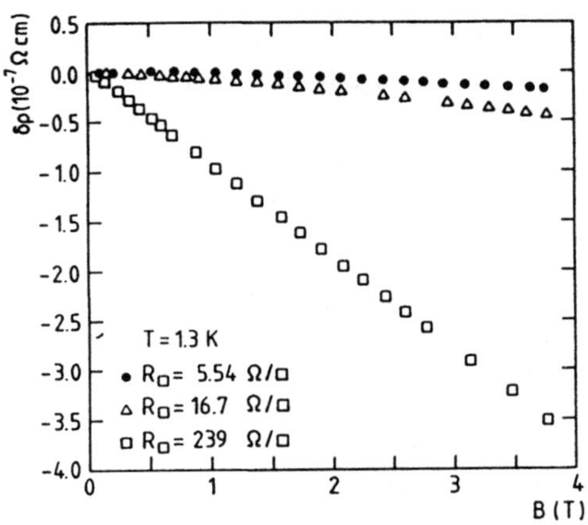

Fig. 5: Magnetoresistivity for CuCr films ($c_{Cr} \simeq 1.0$ at%) with different thickness

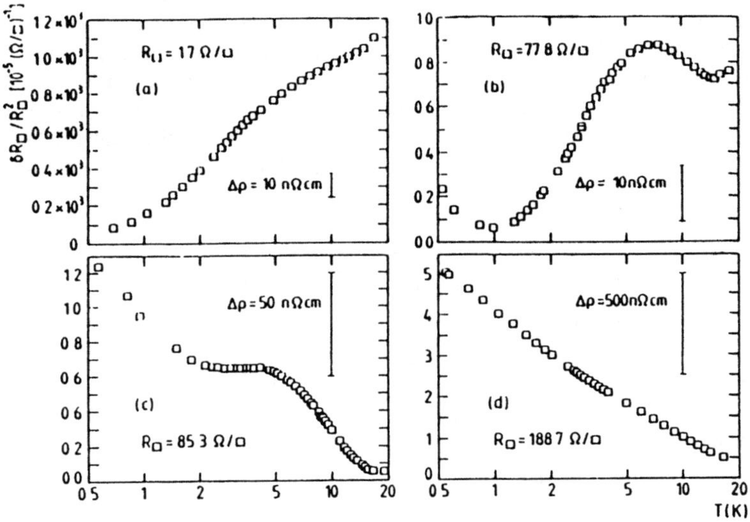

Fig. 6: Temperature dependence of the resistivity in AgMn films ($c_{Mn} \simeq 0.75$ at%) with different thickness

tude of the magnetoresistivity for the thinnest film can not be explained by the Kondo effect. A detailed analysis shows that this magnetoresistance which is larger in a perpendicular field than in a parallel field, can be perfectly described by WEL ! On the other hand, the isotropic negative magnetoresistivity observed for the thinnest film in Fig. 5, can only be explained by the Kondo effect.

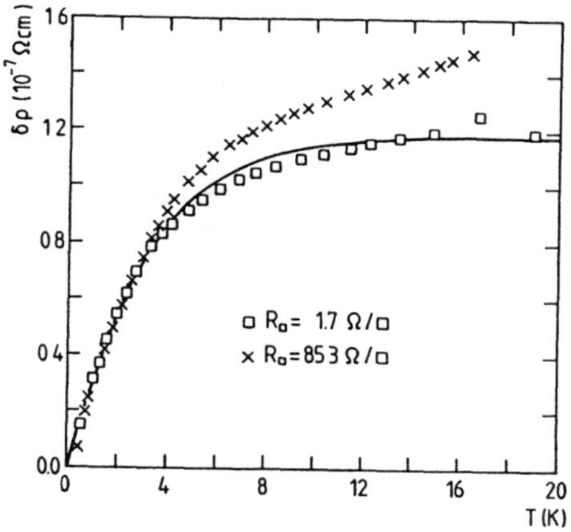

Fig. 7: Temperature dependence of the resistivity for films (a) and (c) shown in Fig. 6, after subtraction of the Coulomb interaction effects

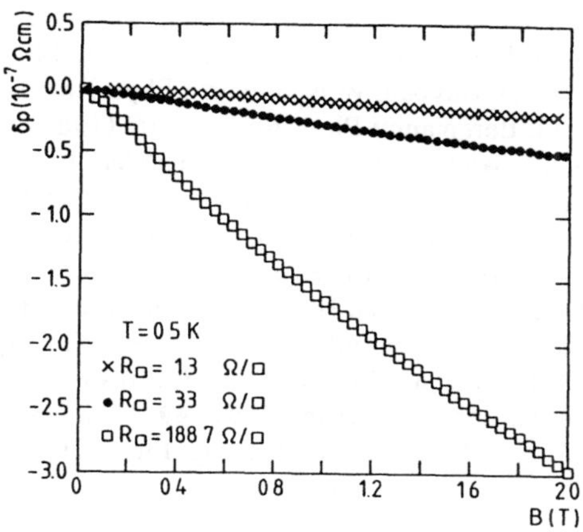

Fig. 8 Magnetoresistivity for AgMn films ($c_{Mn} \simeq 0.75$ at%) with different thickness

5. Conclusions

We have developed a flash evaporation method to prepare homogeneous films of dilute magnetic alloys with a thickness t as small as 4.0 nm. The concentration c of active magnetic scatterers can be determined within 10 %, so that a quantitative comparison with theory is possible.

In very thin CuCr films ($t \simeq 4.0$ nm, $c_{Cr} < 100$ ppm), measuring the low field magnetoresistance caused by the weak electron localization (WEL), allows to calculate the temperature dependence of the magnetic scattering time τ_m. Our experiments clearly indicate that the destruction of the WEL is dominated by the resonant spin-flip scattering in the vicinity of the Kondo temperature T_K. Even when $c \sim 1$ at%, the WEL method can still be used in the thinnest films, provided the single impurity Kondo effect is not too strong. In the thicker films ($t > 10.0$ nm), the bulk properties are recovered for CuCr and AgMn alloys. The isotropic magnetoresistivity and the temperature dependence of the resistivity are dominated by the magnetic scattering. A quantitative description of our experimental results requires, however, that the damping of the RKKY interaction by the elastic defect scattering and the disorder enhanced Coulomb repulsion are taken into account. Finally, our experiments do not indicate the presence of a dimensional crossover in the spin-glass behavior. We also find that the Kondo effect is not affected by the finite film thickness.

Acknowledgements We are much indebted to M. Gijs, P. Santhanam, C. Umbach and R. Webb for helpful discussions. We would also like to thank the Belgian I.I.K.W. for financial support. One of us (C.V.H.) is Bevoegdverklaard Navorser of the Belgian N.F.W.O. .

References

1. G. Gruner, A. Zwadowski : Rep. Prog. Phys. **37**, 1497 (1974)
2. J.A. Mydosh, G.J. Nieuwenhuys: In *Ferromagnetic materials*, ed. by E. P. Wolfarth, Vol.1 (North Holland, Amsterdam 1980) p. 71
3. J.S. Schilling: Adv. Phys. **28** , 657 (1979)
4. P. G. de Gennes : J. Phys. Radium **23**, 630 (1962)

5. P.F. de Chatel: J. Magn. Magn. Mater. **23**, 28 (1981)
6. H. Fukuyama : J. Phys. Soc. Jpn. **51**, 1105 (1982)
7. M.D. Daybell, W.A. Steyert : Phys. Rev. Lett. **20**, 195 (1968)
8. G. Bergmann : Phys. Rep. **1**, 107 (1984)
9. S. Hikami, A.I. Larkin, Y. Nagaoka : Prog. Theor. Phys. **63**,707 (1980)
10. C. Van Haesendonck, M. Gijs, Y. Bruynseraede : In *Localization, Interaction and Transport Phenomena,* ed. by B. Kramer, G. Bergmann, Y. Bruynserade , Springer Ser. Solid-State Sci., Vol. 61 (Springer, Berlin, Heidelberg 1985) p.221
11. R.P. Peters, G. Bergmann, R.M. Mueller : Phys. Rev. Lett. **58**, 1964 (1987)
12. C. Van Haesendonck, J. Vranken, Y. Bruynseraede : Phys. Rev. Lett. **58**, 1968 (1987)
13. U. Larsen : Phys. Rev. B **18**, 5014 (1978)
14. U. Larsen : J. Appl. Phys. **49**, 1610 (1978)
15. E.D. Ramos : J. Low Temp. Phys. **20**, 547 (1975)

Effect of an AC Electric Field on Localization in Thin Metal Films

J. Liu and N. Giordano

Department of Physics, Purdue University,
West Lafayette, IN 47907, USA

1. INTRODUCTION

There has been a good deal of work on the transport properties of disordered systems in recent years. One of the central themes of this work is the crucial importance of phase coherent effects. Effects such as weak localization [1,2], the Aharonov-Bohm effect in multiply connected geometries [3], and universal conductance fluctuations [4] can all be viewed as arising from the interference of electron waves. Processes which destroy or otherwise affect the phase memory of the electrons can thus have a profound influence on transport in these systems. Electron inelastic and spin-spin scattering are well known processes which destroy the phase memory, and both contribute to the so-called phase breaking time, τ_ϕ [1,2]. In addition, it is well established that the application of external fields can also affect phase coherence. The best studied example of this situation involves the magnetoresistance. Because the electron paths are random in size and shape, a magnetic field applied perpendicular to the plane of the path will make an essentially random contribution to the phases of different electrons, and thus destroy the coherent backscattering which gives rise to weak localization [1,2].

Several years ago it was pointed out theoretically [5,6] that an AC electric field could also disrupt the phase coherence. The basic idea is quite simple; the electric field does work on each electron thereby affecting its phase. Since the electron trajectories are essentially random walks, electrons which follow different trajectories will experience effectively random phase changes, and thus phase coherent effects will be quenched. Note that this is *not* the same as simple Joule heating of the electrons; it is essential that the electric field vary with time. The magnitude of the effect produced by an AC field is predicted to depend on the strength and frequency of the field, and it turns out the the effect becomes significant when the frequency, ω, matches the phase breaking time; that is, when $\omega\tau_\phi \sim 1$. Since τ_ϕ is typically of order $10^{-9}-10^{-10}$s (see e.g. [1,2,6,7]), this means that frequencies of order 10^9Hz are required. In this paper we report the observation of this effect in metal films.

2. EXPERIMENTAL METHOD

The experiments described below all employed thin Ag films. The films were prepared by DC sputtering with Ar, and had resistivities of $\sim 25\mu\Omega-$cm. The resistance measurements were performed using standard low frequency (100Hz) bridge techniques. The samples were situated in a resonant cavity such that the electric field was directed in the plane of the film. The cavity was immersed directly in (and filled with) liquid ^{4}He. The main resonant frequency of the cavity was near 8×10^9Hz, and most of the measurements were performed at that frequency. There were also several smaller resonances at nearby frequencies, and similar results were obtained at those frequencies. The microwave source was coupled to the cavity via semi-rigid coaxial cable, whose losses were tolerable at these frequencies. The relative strength of the microwave field was measured using a directional coupler and a diode detector, while the absolute calibration was obtained using measurements described below.

3. RESULTS

Some typical results for the conductance as a function of the strength of the AC field, E, for several different samples at 4.2K are shown in Fig. 1. It is seen that the behavior depends systematically on the value of the sheet resistance, $R_\square$. To interpret the results in Fig. 1 it is important to note that for a metal with a relatively high atomic number, such as Ag, the spin-orbit scattering is strong. This leads to anti-localization [1,2]; i.e., localization yields a *positive* contribution to the conductance, and this contribution becomes larger as the temperature is decreased.

According to the theory, an AC electric field acts as an additional source of phase breaking, and hence should also quench localization. Thus, we would expect that in Ag (and any other system in which anti-localization is dominant) the conductance should become smaller as the magnitude of the AC field is increased. That is, the behavior should be qualitatively similar to that seen when a magnetic field is applied [2]. From Fig. 1, we see that this is precisely what is observed for the samples with $R_\square$ near 100Ω. Samples with much larger values of $R_\square$ display a conductance which increases with E, corresponding to simple Joule heating. It is important to emphasize that the behavior seen for the samples with $R_\square \sim 100\Omega$, i.e., a conductance which decreases as E in increased, *cannot* be caused by Joule heating, since in the limit of small applied fields the conductance increases as the temperature is increased. To the best of our knowledge, the only process which can cause such behavior is the AC effect predicted by Al'tshuler and co-workers [5,6].

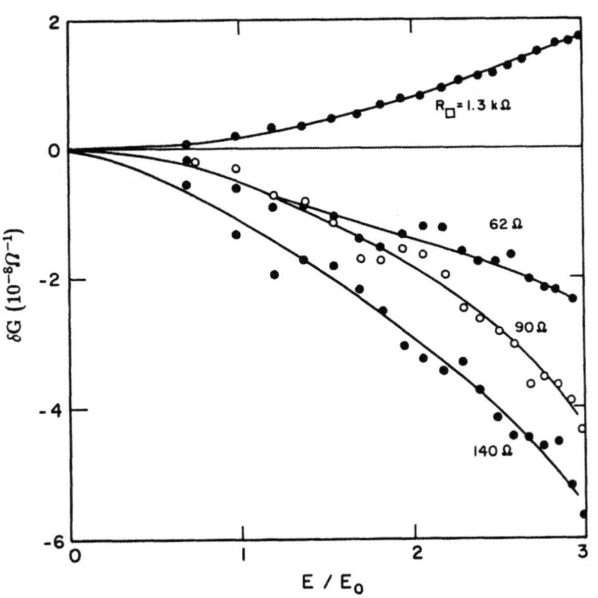

Fig. 1. Conductance change, δG, as function of the strength of the AC electric field at 4.2K for several different samples. The normalizing field here and in Fig. 2, E_o, is $\sim$30V/m. The sheet resistances of the samples are indicated in the figure. The lines are guides to the eye.

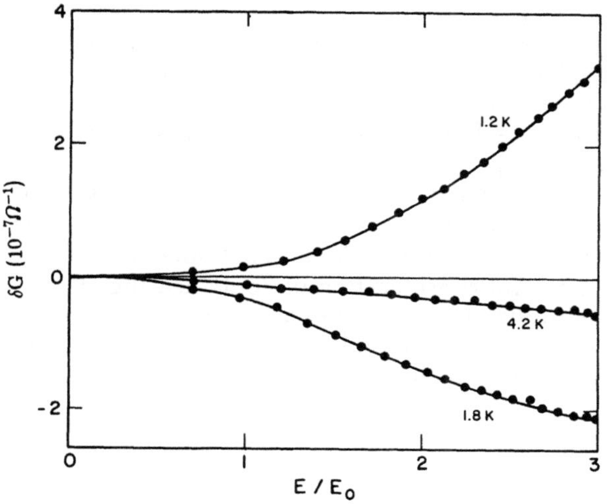

Fig. 2. Conductance change, δG, as a function of the strength of the AC electric field at several different temperatures for a sample with $R_\square=140\Omega$. The lines are guides to the eye.

Figure 2 shows results for a sample with $R_\square=140\Omega$ at several different temperatures. As the temperature is reduced below 4.2K, the AC effect (i.e., the decreasing G as a function of E) first becomes larger. However, at the lowest temperatures Joule heating dominates, and G increases with E. Hence, it is clear that both the AC effect and Joule heating are important for this sample. The magnitudes of the two effects are controlled in large part by the phase breaking time, τ_ϕ, and the electron-phonon scattering time, τ_{ep}, respectively. The phase breaking is due predominantly to electron-electron scattering in these systems [1,6,7], and hence $\tau_\phi \sim T^{-1}$. However, τ_{ep} has a much stronger temperature dependence, typically T^{-p} with $p\sim3$ or 4, depending on the details of the scattering process [8]. Hence, we would expect Joule heating to become relatively more important as the temperature is reduced, in qualitative agreement with the results in Fig 2.

While Joule heating is in some sense a nuisance in these experiments, it can be used to provide a calibration of the magnitude of the AC electric field. For samples in which the Joule heating dominates, one first measures the conductance (at some fixed temperature) as a function of the high frequency field. One can then measure G, in the absence of an AC field, as a function of the low frequency (100Hz) measuring field, which is trivial to calibrate. By comparing the two heating curves (i.e., G as a function of both the high and low frequency fields) one can get an absolute *in situ* calibration of the strength of the high frequency AC field. We have performed a calibration in this way and the result obtained is consistent with room temperature measurements of the field in the cavity, and our best estimates of the Q of the cavity at low temperatures.

A detailed comparison of our results with the theory is complicated by several facts. First, for the general case the theory cannot be evaluated in closed form, but can be expressed only in terms of a complicated definite integral. Second, the theoretical prediction is a sensitive function of τ_ϕ, which has not yet been measured for our samples. Nevertheless, if we assume that electron-electron scattering is the dominant source of phase breaking (as is commonly found in other similar films), then we can use the theory [1,6,7] to estimate τ_ϕ. This yields find $\tau_\phi \approx 9\times10^{-11}$s, for $R_\square=140\Omega$ at 4.2K. We have evaluated the theoretical prediction [9] for this value of τ_ϕ, and find $\delta G\sim6\times10^{-7}\Omega^{-1}$ for E=100V/m. This is about a factor of 10 larger than seen experimentally (compare with Fig. 1). Given the uncertainties in all of the parameters involved, together with the fact that Joule heating may not be completely negligible, we feel that this is reasonable agreement. In particular, we note that changing τ_ϕ by only a factor of two would bring theory and experiment into perfect agreement.[10] We should also note that the behavior seen at high $R_\square$ in Fig. 1 is also well accounted for by the theory. As $R_\square$ is increased, τ_ϕ becomes smaller [6,7] and this in turn makes the AC effect smaller. Hence, for the sample in Fig. 1 with $R_\square$=1.3kΩ, the AC effect is negligibly small, and Joule heating dominates. A detailed analysis of the behavior seen in Fig. 1 will require a careful consideration of Joule heating effects, along with knowledge of τ_ϕ. Magnetoresistance measurements are in progress to determine τ_ϕ for these films, so it should be possible to make a much more detailed comparison with the theory in the near future.

4. DISCUSSION

We have described experiments in which the effect of an AC electric field on localization in two dimensions has been observed. Our results are in reasonably good agreement with the theory [5,6]. The frequency-dependent behavior of three-dimensional systems near the metal insulator transition has been the focus of several recent studies [11]. This behavior is quite interesting, but involves somewhat different physics than does the two-dimensional case considered here. Previous work in this area in two dimensions has been limited to a reported null result for very thin films of Cu and Nb [12], and an investigation of an inversion layer sample [13]. In the latter experiment the behavior in zero field was anomalous, so it is hard to draw any firm conclusions from that work. Thus, there are not really any other experiments which can be compared to our results. It will be interesting to extend our measurements to other materials, including inversion layers, and to higher frequencies.

ACKNOWLEDGMENTS

We thank D. E. Beutler, B. C. Crooker, T. L. Meisenheimer, and G. Ascarelli for many useful discussions. This work was supported in part by the National Science Foundation through grant DMR-8614862.

1. P. A. Lee and T. V. Ramakrishnan, Rev. Mod. Phys. **57**, 287 (1985).
2. G. Bergmann, Phys. Rep. **107**, 1 (1984).
3. S. Washburn and R. A. Webb, Adv. Phys. **35**, 375 (1986).
4. P. A. Lee and A. D. Stone, Phys. Rev. Lett. **55**, 1622 (1985); B. L. Al'tshuler, Pis'ma Zh. Eksp. Teor. Fiz. **41**, 530 (1985) [JETP Lett. **41**, 648 (1985)].
5. B. L. Al'tshuler, A. G. Aronov, and D. E. Khmel'nitzkii, Solid State Comm. **39**, 619 (1981).
6. B. L. Al'tshuler, A. G. Aronov, and D. E. Khmel'nitzkii, and A. I. Larkin, in *Quantum Theory of Solids,* edited by I. M. Lifshitz (Mir Publishers, Moscow, 1982).
7. B. L. Al'tshuler, A. G. Aronov, and D. E. Khmel'nitzkii, J. Phys. C. **15**, 7367 (1982).
8. D. J. Thouless, Phys. Rev. Lett. **39**, 1167 (1977).
9. Equation 3.3.6 in [6].

10. Note also that the effects of spin-orbit coupling are not included in the theory. This would be expected to reduce the magnitude of the theoretical prediction by a factor of two [1,2], which would bring it closer to our experimental results.
11. See, for example, M. A. Paalanen, T. F. Rosenbaum, G. A. Thomas, and R. N. Bhatt, Phys. Rev. Lett. **51**, 1896 (1983); R. J. Deri and T. G. Castner, Phys. Rev. Lett. **57**, 134 (1986).
12. M. E. Gershenzon and V. N. Gubankov, Pis'ma Zh. Eksp. Teor. Fiz. **34**, 32 (1981) [JETP Lett. **34**, 30 (1981)]. It appears that the measurement resolution in this experiment was insufficient to observe the AC effect.
13. S. A. Vitkalov, G. M. Gusev, Z. D. Kvon, and G. I. Leviev, Pis'ma Zh. Eksp. Teor. Fiz. **43**, 145 (1986) [JETP Lett. **43**, 185 (1986)].

Transition from Strong to Weak Localization in Ultrathin Pd Films

J. Kästner[1] *and H. v. Löhneysen*[2]

[1]2. Physikalisches Institut der RWTH Aachen, D-5100 Aachen, Fed. Rep. of Germany*

[2]Physikalisches Institut der Universität Karlsruhe, D-7500 Karlsruhe, Fed. Rep. of Germany

We report on measurements of the electrical resistance $R_\square$ of ultrathin Pd films quench-condensed at 16 K onto quartz substrates in a UHV cryostat and subjected to various annealing stages at temperatures $T_a < 300$ K. With the annealing of the films, $R_\square$ irreversibly decreases to less than a tenth of the initial value. Apparently, initially isolated Pd islands grow and become progressively interconnected with the annealing due to Pd surface diffusion until a 2D percolating cluster is formed, which shows metallic conduction influenced by weak localization.

1. INTRODUCTION

Weak-localization effects have been observed in thin Pd films by a number of authors /1,2/. A "transition" to strongly localized, i.e. insulating behavior can be found when the film becomes very thin. In previous experiments, this transition has been observed for different films where the sheet resistance $R_\square$ was the control parameter /3/. $R_\square$ was varied by unspecified alterations of evaporating conditions and/or growth morphology of the films, and/or thickness. In the present paper we report on the transition from strong to weak localization occurring in single films of Pd which are quench-condensed at cryogenic temperatures and where the transition is achieved by thermal treatment at temperatures up to room temperature.

2. EXPERIMENTAL

The material for the films was evaporated in a UHV cryostat (background pressure $< 10^{-8}$ mbar) from a resistively heated Pd wire (purity 5N) and condensed onto crystalline quartz substrates held at temperatures between 4.2 K and 16 K. The average thickness as measured with a quartz oscillator was about 2 nm for all films. The electrical resistance of the films was measured with a high resolution ac bridge. The films were annealed at various temperatures $T_a < 300$ K. Annealing time was ~ 1 h for $T_a \leq 65$ K and ~ 8 h for $T_a = 120$ K and 265 K. During each annealing step at T_a the resistance decreased irreversibly. Subsequently, the resistance could be traced reversibly up to T_a.

For the sake of clarity and brevity, we report results of one film only. Similar results have been obtained for the other films.

3. RESULTS AND DISCUSSION

Figure 1 shows the sheet resistance $R_\square$ of a Pd film vs. temperature on a log-log plot for different annealing temperatures T_a. The first point to note is the very large irreversible change accomplished by annealing: the value of $R_\square$ (1.5 K) ~

*Present address: CryoVac GmbH & Co KG, D-5210 Troisdorf, FRG

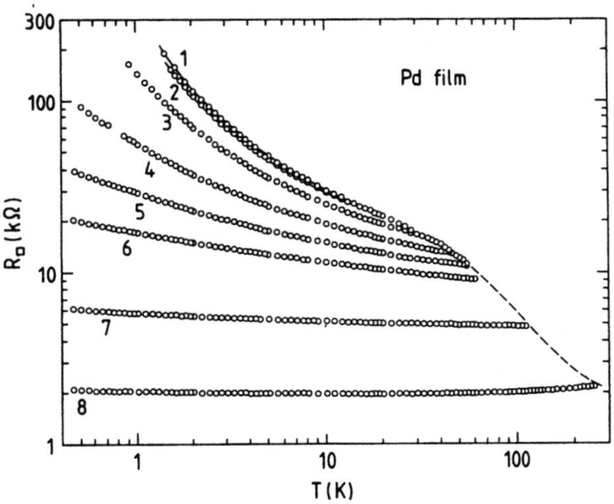

Fig. 1
Electrical sheet resistance $R_{\square}$ vs. temperature T of a Pd film condensed at 4.2 K and subjected to various annealing steps at temperatures T_a; curve 1: T_a = 16 K (as deposited); 2: T_a = 20 K; 3: T_a = 30 K; 4: T_a = 45 K; 5: T_a = 55 K; 6: T_a = 65 K; 7: T_a = 120 K; 8: T_a = 265 K

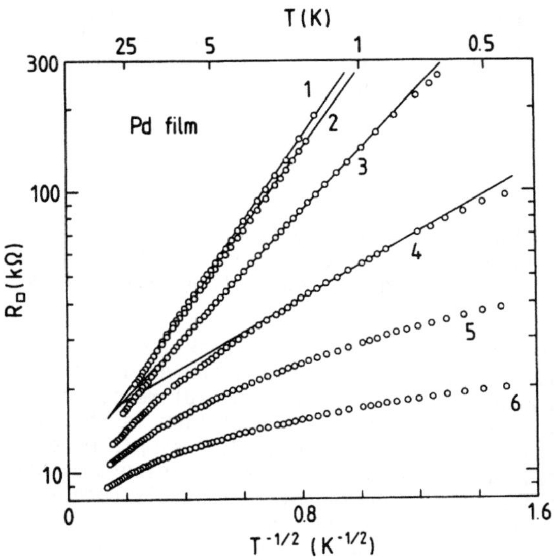

Fig. 2
Data of Fig. 1 (annealing steps 1 to 6) plotted as log $R_{\square}$ vs. $T^{-1/2}$

200 kΩ obtained for the as-deposited film decreases by a factor of 10^2 upon annealing at 265 K. After each intermediate annealing step at a temperature T_a, $R_{\square}(T)$ changes reversibly below T_a. In the following, we shall only consider these reversible $R_{\square}(T)$ curves.

The low-T_a films show a very strong temperature dependence of $R_{\square}$. As can be seen from Figure 2, the data for T_a < 30 K can be represented by $R_{\square} = R_0 \exp(AT^{-1/2})$ over almost an order of magnitude in T. Because of the exponential temperature factor, this resistance behavior can be regarded as that of strongly localized electrons. We note that the $T^{-1/2}$ dependence is expected for electron tunneling between metal grains in granular films /4/ and has been found before in very thin films of Au, Cu, Al, and Sn over a more limited temperature range /5/, where apparently tunneling between islands of the condensed metal occurs.

The temperature dependence for intermediate T_a appears to be rather complex, we therefore discuss first the high T_a-films ($T_a \geq 120$ K). As exemplified in Figure 3

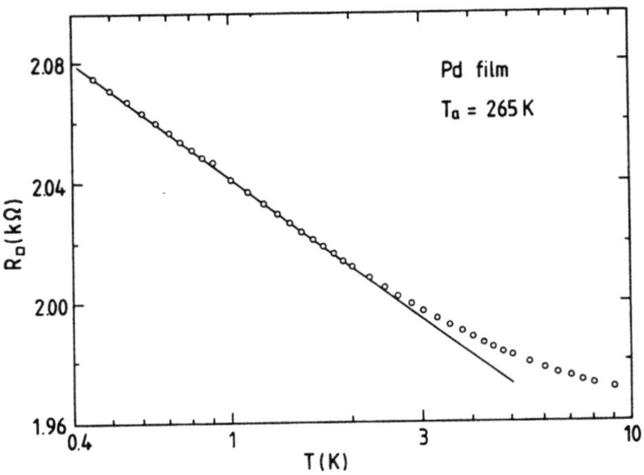

Fig. 3
Resistance $R_\square$ vs. log T of the Pd film annealed at T_a = 265 K.

for T_a = 265 K, the resistance varies as $\Delta R_\square \sim -B \ln(T/T_0)$ which is the familiar behavior for weak localization and/or Coulomb interactions of electrons in 2D films /6/. Weak localization in Pd films has been studied by a number of authors, notably by magnetoresistance measurements /1,2/. Quantitative agreement with weak-localization theory seems to be lacking, which is attributed to the nearly ferromagnetic character of Pd /2/. The coefficient C in the expression

$$\frac{\Delta R_\square}{R_\square^2} = -C \frac{e^2}{2\pi^2 \hbar} \ln \frac{T}{T_0}$$

where T_0 is an arbitrary reference temperature, as determined from our experimental data, is the same for the T_a = 120 K and 265 K films, C = 0.84. This is of the magnitude C = 1 expected theoretically /6/, indicating that our well annealed films show a quasi-uniform 2D resistance behavior.

We now turn to the discussion of the transition from strong to weak localization with increasing T_a. Our data (Fig. 1) indeed show that this transition occurs rather smoothly. For instance, there is no temperature T_a at which $R_\square$ or $dR_\square/dT$ measured at low temperatures show a pronounced feature. An intriguing interpretation is that a percolation - localization crossover /7/ occurs as T_a is increased. A simple model would be that small and initially isolated Pd islands with electron transport occurring only through tunneling, gradually become interconnected by Pd channels because surface diffusion of Pd sets in when the films are annealed. In the films of the intermediate T_a region, different transport processes may be present in the irregular 2D network: tunneling as discussed before and "metallic" transport through 1D channels, having a stronger T dependence than 2D weak localization. Close to percolation, localization on a fractal may also become important. When a percolating metallic cluster is formed after annealing at a sufficiently high T_a, 2D weakly localized resistance behavior is observed.

Although no electron-microscopic study of quench-condensed Pd films is available, our interpretation is compatible with a study performed on thin (< 3 nm) quench-condensed films annealed at room temperature which shows a continuous structure /8/. Resistivity measurements on these films appear to be in qualitative accord with our data.

In conclusion, our experiments show the transition from strong to weak localization in a single film, the only parameter being the annealing temperature. The data suggest a gradual growth and progressive interconnection of initially isolated Pd islands with annealing until a 2D percolating cluster is formed, which shows metallic conduction influenced by weak localization and eventually also by Coulomb interaction effects.

ACKNOWLEDGEMENT

This work was supported by Deutsche Forschungsgemeinschaft through Sonderforschungsbereich 125 Aachen-Jülich-Köln.

REFERENCES

1. R. S. Markiewicz, C. J. Rollins: Phys. Rev. B 29, 735 (1984)
2. G. Bergmann, G. Mathon: Solid State Commun. 55, 512 (1985)
3. M. J. Burns, W. C. Mc Gimms, R. W. Simon, G. Deutscher, P. M. Chaikin: Phys. Rev. Lett. 47, 1620 (1981)
4. B. Abeles, Ping Sheng, M. C. Coults, Y. Arie: Adv. Phys. 24, 407 (1975)
5. R. C. Dynes, J. P. Garno, J. M. Rowell: Phys. Rev. Lett. 40, 479 (1978)
6. For a review, see G. Bergmann, Phys. Rep. 107, 1 (1984)
7. A. Palevski, G. Deutscher: Phys. Rev. B 34, 431 (1986)
8. G. Dumpich, H. Kristen, E. F. Wassermann: Z. Physik B 51, 251 (1983)

Electron Transport Properties of a 100 Å Crystalline In_2O_{3-x} Film

M. Ben-Shlomo and R. Rosenbaum

School of Physics and Astronomy, Raymond and Beverly Sackler Faculty of Exact Sciences, Tel-Aviv University, Ramat-Aviv, 69978, Israel

1. INTRODUCTION

Crystalline In_2O_{3-x} is an interesting material. Thin films can be prepared by evaporating In_2O_3 in a partial oxygen atmosphere of $5x10^{-4}$ mm/Hg upon a heated 160 C substrate. The evaporation rate and partial oxygen pressure determine the electron carrier concentration per unit volume N. For the 100 A film studied here, N is on the order of $9x10^{19}/cm^3$ and is smaller by a factor of one thousand compared to the concentration found in most metals. The small value of N has important consequences: a) the plasma frequency falls in the near infrared, and these films are transparent in the visible; b) since the Hall voltage scales inversely with N, one can measure the Hall voltage with sufficient accuracy in order to detect any possible temperature dependence in the Hall constant data R_H; and c) the electron screening constant F takes on a rather large value close to one. The small value of N gives rise to less electron screening and, hence, to larger electron-electron interaction contributions to these transport measurements. The interaction processes in In_2O_{3-x} contributed to 1) a ln(1/T) temperature dependence in the Hall constant data R_H, 2) a negative magnetoconductance to the high field, low temperature, magnetoconductance data, and 3) a major contribution to the resistance versus temperature data in zero and finite magnetic fields. However, weak localization effects were also important. Weak localization dominated the low field magnetoconductance data and also gave an important contribution to the zero-field resistance data. Since both weak localization and electron-electron interaction effects are important in In_2O_{3-x}, the 2d theories for these processes are briefly summarized.

2. THEORETICAL BACKGROUND

2.1 2d Weak Localization Theory Using Perpendicular Magnetic Fields

For thin 2d films exhibiting weak localization (WL), the general case of perpendicular magnetoconductance (MC) was derived by HIKAMI, et al. /1/ and further studied by MAEKAWA and FUKUYAMA /2/. BERGMANN /3,4/ was one of the first to extract spin-orbit τ_{so}, magnetic scattering τ_s, and inelastic scattering $\tau_i(T)$ times using this theory. The reader is referred to reference /5/ for the WL expressions for the magnetoconductance, $\Delta\sigma_{\square,WL}$, and for the conductivity $\sigma_{\square,WL}$.

If weak localization effects only are present, FUKUYAMA /6/ has shown theoretically that weak localization does not contribute any temperature dependence to the Hall constant R_H. OVADYAHU and IMRY /7/ and recently TOUSSON and OVADYAHU /8/ have published strong experimental evidence that there is no temperature dependence of R_H in In_2O_{3-x} films measured in very small magnetic fields of 0.05 Tesla or less; such small fields do not suppress the weak localization processes. There is no theory yet on the temperature dependence of the Hall constant when both weak localization and electron-electron interaction effects are present simultaneously.

2.2 2d Electron-Electron Interaction Theory

Electron-electron interactions play a dominant role in metallic systems which have small electron carrier concentrations $N = 1/eR_H$ since there is less screening of the

Coulomb potential as N decreases. All the interaction formulae scale with a single parameter, the electron screening constant F. According to ALTSHULER, et al. /9,10/, $F \cong 1-4 l_s k_F/\pi$ for $l_s k_F<1$ where k_F is the Fermi vector = $(3\pi^2/eR_H)^{1/3}$ and l_s is the electron screening length = 0.0154 $R_H^{-1/6}$. The theoretical screening constant F is related to the experimental 2d screening constant F_{2d} through the expression $F_{2d} = 8(1+F/2)\ln(1+F/2)/F - 4$, /14/. For In_2O_{3-x} where N is small and on the order of $9\text{x}10^{19}/\text{cm}^3$, F is predicted to be 0.7.

The condition for using the 2d interaction formulae is that the interaction length $l_T = (D\hbar/k_BT)^{1/2}$ be greater than the film thickness d. For a 100 Å film, temperatures below 10 K satisfy the 2d criterion. There are two electron-electron interaction contributions. One comes from an orbital effect of the magnetic field and has been discussed by LARKIN /11/, ALTSHULER et al. /12/, and FUKUYAMA /13/. This orbital contribution was negligible as our film exhibited no superconducting transition. The second interaction contribution, arising from spin splitting of conduction electron energies, plays the dominant role in the transport properties of In_2O_{3-x}. The theory has been developed by ALTSHULER, et al. /9,10/; the most recent developments appear in the review paper of LEE and RAMAKRISHNAN /14/. The reader is referred to reference /14/ for the expressions for the MC, $\Delta\sigma_{\square,int}$ and the conductivity $\sigma_{\square,int}$.

ALTSHULER, KHMEL'NITZKII, LARKIN and LEE /10/ had made the surprising prediction that the Hall constant R_H should increase at twice the rate of the resistance increase at decreasing temperatures, provided that the electron-electron interactions are only present and that weak localization effects are either absent or are suppressed by an applied magnetic field. BERGMANN /15,16/ has elaborated upon the Hall constant expression by inserting the theoretical conductivity results of LEE and RAMAKRISHNAN /14/ into the formula, and BERGMANN /15,16/ derived and also observed the ln(1/T) temperature dependence of R_H. The experimental 2d screening constant F_{2d} is obtained from the slope of the R_H data.

3. EXPERIMENTAL DATA AND RESULTS

The experimental screening constant F_{2d} can be determined directly from the Hall constant data R_H; details can be found in the LT18 Proceedings /17/. For this film, F_{2d} = 0.66 corresponding to the theoretical screening constant F of 0.75, in good agreement with the theoretical prediction of F = 0.7.

The MC data in the He^3 temperature region is shown in Fig. 1. We have taken MC data up to 37 K but only the low temperature data is shown. At low magnetic fields, the MC data is always positive, as a result of the very weak spin-orbit scattering process; there is still no explanation for this. In the low field region below 1 Tesla, the MC increases as the temperature is decreased; this results from the temperature dependence of the inelastic scattering time $\tau_i \propto 1/T$. The low field data has been analyzed using only the WL theory of HIKAMI, et al. /1/. Values of τ_{so}, τ_s, and $\tau_i(T)$ are $\tau_{so} = 7.6\text{x}10^{-11}$ sec, $\tau_s = 7.3\text{x}10^{-11}$ sec, and $\tau_i = 133\text{x}10^{-12}/T$ in sec. The inelastic scattering time seems to be well described by the Nyquist formula of ALTSHULER, et al. /18/.

The most outstanding feature of the MC data occurs in the highest fields and can best be seen in the 0.5 K data where the MC from interactions is strong and negative. One will observe in Fig. 1 that the 0.5 K data is smaller in magnitude than the 1.07 K data, crossing the 1.07 K data already at B = 3 Tesla. The interaction MC, $\Delta\sigma_{\square,int}$, was determined by 1) calculating the MC from the WL theory at magnetic fields higher than those used in the low field fitting region, and 2) substracting from the experimental MC data the calculated WL MC contribution. The results for $\Delta\sigma_{\square,int}$ are plotted in Fig. 2 as a function of the dimensionless parameter $f = g\mu_B B/k_B T$. In Fig. 2 values of T range from 0.5 K to 4.2 K and values of B range from 1 to 5 Tesla. According to the interaction MC theory /9.10,14/, different combinations of T and B which give rise to the same value of f should yield the same value of $\Delta\sigma_{\square,int}$. Experimentally, this is not observed in Fig. 2 and might result from our inability to

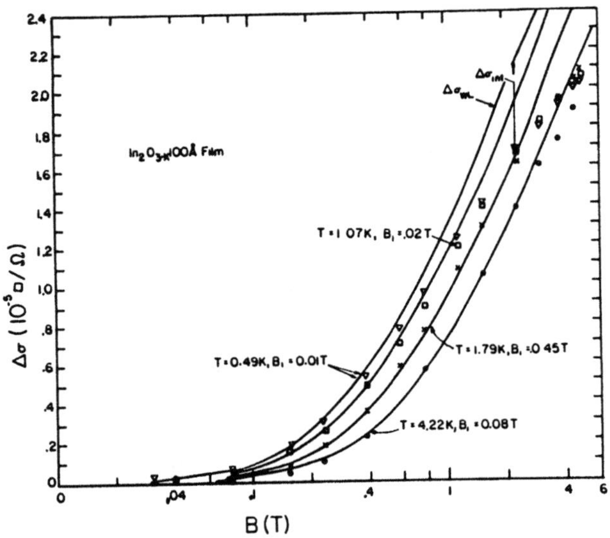

Fig. 1: MC data versus magnetic field.

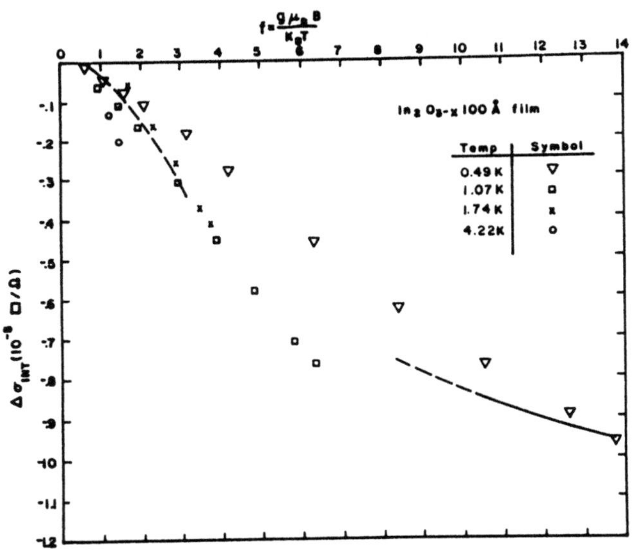

Fig. 2: Interaction MC versus the parameter $f = g\mu_B B/k_B T$.

measure the MC with sufficient accuracy to extract out the small values of $\Delta\sigma_{\square,int}$ above 1 K; our measuring accuracy in the MC data is $\pm$ 3%. Another explanation for the discrepancies might come from fitting difficulties of the WL theory to the low field MC data where more than one unique set of fitting parameters can be found to give equivalently good fits to the data. However, the 0.5 K data appears to be in good agreement with the theory where the lines appearing in Fig. 2 correspond to the small and large f limits of the MC theory /14/. The only parameter in the interaction MC theory is the 2d electron screening constant F_{2d}, known earlier from the Hall constant data to be F_{2d} = 0.66.

The resistance versus temperature (R vs T) data for zero and for a 1.95 Tesla field is shown in Fig. 3. If interactions were not present, the 1.95 T data would not have any temperature dependence since the field is sufficiently strong to suppress the

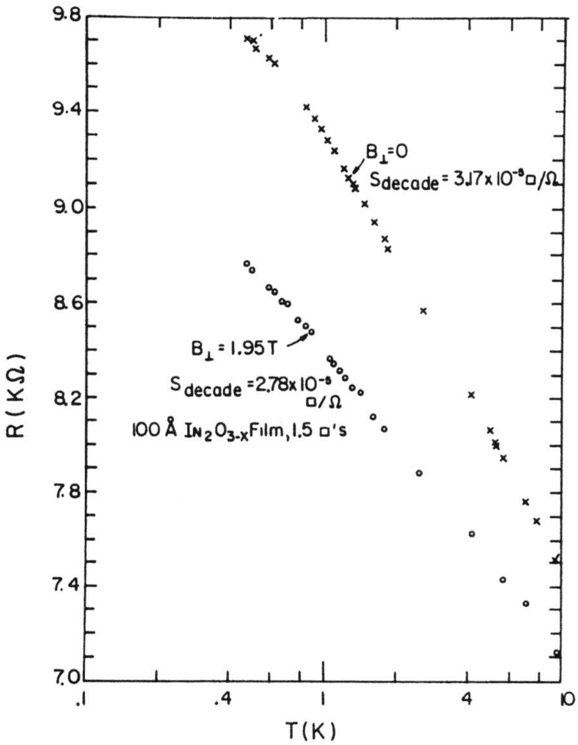

Fig. 3: Resistance versus temperature in a zero and in a 1.95 Tesla field.

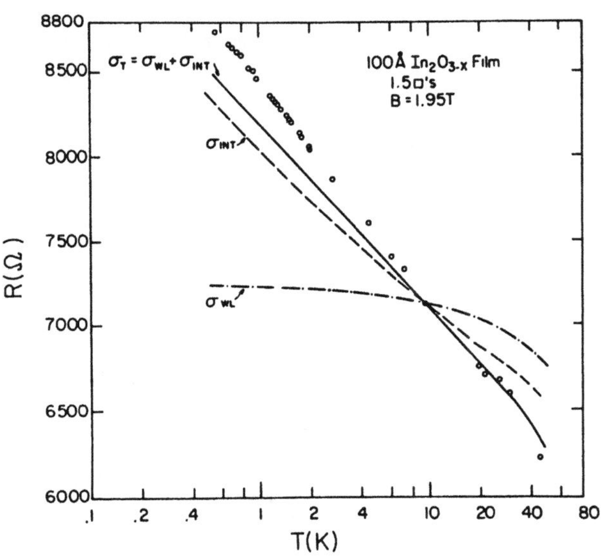

Fig. 4: Theoretical fits to the 1.95 T resistance versus temperature data.

temperature dependence of the weak localization contribution. Yet, the 1.95 T data exhibits a strong temperature dependence, arising from electron-electron interactions. Fig. 4 shows fits of the WL conductivity theory, of the interaction conductivity theory, and of the combined contribution of the two processes to the 1.95T resistance data versus temperature. From Fig. 4, the 1.95 T field suppressed the temperature dependence of the WL contribution. Interestingly, the 2d electron screening constant F_{2d} has little influence upon the fitting procedure. Again, F_{2d} = 0.66 was used.

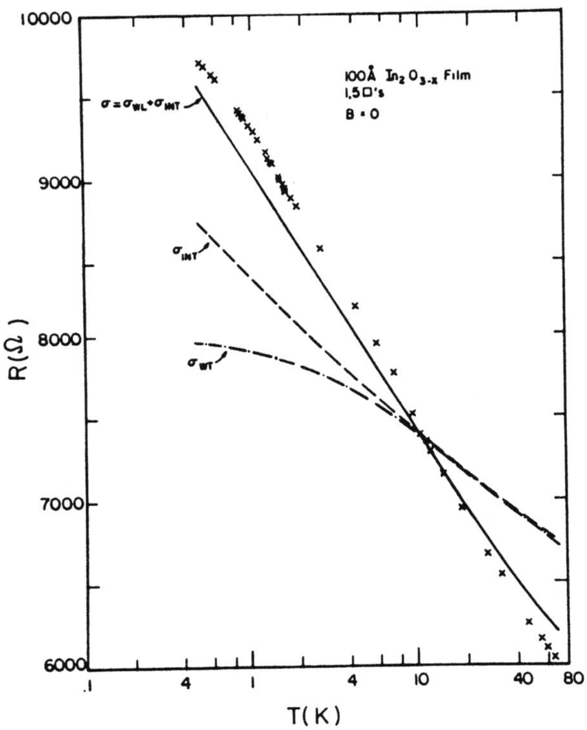

Fig. 5: Theoretical fits to the zero field resistance versus temperature data.

Fig. 5 shows fits of the WL conductivity theory, of the interaction conductivity theory, and of the combined contribution of the two processes to the zero field resistance data. The magnitude of the interaction contribution was fixed by the 2d electron screening constant, $F_{2d} = 0.66$, known from the Hall constant data. The WL theory yielded an important contribution; the three parameters in this theory, the inelastic scattering time $\tau_i(T)$, the spin-orbit scattering time τ_{so}, and the magnetic scattering time τ_s were known from the fits to the low field MC data of Fig. 1. Agreement is quite good.

4. SUMMARY

Thin non-superconducting, crystalline, In_2O_{3-x} films exhibit a beautiful and rich interplay of weak localization and electron-electron interaction effects. In_2O_{3-x} is an exciting metallic system in that the Hall constant can be measured accurately and with moderate ease, allowing for the independent determination of the electron screening constant F_{2d}. Once F_{2d} is known, the remaining transport properties arising from interactions are uniquely determined; and the interaction predictions can be accurately tested against the experimental results. For In_2O_{3-x}, the agreement is impressive.

1. S. Hikami, A. Larkin, and Y. Nagaoka, Prog. Theor. Phys. 63, 707 (1980).
2. S. Maekawa and H. Fukuyama, J. Phys. Soc. Jpn. 50, 2516 (1981).
3. G. Bergmann, Phys. Rev. Lett. 48, 1046 (1982).
4. G. Bergmann, Phys. Rev. Lett. 49, 162 (1982).
5. Y. Fehr, S. May-tal, and R. Rosenbaum, Phys. Rev. B33, 6631 (1986).
6. H. Fukuyama, J. Phys. Soc. Jpn. 49, 644 (1980).
7. Z. Ovadyahu and J. Imry, Phys. Rev. B24, 7439 (1981).
8. E. Tousson and Z. Ovahyahu, Solid State Commun. 60, 407 (1986).

9. B. L. Altshuler, A.G. Aronov, and P. A. Lee, Phys. Rev. Lett. 44, 1288 (1980).
10. B. L. Altshuler, D. Khmel'nitskii, A. I. Larkin, and P. A. Lee, Phys. Rev. B22, 5142 (1980).
11. A. I. Larkin, JETP Lett. 31, 219 (1980).
12. B. L. Altshuler, A. G. Aronov, D. E. Khmelnitskii, and A. I. Larkin, Sov. Phys. - JETP 54, 411 (1981).
13. H. Fukuyama, J. Phys. Soc. Jpn. 50, 3407 (1981).
14. P. Lee and T. Ramakrishnan, Rev. Modern Phys. 57, 287 (1985).
15. G. Bergmann, Solid State Commun. 49, 775 (1984).
16. G. Bergmann, Phys. Reports 107, 1 (1984).
17. M. Ben-Shlomo and R. Rosenbaum, "Temperature Dependent Hall Constant Data in a 100 Å Crystalline In_2O_{3-x} Film" to be published in the Proc. of LT18.
18. B. L. Altshuler, A. G. Aronov, and D. E. Khmelnitsky, J Phys. C15, 7367 (1982).

Electron Localization in Heavily Doped Polyacetylene

F. Komori, F. Okamoto, and S. Ikehata

Department of Physics, Faculty of Science, University of Tokyo, Hongo 7-3-1, Bunkyo-ku, Tokyo 113, Japan

1. INTRODUCTION

Doped polyacetylene shows insulator-to-metal transition with increasing concentration of dopants. At present, however, metallic conduction in polyacetylene has been reported only in AsF_5 doped samples [1,2], and the presence of localization effect was suggested in them. In the present paper, we will show metallic conduction of the polyacetylene films doped heavily with ClO_4: we measured the conductivity at low temperature and in magnetic field. Then we will analyze and discuss the results by the localization and interaction effects in disordered metals. Heavily doped polyacetylene is actually a disordered metal, and the weak localization can be a tool for study of the electronic state.

2. SAMPLES

Cis-polyacetylene films were synthesized by the the method of ITO et al.[3] and were thermally isomerized to trans-polyacetylene. Doping was carried out by electrochemical method [4] in the electrolysis of 0.5 M $Bu_4N^+ClO_4^-$ in $CHCl_2$. The concentration of the dopant is estimated to be 10 % by the weight uptake. The doped film was cut into a rectangular shape and the conductance was measured by conventional dc four terminal method with silver paste as electrodes. The conductivity of the film, which we call the No.1 sample, at room temperature is about 25 kS/m. In order to discuss clearly the effect of randomness on the electronic transport, we also studied a resistive film which was lightly exposed to air. The conductivity of this film (No.2) is 10 kS/m at room temperature.

3. RESULTS

The temperature dependence of the conductivity between 0.5 K and 77 K is shown in Fig. 1 for both the films. We note that the conductivity is well expressed as $A + B\sqrt{T}$ with constants A and B for the films in this temperature range. Despite the large difference between the values of A, the value of B of the No.1 is only about twice as large as that of the No.2.

In the AsF_5 doped polyacetylene [2,3], temperature dependence of resistance was observed to be logarithmic. By plotting our data in the similar way, we indeed found that the sample No.1 shows the similar dependence. Nevertheless, the temperature range of the logarithmic dependence in the resistivity, which is almost the same as in the AsF_5 samples, is narrower than that of square root in the conductivity. Furthermore in the No.2 sample, the logarithmic dependence is no longer observed. These facts suggest that the latter dependence is intrinsic in heavily doped polyacetylene at these temperatures.

Figure 2(a) shows the magnetoconductivity $\Delta\sigma(H) = \sigma(H) - \sigma(0)$ of the No.1 film at several temperatures below 4.2 K. It is positive at high temperature or in low magnetic field, and is negative at low temperature or in high field. In addition, square root dependence on the field in the negative magnetoconductance is observed

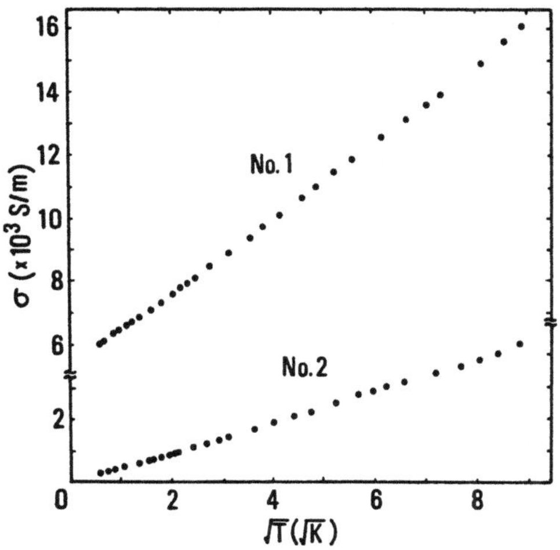

Fig. 1 Conductivity of two polyacetylene samples as a function of square root of temperature below 77 K. The No.1 sample is a clean film and the No.2 is a resistive film lightly exposed to air.

in high field below 2 K with almost the same coefficient. At the lowest temperature, however, it has a trend to saturation above 4 T. The magnetoconductance of the No.2 sample is always negative below 4.2 K as shown in Fig. 2(b). The square root dependence in high field and the trend of the saturation are also observed in this sample.

4. DISCUSSION

The square root dependence of the conductivity on temperature and magnetic field is observed in many doped semiconductors [5] and thick resistive metal films [6]. These are well explained by the localization and interaction effects in three-dimensional weakly localized metals. The experimental results of the doped polyacetylene films qualitatively resemble these effects although they have been studied as one-dimensional conductors. Because the polyacetylene film is a three dimensional network of fibrils [7], the theory of the above effects, which are insensitive to the microscopic structure, can be applied to them by regarding the whole films as three-dimensional conductors. Thus we next compare the results with the theory in detail.

4.1 THEORIES

The localization contribution to the conductivity is theoretically given [8] as

$$\sigma_L = \frac{e^2}{2\pi^2\hbar} F\left(\frac{c\hbar}{4\tau_\phi DeH}\right)\sqrt{\frac{eH}{c\hbar}} + \frac{e^2}{2\pi^2\hbar}\sqrt{\frac{1}{D\tau_\phi}} \quad . \tag{1}$$

Here, D is the diffusion constant, τ_ϕ the phase-memory time defined as $\tau_\phi^{-1} = \tau_\varepsilon^{-1} + 2\tau_s^{-1}$ with inelastic scattering time, τ_ε and spin scattering time, τ_s. The function F(x) is defined in ref. 8 and has asymptotic forms as

$$F(x) = 0.605 - 2x \qquad \text{for } x \ll 1$$
$$= x^{-3/2}/48 \qquad \text{for } x \gg 1.$$

We neglected the spin-orbit scattering in (1) because the spin-orbit interaction is small in light elements such as carbon and hydrogen. This theory can be applied when the conduction over the characteristic length, $\sqrt{D\tau_\phi}$, is three dimensional.

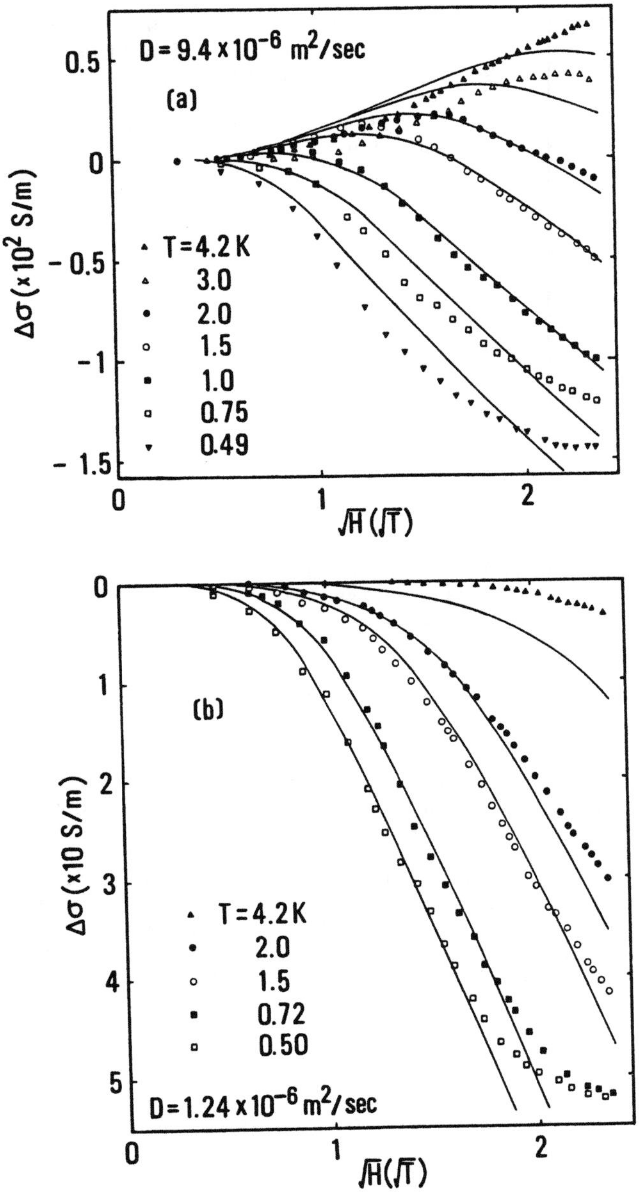

Fig. 2 Magnetoconductivity as a function of square root of magnetic field at various temperatures for the No.1 (a) and the No.2 (b) samples. Solid lines are theoretical curves of the magnetoconductivity (1,3) fitted to the experimental data with the parameters given in the text and the diffusion constants calculated from the conductivity and given in the figures.

In the interaction theory [9,10], the conductivity depends on temperature as

$$\sigma_I(T) = \frac{e^2}{2\pi^2\hbar}\, 0.46\left(\frac{4}{3} - F\right)\sqrt{\frac{k_BT}{D\hbar}} \quad . \tag{2}$$

Here, F is an average of the Coulomb interaction and take a value of the order of 1. The magnetoconductivity by this mechanism is given as

$$\sigma_I(H) = \frac{e^2 F}{4\pi^2\hbar}\sqrt{\frac{k_BT}{2D\hbar}}\, G\left(\frac{g\mu_B H}{k_BT}\right) \quad . \tag{3}$$

Here, g is g-factor, and G(x) is given in ref. 9 with asymptotic forms as

$$G(x) = \sqrt{x} - 1.3 \quad \text{for } x \gg 1$$
$$= 0.053x^2 \quad \text{for } x \ll 1.$$

We neglected the interaction contribution from particle-particle channels [10] for simplicity. The magnetoconductivity is negative and the magnitude increases with decreasing temperature. Square root dependence on H appears when H is much larger than $k_BT/g\mu_B$. This theory can be applied when the conduction over another characteristic length, $\sqrt{k_BT/\hbar D}$, is three dimensional. The extra conductivity due to both the effects can be simply added in the first order approximation.

4.2 MAGNETOCONDUCTANCE

Because the magnetoconductance gives much information on the electronic conduction, we will discuss this first. With the common values of the parameters $\tau_\phi = 10^{-10}$ sec, g=2 and F=1 in eqs. 1 and 3, we can reproduce qualitatively the experimental results of the magnetoconductance in both films as shown in Figs.2. Here, we divide the theoretical values by 4 in Fig. 2a and by 33 in Fig. 2b to compare the results in the same figures.

The difference between these two sets of theoretical curves arises from that of the diffusion constant calculated from the conductivity at 4.2 K using the free electron model. The magnitude of positive magnetoconductivity due to the localization effect depends weakly on the diffusion constant, while that of negative magnetoconductivity due to the interaction increases with decreasing the diffusion constant, ie. conductivity. This is the reason why the positive magnetoconductance appears only in the conductive film. Furthermore, by taking a reasonable value of g-factor, we can explain variation of magnetoconductance with temperature. The negative magnetoresistance observed in the AsF_5 doped films [1,2] can be also interpreted with the same mechanism; localization effect is dominant because these films are highly conductive.

The better fitting can be achieved if we allow τ_ϕ to vary. The obtained value of τ_ϕ decreases from 1.1×10^{-10} sec to 4.6×10^{-11} sec for the No.1 film and from 4.4×10^{-10} sec to 1.0×10^{-10} sec for the No.2 film with decreasing temperature. This dependence implies that the spin-flip scattering is dominant in τ_ϕ because τ_ε should increase with the decrease of temperature. The presence of the spin-flip scattering is consistent with the magnetic susceptibility which rapidly increases below 40 K while it is Pauli paramagnetic at high temperature. However, the origin of this increase is not clear.

In spite of the above success, qualitative agreement with the theory is not so satisfactory as in the case of the well studied disordered metals [5,6]; for example, the saturation of negative magnetoconductance is not expected in the theory. This can be ascribed partly to the fibril structure of the polyacetylene film [7] because microscopic disordedness reduces the range of the magnetic field where the theories can be applicable. The quantitative disagreement is also ascribed to the complicated structure; the values of the parameters D and F, which are responsible for the magnitude of the effects, can not be determined simply from the conductivity if the electron conducts inhomogeneously.

4.3 TEMPERATURE DEPENDENCE

As for the temperature dependence, the interaction effect is expected to dominate the localization because τ_ϕ is weakly dependent on T as we discussed above. Actually, $\sqrt{T}$ term is observed corresponding with (2). Quantitatively, however, the magnitude of this term in the No.1 sample is several times larger than the theoretical value calculated from (2) with D and F. To improve the agreement, we should take smaller value of D and/or F. Here, we again encounter the ambiguity of these parameters in the quantitative analysis. In the No.2 sample, the quantitative disagreement is so large that it seems difficult to explain the conduction only by the interaction effect.

5. CONCLUSION

We have shown that the electronic conduction in heavily doped polyacetylene at low temperature is qualitatively understood as the localization and the interaction effects in three-dimensional disordered metals. Here we treat the film as homogeneous three-dimensional network of metallic fibrils. The estimated values of characteristic lengths of these effects from D are several hundred A and slightly larger than the diameter of the fibril [7], suggesting consistency of the analysis. However, because the system is microscopically inhomogeneous, further understanding of the structure of the doped films should be necessary to discuss the conduction more precisely and quantitatively.

REFERENCES

1. J.F. Kwark, T.C. Clarke, R.F. Greene and G.B. Street: Solid State Commun. 31 (1979) 355.
2. C.M. Gould, D.M. Bates, H.M. Bozler, A.J. Heeger, M.A. Druy and A.G. MacDiarmid: Phys. Rev. B23 (1981) 6820.
3. T.Ito, H. Shirakawa and S. Ikeda: J. Polym. Sci. Poly. Chem, Ed.12 (1974)11.
4. P.J. Nigrey, D. Maclennes, D.P. Nairns, A.G. MacDiarmid and A.J. Heeger: J. Electrochem. Soc. 128 (1981)1651.
5. for review, Y. Ootuka and A. Kawabata: Prog. Theore. Phys. Supple. 84 (1985) 249.
6. for review, S. Kobayashi and F. Komori: Prog. Theore. Phys. supple. 84 (1985) 224.
7. J.C.W. Chien: Polyacetylene (Academic Press, London, 1984).
8. A. Kawabata: J. Phys. Soc. Jpn. 49 (1980)628.
9. P.A. Lee and T.V. Ramakrishnun: Phys. Rev. B 26 (1982) 4009.
10. Y. Isawa, K .Hoshino and H. Fukuyama: J. Phys. Soc. Jpn. 51 (1982)3262.

Low Temperature Limit of Quantum Corrections to the Transport Properties of a Disordered CuTi Alloy

P. Lindqvist and Ö. Rapp

Solid State Physics, The Royal Institute of Technology, S-10044 Stockholm, Sweden

1. INTRODUCTION

The theories of quantum corrections to classical transport properties have successfully explained several qualitative features of the resistivity and magnetoresistance in a number of glassy metal alloy systems. In many cases, however, there are notable discrepancies between quantitative results. For example, rather different values for the spin-orbit scattering time, τ_{so}, in CuTi alloys have been obtained in various investigations [1] and a range of exponents, within about a factor of 6, for the temperature dependence of the inelastic scattering (phase breaking) time τ_i in CuY alloys have been reported by different authors [2].

One factor which has contributed to such difficulties is no doubt the complexity of the theoretical formulations [3] and of the fitting procedures. Besides τ_i and τ_{so} which are often used as fitting parameters, one must also know the diffusion constant D, the resistivity ρ, the screening factor F_σ, and the net electron-phonon interaction constant λ_e, to account quantitatively for the observations. Furthermore, parameters have sometimes been introduced to scale the observed magnetoresistance such as a factor 1.2 [4] or a fitting parameter α in the range 1-1.5 [5].

In the present paper we present an initial analysis of our measurements of the resistance and magnetoresistance of some Cu-rich CuTi alloys. These measurements cover an extended temperature range down to the mK region and we can extrapolate the field dependence of the magnetoresistance to the T=0 K limit. In this limit the expressions for some of the contributions to $\rho(B)$ simplify considerably and we can compare the theoretical predictions with the observations.

2. EXPERIMENTAL DETAILS

Starting materials of nominal purities 99.9999 wt% Cu and 99.97 wt% Ti (from Materials Research Corporation, Marz grade) were repeatedly melted in an arc-furnace under reduced pressure of Ar. Ribbons were produced by melt-spinning in He gas. The alloys were found to be X-ray amorphous. The density was determined by weighing in methanol. For $Cu_{65}Ti_{35}$ the result was 7.03 g/cm^3. This permitted fairly accurate measurements of the resistivity with the result corresponding to 182±5 μΩcm at 4.2 K for this alloy.

Temperatures below 1 K were obtained in a dilution refrigerator. The magnetoresistance was measured in a superconducting solenoid with slow sweeping rates of the magnetic field. Below 100 mK this rate approached 2 hours/Tesla to avoid eddy current heating of the heat exchangers.

3. RESULTS AND ANALYSIS

3.1 Experimental Results

The magnetoresistance of $Cu_{65}Ti_{35}$ at several temperatures is shown in Fig. 1 versus $\sqrt{B}$. It can be seen that there is a field range over which the magnetoresistance is linear in $\sqrt{B}$ which increases with decreasing temperature. At 45 mK, this linearity extends from 0.15 to 4 T.

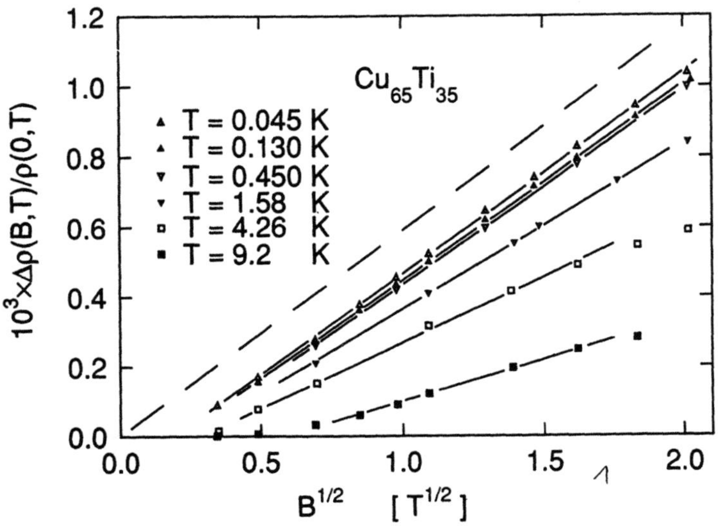

Figure 1 Magnetoresistance vs $\sqrt{B}$ at different temperatures for disordered $Cu_{65}Ti_{35}$. Equation (5) is shown by the dashed line

The zero temperature limit can be obtained by extrapolation. We find empirically that the slopes of the straight lines in Fig. 1 are approximately proportional to $\sqrt{T}$. This is illustrated in Fig. 2. Allowing also for some different extrapolation schemes, we find at T=0 K the slope $(0.62\pm0.04)\times10^{-3}$ $(\text{Tesla})^{-1/2}$ which will be compared to theoretical estimates below. From the results of HOWSON and GREIG [6] and our value of ρ, one obtains 0.64×10^{-3} $T^{-1/2}$ at 150 mK for $Cu_{65}Ti_{35}$.

The contribution to the magnetoresistance from interaction effects in the particle hole channel [3] contains the screening factor F_σ. In the low temperature limit where interaction effects dominate in the temperature dependence of the resistivity, $\rho(T)$, one can obtain F_σ from the slope of ρ vs $\sqrt{T}$. This is shown in Fig. 3. For temperatures below 150 mK we observe saturation in $\rho(T)$, which is common in many metallic glasses. With

$$F_\sigma = \frac{2}{3}\left[\frac{4}{3} + \frac{\Delta\rho}{\rho^2\sqrt{T}}\,\frac{\sqrt{2}\,4\pi^2\hbar}{1.3\,e^2}\,\sqrt{(\hbar D/k_B)}\right] \tag{1}$$

from [3], $D=(3\pm0.3)\times10^{-5}$ m^2/s from the specific heat measurements of MOODY and NG [7] and our resistivity result, and from the slope of $\Delta\rho/\rho\sqrt{T} = (-5.3\pm0.3)\times10^{-4}$ $K^{-1/2}$ in Fig. 3, we obtain $F_\sigma = 0.37\pm0.07$.

The electron-phonon interaction was neglected in the evaluation of the density of states from the specific heat results since arguments will be given below that $\lambda\approx0$ is a reasonable assumption for this alloy. Our value of F_σ is in agreement with the results of HICKEY and coworkers [8] and smaller than those obtained by SCHULTE and FRITSCH [9].

Nevertheless, caution must be exercised when using such results for F_σ. In many metal glasses where D is much larger than in CuTi, the slope of $\rho(T)$ vs $\sqrt{T}$ is similar to that in CuTi alloys, which may lead to unphysical negative values of F_σ and it is not known how accurate (1) is. Furthermore, since the temperature range over which $\rho(T)$ is linear in $\sqrt{T}$ is limited for the present alloy, it is not completely clear that we can reach the region where a $\sqrt{T}$ term dominates the resistivity. If weak localization effects must be simultaneously included, the coefficient of the $-\sqrt{T}$ term will increase, perhaps by 30%, and such a change could decrease F_σ by 80%.

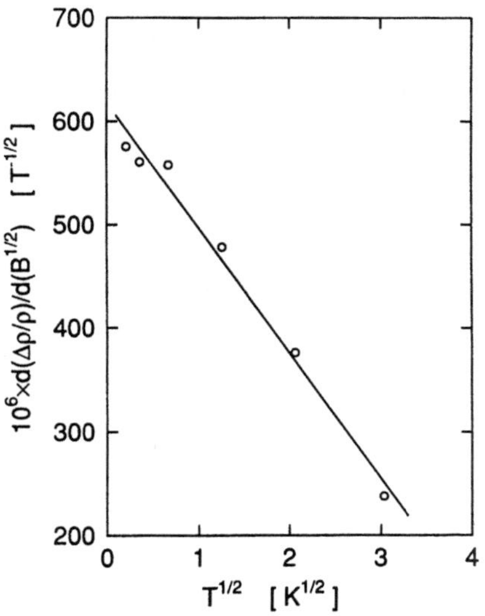

Figure 2 The slopes of the straight lines in Fig. 1 on a $\sqrt{T}$ scale

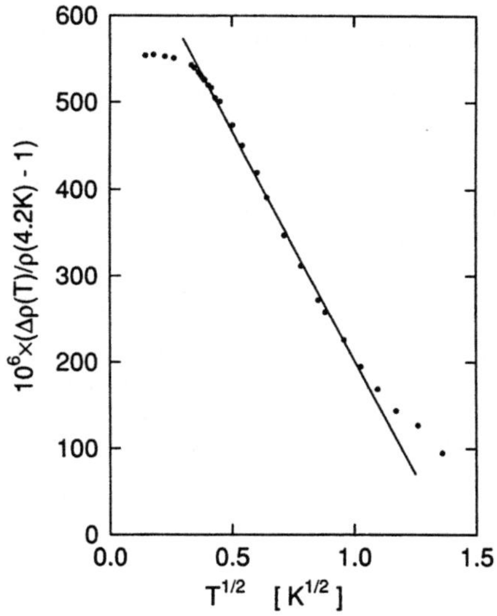

Figure 3 The resistance vs $\sqrt{T}$ for disordered $Cu_{65}Ti_{35}$

3.2 Analysis

Three contributions to the magnetoresistance will be considered [3,10,11], i.e. those from weak localization, interaction effects in the particle-hole channel, and interaction effects in the particle-particle channel. In the limit T=0 K all three contributions are proportional to $\sqrt{B}$. This follows immediately for the interaction terms and the constant values for the functions $g_3(x)$ [3] and $\varphi_3(x)$ [10] for large values of the argument. For the contribution from weak localization in the formulation of ALTSHULER and coworkers [10], it is readily seen that at low enough temperatures we always have $\tau_i \gg \tau_{so}$ and a constant value of the function $f_3(x)$ so that $\Delta\rho \sim \sqrt{B}$ for a field range which approaches B=0 when $T \to 0$. It can be noted that in the FUKUYAMA-HOSHINO theory [11], which includes spin-orbit interaction and Zeeman splitting, the magnetoresistance at T=0 K can be linear in $\sqrt{B}$ over an extended field range, provided τ_{so} has a suitable value, and that this slope can be much larger than that derived from (2) below. Equation (2) can be recovered from this theory [11] only in the limit T=0 and small B.

Collecting these results we obtain

$$\Delta\rho/\rho = \frac{0.605\, e^2}{4\pi^2 \hbar} \left(\frac{e}{\hbar}\right)^{1/2} \rho \sqrt{B} \quad \text{(weak localization)} \tag{2}$$

$$+ \frac{e^2}{4\pi 2 \hbar} \sqrt{(g\mu_B/2hD)}\, F_\sigma\, \rho \sqrt{B} \quad \text{(interaction, particle-hole)} \tag{3}$$

$$+ \frac{1.9\, e^2}{2\pi^2 \hbar} \left(\frac{e}{\hbar}\right)^{1/2} g\, \rho \sqrt{B} \quad \text{(interaction, particle-particle)}\ . \tag{4}$$

In (4), g is a coupling function which determines the sign of this contribution and is positive for net repulsive interaction and negative for net attractive interaction. The present alloy is outside of the superconducting region, implying at most a small attractive electron-phonon interaction λ. The concentration dependence of λ is not known. One may assume that glassy metals with Ti have qualitative similarities to those with Zr. In ZrCu, λ decreases faster than linearly with Zr concentration [12] and in ZrCo alloys, λ approaches 0 at the superconducting phase boundary [13]. Therefore we find it reasonable to assume that in $Cu_{65}Ti_{35}$ one has $\lambda<\mu^*$ (= the repulsive part of the interaction), that the net interaction $\lambda_e = \lambda-\mu^*$ is negative and the contribution to the magnetoresistance positive. With $\lambda \approx 0$ we obtain $g \approx 0.05$. If λ has a small non-zero value g can be somewhat smaller. Furthermore, strong spin-orbit scattering may reduce the magnitude of g and sometimes a factor 1/4 is inserted in (4). Therefore our estimate of g is probably an upper limit.

At T=0 K we thus obtain from (2), (3), and (4) the following estimates for the slope of the magnetoresistance (with B in Tesla)

$$\Delta\rho/\rho_0/\sqrt{B} = [(0.27\pm0.01) + (0.23\pm0.07) + (0.09\pm0.02)]\times10^{-3} = (0.59\pm0.07)\times10^{-3}. \quad (5)$$

This result is shown by the dashed line in Fig. 1.

There is reasonable agreement between this estimate and our experimental result of $(0.62\pm0.04)\times10^{-3}\ T^{-1/2}$. In conclusion we find that the magnetoresistance of disordered $Cu_{65}Ti_{35}$ at T=0 K seems to be quantitatively well accounted for by the sum of the contributions from weak localization in the formulation by ALTSHULER and coworkers [10] and those from interaction effects in the particle-hole and particle-particle channels.

ACKNOWLEDGEMENTS

We are grateful to L. Arnberg, The Swedish Institute for Metals Research, for his help in melt-spinning the alloy. Part of this work has been financed by Naturvetenskapliga Forskningsrådet.

REFERENCES

1. M.A. Howson, B.J. Hickey, C. Shearwood: J. Phys. F16, L175 (1986)
2. Ö. Rapp, Y.-F. Li: submitted to Phys. Rev. B
3. e.g. P.A. Lee, T.V. Ramakrishnan: Rev. Mod. Phys. 57, 287 (1985)
4. S.J. Poon, K.M. Wong, A.J. Drehman: Phys. Rev. B31, 1668 (1985)
5. J.B. Bieri, A. Fert, G. Creuzet, A. Schul: J. Phys. F16, 2099 (1986)
6. M.A. Howson, D. Greig: J. Phys. F16, 989 (1986)
7. D.E. Moody, T.K. Ng: Physica 126B, 371 (1984)
8. B.J. Hickey, D. Greig, M.A. Howson: J. Phys. F16, L13 (1986)
9. A. Schulte, G. Fritsch: J. Phys. F16, L55 (1986)
10. B.L. Altshuler, A.G. Aronov, A.I. Larkin, D.E. Khmelnitskii: Sov. Phys. JETP 54, 411 (1981)
11 H. Fukuyama, K. Hoshino: J. Phys. Soc. Japan 50, 2131 (1981)
12. Ö. Rapp: Phys. Rev. B34, 2878 (1986)
13. M. Flodin, L. Hedman, Ö. Rapp: Phys. Rev. B34 4558 (1986)

Localization of Electrons and Inelastic Relaxation Time in Metal Films

Y.F. Komnik, E.I. Bukhshtab, and A.V. Butenko

Institute for Low Temperature Physics and Engineering,
UkrSSR Academy of Sciences, Kharkov 310164, USSR

The study of quantum interference effects in low temperature conductivity of quasi-two-dimensional conducting systems, particularly of thin semimetal films /1-5/, has revealed a good agreement of experimental results with the theory of weak localization of electrons (WL) and electron-electron interaction (EEI). This allows using the WL and EEI effects as a tool for investigating the mechanisms of electron scattering in thin metal films. The magnitude of WL effect depends on the phase relaxation time of the electron wave function τ_φ which is governed by inelastic relaxation processes in metals.

At sufficiently low temperatures the main channel of inelastic scattering is the electron-electron one. Reference 6 shows that in this case

$$\tau_\varphi^{-1} = \frac{\pi kT}{\hbar} \frac{e^2 R_\square}{2\pi^2\hbar} \ln \frac{\pi\hbar}{e^2 R_\square}, \tag{1}$$

where $R_\square$ is the resistance of film square section. Experimental data for τ_φ obtained from the WL effect are described by the $\tau_\varphi^{-1} \sim T$ relation in films of bismuth /2/, antimony /5/, aluminium /7/, silver /8/. References 4 and 5 show that the numerical values of τ_φ agree with Eq. (1).

With the increase of temperature there is a transition to the relation $\tau_\varphi^{-1} \sim T^2$. At this, as shown in Ref. 5, there occurs a change of τ_φ dependence on film thickness, i.e. on $R_\square$ or τ (the elastic relaxation time). The main mechanism of inelastic scattering at such temperatures is probably the electron-phonon one. The $\tau_{eph} \sim T^2$ relation is characteristic of semimetals at low temperatures. Theoretical calculation /9/ of the phase relaxation time at electrons scattering by transverse phonons yields $\tau_\varphi \sim T^{-2}$ for "high" ($\vartheta_D \gg kT \gg \hbar S_T/l$) and $\tau_\varphi \sim T^{-4}$ for "low" temperatures ($kT \ll \hbar S_T/l$, S_T is the transverse sound velocity, l is the mean free path of electrons). The value of $\hbar S_T/l$ for Bi films is ~ 1 /K/.

If the inelastic scattering is due to both mechanisms at a time, the $\tau_\varphi(T)$ dependence has the form

$$\tau_\varphi^{-1}(T) = AT + BT^2. \tag{2}$$

The experimental study of WL effect makes it possible to use the magnetic field dependences of conductivity for directly determining the value of $D\tau_\varphi$ where D is the electron diffusion coefficient which should be considered constant in the temperature range under study (the temperature-dependent part of resistance is about 10^{-3}).

By plotting $(\tau_\varphi \cdot T)^{-1}$ or $(D\tau_\varphi \cdot T)^{-1}$ as a function of T, according to Eq. (2) we should obtain a straight line crossing the ordinate axis at point A with slope B. However, the analysis of experimental curves has demonstrated that at sufficiently low temperatures there exists a straight horizontal section and only at a certain temperature T_0 the $(\tau_\varphi \cdot T)^{-1}$ increases and subsequently turns into a straight line with slope B.

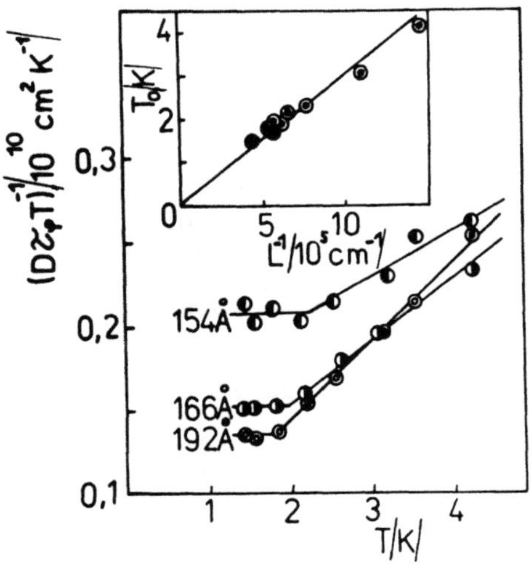

Fig. 1

The figure shows the $(D\tau_\varphi \cdot T)^{-1}$ dependence on T for three film samples of bismuth-antimony alloy of various thicknesses to demonstrate "freezing-out" of the $\tau_\varphi^{-1} \sim T^2$ contribution. At low temperatures, the experimental curves are horizontal, the quadratic dependence $\tau_\varphi^{-1} \sim T^2$ manifests itself at higher temperatures. The temperature T_0 of experimental curves inflection decreases with the increase of film thickness.

The observed peculiarity of electron-phonon scattering can be attributed to the fact that at low temperatures the thermal phonon wavelength $\lambda_{ph} = 2\pi\hbar S/kT$ (S being the sound velocity) exceeds the film thickness L. In this case only two-dimensional vibrations are possible for a film not bound to its substrate. The dependence $\tau_\varphi^{-1} \sim T$ can be expected at inelastic scattering of electrons by surface (two-dimensional) phonons existing at any temperature. The contribution of these processes can account for the fact that, as a rule, the experimental values of τ_φ at low temperatures are lower (about twofold) than those calculated by Eq. (1) /4,5/. In a real film, the phonon spectrum is determined by an acoustic mismatch between the film and the substrate which is noticeable for bismuth films deposited on mica and glass since the velocities of sound in these materials differ by a factor of two or three.

The temperature T_0 at which $\lambda_{ph}/2 = L$ is defined by

$$T_0 = \pi\hbar S/k\,L\,. \qquad (3)$$

Inset to the figure shows T versus 1/L which is a straight line passing through the origin of coordinates. According to Eq. (3), the slope of this curve is governed only by the sound velocity S equal to $1.4 \cdot 10^5$ cm/s which is in good agreement with the transverse sound velocity in bulk bismuth crystals. It should be noted that the points experimentally obtained by us for bismuth and bismuth-antimony films, as well as the results for bismuth films reported elsewhere /2,3/, lie on the same straight line.

Similar curves plotted for antimony films yield $S = 2 \cdot 10^5$ cm/s, for beryllium films - $S = 9.6 \cdot 10^5$ cm/s which is in good agreement with the known values of transverse sound velocity.

Having analysed the experimental data on temperature dependence of time τ_φ in the films of aluminium /7/ and silver /8/, we obtained the curves similar to those shown in the figure. Namely, in these films the dependence $\tau_\varphi(T)$ exhibits the same peculiarity as in the semimetal films at T values which are also dictated by the film thickness and sound velocity in compliance with Eq. (3).

CONCLUSION

Thus, the study of inelastic scattering of electrons in semimetal films by measuring the localization quantum corrections to conductivity made it possible to detect the peculiarity in temperature dependence of the electron-phonon scattering, i.e. the "freezing-out" of the $\tau_\varphi^{-1} \sim T^2$ contribution below the temperature T_0 at which the phonon wavelength becomes equal to the film thickness.

REFERENCES

1. Yu.F. Komnik, E.I. Bukhshtab, V.V. Andrievskii, A.V. Butenko: J. Low Temp. Phys. 52, 317 (1983)
2. F. Komori, S. Kobayashi, W. Sasaki: J. Phys. Soc. Jap. 52, 368 (1983)
3. A.K. Savchenko, A.S. Rylik, V.N. Lutskii: ZhETF 85, 2210 (1983)
4. E.I. Bukhshtab, A.V. Butenko, Yu.F. Komnik, V.V. Pilipenko: Solid State Commun. 53, 347 (1985).
5. A.V. Butenko, E.I. Bukhshtab: ZhETF 88, 1053 (1985)
6. B.L. Altshuler, A.G. Aronov, D.E. Khmel'nitskii: J. Phys. C15, 7367 (1982)
7. M.E. Gershenson, V.N. Gubankov, Yu.E. Zhuravlev: ZhETF 85, 287 (1983)
8. M.E. Gershenson, V.N. Gubankov, Yu.E. Zhuravlev: ZhETF 83, 2348 (1982)
9. J. Rammer, A. Schmid: Phys. Rev. B34, 1352 (1986)

Thermoelectric Power and Coulomb Gap in Inhomogeneous Thin Bismuth Films

J.J. Lin, J.L. Cohn, and C. Uher

Department of Physics, University of Michigan, Ann Arbor, MI 48109, USA

Abstract

The resistance and thermoelectric power of very thin bismuth films have been measured in the temperature range 0.3-10K. A single sample was evaporated in an oxygen atmosphere of 50 mtorr and thereafter oxidized in steps to "tune" the sheet resistance, $R_\square$, from 1-60 kΩ/□. The thermopower for low $R_\square$ decreases linearly with decreasing temperature as is typical for a metal. For $R_\square > 24$kΩ/□ the thermopower diverges rapidly, reaching values at 0.3K which are ~4 orders of magnitude larger than those for low $R_\square$. This behavior suggests that there is a gap opening at the Fermi level, presumably due to a strong depletion of the density of states arising from electron-electron interactions. Surprisingly, the resistances in all cases exhibit temperature dependences characteristic of electron-electron interaction and localization effects in the weak disorder limit, with a ℓnT dependence for low $R_\square$ changing gradually to $1/\sqrt{T}$ for high $R_\square$. This latter behavior suggests the presence of quasi-one-dimensional conducting channels in the films. Preliminary TEM micrographs indicate that high $R_\square$ films are characterized by a labyrinth pattern of multiply-connected metallic grains with nonpercolating insulating inclusions.

1. Introduction

Considerable progress has been made in the past several years in the understanding of transport in disordered systems [1,2]. In particular, the theories of localization [3] and electron-electron interactions [4] have been very successful in predicting the behavior of weakly disordered two-dimensional (2D) thin films and very narrow wires (1D). Less well understood is the nature of transport for intermediate disorder with mixed dimensionality. In addition to the intrinsic disorder of the conducting medium, very thin metal films may develop larger scale cracks or insulating inclusions that represent internal boundaries. These "morphological" inhomogeneities may further restrict the dimensionality of the system and thus be important in determining transport properties. Much of the current understanding of disordered systems has come from the great body of experimental work on resistance and magnetoresistance in homogeneous systems [1,2], while thermoelectric effects have been largely ignored. To our knowledge only one experimental study has been undertaken thus far to address the thermopower in strongly disordered thin metal films[5]. Because the thermopower is very sensitive to the density of states at the Fermi level, its measurement provides important information complementary to that obtained from measurements of resistance. Of particular interest in this regard is the role played by electron-electron interactions which may lead to the formation of a Coulomb gap [6].

In this paper we report measurements of low-temperature thermopower in very thin bismuth films whose electrical characteristics in the weak disorder regime are well established [7,8]. By controllably oxidizing a given sample we have observed significant changes in the temperature dependence of resistance and thermopower with increasing sheet resistance, $R_\square$. The thermopower data clearly indicate the opening of an energy gap for films with $R_\square > 24$KΩ/□. Surprisingly, the conduction in these high $R_\square$ films is not activated, but rather exhibits a temperature dependence consistent with one-dimensional electron-electron interaction and localization effects in the limit of weak disorder. These results suggest that the morphology of the conducting medium is very important.

2. Theory

In the weak disorder regime, localization [3] and electron-electron interaction [4] effects give rise to a resistance that increases with decreasing temperature as ℓnT and $1/\sqrt{T}$ in 2D and 1D, respectively. With increasing disorder, the scaling theory of localization predicts a gradual approach to exponential localization where the resistance rises exponentially with an inverse power of T with decreasing temperature.

The thermopower, S, has not been treated theoretically for the case of intermediate disorder, however its temperature dependence in limiting cases is understood. For weakly disordered thin metal films [9] the Boltzmann result, $S \propto (k/e)[N(E_F)kT]$, is valid as long as the density of states at the Fermi level, $N(E_F)$, remains finite. In systems with an energy gap, e.g. semiconductors, electrons transport more than thermal energy, and S increases as $T \rightarrow 0$ [5,10]. This latter behavior is particularly relevant for systems with stronger disorder where electron-electron interaction effects are enhanced and may be sufficient to open a gap at the Fermi energy [6].

3. Experiment

A sample was prepared in a standard four-probe bridge pattern by evaporation of 99.9999% purity Bi in an oxygen atmosphere of 50 mtorr onto glass substrates held at room temperature, with a deposition rate of ~1Å/sec. Following the deposition of nominally 50Å Bi, the conduction channel of the bridge was coated with ~250Å SiO as a protective layer to inhibit rapid oxidation and sample deterioration in ambient conditions. The as-prepared film had a residual resistivity, $\rho(4K) \sim 2800\ \mu\Omega$-cm. Electrical contact to the bridge was made with superconducting wires(T_c=9.5K) indium soldered to 5000Å thick Bi pads. The sample was mounted with one end thermally anchored to the mixing chamber of a dilution refrigerator and the other affixed with a resistive heater. The temperatures of these "cold" and "hot" ends of the sample were monitored by calibrated germanium and carbon glass resistance thermometers attached to the back of the substrate. Thermopower and DC resistance measurements were performed on successive low-temperature runs following *in-situ* oxidation treatments at room temperature to "tune" the sheet resistance. For the thermopower measurements, temperature gradients were typically ~10% of the average temperature of the sample. Signals were monitored on a digital nanovoltmeter with amplified output connected to a chart recorder.

4. Results and Discussion

Figure 1 shows low-temperature thermopowers with values of $R_\square$ at 4.2K indicated for each curve. The thermopower is negative, indicating a dominant contribution from electron diffusion. Because electron mean free paths are severely limited in disordered thin films, the phonon drag effects characteristic of pure metal [11] are substantially reduced, and we interpret the low-temperature thermopower as reflecting the carrier energy distribution. The curves for $R_\square$ = 1.26 and 8.81K$\Omega/\square$ are virtually indistinguishable, and follow a linear dependence on temperature to below 0.7K where the signal becomes too small to measure. For $R_\square$ =23.3k$\Omega/\square$, the thermopower decreases approximately linearly with temperature before diverging rapidly at ~2K. This divergence occurs less rapidly and at higher temperatures with increasing $R_\square$. We conclude that a gap is opening at the Fermi level, presumably due to a strong depletion of the density of states arising from electron-electron interactions [5,6]. Interestingly, the film with $R_\square$(4.2K)=23.3k$\Omega/\square$ has $R_\square$ very close to the "crossover" value ($R_{2D} \equiv 2\pi\hbar/e^2 \approx 26k\Omega/\square$) defining the clas-

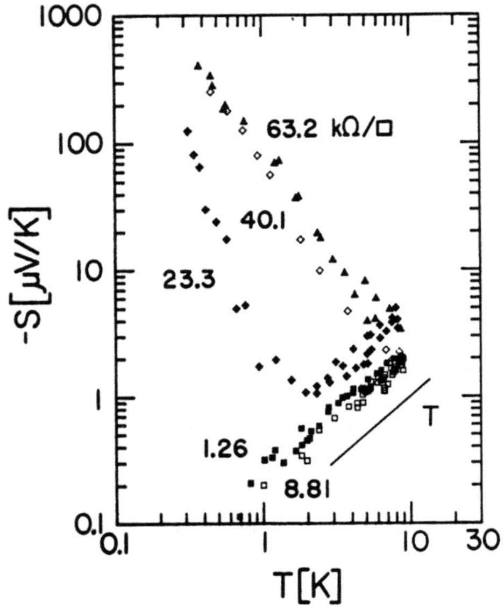

Fig. 1 Log-log plot of -S versus T. Labels indicate values of $R_{\square}$ at 4.2K.

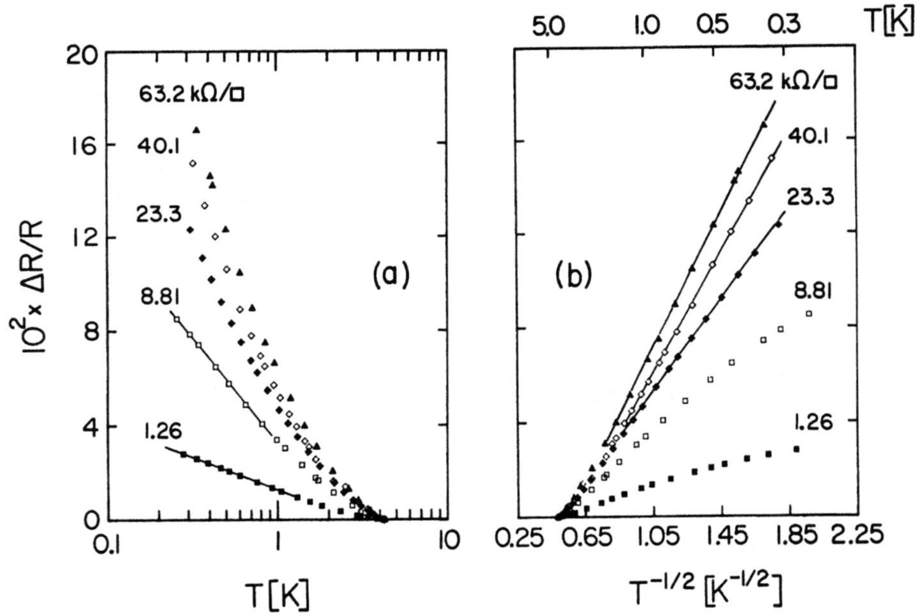

Fig. 2 Relative resistance rise, $\Delta R/R$ (normalized to 4.2K), versus (a) ℓnT and (b) $1/\sqrt{T}$. Labels indicate values of $R_{\square}$ at 4.2K and lines are guides to the eye.

sical picture of 2D "metal-insulator" transitions. As the system is driven toward stronger disorder (increasing $R_{\square}$), we expect the gap to become larger, and this is manifested as an increase in the temperature at which S diverges. These results are in qualitative agreement with a previous study [5] on Pd and Pd-Au films. There are several surprising features of our data that deserve particular attention. As mentioned above, the measurements of Fig. 1 clearly indicate a more rapid divergence of S for $R_{\square}$ = 23.3kΩ/□ than for higher $R_{\square}$. In addition, these divergencies are far more rapid (approximately ∝ $1/T^4$ for $R_{\square}$ = 23.3kΩ/□) than, for example, in semiconductors where $S \propto E_{gap}/kT$ for $E_{gap} >> kT$ [10]. We have recently observed this same trend in measurements performed on a second sample with $R_{\square}$ tuned in smaller increments near 20 kΩ/□. These results will be discussed elsewhere [12].

In apparent contrast to the behavior of the thermopower, the presence of an energy gap is not evidenced in the temperature dependence of the resistance for high $R_{\square}$. In Fig. 2(a), for low $R_{\square}$ the resistance rises as ℓnT, in good agreement with previous studies of thin film bismuth in the weak disorder regime [7,8]. This temperature dependence changes gradually to a $1/\sqrt{T}$ behavior for high $R_{\square}$ [Fig. 2(b)], consistent with interaction and weak localization effects observed in narrow wires [13] and inhomogeneous films with mixed dimensionality [14,15]. We find that the resistance ratio, R(300K)/R(4K), is 0.91±0.04 for all films, indicating that the intrinsic disorder in the conducting medium is essentially unchanged by the oxidation. The quasi-one-dimensional behavior of the resistance is, no doubt, associated with the details of the film morphology. TEM micrographs for high $R_{\square}$ films indicate a pattern of multiply-connected metallic grains separated by insulating (oxide) inclusions [12]. Evidently the oxidation treatment causes the insulating portions to enlarge, thus constricting the conduction channels and reducing the effective system dimensionality.

5. Conclusions

We have observed the onset of rapid divergencies in the thermopower of very thin bismuth films for sheet resistance exceeding $R_{\square}$ ~24kΩ/□. These results clearly indicate the opening of an energy gap at the Fermi level for high $R_{\square}$, apparently a consequence of enhanced electron-electron interaction effects in the presence of increasing disorder. At the same time, in the range of temperatures where S diverges, the resistance behaves as in a weakly disordered quasi-1D metal. The simultaneous observation of these seemingly contradictory behaviors is associated, in part, with the inhomogeneous morphology of the high $R_{\square}$ films.

This work was supported by NSF-Low Temperature Physics-Grant DMR-8508392.

References

1. G. Bergmann, Phys. Rep. 107, 1 (1984).
2. P.A. Lee and T.V. Ramakrishnan, Rev. Mod. Phys. 57, 287 (1985).
3. E. Abrahams, P.W. Anderson, D.C. Licciardello, and T.V. Ramakrishnan, Phys. Rev. Lett. 42, 673 (1979).
4. B.L. Al'tshuler, A.G. Aronov and P.A. Lee, Phys. Rev. Lett. 44, 1288 (1980); B.L. Al'tshuler, D. Khmel'nitzkii, A.I. Larkin, and P.A. Lee, Phys. Rev. B 22 5142 (1980).
5. M.J. Burns, W.C. McGinnis, R.W. Simon, G. Deutscher, and P.M. Chaikin, Phys. Rev. Lett. 47, 1620 (1981); M.J. Burns, and P.M. Chaikin, Phys. Rev. B 27, 5924 (1983).
6. A.L. Efros and B.I. Shklovskii, J. Phys. C 8, L49 (1975); A.L. Efros, N. Van Lien, and B.I. Shklovskii, Solid State Commun. 32, 851 (1979).
7. P.H. Woerlee, G.C. Verkade, and A.G.M. Jansen, J. Phys. C 16, 3011 (1983).
8. F. Komori, S. Kobayashi, and W. Sasaki, J. Phys. Soc. Jpn. 52, 368 (1983).
9. V.V. Afonin, Yu M. Gal'perin, and V.L. Gurevich, Sov. Phys. JETP 60, 194 (1984).
10. N.F. Mott and E.A. Davis, Electronic Processes in Non-Crystalline Materials (Clarendon, Oxford, 1979).
11. C. Uher and W.P. Pratt, Jr., J. Phys. F 8, 1979 (1978).
12. J.L. Cohn, J.J. Lin, and C. Uher, (to be published).
13. D.E. Beutler, Ph.D. thesis, Purdue University, 1986 (unpublished).
14. Y. Koike, M. Okamura, and T. Fukase, J. Phys. Soc. Jpn. 54, 3018 (1985).
15. A. Palevski and G. Deutscher, Phys. Rev. B 34, 431 (1986).

Comparison of Localization Effects in Granular Magnesium and Granular Aluminum. Superlocalization?

M. Kunchur, P. Lindenfeld, and W.L. McLean

Serin Physics Laboratory, Rutgers University,
Piscataway, NJ 08855-0849, USA

When magnesium is evaporated in the presence of a small amount of oxygen a mixture of metallic grains and oxide is formed, similar in many respects to granular aluminum. Since magnesium is not superconducting the normal-state properties can be studied more directly, in the absence of the complications which in aluminum are associated with superconductivity.

In this contribution we show that with changing oxygen content granular magnesium undergoes a metal-insulator transition as does granular aluminum. We present results on the temperature and magnetic-field dependence of the resistance which show differences between the two materials which can, in part, be ascribed to the presence of superconductivity in aluminum.

The main motivation, however, was to search for the presence of a slope change on the graph of log R against $T^{-1/2}$ which is present in granular aluminum and has been interpreted in several granular metals as the result of superconductivity in the grains [1-3]. Recently it has been suggested that it may be the result of "superlocalization", i.e. the combined result of localization and percolation [4,5]. We find that such a slope change is present in granular magnesium, but smaller and not accompanied by the very large negative magnetoresistance which is found in granular aluminum.

Fig. 1 shows a graph of log R against $T^{-1/2}$ for six granular Mg specimens made at the same time by electron-beam evaporation on water-cooled substrates in the presence of a small amount of oxygen. Preliminary TEM measurements show a grain structure with a characteristic size of about 70Å (compared to about 30Å for aluminum). Deposition on liquid-nitrogen cooled substrates does not cause a substantial difference in the transport properties.

We characterize the specimens by the ratio $r = R_{4.2}/R_{RT}$ which is better known than the room-temperature resistivity ρ_{RT}. The specimens on Fig. 1 have values of r from 2.5 to 18. The corresponding values of ρ_{RT} range from 0.02 Ωcm to 0.3 Ωcm. On this graph the metal-insulator transition is the boundary between specimens for which the curve is concave downward and saturates at low temperatures, and specimens for which the resistance continues to increase and eventually diverges as the temperature is lowered. The lines for the specimens with r = 2.5, 3.0, and 4.0 curve downwards and therefore represent metallic behavior. The curves for the other specimens represent insulating behavior.

The lines for the more resistive samples exhibit a definite slope change ("kink") similar to that which is also found in granular aluminum and other granular metals. The temperature at which the kink occurs changes from 0.23 for the specimen with r = 6.1 to 0.91 for r = 18.

Fig. 2 shows the magnetic-field dependence for three of the Mg specimens at 0.6K and one aluminum specimen. At low fields the magnetoresistance of the Mg samples is positive, as expected from spin-orbit scattering. In aluminum this contribution is masked by superconductivity which, however, disappears near the critical field $H_{c2}(0)$ of 3.6T [6]. Indeed at higher fields the slope of the magnetoresistance curves is negative and of comparable magnitude in corresponding specimens of both materials when they are in the metallic region far from the

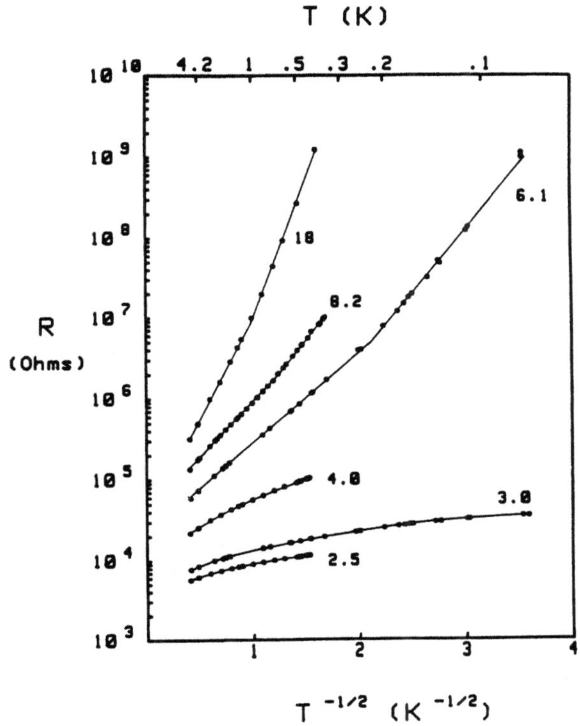

Fig. 1. Log R vs. $T^{-1/2}$ for six granular magnesium specimens. The number on each curve is the ratio $r = R_{4.2}/R_{RT}$

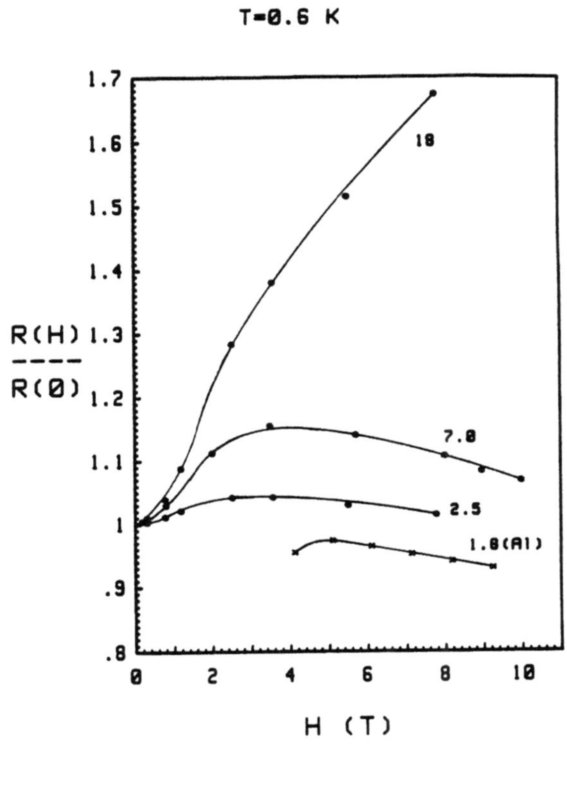

Fig. 2. The magnetoresistance of 3 magnesium specimens and one aluminum specimen (lowest curve) at 0.6K. The number on each curve is the resistance ratio r

metal-insulator transition. Near the metal-insulator transition and in the insulating region the magnetoresistances for the two materials are quite different.

In aluminum a field of about 4T causes the resistance to drop by factors of up to several hundred at about 0.5K and causes the kink to disappear [3,7]. This result and similar ones in other granular metals [2] have therefore been interpreted in terms of the destruction of superconductivity in the grains or in clusters of grains. The superconductivity in the isolated grains or clusters is accompanied by a decreased density of states and hence a higher resistivity [1,2]. The absence of a comparably large effect in magnesium is consistent with this interpretation.

An alternative interpretation incorporating both localization and percolation effects was recently put forward by Deutscher [4] and by Deutscher, Levy, and Souillard [5]. They derive a temperature variation where log R is proportional to $T^{-3/7}$. The exponent is too close to 1/2 for the present experiment to rule out one or the other definitely. Since a magnetic field of 9T should have a large effect on any localization phenomenon a large magnetoresistance is again expected.

A factor which affects the suggested effect is the grain size. It is shown in Ref. 5 that the quantity T_o in the expression $R = R_o \exp(T_o/T)^\alpha$ is inversely proportional to the volume of a grain. For localization effects to be important near the percolation threshold on a percolating cluster of grains it is also necessary that the inelastic scattering length be large compared to the grain size. For these reasons the magnitude of the effect should be smaller in Mg than in Al.

In specimens that are sufficiently far from the metal-insulator transition on the metallic side the magnetoresistances in Al and Mg (beyond H_{c2} for Al) are close, so that the characteristic lengths for inelastic and spin-orbit scattering in the two materials are also close. To illustrate this point we show the spin-orbit scattering length as a function of r on Fig. 3. The graph is consistent

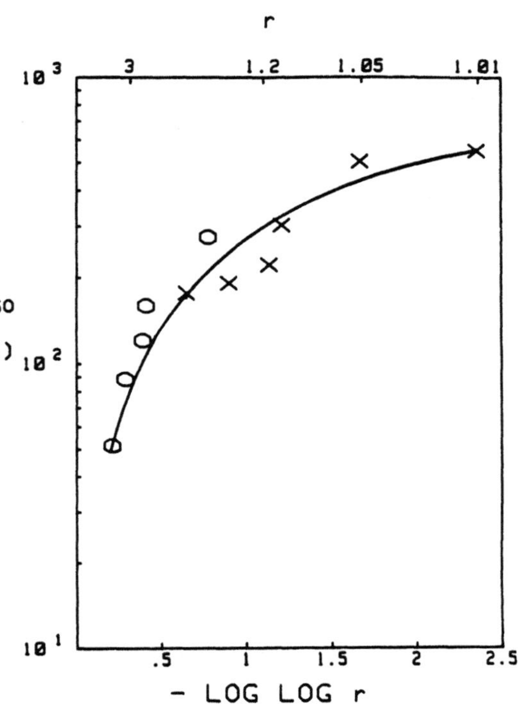

Fig. 3. The variation of the spin-orbit scattering length L_{SO} with the resistance ratio r. The log-log scale is chosen so as to expand the region near r = 1

with similar behavior as a function of r (and resistivity) in the two cases. It illustrates incidentally that the spin-orbit scattering may be significant for large values of r, i.e. for large amounts of elastic scattering, in spite of the low values of Z of magnesium and aluminum.

We conclude that the normal-state behavior of granular magnesium and aluminum is similar on the metallic side far from the metal-insulator transition, but that there are large differences on the insulating side. The results are consistent with the presence of superlocalization effects in both materials, larger in aluminum because of the smaller grain size, and present in aluminum in addition to effects related to local superconductivity.

We are grateful to G. Deutscher for many discussions and suggestions. This work was supported by the National Science Foundation under grant No. DMR 85-11982.

References

1. C.J. Adkins, J.M.D. Thomas, and M.W. Young: J. Phys. C: Solid State Phys. 13, 3427 (1980)
2. B.I. Belevtsev and A.V. Fomin: Fiz. Nizk. Temp. 12, 103 (1986) [Sov. J. Low Temp. Phys. 12, 60 (1986)]
3. M. Kunchur, Y.Z. Zhang, P. Lindenfeld, W.L. McLean, and J.S. Brooks: unpublished
4. G. Deutscher: Phil. Mag.: in press
5. G. Deutscher, Y. Levy, and B. Souillard: unpublished
6. T. Chui, P. Lindenfeld, W.L. McLean, and K. Mui: Phys. Rev. B 24, 6728 (1981)
7. H.K. Sin, P. Lindenfeld, and W.L. McLean: Phys. Rev. B 30, 4067 (1984)

Analysis of Magnetoresistance Data on Aluminum

J. Romijn[1], *J.G.F. Kablau*[1], *T.M. Klapwijk*[2], *and J.E. Mooij*[1]

[1]Department of Applied Physics and Centre for Submicron Technology, Delft University of Technology, P.O. Box 5046, NL-2600 GA Delft, The Netherlands

[2]Department of Applied Physics, University of Groningen, Nijenborgh 18, NL-9747 AG Groningen, The Netherlands

We investigate whether by adding the theory of Brenig instead of the usual theory of Larkin to the theory of weak-localization leads to a better agreement of the magnetoresistance data of superconducting materials at magnetic fields larger than $\{2k_BT/(\pi eD)\}\ln(T/T_c)$. Comparison of the values of the inelastic scattering times obtained with the two methods showed differences of only a few percent. The magnetic field region where a reasonable description of the magnetoresistance data by including the Maki-Thompson term of Brenig is achieved increased with a factor of four. The self-energy contributions appear to be about a factor of two too large.

1. Introduction

The inelastic scattering rate of electrons τ_{in}^{-1} is determined from magnetoresistance data. Weak localization as well as electron-electron interactions contribute to the conductivity. For superconducting metals above the critical temperature T_c, the superconducting fluctuations play a dominant role and have to be included as well. Usually this Maki-Thompson correction as calculated by LARKIN [1] is employed, however his expression is restricted to a small magnetic field B and a limited temperature T according to

$$\hbar/\tau_{in},\ 4\,DeB \ll (8k_BT/\pi)\ln(T/T_c) \tag{1}$$

with D the diffusion constant. A general description has been given by BRENIG [2]. He calculated the contributions to the conductivity by electron-electron interactions in the particle-particle channel (Cooperon) containing the superconducting fluctuations (i.e. the Maki-Thompson correction and the Aslamazov-Larkin correction) and the self-energy corrections. This theory is restricted to magnetic fields with $B \ll \hbar/(el^2)$, which is a weak limitation. Here l is the mean free path. The self-energy corrections are the generalized forms of the g_4-process of FUKUYAMA [3]. The correction to the conductivity of the electron-electron interactions in the particle-hole channel (Diffuson) is calculated by LEE and RAMAKRISHNAN [4].

We studied the analysis of magnetoresistance data in a magnetic field ranging from −1.5 T to 1.5 T and are interested whether the Brenig expressions provide a better description of the experimental data at fields beyond the limitation of Larkin. The temperature regime studied is not too close to T_c ($T/T_c-1 \gg 0.1$) where the Aslamazov-Larkin correction can be neglected. In the investigated magnetic field region the contribution in the diffusion channel is negligible as well.

2. Theory

For the corrections to the conductivity due to weak localization in the two-dimensional (2D) limit, $d \ll (D\tau_i)^{1/2}$ for samples with thickness d, the expressions of HIKAMI et al. [5] have been applied. With $\sigma_o = e^2/(2\pi^2\hbar)$, the magnetoconductance change $\Delta\tilde{\sigma}_x = \{\Delta\sigma_x(B,T) - \Delta\sigma_x(0,T)\}/\sigma_o$ and ψ the digamma-function, the contribution due to inelastic scattering is

$$\Delta\tilde{\sigma}_{WL}^{in} = \psi(1/2 + \gamma_{in}/\alpha) - \ln(\gamma_{in}/\alpha) \tag{2a}$$

and due to spin-orbit scattering is

$$\Delta\sigma^{so}_{WL} = {}^3/_2\left[\{\psi(^1/_2 + \gamma_*/\alpha) - \ln(\gamma_*/\alpha)\} - \{\psi(^1/_2 + \gamma_{in}/\alpha) - \ln(\gamma_{in}/\alpha)\}\right] . \quad (2b)$$

Here $\alpha = 4DeB/(2\pi k_BT)$ and $\gamma_x = (\hbar/\tau_x)/(2\pi k_BT)$. The characteristic time τ_* is defined in terms of the inelastic scattering time τ_{in} and the spin-orbit scattering time τ_{so} according to $\tau_*^{-1} = \tau_{in}^{-1} + {}^4/_3\tau_{so}^{-1}$. The finite spin scattering is expected to be negligible in aluminum [6].

The Brenig expressions for the contribution to the electron-electron interactions are more complicated. All terms consist basically of a sum of integrals containing digamma-functions with complex arguments. At high magnetic fields the conductivity correction vanishes. The following quantities are defined (ι is an integral value):

$$\eta_\iota = \alpha(\iota + {}^1/_2) + \gamma_{in} \quad (3a)$$

$$\psi_\iota(z) = \psi\{{}^1/_2(1+\eta_\iota) + z\} \quad (3b)$$

$$\Gamma_\iota(z) = [\ln(T/T_c) + \psi_\iota(z) - \psi(^1/_2)]^{-1} . \quad (3c)$$

The Maki-Thompson contribution as calculated by Brenig is given by

$$\Delta\sigma_{MT} = {}^1/_2\left[\alpha \sum_\iota \beta(T,\eta_\iota)/\eta_\iota - \int_{\gamma_{in}}^{\infty} d\eta\, \beta(T,\eta)/\eta\right] \quad (4)$$

with

$$\beta(T,\eta_\iota) = -\int dx \frac{2\pi}{\sinh^2 2\pi x}\left[\mathrm{Im}\, \psi_\iota(ix)\right]\left[\mathrm{Im}\, \Gamma_\iota(ix)\right] . \quad (5)$$

The second term on the right hand side in eq. 4 is the conductivity correction in the limit of zero magnetic field obtained by changing from a summation to an integration ($\alpha\iota \rightarrow \eta$).

3. Analysis

The expressions of Brenig are not suitable for a straightforward fit. We started with applying the usual method to obtain a good prediction of the scattering times using the Larkin expression

$$\Delta\sigma_{MT} = \beta(T,0)\{\psi(^1/_2 + \gamma_{in}/\alpha) - \ln(\gamma_{in}/\alpha)\} . \quad (6)$$

Here it is essential to use eq. 5 with $\eta_l = 0$ for the β-parameter of Larkin rather than applying one of the limiting cases $\beta(T) \rightarrow (\pi^2/4)\ln^{-1}(T/T_c)$ for $T \rightarrow T_c$ or $\beta(T) \rightarrow (\pi^2/6)\ln^{-2}(T/T_c)$ for $T >> T_c$, otherwise deviations of a factor of two may be obtained for the inelastic scattering time.

The simplex-method [7] has been applied for the fit-procedure. Besides the inelastic and the spin-orbit scattering time the zero magnetic field and the minimum or maximum of the conductance at zero field are fitted as well. The current through the coil of the superconducting magnet where the field is equal to zero is not well known due to remanence of the magnet. Further the resistance change must be scaled to zero at zero magnetic field.

For a set of values of the inelastic scattering time round the predicted value, the Maki-Thompson contribution according to eq. 4 is calculated. The integrals are computed with a five-point Gaussian-Legendre quadrature formula. The digamma-functions are approximated with a finite series using the asymptotic expansions and the recurrence relations. Improved values for the scattering times are now found by fitting again with the simplex method, however, with the Maki-Thompson contribution of Larkin substituted by an interpolated value of the calculated numbers from the expression of Brenig.

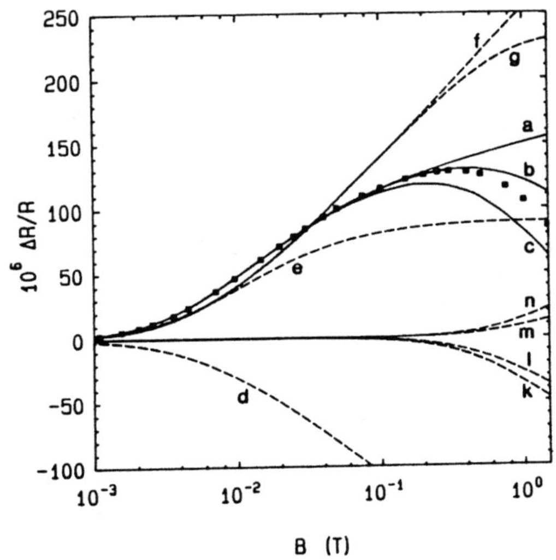

Fig. 1. The normalized magnetoresistance of a 2D-aluminum sample at 4.08 K. $R_{\Box} = 3.13\ \Omega$, $T_c = 1.71$ K and $D = 5.75 \times 10^{-4}\ m^2/s$. The squares are the experimental data. The solid lines are the theoretical curves obtained with the Larkin expression and the Brenig expression as indicated. The dashed lines are the various corrections.

a $= \Delta\tilde{\sigma}_{WL} + \Delta\tilde{\sigma}_{MT}$ (L)
b $= \Delta\tilde{\sigma}_{WL} + \Delta\tilde{\sigma}_{MT}$ (B)
c = full theory
d $= \Delta\tilde{\sigma}^{in}_{WL}$
e $= \Delta\tilde{\sigma}^{so}_{WL}$
f $= \Delta\tilde{\sigma}_{MT}$ (L)
g $= \Delta\tilde{\sigma}_{MT}$ (B)
k $= \Delta\tilde{\sigma}_1$
l $= \Delta\tilde{\sigma}_2$
m $= \Delta\tilde{\sigma}_3$
n $= \Delta\tilde{\sigma}_4$

L ≡ Larkin, B ≡ Brenig

A typical result is given in Fig. 1 for a sample with sheet resistance $R_{\Box} = 3.13\ \Omega$, $T_c = 1.71$ K and $D = 5.75 \times 10^{-4}\ m^2/s$ at a temperature of 4.08 K, showing the difference between applying the expression of Larkin (curve a) and the expression of Brenig (curve b). The total resistance change is

$$\frac{\Delta R}{R} = -R_{\Box}\sigma_0 \sum_x \Delta\tilde{\sigma}_x . \tag{7}$$

The inelastic scattering time is 3.0×10^{-10} s and the spin-orbit time is 9.5×10^{-11} s. The numerical values of the scattering times found this way differ a few percent from the values obtained with the Larkin expression. A significant discrepancy of the description at high magnetic fields is seen and the range of fields where magnetoresistance data can be fitted is still limited. To determine whether the deviations may be explained by the self-energy contribution, the terms as given by Brenig (eq. 52, 57, 59 and 64 in ref. 4) have been calculated. At relatively low fields i.e. where the magnetoresistance approaches the value at zero field, the convergence of the summations in the terms is very slow. In this case the sum has been approximated by an integral (k is an integral value and h is the small parameter)

$$\sum_{k=0}^{\infty} F(a+kh) \approx h^{-1} \int_a^{\infty} F(t)dt + {}^1\!/\!_2 F(a). \tag{8}$$

The contributions of the self-energy terms are shown in Fig. 1 indicated with $\Delta\tilde{\sigma}_1$ to $\Delta\tilde{\sigma}_4$ as well as the curve for the full theory. Clearly no improved agreement is achieved by including the self-energy contributions. Satisfactory agreement with the experimental results would be obtained if the self-energy correction is about a factor of two smaller than calculated by Brenig.

4. Discussions

EILER [8] has calculated the magnetic field dependence of the inelastic scattering time. This effect is well described by an effective inelastic scattering time and does not influence the description of the magnetoresistance data at relatively high fields. By BRENIG et al. [9] an enhanced inelastic scattering rate induced by the superconducting fluctuations has been calculated, which is temperature and magnetic field dependent. At temperatures close to T_c a significant effect is expected. Experimental data for the inelastic scattering rate are in good agreement with this theory [10]. In the temperature range studied, the influence of the magnetic field will be very small. For the sample shown in Fig. 2 the maximum effect is only 3% and can be neglected.

The computations applying the expressions of Brenig are extremely time consuming and cannot be used on a regular basis. Concerning the computed values of the inelastic scattering times, no significant differences compared to the values obtained with the Larkin-expression are observed. If the expression of Larkin is applied it is essential to use the full-expression for the β-parameter. Including the Maki-Thompson contribution as given by Brenig, the magnetic field region where a reasonable description of the magnetoresistance data can be obtained, improves with a factor of four.

Acknowledgement

We thank Dr. Brenig for making some graphs of his numerical results available to us.

References

1. A.I. Larkin, Pisma Zh. Eksp. Teor. Fiz. 31, 239 (1980) [JETP Let.. 31, 219 (1980)]
2. W. Brenig, J. of Low Temp. Phys. 60, 297 (1985)
3. H. Fukuyama, J. Phys. Soc. Jpn, 50, 3407 (1981)
4. P.A. Lee and T.V. Ramakrishnan, Phys. Rev. B 26, 4009 (1982)
5. S. Hikami, A.I. Larkin and Y. Nagaoka, Prog. Theor. Phys. 63, 707 (1980); S. Maekawa and H. Fukuyama, J. Phys. Soc. Jpn, 50, 2516 (1981)
6. Y. Bruynseraede, C. van Haesendonck, M. Gijs and G. Deutscher, Phys. Rev. Lett. 50, 277 (1983)
7. J.A. Nedlar and R. Mead, Computer Journal 7, 308 (1965); M.S. Caceci and W.P. Cacheris, Byte 9, 340 (1984)
8. W.Eiler, Solid State Comm. 56, 917 (1985)
9. W Brenig, M. Chang, E.Abrahams and W. Wölfle, Phys. Rev. B, 31, 7001 (1985)
10. J.M. Gordon and A.M. Goldman, Phys. Rev. B, 34, 1500 (1986)

Electron-Phonon Scattering Rates in Dirty Three-Dimensional Aluminum Films

J. Romijn[1], *R.M.G. Boesten*[1], *T.M. Klapwijk*[2], *and J.E. Mooij*[1]

[1]Department of Applied Physics and Centre for Submicron Technology, Delft University of Technology, P.O. Box 5046, NL-2600 GA Delft, The Netherlands

[2]Department of Applied Physics, University of Groningen, Nijenborgh 18, NL-9747 AG Groningen, The Netherlands

Inelastic scattering rates for electrons in three-dimensional aluminum films have been measured in a regime where the electron-phonon scattering dominates the electron-electron scattering. The mean free path has been varied and the temperature dependence of the electron-phonon scattering rate studied. A cross-over from T^3 to T^2 dependence has been observed. The data fail to yield a T^4 dependence as predicted by Schmid for very dirty material.

1. Introduction

The distribution of electrons over the energies in metals is established by electron-phonon and electron-electron scattering. At sufficiently high temperatures electron-phonon scattering dominates. Here, the dimensionality of the sample depends on the ratio of the characteristic phonon wavelength λ_{ph} and the thickness of the sample d. For clean metals where the mean free path l is larger than the phonon wavelength, $l/\lambda_{ph} > 1$, the theory predicts an electron-phonon collision rate of $(k_BT/\hbar)^3/\omega_D^2$ [1]. Here ω_D is the Debije-frequency. The theory agrees with the experimental results for such films. For dirty films, $l/\lambda_{ph} < 1$, several authors have calculated a different temperature dependence of the electron-phonon scattering rate. TAKAYAMA [2] and KAGAN and ZHERNOV [3] predict a rate proportional to T^2. SCHMID [4] predicts a much more complicated dependence on temperature ranging from T^2 to T^4 depending on the mean free path. We investigated dirty three-dimensional films in a regime where electron-phonon scattering is expected to dominate. The inelastic scattering times are determined from magnetoresistance data.

2. Experiments

The aluminum films are prepared by e-gun evaporation of 99.999%-pure aluminum onto room temperature thermally oxidized silicon substrates. The variations of the sheet resistance are achieved by evaporation in an oxygen atmosphere of about 10^{-4} Pa. Standard photo-lithography with lift-off has been used to pattern the films in a four-terminal geometry. The length of the films is 5 mm and the width 50 µm. The samples investigated have a thickness of 100 nm with a sheet resistance $R_\square$ ranging from 3 Ω to 10 Ω. An a.c.-driven double Kelvin bridge has been used to measure the magnetoresistance in a magnetic field range of -1.5 T to +1.5 T and a temperature range of 4 K to 15 K. Employing a bias current of 5 µA rms a resolution better than 1 ppm has been achieved. The stability of the circuit is better than 3 ppm over an hour. The magnetic field is driven with a resolution of 0.5×10^{-4} T in the vicinity of the zero field value. The superconducting magnet showed a remanent magnetic field of about 6×10^{-3} T. The zero magnetic field is determined from the symmetry with the field of the magnetoresistance.

3. Analysis of magnetoresistance data

Weak localization, superconducting fluctuations and electron-electron interactions influence the conductivity of thin films. As the films are two-dimensional with respect to weak-localization, the expressions of HIKAMI et al. [5] have been applied. With ψ the digamma-function, $\sigma_0 = e^2/(2\pi^2\hbar)$ and $\Delta\tilde{\sigma}_x = \{\Delta\sigma_x(B,T) - \Delta\sigma_x(0,T)\}/\sigma_0$, the contribution due to inelastic scattering is given by

$$\Delta\sigma^{in}_{WL} = \psi(1/2 + \gamma_{in}/\alpha) - \ln(\gamma_{in}/\alpha) \quad (1)$$

and for the contribution due to spin-orbit scattering by

$$\Delta\sigma^{so}_{WL} = 3/2\left[\{\psi(1/2 + \gamma_*/\alpha) - \ln(\gamma_*/\alpha)\} - \{\psi(1/2 + \gamma_{in}/\alpha) - \ln(\gamma_{in}/\alpha)\}\right] . \quad (2)$$

Here $\alpha = 4DeB/(2\pi k_B T)$ and $\gamma_x = (\hbar/\tau_x)/(2\pi k_B T)$, with D the diffusion constant, B the magnetic field and T the temperature. The characteristic time τ_* is defined in terms of the inelastic scattering time τ_{in} and the spin-orbit scattering time τ_{so} according to $\tau_*^{-1} = \tau_{in}^{-1} + 4/3\tau_{so}^{-1}$. The spin scattering is expected to be negligible in aluminum [6].

The superconducting fluctuations are described by the contributions of Maki-Thompson and Aslamazov-Larkin. As we have measured in a temperature regime not to close to T_c the Aslamazov-Larkin fluctuations are neglected. For the Maki-Thompson fluctuations we applied the expression of LARKIN [7]

$$\Delta\sigma_{MT} = \beta(T)\{\psi(1/2 + \gamma_{in}/\alpha) - \ln(\gamma_{in}/\alpha)\} . \quad (3)$$

His expression is valid in a magnetic field region satisfying $4DeB << (8k_BT/\pi)\ln(T/T_c)$. For the β-parameter we applied the expression in an integral form as described by BRENIG [8] rather than one of the limiting cases $\beta(T) \rightarrow (\pi^2/4)\ln^{-1}(T/T_c)$ for $T \rightarrow T_c$ or $\beta(T) \rightarrow (\pi^2/6)\ln^{-2}(T/T_c)$ for $T >> T_c$

$$\beta(T) = -\int dx \frac{2\pi}{\sinh^2 2\pi x}\left[\mathrm{Im}\,\psi(1/2+ix)\right]\left[\mathrm{Im}\{\ln(T/T_c) + \psi(1/2+ix) - \psi(1/2)\}^{-1}\right] . \quad (4)$$

For small magnetic fields $B < \pi k_BT/(2eD)$ the corrections due to the electron-electron interactions, i.e. the interactions in the particle-particle channel as well as in the particle-hole channel, can be neglected. We used the simplex-method [9] for a straightforward fit with four parameters. Besides the inelastic and the spin-orbit scattering time the zero magnetic field and the minimum or maximum of the magnetoresistance are fitted as well. The magnetic field has been restricted to below $\pi k_BT/(4eD)$. The total resistance change is then defined by

$$\frac{\Delta R}{R} = -R_\square\sigma_0(\Delta\sigma^{in}_{WL} + \Delta\sigma^{so}_{WL} + \Delta\sigma_{MT}) . \quad (5)$$

4. Experimental results

The critical temperature of the film has been determined from the resistive transition as a function of temperature. The mean free path l and the diffusion constant $D = v_F l/3$ in our films are calculated from the resistivity ρ assuming a ρl-product of $4 \times 10^{-16}\ \Omega m^2$ for aluminum. The resistivity follows from the geometry of the sample and the resistance at 4.2 K. The Fermi-velocity v_F is 1.36×10^6 m/s. The phonon wavelength λ_{ph} is equal to $2\pi/q_{ph}$ where the phonon wavevector q_{ph} is determined by the sound velocity v_s via $q_{ph} = 4\pi k_BT/(\hbar v_s)$. For longitudinal waves in aluminum v_s is 6.4×10^3 m/s. An overview of the sample parameters is given in Table I. The samples are in the dirty limit ($l/\lambda_{ph} < 1$) and are three-dimensional with respect to the phonon wavelength ($d/\lambda_{ph} = 17$ at 4.2 K).

Table I. Sample parameters at 4.2 K. The length of the sample is 5 mm, the width 50 μm and the thickness 100 nm.

Sample	$R_\square$ [Ω]	T_c [K]	l [nm]	D [m^2/s]	$k_F l$	l/λ_{ph}	τ_{so} [ps]
A	3.13	1.85	1.3	5.9×10^{-4}	23	0.22	95
B	3.13	1.71	1.3	5.8×10^{-4}	22	0.22	90
C	4.75	1.99	0.84	3.8×10^{-4}	15	0.14	100
D	10.0	2.11	0.40	1.8×10^{-4}	7.0	0.07	120

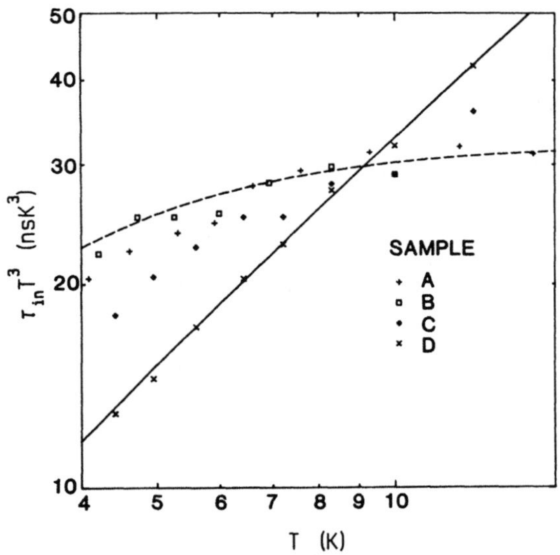

Fig.1. The inelastic scattering times scaled with T^3 as a function of temperature T. The dashed line is the best fit for samples A and B according to $\tau_{in}^{-1} = a_1T + a_3T^3$. The solid line is the best fit for sample D according to $\tau_{in}^{-1} = b_1T + b_2T^2$.

The spin-orbit scattering times as determined experimentally are presented in Table I. Above about 10 K the times tend to decrease with increasing temperature. ABRIKOSOV and GOR'KOV [10] predict a spin-orbit scattering rate proportional to the elastic scattering rate τ^{-1} as $\tau_{so}^{-1} = \tau^{-1}(Z/137)^4$, with Z the atomic number ($Z_{Al} = 13$). We find spin-orbit scattering times which are an order of magnitude too large, presumably due to the oxygen in the aluminum [12]. The results for the inelastic scattering times are scaled with T^3 and shown in Fig. 1 as a function of temperature.

5. The electron-phonon scattering time

The inelastic scattering rate contains contributions of the electron-electron scattering and electron-phonon scattering according to $\tau_{in}^{-1} = \tau_{e\text{-}e}^{-1} + \tau_{e\text{-}ph}^{-1}$. The electron-electron scattering rate has been calculated by various authors [11]. They all find the following expression for quasi two-dimensional films with differences expressed in a factor κ of order unity:

$$\gamma_{e\text{-}e} = \frac{\hbar/\tau_{e\text{-}e}}{2\pi k_B T} = R_{\square}\sigma_o \ln\left(\frac{\kappa}{2\pi R_{\square}\sigma_o}\right). \tag{6}$$

In our films $\tau_{e\text{-}e}^{-1}$ is smaller than $\tau_{e\text{-}ph}^{-1}$ but can not be neglected in the analysis [12]. As the electron-electron scattering rate is proportional to T and the electron-phonon scattering rate is expected to be proportional to T^3 in clean films and T^2 in dirty films, we fitted τ_{in}^{-1} to $a_1T + a_3T^3$ and to $b_1T + b_2T^2$. The results of the cleaner films A and B can reasonably be described with the cubic relation with $a_1 = 2.1 \times 10^8\ s^{-1}K^{-1}$ and $a_3 = 3.1 \times 10^7\ s^{-1}K^{-3}$. The theoretical value [12] for a_3 is a factor of 3.5 below this fitted value. The dirtiest film D can well be described with the quadratic expression with $b_1 = 2.4 \times 10^8\ s^{-1}K^{-1}$ and $b_2 = 2.8 \times 10^8\ s^{-1}K^{-2}$. It is not possible to describe the inelastic scattering rate of sample C with an intermediate value of the mean free path with either of the two expressions. The fitted values for a_1 and b_1 agree within a factor of two with the theory (eq. 6).

For the electron-phonon scattering rate Schmid predicts a complicated dependence on temperature and mean free path

$$\gamma_{e\text{-}ph} = \frac{\hbar/\tau_{e\text{-}ph}}{2\pi k_B T} = \frac{\Lambda}{2\pi}\varepsilon^2 B(\varepsilon\lambda), \tag{7}$$

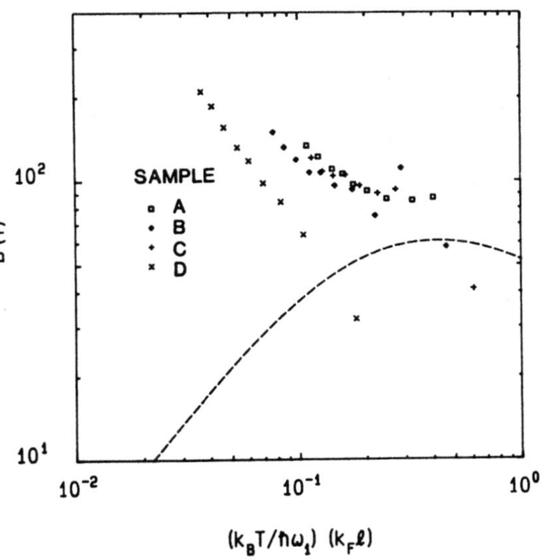

Fig.2. Comparison of the experimental data with the theory of Schmid. The inelastic scattering time has been scaled according to $B(T) = \hbar^3\omega_1{}^2/\{\Lambda\tau_{in}(k_BT)^3\}$. The dashed line is the theoretical curve.

where $\varepsilon = k_BT/(\hbar\omega_1)$ and $\lambda = k_Fl$. The function $B(\varepsilon\lambda)$ is shown in Fig. 2 (dashed line). In aluminum for $\omega_1 = v_sk_F$ a value of $1.1 \times 10^{14}\ s^{-1}$ is estimated using $k_F = (2/3\sigma_o\rho l)^{-1/2}$ and following Schmid Λ=2. With these values and assuming $\tau_{in}{}^{-1} \approx \tau_{e\text{-}ph}{}^{-1}$ the data have been plotted in Fig. 2 using $B(T) = \hbar^3\omega_1{}^2/\{\Lambda\tau_{in}(k_BT)^3\}$. The magnitude of the values of the experimental data agrees reasonably with the theory. However, our data fail to yield the cross-over to a T^4 dependence of the electron-phonon scattering rate. The obtained spin-orbit scattering times indicate a larger ρl-product than estimated as well as a tendency of a decreasing ρl -product with increasing mean free path. As a consequence, the values of k_F and ω_1 will be smaller and the mean free path larger. Our data in Fig.2 will shift to the right.

6. Conclusions

We have measured the inelastic scattering rate in aluminum films where the rate is dominated by electron-phonon collisions. The mean free path of the samples has been varied in a range where a cross-over from T^3 to T^2 dependence of the electron-phonon scattering is observed. Comparison of the rates with the prediction of Schmid does not show a cross-over to a T^4 dependence. This may be due to an underestimated value for the mean free path. The measured electron-electron scattering rates are in good agreement with the theoretical prediction.

References

1. W.E. Lawrence and A.B. Meador, Phys. Rev. B 18, 1154 (1978)
2. H. Takayama, Z. Phys. 263, 329 (1973)
3. Yu. Kagan and A.P. Zhernov, Zh. Eksp. Teor. Fiz. 50, 1107 (1966) [Sov. Phys.-JETP 23, 737 (1966)]
4. A. Schmid, in Localization, Interaction and Transport Phenomena, Vol. 61 of Springer Series in Solid State Sciences, ed. B. Kramer, G. Bergmann and Y. Bruynseraede (Springer Verlag, Berlin, 1985), p. 212
5. S. Hikami, A.I. Larkin and Y. Nagaoka, Prog. Theor. Phys. 63, 707 (1980)
6. Y. Bruynseraede, C. van Haesendonck, M. Gijs and G. Deutscher, Phys. Rev. Lett. 50, 277 (1983)
7. A.I. Larkin, Pis'ma Zh. Eksp. Teor. Fiz. 31, 239 (1980) [JETP Lett. 31, 219 (1980)]
8. W. Brenig, J. of Low Temp. Physics. 60, 297 (1985)

9. J.A. Nedler and R. Mead, Computer Journal 7, 308 (1965); M.S. Caceci and W.P. Cacheris, Byte 9, 340 (1984)
10. A. Abrikosov and L.P. Gor'kov, Zh. Eksp. Teor. Fiz. 42, 1088 (1962) [Sov. Phys.-JETP 15, 752 (1962)]
11. B.L. Altschuler, A.G. Aronov and D.E. Khmel'nitskii, J. Phys. C: Solid State Phys., 15, 7367 (1982); H. Fukuyama, J. of Phys. Soc. of Jpn, 53, 3299 (1984); W. Eiler, J. of Low Temp. Phys. 56, 481 (1984)
12. P. Santhanam, S. Wind and D.E. Prober, Phys. Rev. B. 35, 3188 (1987)

Superconducting Fluctuations and Localization in Two-Dimensional Al-Ge

L.R. Tessler and G. Deutscher

Department of Physics and Astronomy, Tel Aviv University, 69978 Ramat Aviv, Israel

1. Introduction

Magnetoconductivity (MC) in the presence of superconducting fluctuations in disordered two-dimensional thin films has been extensively studied in the last few years, especially after Larkin made explicit the connection between weak localization (WL) and the fluctuations [1]. His original results were however limited to temperatures not too close to T_c and low magnetic fields. The developments of Lopes dos Santos and Abrahams [2] generalized Larkin's results removing part of these limitations. They also predicted the saturation of the MC close to T_c and at high enough fields. In these conditions they predicted the MC due to Maki-Thompson (MT) fluctuations to have the form:

$$\Delta\sigma_{MT}(T,H) = -\sigma_{MT}(T,H=0) + \frac{e^2}{8\hbar}\frac{\pi c k_B T}{DeH} \tag{1}$$

In the present paper we present measurements of MC on two-dimensional granular aluminum-amorphous germanium thin films, including in the saturation region. The results obtained show good agreement with the theory.

2. Materials and Methods

The samples were prepared by electron-beam co-evaporation of germanium and aluminum from separate sources onto glass substrates kept at room temperature, in a vacuum better than 5×10^{-6} torr. Care was taken to ensure that no heavy or magnetic elements contaminated the films. All the samples were approximately 200 Å thick, 2 mm wide and 15 mm long. The aluminum to germanium ratio in the samples was controlled by varying the relative evaporation rates, which were kept between 1 and 4 Å/sec. The films consist of granular aluminum embedded on an amorphous germanium matrix. The MC measurements took place on a liquid helium cryostat equipped with a magnet able to produce fields up to 12 kOe. A conventional four-terminal DC system with currents of 1 μA was used. Table I lists the relevant parameters of the samples. $R^{4.2}_{sq}$ stands for the resistance per square at 4.2 K.

Table I. Sample parameters

Sample	% Al	Thickness [Å]	$R^{4.2}_{sq}$ [Ω]	T_c [K]	τ_{so} [sec]	D $\left(\frac{cm^2}{s}\right)$
A	75	270	11.3	1.85	3.2×10^{-11}	8.5
B	66	204	65.7	2.01	2.5×10^{-11}	6.5
C	40	177	220	1.27	7.4×10^{-12}	2.2

3. Results

All the MC measurements were performed with the magnetic field perpendicular to the film. In samples A and B the value of T_c obtained from fitting the experimental data to the theory agreed well with the measured ones. In sample C T_c was not reached, so we adopted the value obtained from the fits. In Fig. 1 we represent $\sigma = 1/R_{sq}$ as a function of the applied magnetic field at different temperatures for sample C. Note that at small fields σ *decreases* with the temperature, while for fields above around 3 k Oe it *increases*. Above 3 kOe σ depends logarithmically on the temperature, which is characteristic of a two-dimensional normal metal. This behavior was found in all samples.

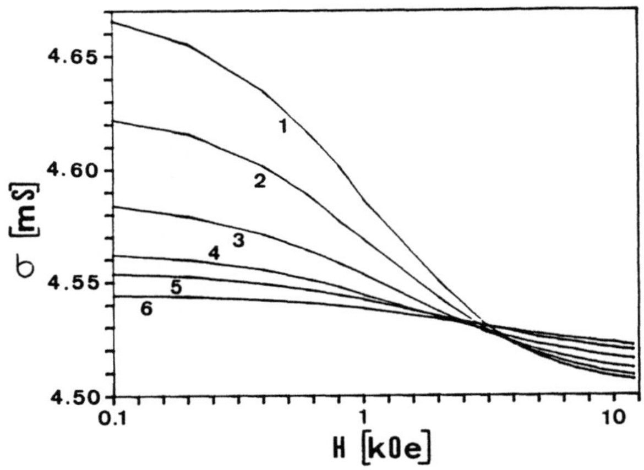

Figure 1. Conductivity σ as a function of magnetic field in sample C at different temperatures: 1=1.55 K, 2=1.7 K, 3=2.0 K, 4=2.5 K, 5=3.0 K, 6=4.2 K

The MC saturates at high enough fields. In order to obtain the characteristic inelastic and spin-orbit fields (H_i and H_{so}) and the diffusion constant we proceeded as follows:

i) We first fitted the MC at low fields and at temperatures not too close to T_c to the expression

$$\Delta\sigma_{MT,WL}(T,H) = \frac{e^2}{2\pi^2\hbar}\left\{f_1(T,H) - \beta\left(\frac{T}{T_c}\right)\left[\psi\left(\frac{H_i}{H}+\frac{1}{2}\right) - \ln\left(\frac{H_i}{H}\right)\right]\right\} \tag{2}$$

$$f_1(T,H) = \frac{3}{2}\,\psi\left(\frac{H_2}{H}+\frac{1}{2}\right) - \frac{1}{2}\,\psi\left(\frac{H_i}{H}+\frac{1}{2}\right) - \ln\left[\frac{H_2^{3/2}}{H\,H_i^{1/2}}\right] \tag{3}$$

where ψ is the di-gamma function, β is the function defined in [2], $H_2(T) = H_i(T) + H_{so}$. From these fits we obtained H_{so} and $H_i(T)$. This procedure was repeated for different temperatures. As no measurements were made too close to T_c we made the reasonable approximation $\beta\ (T/T_c,\delta) \simeq \beta\ (T/T_c,\delta=0)$ (as in [3]). We did not consider the contributions of the Aslamazov-Larkin fluctuations and of the ghost critical field H^*_{c2} [4]. Taking these contributions in account did not improve the fittings and only made them less reliable by introducing extra parameters.

ii) Knowing $H_{i(T)}$ and H_{so} we calculated the WL contribution to the MC at temperatures close to T_c and obtained the MT contribution to the MC:

$$\Delta\sigma_{MT} = \Delta\sigma_{MT,WL} - \Delta\sigma_{WL}. \tag{4}$$

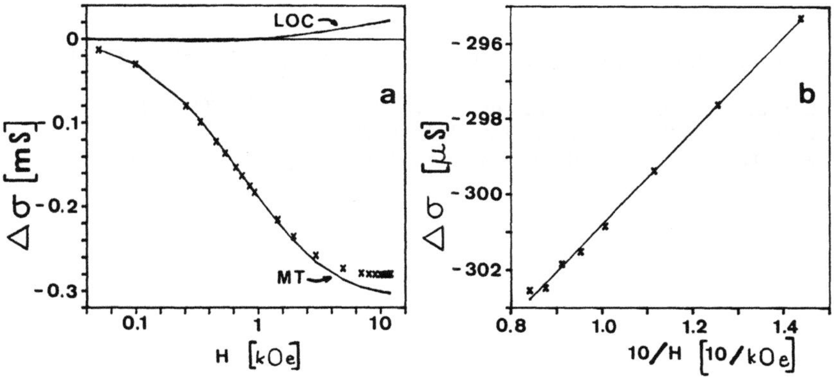

Figure 2. a) Experimental points, Localization and Maki-Thompson contributions to the magnetoconductance in sample B. b) $\Delta\sigma_{MT}$ as a function of 1/H at 2.5 K

iii) We plotted $\Delta\sigma_{MT}$ as a function of 1/H in the saturation region and obtained D according to (1) by linear regression. In Fig. 2 we illustrate this procedure for sample B. Fig. 2a shows the various $\Delta\sigma$ as a function of log H, and Fig. 2b show the $\Delta\sigma_{MT}$ against 1/H in the saturation region. The values of $\tau_{so}=\hbar c/4DeH_{so}$ and D for all samples are listed in Table I.

4. Discussion

We were able to test the validity of (1) for the first time. The system aluminum-amorphous germanium is very appropriate to do this, as the MC saturates at relatively low fields. We used (1) to calculate the diffusion constant D. The values of D calculated this way in samples A and B were compared to the values obtained from the derivative of H_{c2} in relation to the temperature at T_c, and they were found compatible within the experimental uncertainty. The diffusion constant decreased with increasing disorder in the films, as expected.

We presented with details the procedure used to obtain D. This method can be applied to any two-dimensional film that presents superconducting fluctuations. In order to apply this method the MC should be measured at high enough fields and at temperatures close to T_c.

In the samples studied the spin-orbit scattering time τ_{so} decreases with the concentration of germanium in the sample. In the temperature range studied τ_{so} was always smaller than τ_i. This can possibly be attributed to an enhancement of spin-orbit scattering scattering on the aluminum-amorphous germanium interfaces.

Acknowledgements

This work was partially supported by the Oren Family Chair for Experimental Solid State Physics and by the US-Israel Binational Science Foundation. One of the authors (L.R.T.) is the holder of a Helmuth Heineman Scholarship.

References

1. A. I. Larkin, Pis'ma Eksp. Theor. Fiz. 31, 239 (1980) [JETP Lett. 31, 219 (1980)]
2. J. M. B. Lopes dos Santos and E. Abrahams, Phys. Rev. B31, 172 (1985)
3. J. M. Gordon and A. M. Goldman, Phys. Rev. B34, 1500 (1986)
4. A. Kapitulnik, A. Palevski and G. Deutscher, J. Phys. C18, 1305 (1984)

Electron Inelastic Scattering and the Upper Critical Field in Thin Nb Films and Wires

M. Hikita, Y. Tajima, and T. Tamamura

NTT Electrical Communications Laboratories, Tokai, Ibaraki 319-11, Japan

1 Introduction

The latest microelectronics technologies such as thin film deposition and microfabrication have given considerable impetus to experimental investigation the localization in two- and one-dimensional systems. Recent weak localization theories successfully explain the anomalous magnetoconductance of thin films in simple metals like Mg, Cu, Ag, Au, and Al [1,2,3]. In these studies, the temperature dependence of the inelastic scattering time (τ_ϕ) was found to be expressed by the formula $T^{-\alpha}$. However, the values of the coefficient α experimentally obtained were ambiguous, ranging from 1.6 to 3.0. This situation is particularly complex in superconducting transition metals such as Nb [4,5], because the fluctuation effects must also be taken into account.

In this paper, we report on a systematic study of the τ_ϕ in thin and high quality Nb films with different thickness in order to provide a new quantitative evaluation method for quantum transport phenomena. We also studied the upper critical field (H_{c2}) and localization in fine Nb wires thinner than 0.1μm fabricated by nanometer electron-beam (EB) lithography and dry-etching techniques.

2 Film and Wire Preparation and Measurement

Thin Nb films and wires were carefully fabricated to avoid contamination and damage. Si wafers used as a substrate were coated with silicon nitride by CVD. High vacuum EB evaporation was used for very thin Nb deposition. The substrate surface was cleaned by Ar ion sputtering, then Nb films with thicknesses of 25, 30, 50, 80 Å were deposited at an almost constant deposition rate of 0.1nm/s at room temperature. Thin Si films of 25Å thickness were successively deposited on Nb films to protect the Nb surface from the atmosphere. The pressure was maintained at about $1-3\times10^{-7}$Torr during deposition. By these procedures, the oxidation of very thin Nb films can be virtually removed, which was proved by the resistivity measurements.

Thin Nb wires were patterned by finely focused EB exposure of a silicone-based negative resist / AZ-type photoresist 2-layer system [6]. Nb etching was carried out by reactive ion etching using $CBrF_3$ gas, then the resist was removed by washing the wafer in methylethylketone.

The magnetoconductance and H_{c2} were measured by a four-terminal AC(73Hz) bridge and a four-terminal AC method, respectively, in a magnetic field applied perpendicular to the direction of Nb surface.

3 Results and Discussion

3.1 Calculation of τ_ϕ

The temperature dependent τ_ϕ was calculated from the following formula considering weak localization (wl), Maki-Tompson fluctuation (MT) and Aslamazov-Larkin fluctuation (AL):

$$\Delta G(H,T) = \Delta G_{wl}(H,T) + \Delta G_{MT}(H,T) + \Delta G_{AL}(H,T) \tag{1}$$

$$\Delta G_{wl}(H) = G_{00}(3/2F(a1) - 1/2F(a2)) \tag{2}$$

$$\Delta G_{MT}(H) = -\beta \Delta G_{wl}(H, H_{so} = 0) \tag{3}$$

$$F(x) = \psi(1/2 + x) - \psi(x)$$

$$\psi : digamma\ \ function$$

$$a_1 = H_\phi + 4/3H_{so},\ a_2 = H_\phi,\ H_x = \Phi_0/(4\pi D\tau_x),$$

$$D = 1/3(v_F l),\ G_{00} = e^2/2\pi^2 h = 1.235 \times 10^{-5}\Omega^{-1}.$$

The conventional equation for two dimensions from [7] was used as $G_{AL}(H)$. The diffusion constant D was estimated from the dH_{c2}/dT measurement near T_c. The H_{so} of Nb film is not sensitive to the value of H_ϕ , though H_{so}($\tau_{so} = 8 \times 10^{-13}$,ref.[8]) is constant throughout the present paper. In fact, magnetoconductance of the Nb film is positive, in other words, $H_\phi \ll H_{so}$.

3.2 Thin Nb films

All thin Nb films prepared show superconductivity, and the measured T_c and the estimated diffusion constant D, which are summarized in the Table in Fig.1, vary with Nb film thickness d. The temperature dependences of the calculated τ_ϕ for each Nb film are also shown in Fig.1.

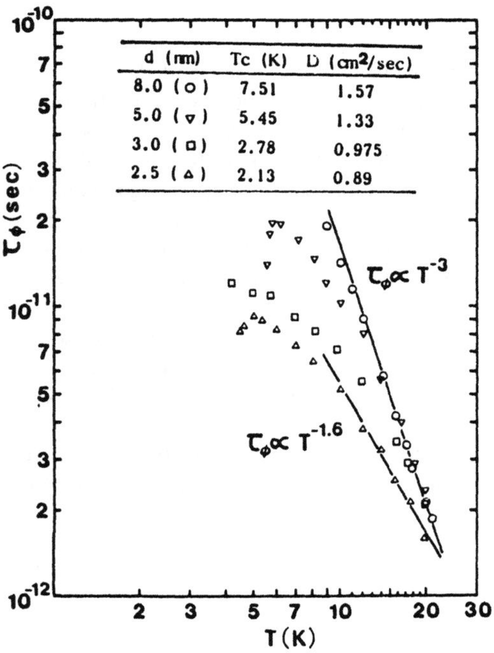

Fig.1 Temperature dependence of τ_ϕ in thin Nb films

In a temperature region above 10K, the value of τ_ϕ appears to follow a relation $T^{-\alpha}$. The αs change from -3 to -1.6 with decreasing film thickness. These curves can be well fitted to a function of the form

$$1/\tau_\phi = C_0 + C_1T + C_3T^3. \tag{4}$$

A similar form was found in several cases with Al [9,10], Nb [5], Pd [11] films, and C_0, C_1 and C_3 were explained as coefficients of scattering times of magnetic-impurity, electron-electron $(e-e)$ and electron-phonon $(e-p)$ interactions, respectively. However, the the values of C_0, C_1 and C_3 obtained from the experiments are in poor agreement with the quantitative explanation in dirty metals [5,8].

Parameter fitting results are compared with theories in Fig.2. As the theoretical C_1 value (C_{1th}), Fukuyama and Abrahams [12] derived the formula

$$1/\tau_{e-e} = (k_BTe^2/2\pi\hbar^2)R_\square ln(\pi\hbar^2/e^2R_\square) = C_{1th}T. \tag{5}$$

C_{1th}, which can be determined by measuring $R_\square$, showed an excellent agreement with our C_1 data as shown in Fig.2. On the other hand, the $e-p$ scattering time depends on the phonon dimensionality of the film. In the 3D limit, Lawrence and Meador [13] obtained the formula

$$1/\tau_{e-p} = 14\pi\zeta(3)\lambda\omega_D(T/T_D)^3 = C_{3th}T^3, \tag{6}$$

where D is the Debye frequency, λ is the $e-p$ coupling parameter and T_D is the Debye temperature. In the 2D dirty limit, Raffy et al.[11] found an expression modified from the 2D clean limit case:

$$1/\tau_{e-p} = 14\pi^2\zeta(2)(k_B\lambda lT^3/\hbar dT_D^2) = C_{3th}T^3. \tag{7}$$

As the phonon dimensionality of the film with d smaller than 8nm is estimated to be 2 in the present measurement, we should adopt the 2D dirty limit for τ_{e-p} evaluation. At $d\approx$8nm, C_{3th} obtained from (7) agrees with the observed C_3, where the following values are used: $\lambda = 1$, T_D=277K, $l = 3D/v_F$, $v_F = 2.7 \times 10^7$cm/s [5,14]. Eq.(7) indicates that C_{3th} increases with decreasing d as d^{-1}, however, C_3 of our thin films is not sensitive to d. This discrepancy could be caused by a structural difference between an ideal free-standing film and the actual film which is sandwiched between the substrate and the thin Si oxide. When this consideration is taken into account, we can conclude that the observed C_3 is in fair agreement with C_{3th} in the 2D dirty limit.

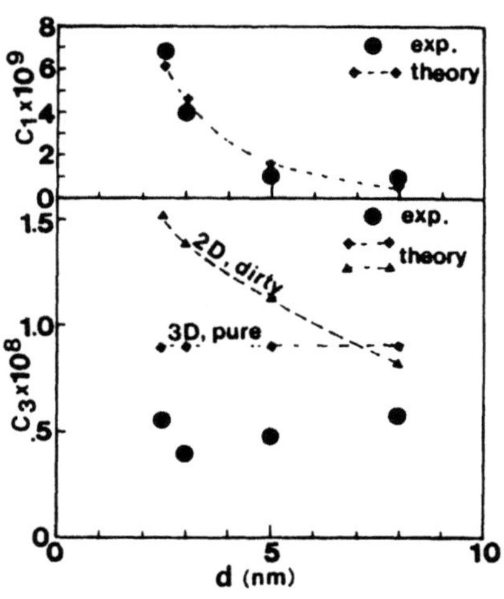

Fig.2 $C1(e-e)$ and $C3(e-p)$ values vs. Nb film thickness

It is worthwhile noting that τ_ϕ in Fig.1 shows an inverse temperature dependence in a temperature region just above T_c. This phenomenon had not been reported previously in Nb films [4,5], and was only observed in Al films [15]. This origin was explained in terms of divergent pair-breaking near T_c[16], but a future study will be necessary to clarify whether Nb films show the same behavior as Al or not.

3.3 Thin wires

Figure 3 shows a SEM picture of fabricated Nb wires (W=100nm, L=5μm, d=8nm) for 4-terminal measurement. The temperature dependence of H_{c2} in thin Nb wires is compared with that in Nb film(W=50μm, d=8nm) in Fig.4. The H_{c2} of the film decreases linearly with increasing temperature. On the other hand, the H_{c2} of the wire drops sharply near T_c and shows substantial deviation from the linear dependence. The measurements with the same wire sample with different current flows, 1nA (2.5×10^2A/cm^2) and 18nA (2.25×10^3A/cm^2), resulted in the same H_{c2} values, indicating that anomalous H_{c2} behavior of thin wires is not due to the heating and critical current effects, but is intrinsic. These results indicate, as the inelastic and elastic scattering lengths in this wire are shorter than the width of the wire, this wire can be regarded as a two-dimensional conductor. This is supported by the fact that dH_{c2}/dT in the linear region is almost the same as that in the film. Thus this anomaly could be due to a superconducting dimensionality change from 2D to 1D near T_c. The magnetoconductance of the wire above T_c was also measured, but was not well fitted by either 1D or 2D theories[17]. These H_{c2} and magnetoconductance anomalies in Nb wires will be studied in the future.

4 Conclusion

Quantum transport phenomena of very thin Nb films and wires fabricated with attention to avoid damage and contamination were studied. The temperature dependence of τ_ϕ changes from $T^{-1.6}$ to T^{-3} with film thickness from 2.5nm to 8nm. The $1/\tau_\phi$ can be well fitted to the function $C_0 + C_1 T + C_3 T^3$, where C_0 corresponds to the inverse of the magnetic impurity scattering time. The C_1 shows excellent agreement with a theoretical coefficient of the electron-electron scattering time. The C_3 can be explained as the coefficient of the electron-phonon scattering time in 2D dirty limit. The anomalous behavior of H_{c2} is also found in narrow (0.1μm)2D thin Nb wire.

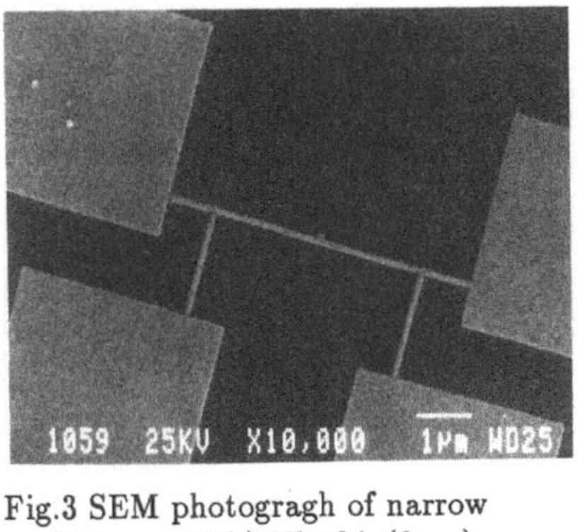

Fig.3 SEM photogragh of narrow (0.1μm width) Nb thin(8nm) film for measurement

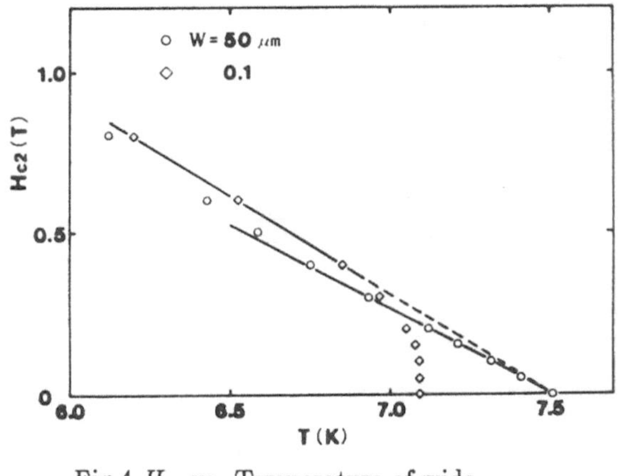

Fig.4 H_{c2} vs. Temperature of wide (50μm width) and narrow (0.1μm width) Nb thin(8nm) film

The authors would like to thank Dr. S.Kurihara for the helpful discussions, and Mr. M.Kakuchi for assistance in fabricating Nb wires.

References

1. G.Bergmann, Phys.Rep.,107, 1(1984).
2. P.A.Lee and T.V.Ramakrishnan, Rev.Mod.Phys., 57, 287(1985).
3. S.Kobayashi and F.Komori, Prog.Theor.Phys., 84 suppl., 224(1985).
4. M.E.Gershenzon, V.N.Gubankov and E.Zhuravlev, Sov.Phys.JETP, 58, 167(1983).
5. B.J.Dalrymple, S.A.Wolf, A.C.Ehrlich and D.J.Gillespie, Phys.Rev.B, 33, 7514(1986).
6. M.Kakuchi, M.Hikita and T.Tamamura, Appl.Phys.Lett., 48, 835(1986).
7. E.Abrahams, R.E.Prange and M.J.Stephan, Physica, 55, 230(1971).
8. C.Van Haesendonck, M.Gijs and Y.Bruynseraede, Localization, Interaction and Transport Phenomena, ed B.Kramer, G.Bergmann and Y.Bruynseraede (Springer, 1985) p.221.
9. J.M.Gordon, C.J.Lobb and M.Tinkham, Phys.Rev.B, 28, 4046(1983).
10. P.Santhanam and D.E.Prober, Phys.Rev.B, 29, 3733(1984).
11. H.Raffy, L.Dumoulin, P.Nedellec and J.P.Burger, J.Phys.F, 15, L37(1985).
12. H.Fukuyama and E.Abrahams, Phys.Rev.B, 27, 5976(1983).
13. W.E,Lawrence and A.B.Meador, Phys.Rev.B, 18, 1154(1978).
14. H.R.Kerchner, D.K.Christen and S.T.Sekula, Phys.Rev.B, 24, 1200(1981).
15. J.M.Gordon, C.J.Lobb and M.Tinkham, Phys.Rev.B, 29, 5232(1984).
16. S.Maekawa, Localization, Interaction and Transport Phenomena, ed B.Kramer, G.Bergmann and Y.Bruynseraede (Springer, 1985) p.130.
17. D.E.Prober, S.Wind and P.Santhanam, Localization, Interaction and Transport Phenomena, ed B.Kramer, G.Bergmann and Y.Bruynseraede (Springer, 1985) p.148.

Part V

Weak Localization Effects: Space-Charge Layers

Higher Subband Effect on the Spin-Orbit Interaction in a 2D System in InAs n-Inversion Layers

Y. Kawaguchi[1], *I. Takayanagi*[2], *and S. Kawaji*[2]

[1]Gakushuin Women's Junior College, 3-20-1 Toyama, Shinjuku-ku, Tokyo 162, Japan

[2]Department of Physics, Gakushuin University, 1-5-1 Mejiro, Toshima-ku, Tokyo 171, Japan

1. Introduction

The two-dimensional system in n-inversion layers of InAs crystals is a good system for investigating the effect of higher subbands on the spin-orbit interaction for the following reasons: (1) conduction electrons have a large g-factor ($g \sim -15$), and (2) owing to their light effective mass ($m^* \sim 0.023m_e$), electrons populate in the first excited subband at a low electron concentration. The electron concentration is estimated around $1.5\times10^{16}m^{-2}$ with the triangular potential approximation.

We have observed a remarkable negative magnetoconductance effect in the n-inversion layers with the electron concentration N_s of $1.5\times10^{16}m^{-2}$ at the SiO_2-InAs interface below 77K [1]. The negative magnetoconductance observed in both perpendicular and parallel fields ($B_\perp$ and $B_{/\!/}$) is explained by the weak localization theory with the spin-orbit interaction. The magnetoconductance $\Delta\sigma$ in $B_{/\!/}$ was well reproduced by the theory of MAEKAWA and FUKUYAMA[2]. The fitting of the theory to the data gives a relaxation time of spin-orbit scattering τ_{so} of 8×10^{-13}s and inelastic scattering τ_ε of 2×10^{-11}s at 1.1K [1]. However, the theory is not able to reproduce the experimental $\Delta\sigma$ in $B_\perp$.

In this paper we will show results for the N_s dependence of $\Delta\sigma$ in n-type inversion layers of InAs-MISFETs. Negative $\Delta\sigma$ has also been observed in both $B_\perp$ and $B_{/\!/}$. The magnitude of negative $\Delta\sigma$ in $B_\perp$ has a dip at around N_s=$1.8\times10^{16}m^{-2}$. The magnitude of negative $\Delta\sigma$ in $B_{/\!/}$ has a bend at the same N_s where the $\Delta\sigma$ in $B_\perp$ shows a dip. These results will be explained by introducing an intersubband effect qualitatively.

2. Experimental

Hall-bar-type MISFETs were fabricated on the B-face of the (111) surface of p-InAs crystals. The acceptor concentration of the crystal is $7\times10^{22}m^{-3}$. A silicon dioxide film for the gate insulator was prepared from SiH_4 and O_2 by chemical vapor deposition. The film thickness was about 1500Å. The length and width of the channel were 551 and 276μm , respectively. The channel had some conduction at Vg=0, where $N_s = 3.6\times10^{16}m^{-2}$ and the mobility $\mu_H = 0.23m^2/Vs$ at T=1.1K. However, the current passing through the outer part of the channel was eliminated by removing the SiO_2 film from it. The ratio of the conductivity in the channel to that out of the channel was more than one hundred[1].

By varying Vg, the electron mobility goes up from 0.29 $m^2/V\,s$ at $N_s = 5\times10^{15}m^{-2}$ to $0.38m^2/V\,s$ at $N_s = 1.3\times10^{16}m^{-2}$, and comes down to 0.20 $m^2/V\,s$ at $N_s = 5.5\times10^{16}m^{-2}$ at 1.1 K. The mobility μ_H and N_s are estimated from the Hall coefficient R_H and the conductivity $\sigma(0)$ by $\mu_H = R_H \cdot \sigma(0)$ and $N_s = 1/e R_H$. The N_s increases proportionally with increasing gate voltage up to $N_s \sim 3\times10^{16}m^{-2}$. At around $N_s = 3\times10^{16}m^{-2}$ the coefficient of proportionality reduces by 15 %.

3. Results and Discussion

The magnetoconductance has been measured up to a magnetic field B of 1.0T as a function of N_s at temperatures 1.1 and 4.2K. The magnetoconductance in $B_\perp$ has a minimum value at a field around 0.1T. The magnetoconductance in $B_{//}$ decreases monotonically as the field increases. Figure 1 shows the magnetic field dependence of $\Delta\sigma$ in $B_\perp$ and $B_{//}$ in the low field region. Both $\Delta\sigma$ are negative in the whole magnetic field region observed. These $\Delta\sigma$ are fairly well explained by the weak localization theory including the spin-orbit interaction with suitable values of parameters τ_ε and τ_{so}. Parameters τ_ε and τ_{so} are listed in Table 1. These parameters are obtained by fitting Maekawa and Fukuyama's theory to the experimental $\Delta\sigma$ in $B_{//}$ in the electric quantum limit.

The N_s dependence of $|\Delta\sigma_\perp|$ and $|\Delta\sigma_{//}|$ in a magnetic field of 0.1T are shown in Fig. 2. There is a considerable difference between the N_s dependence of $|\Delta\sigma_\perp|$ and $|\Delta\sigma_{//}|$. In the $|\Delta\sigma_\perp|$ versus N_s curve there is a dip in the vicinity of N_s= $1.8\times10^{16}m^{-2}$. In the $|\Delta\sigma_\perp|$ versus N_s curve, however, there is only a bend at the same N_s where a dip is observed in $|\Delta\sigma_\perp|$ versus N_s.

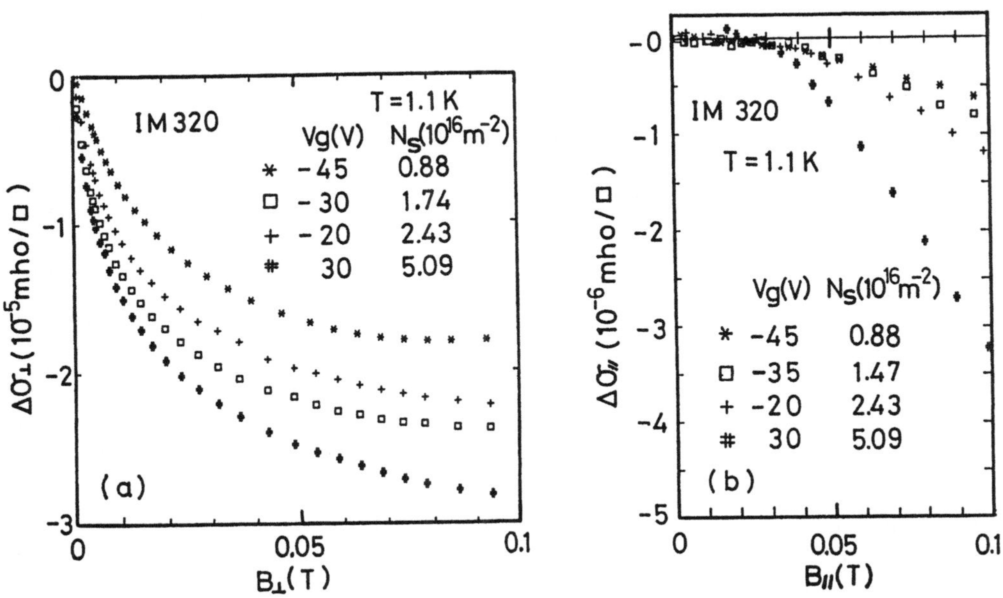

Fig. 1 Magnetic field dependence of $\Delta\sigma$ in perpendicular (a) and parallel (b) field

Table 1. Electron concentration dependence of relaxation times

N_s ($10^{16}m^{-2}$)	τ_0 (ps)	τ_ε (ps)	τ_{so} (ps)	K
0.45	0.038	42	1.5	1.1
0.80	0.048	45	1.3	1.2
1.2	0.050	60	0.9	1.2
1.5	0.049	62	1.0	1.2

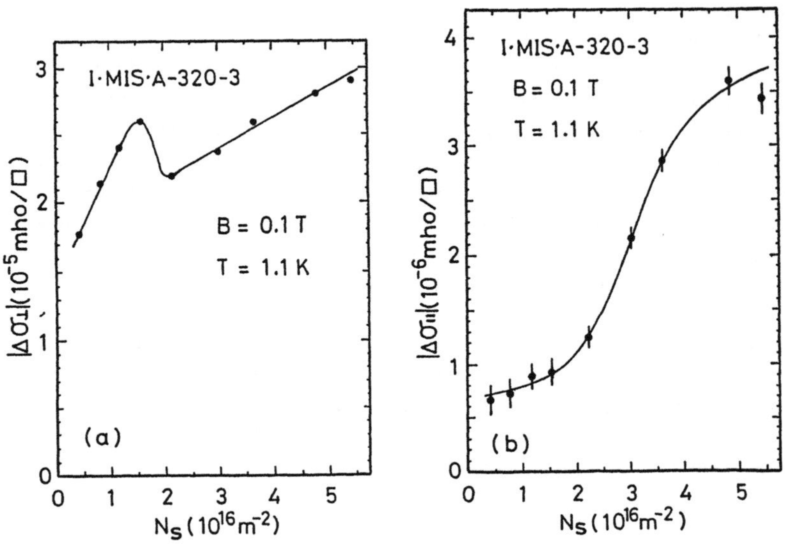

Fig. 2 Electron concentration dependence of $|\Delta\sigma_\perp|$ (a) and $|\Delta\sigma_{/\!/}|$ (b)

In the lower field limit, $\Delta\sigma_\perp$ and $\Delta\sigma_{/\!/}$ are expressed as follows[2]:

$$\Delta\sigma_\perp = \frac{e^2}{2\pi^2\hbar}\left[\left\{\frac{3}{(1+R)^2}-1\right\}\frac{(a\,\tau_\varepsilon)^2}{48} - \frac{1}{4}\left\{R\left(1+\frac{1}{1+R}\right) - \log\left(1+R\right)\right\}\gamma\right] , \quad (1)$$

$$\Delta\sigma_{/\!/} = -\frac{e^2}{16\pi^2\hbar}\left\{R\left(1+\frac{1}{1+R}\right) - 2\cdot\log\left(1+R\right)\right\}\gamma , \quad (2)$$

where $R = 4\frac{\tau_\varepsilon}{\tau_{so}}\cdot K$, $K = \frac{1/\tau_0 + 3/\tau_{so}}{1/\tau_0 - 1/\tau_{so}}$, $\gamma = \left(\frac{g\mu_B\tau_{so}}{2\hbar}\cdot K\right)^2$ and $a = \frac{4DeB}{\hbar} = \frac{4E_F\mu B}{\hbar}$.

In the present case, $\tau_\varepsilon/\tau_{so}$ has a value between 28 and 71 and K is constant at a value of about 1.2 as N_s varies from 4.5×10^{15} to $2.2\times10^{16}m^{-2}$. The first term in the right hand side of (1) is dominant for the negative $\Delta\sigma$ in $B_\perp$. In the right hand side of (2), the logarithmic term does not have a significant contribution to the negative $\Delta\sigma$ in $B_{/\!/}$. The expressions for the $\Delta\sigma$ reduce to

$$\Delta\sigma_{\perp} = -10.0 \left[E_F(\mathrm{eV})\mu(\mathrm{m^2/V\cdot s})\tau_{\varepsilon}(\mathrm{ps})B(\mathrm{T})\right]^2 \ [\mathrm{mho/sq.}] \ , \tag{3}$$

$$\Delta\sigma_{//} = -26.8\cdot 10^{-6}\, K^3 \tau_{\varepsilon}(\mathrm{ps})\tau_{so}(\mathrm{ps}) \left[B(\mathrm{T})\right]^2 \ [\mathrm{mho/sq.}] \ . \tag{4}$$

The magnetoconductance in $B_{\perp}$ is dependent on the square of μ and E_F. On the other hand, $\Delta\sigma_{//}$ is independent of μ and E_F.

In the following discussion we assume that the magnetoconductance is the sum of $\Delta\sigma_i$ in each subband. As $\Delta\sigma_{\perp}$ is weighted by E_F, the observed $\Delta\sigma_{\perp}$ is contributed mainly by the ground subband. In the N_s region higher than the dip in $|\Delta\sigma_{\perp}|$ versus N_s curve, a slope of $|\Delta\sigma_{\perp}|$ versus N_s curve is less than a quarter of that in the N_s region lower than the dip. As the higher subband is occupied by electrons in the high N_s region, an increasing rate of E_F is less than the low N_s region. The dip is observed at an N_s where the slope changes. As electrons populate the higher subband, the intersubband scattering effect is expected in the carrier mobility . The calculation by MORI and ANDO[3] predicted that the mobility had a sharp decrease when the Fermi energy crossed the energy of the higher subband bottom. The dip observed in the $|\Delta\sigma_{\perp}|$ versus N_s curve shows the energy of the higher subband bottom. Yamaguchi reported that the electrons populate the higher subband at $N_s=1.5\times10^{16}\mathrm{m}^{-2}$ from the Shubnikov-de Haas experiment [4].

The magnetoconductance in the high N_s region is contributed by both ground and higher subbands. The magnitude of $|\Delta\sigma_{//}|$ in the vicinity of $N_s=5\times10^{16}\mathrm{m}^{-2}$ is almost four times as large as that in the low N_s region. It implies that the product of τ_{ε} and τ_{so} in the higher subband is larger than that in the ground subband.

In conclusion, the energy level of the first excited subband is estimated by the magnetoconductance measurement at low magnetic fields.

References.

1. Y. Kawaguchi, I. Takayanagi, S. Kawaji: J. Phys. Soc. Jpn. 56 (1987) 1293
2. S. Maekawa, H. Fukuyama: J. Phys. Soc. Jpn. 50 (1981) 2516
3. S. Mori, T. Ando: Phys. Rev. B19 (1979) 6433
4. E. Yamaguchi: Phys. Rev. B32 (1985) 5280

Magnetotransport Studies of Localization and Interaction Effects in HgCdTe Two-Dimensional Electron Gas

E.C. Palm[1], *P.S. Kobiela*[1], *W.P. Kirk*[1], *R.A. Schiebel*[2], *and M.A. Reed*[2]

[1]Department of Physics, Texas A&M University, College Station, Texas, 77843, USA

[2]Central Research Laboratories, Texas Instruments, Dallas, TX 75265, USA

The two-dimensional electron gas (2-DEG) formed in the inversion layer of $Hg_{.78}Cd_{.22}Te$ MISFETS has been probed with transverse magnetoresistance ρ_{xx} and Hall resistance ρ_{xy} measurements. The samples were fabricated with a gate structure such that the electronic states could be varied from a mainly localized regime to one in which they were predominantly extended states. The carrier density in the 2-DEG simultaneously increased over an order of magnitude. With fixed temperature the magnetoresistance could be changed from a strongly weak localized behavior to a regime where quadratic behavior dominated. Since the quadratic behavior showed no temperature dependence and the Shubnikov-de Haas oscillations displayed behavior of two conduction bands, the quadratic magnetoresistance is interpreted as evidence of two subbands of conducting electrons. At all carrier densities, evidence for spin-orbit scattering was present. At low carrier densities an unusual asymmetry effect with respect to the magnetic field was observed. This asymmetry is linear in nature and is interpreted in terms of the transition from localized to extended states with increasing carrier density. Measurements were performed from 32 mK to 4 K and at low fields (<1.0 T). Since the HgCdTe system is rather unique (e.g. low effective mass, $m^*=0.006\ m_0$, and large g-factor, $g^*=94$), it offers an excellent opportunity to study characteristic scattering and interaction effects. In addition, the gate structure allowed the sensitivity of the system to be tuned over a wide range of different phenomena from weak localization to classical magnetotransport.

1. EXPERIMENTAL PROCEDURES

1.1 MISFET Sample Structure

The sample had a substrate of p-type $Hg_{.78}Cd_{.22}Te$ with a gate above an insulating layer. By applying a positive voltage to the gate a two-dimensional electron gas was formed at the interface. The carrier concentration could be varied from 4.7×10^{10} cm^{-2} to more than 9.0×10^{11} cm^{-2}. The conduction channel was 107 μm by 406 μm. Contact was made to the 2-DEG with six ion-implanted diodes, which provided a source, drain, and four voltage probes in the standard Hall configuration. An additional independent gate circumscribed the perimeter of the channel to eliminate unwanted inversion layers outside the conduction channel. This guard ring was above an additional insulating layer and was about three times farther from the channel than the gate.

1.2 Measurement Technique

The sample was kept in a light tight container prior to being cooled in a dilution refrigerator. Measurements were performed between 32 mK and 4 K at low magnetic fields (< 1 T). The fields were applied perpendicular to the sample and swept from negative to positive values. At regular intervals an excitation current of 500 nA was pulsed forward for 500 ms while voltages were measured simultaneously with digital voltmeters. The current was then pulsed in the reverse direction and the voltages recorded. For each recording interval, positive and negative voltage measurements were averaged to eliminate thermal EMF's and transient noise effects. Since the pulses were short enough in duration and low enough in amplitude, Joule heating did not cause observable departure from thermal equilibrium.

2. RESULTS AND DISCUSSION

2.1 Localized Regime

At low gate and negative gate voltages, the sample was highly resistive ($10^5\ \Omega$) (as shown in Fig. 1a) and displayed fluctuations in magnetoresistance similar to those seen in submicron size gold, silicon, and GaAs structures [1,2,3,4]. These fluctuations were reproducible as functions of magnetic field but appeared to be random in nature. In the aforementioned systems, similar universal conductance fluctuations have been ascribed to phase coherent conduction effects due to distances between the voltage contacts being comparable to the phase coherent length L_ϕ of the sample [5]. The sample size in which these fluctuations have been seen are much smaller than our sample size, but if L_ϕ in HgCdTe is on the order of 100 μm then these same effects would be important here also. With our present experimental method there is too much noise in the system to make reliable measurements. Efforts are currently underway using lock-in amplifier techniques to measure ρ_{xx} at lower levels of extraneous noise.

Other features in this regime are sharp transitions between different resistance states as shown by the data in Fig. 2b. These transitions occur at certain temperature, magnetic field, and gate voltage combinations. Similar transitions have been seen in Silicon MOSFETS [6,7] and have been attributed to traps in the insulator as the interface traps and releases electrons or groups of electrons. Again

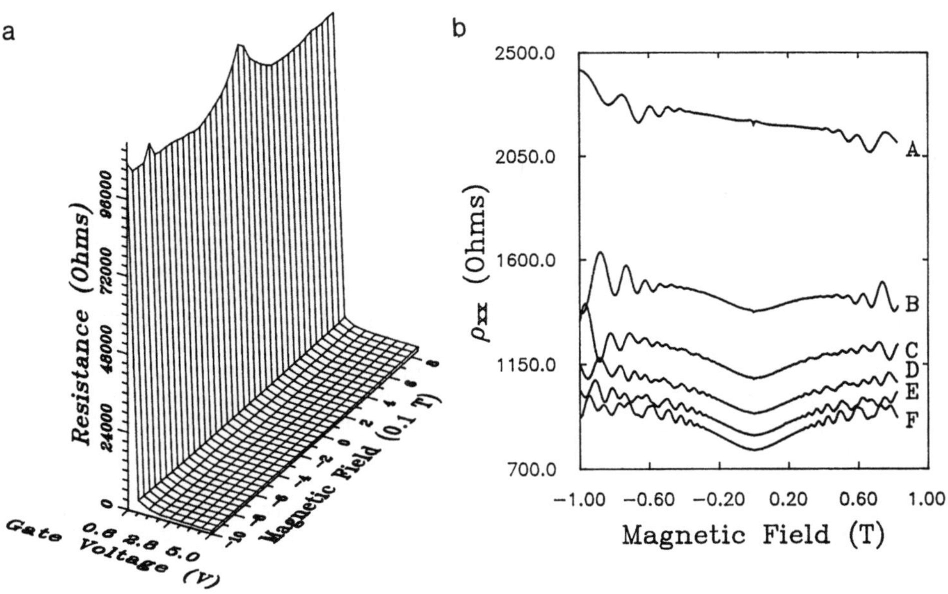

Figure 1a shows the sharp increase in magnetoresistance at low gate voltages, between 0.65 V and 5.00 V. Figure 1b shows the magnetoresistance ρ_{xx} (magnified 20x Fig. 1a) at higher gate voltages: (A) 1.5 V, (B) 2.0 V, (C) 2.5 V, (D) 3.0 V, (E) 3.5 V, (F) 4.0 V.

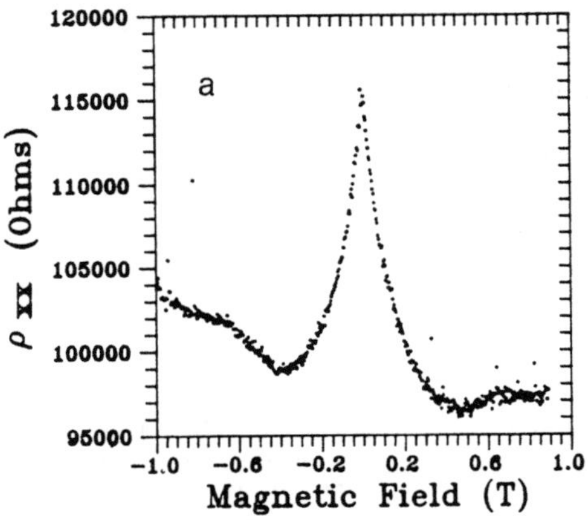

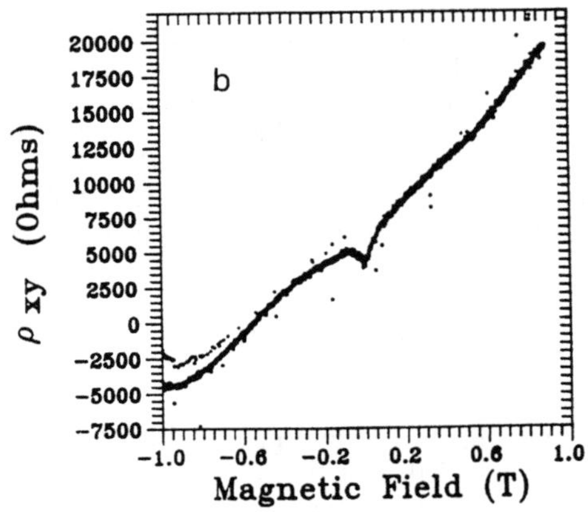

Figures 2a & b show the transverse ρ_{xx} and Hall ρ_{xy} resistances respectively. Data were recorded at 250mK with a gate voltage of 0.65 V, which gave a maximum weak localization behavior.

an effort to further study this phenomena with a different experimental technique is underway.

2.2 Weak localization

As the gate voltage was increased from low values, one moved into a regime where the magnetoresistance was dominated by weak localization. The size of the weak localization peak and the dramatic changes in the magnetoresistance with gate voltage are illustrated in Figs. 1a & 2b. This indicates conduction of a free electron gas in the presence of a high degree of disorder. When the gate voltage was increased further, the weak localization decreased due to an increasing elastic scattering time and a corresponding decrease of the disorder in the system.

A interesting feature about the data is the pronounced asymmetry in the transverse magnetoresistance. This can be explained if some regions of the interface contain localized electronic states and other regions contain extended states. Then as the gate voltage increases, the regions with extended states grow and join to form a primary conduction path. Under these circumstances it would not be unlikely for a conduction path to form in the sample with diagonal components. Such a path, nonparallel to the conduction channel, would cause a mixing of the classical Hall effect and the magnetoresistance. This mixing would lead to a non-zero Hall voltage when the magnetic field equals zero. This was observed and in addition the non-zero Hall voltage decreased with a corresponding decrease in asymmetry as the gate voltage was increased. Figures 2a & b show the magnetoresistance and the Hall resistance when the weak localization is most pronounced. One can clearly see asymmetry in the transverse resistance as well as the weak localization effects superimposed on the linear Hall resistance. A full range of guard ring voltages have been used to try and affect the asymmetry. No significant changes were observed thus indicating the asymmetry is not associated with the perimeter of the conduction channel but instead involves a process within the channel.

The data were fitted to the theory of HIKAMI et. al. [8] employing later corrections [9,10]. The spin orbit scattering time remained fairly constant at about $3x10^{-11}$ s over the range of applied gate voltages. The elastic scattering time varied from about $3x10^{-15}$ s at carrier concentrations of $4.7x10^{10}$ cm^{-2} to $8x10^{-15}$ s at the higher carrier concentrations. The inelastic time decreased for increasing carrier concentrations which is inconsistent with a predominantly electron-electron inelastic scattering mechanism at these temperatures. This is possibly due to many body correlations between the electrons. Future measurements in a parallel field should provide more insight to this mechanism. The temperature dependence of these measurements will be reported in the future.

2.3 Classical Magnetotransport

Increasing the gate voltage further removed all traces of weak localization, although a slight dip due to spin-orbit scattering when the magnetic field was zero was present at all high gate voltages. The ρ_{xx} data display a quadratic resistance which arises from two distinct subbands participating in the conduction process. A beating pattern in the Shubnikov-de Haas oscillations (Fig. 1b) also indicates the presence of these subbands. The quadratic behavior is independent of the temperature, indicating that the bands must both be electron states and not electron-hole states. The analysis is complicated by the presence of two bands, thus doubling the number of physical quantities to calculate. By measuring the Hall coefficient for $\omega_c\tau << 1$ and $\omega_c\tau >> 1$ as well as fitting the transverse magnetoresistance, we gain information on the carrier concentrations and mobilities of each band. However, the mobilities of the two bands must be independently determined in order to complete the analysis. This will be accomplished by measuring the temperature dependence of the Shubnikov-de Haas oscillations and then fitting the amplitude of these oscillations to theory.

3. CONCLUSION AND ACKNOWLEDGMENTS

3.1 Conclusion

The HgCdTe MISFET offers a unique opportunity to study electron scattering and interfacial interactions in the 2-DEG. We have shown that the electronic states at the interface can be changed dramatically from a localized regime to a classical free electron gas by using a gate voltage to change the interfacial potential well conditions of the 2-DEG. The magnetotransport data correspondingly displayed a broad variety of different effects. The effect of having two subbands complicates the analysis but this can be handled. Future experiments will look at samples with different geometries and sizes. Additional analysis of the localized regime with different experimental techniques as mentioned earlier are underway.

3.2 Acknowledgments

The authors wish to thank Wieslaw Szott for his help with the calculations and useful discussions. They also wish to acknowledge with thanks financial support in part from the National Science Foundation grant DMR 8405197, Texas Instruments, the Texas A&M University Board of Regents Advanced Materials Program, and the CEMR.

References

1. A. D. Benoit, S. Washburn, C. P. Umbach, R. B. Laibowitz, and R. A. Webb, Phys. Rev. Lett. 57, 1765 (1986).
2. A. Benoit, C. P. Umbach, R. B. Laibowitz, and R. A. Webb, Phys. Rev. Lett. 58, 2343 (1987).
3. W. J. Skocpol, P. M. Mankiewich, R. E. Howard, L. D. Jackel, and D. M. Tennant, Phys. Rev. Lett. 56, 2865 (1986).
4. W. J. Skocpol, P. M. Mankiewich, R. E. Howard, L. D. Jackel, and D. M. Tennant, Phys. Rev. Lett. 58, 2347 (1987).
5. M. Büttiker, Phys. Rev. Lett. 57, 1761 (1986).

6. K. S. Ralls, W. J. Skocpol, L. D. Jackel, R. E. Howard, L. A. Fetter, R. W. Epworth, and D. M. Tennant, Phys. Rev. Lett. 52, 228 (1984).
7. M. J. Kirton and M. J. Uren, Appl. Phys. Lett. 48, 1270 (1986).
8. S. Hikami, A. I. Larkin, and Y. Nagaoka, Prog. Theor. Phys. 63, 707 (1980).
9. S. Maekawa and H. Fukuyama, J. Phys Soc. Jpn. 50, 2516 (1981).
10. G. Bergman, Phys. Rep. 107, 1 (1984).

Localization and Proximity Effect in InAs Inversion Layer Coupled Superconducting Junction

J. Nitta and T. Kawakami

NTT Electrical Communications Laboratories,
Musashino-shi, Tokyo 180, Japan

1. Introduction

It is well known that a two-dimensional electron gas(2DEG) is formed on the surface inversion layer of p-type InAs. The localization effect has already been confirmed in the inversion layer of InAs because of its low dimensionality[1],[2]. Recently, a supercurrent has been observed in an Nb-(p-type InAs)-Nb coupled junction[3] due to the superconducting proximity effect. According to the proximity effect theory[4], the Cooper pair correlation function exponentially decays in a normal metal of a superconductor-normal metal-superconductor(SNS) junction. The decay length is given by the coherence length of the normal metal. The quantum correction to the coherence length due to the interaction associated with Anderson localization has been theoretically examined by Fukuyama-Maekawa(FM)[5]. The purpose of this paper is to clarify the localization effect in the proximity effect system of the InAs inversion layer coupled superconducting junctions.

2. Experiment

The schematic structure of the fabricated junction is shown in Fig. 1. Two superconducting Nb electrodes with a thickness of 100 nm were coupled with the inversion layer on a p-type InAs substrate. The substrate used in this experiment was (100)-oriented, Zn-doped single crystal with 2.6×10^{17} cm^{-3} acceptor concentration. After fabricating a conventional junction[3], an anodic oxide layer with a thickness of about 20 nm was formed on the surface of InAs channel to get a good conductive surface layer. The coupling length L, the separation between the Nb electrodes, was made on the same chip in the range from 0.2 µm to 5.0 µm. The electrode width W was designed to be 80 µm or 150 µm, which is much longer than L, to reduce a surface leak current out of the channel. The resistance values of the junctions at 4.2 K were approximately linear in the coupling length L. This fact shows that the electric current did not flow through the p-type bulk and the surface leak current was negligible. Under the assumption that the quantum correction to the resistance due to localization is small, $N_s\mu$ products can be estimated to be 4.0×10^{15} (1/Vs) and 2.3×10^{15} (1/Vs) for chip #1 and #2, respectively, where N_s and μ are the carrier concentration and the mobility of the inversion layer.

3. Result and Discussion

(i) Temperature dependence of normal resistance R_n

As is well known, the quantum correction to resistivity in disordered 2DEG systems gives a logarithmic increase with a decrease in temperature. However, the normal resistance R_n in proximity systems is expected to decrease because of an increase in the coherence length with a decrease in temperature. The coherence length is given in the 2DEG systems as

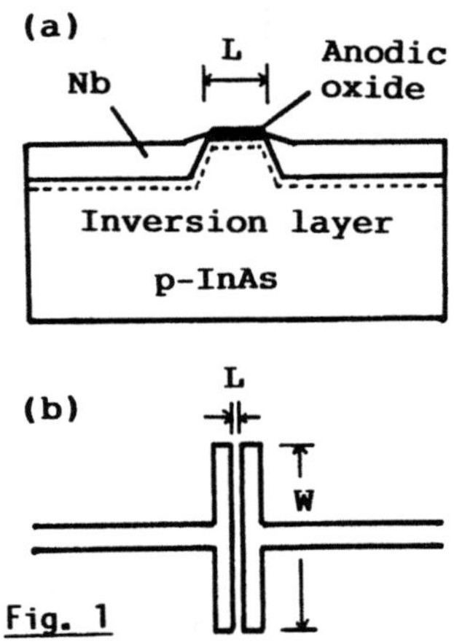

Fig. 1

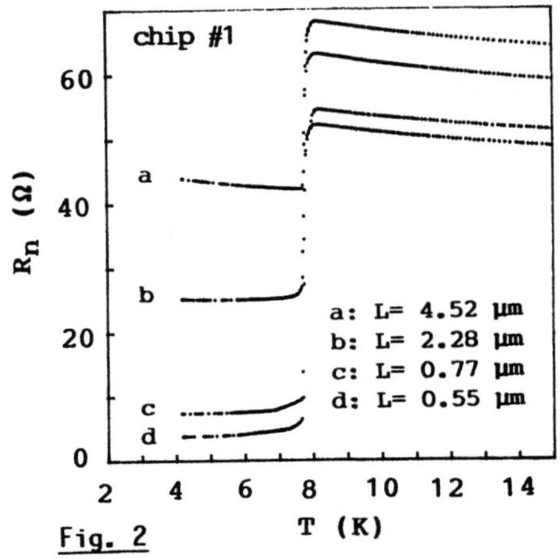

Fig. 2

Fig. 1 Schematic of the junction structure: (a) cross-sectional view of the junction, (b) electrode configuration

Fig. 2 Temperature dependence of normal metal resistance for various coupling lengths L junctions

$$\xi_{NO}(T)=(\hbar^3 N_s \mu/2k_B T e m^*)^{1/2}, \quad (1)$$

where m^* is the carrier effective mass, and T is the temperature.

The normal resistance R_n was measured for junctions with various coupling lengths L. The temperature dependences of R_n for chip #1 are shown in Fig. 2. These properties were found to depend on the coupling length L. R_n for L> 2 µm increased gradually as temperature decreased, while R_n for L< 1 µm decreased as temperature decreased. The L> 2 µm junctions are thought mainly to be affected by localization. As will be discussed below, the effect of localization was also confirmed on these L> 2 µm junctions from the magnetoconductance. The proximity effect is not strongly affected because L is much longer than the coherence length ξ_{NO}(4.2 K)= 0.1 µm. The proximity effect predominantly affects L< 1 µm junctions, where the effective coupling length is reduced by the coherence length.

(ii) Magnetoconductance

The distinctive feature of localization appears when measuring magnetoconductance properties. Under magnetic field, the time-reversal symmetry for the quantum interference is destroyed. As a result of this fact, a positive-magnetoconductance was observed in various disordered systems. For InAs crystals, a negative magnetoconductance was also observed because of a strong spin-orbit interaction. In proximity effect systems, the induced pair potential is expected to also be destroyed by the field. This leads to a negative-magnetoconductance.

The magnetoconductance in the perpendicular field is plotted in Fig. 3 for chip #1. It was also found that the magnetoconductance depended on L. For L> 2 µm junctions, the magnetoconductance properties consist of three components. At a low magnetic field, the conductance decreased due to the anti-localization effect caused by the strong spin-orbit interaction. At an intermediate field from 0.1 T to 0.3 T, the conductance increased as the field increased due to the localization effect. Above 0.3 T, the junction showed the classical magnetoconductance properties. Figure 4 shows a comparison between the data for L= 4.52 µm junction and the Hikami, Larkin and Nagaoka(HLN) theory[6]. Here, the spin-orbit scattering time τ_{so} and the inelastic scattering time τ_ε are treated as fitting parameters,

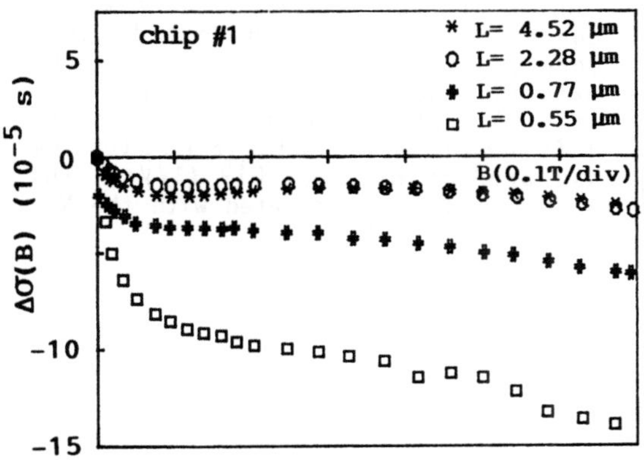

Fig. 3

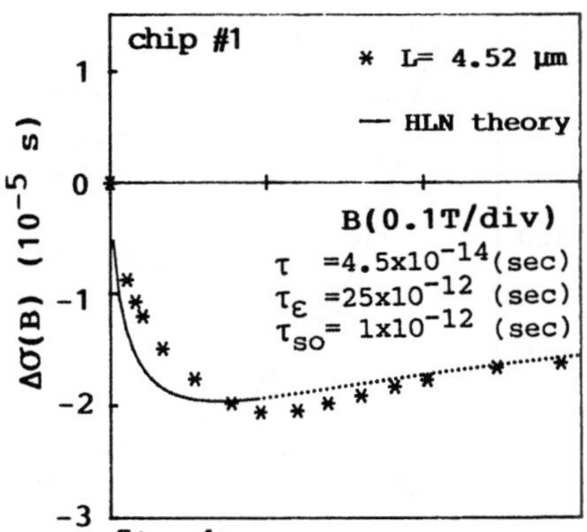

Fig. 4

Fig. 3 Magnetoresistance in perpendicular field for various coupling lengths L junctions

Fig. 4 Comparison between the HLN theory and the experiment. The dotted line corresponds to the region where the theory cannot hold.

and the $N_s\mu$ product was kept to be 4.0×10^{15} (1/Vs). The experimental result did not quantatively but qualitatively agree with the theory. The magnetic field of a minimum conductance obtained by the theory is lower than that of the experiment. The same discrepancy has been pointed out[2] in the InAs inversion layer.

The degree of negative magnetoconductance became large as L became short. For L< 1 μm junctions, anomalously large negative magnetoconductance properties were observed. The large negative magnetoconductane can not be explained in terms of localization. It is thought that the destruction of the proximity effect in the magnetic field was predominantly observed in short L junctions. One possible reason for the destruction can be attributed to the pair breaking effect.

(iii) Temperature dependence of supercurrent I_c

Recently, FM[5] showed that the coherence length was reduced due to localization. According to their theory, the coherence length of the 2DEG proximity system is modified as

$$\xi_N(T)=\xi_{N0}(T)(1+\lambda\ln\pi k_BT\tau/\hbar)^{1/2}(1+\lambda\ln k_BT\tau/\hbar)^{1/2}, \qquad (2)$$

where λ is the randomness parameter and is given as $\lambda=\hbar/2\pi\varepsilon_F\tau$, τ is the elastic scattering time, and ε_F is the Fermi energy. Therefore, the temperature dependence of I_c is expected to be modified, because it is expressed by using the coherence length as[7]

$$I_c\propto\Delta_n(T)^2(1/\xi_{N0})[\cosh^2(L/2\xi_{N0})]^{-1}, \qquad (3)$$

where $\Delta_n(T)$ is the induced pair potential at the SN boundary.

In this Nb/2DEG/Nb system, a supercurrent was observed for L= 0.27 μm junction of chip #2. The temperature dependence of the supercurrent is shown in Fig. 5. The solid and dotted lines represent the calculated curves on the basis of the FM theory and the existing theory[7], respectively. The experimentally obtained result can be fitted both by the FM theory and the existing theory, treating N_s and μ as fitting parameters. The $N_s\mu$ products evaluated from the FM theory and the existing theory are 1.06×10^{15} (1/Vs) and 0.86×10^{15} (1/Vs) as the best fit values, respectively. These $N_s\mu$ products correspond to ξ_N(4.2 K)=0.041 μm and ξ_{N0}(4.2

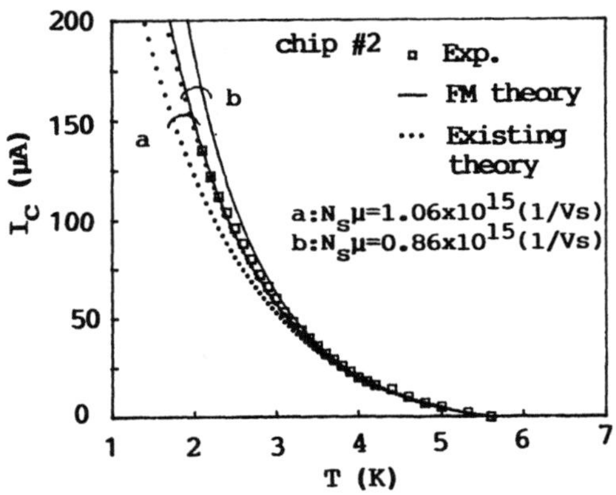

Fig. 5 Temperature dependence of supercurrent. The solid and dotted lines correspond to the FM theory and the existing theory, respectively. (a) A pair of solid and dotted lines on the left hand side are calculated with the $N_s\mu$ product of 1.06×10^{15} (1/Vs). (b) Another pair are estimated with the $N_s\mu$ of 0.86×10^{15} (1/Vs).

K)=0.050 μm, respectively. Here, m was assumed to be 0.024. However, the products are about half 2.3×10^{15} (1/Vs), obtained from the sheet resistance. One reason for this discrepancy may be attributed to the effective mass. It is reported that the effective mass in 2DEG of InAs-GaSb superlattice is enhanced as the Fermi energy increases[8]. Although the effective mass in the p-InAs inversion layer has not yet been obtained experimentally, the FM theory and the existing theory under an assumption of $m^*=0.063$ can be fitted to the experiment with the $N_s\mu$ product of 2.3×10^{15} (1/Vs) obtained from the sheet resistance.

According to the FM theory, the pair amplitude is also reduced due to localization. In the above calculation, this effect was not taken into account. The correction to the amplitude yields a smaller $N_s\mu$ product. If the $N_s\mu$ product obtained from the sheet resistance is true, the randomness parameter λ becomes 0.034. It may be difficult to observe a clear localization effect on temperature dependence of I_c under low λ conditions.

4. Summary

The InAs inversion layer coupled superconducting junctions has been studied from the view point of localization. The temperature dependences of normal resistance R_n and magnetoconductance are found to depend on the coupling length L. The L> 2 μm junctions are affected mainly by localization. The proximity effect is predominant in L< 1 μm junctions. The temperature dependences of supercurrent I_c can be fitted both by the existing theory and the FM theory. The clear quantum correction to the coherence length due to the localization has not yet been observed.

Acknowledgement

The authors would like to thank Prof. H.Fukuyama and Prof. S.Maekawa for their valuable discussions and comments. They also express their thanks to Drs. Y.Kato, T.Kimura, T.Izawa and T.Yamada for their encouragement, Drs. H.Takayanagi and E.Yamaguchi for their helpful discussions, and T.Kawashima for performing the electron beam lithography.

References

1. E. Yamaguchi; Phys. Rev. **B32** 5280 (1985)
2. Y. Kawaguchi, I. Takayanagi and S. Kawaji; J. Phys. Soc. Jpn. **56** 1293 (1987)
3. H. Takayanagi and T. Kawakami; Phys. Rev. Lett. **54** 2249 (1985)

4. P. G. de Gennes; Rev. Mod. Phys. **36** 225 (1964)
5. H. Fukuyama and S. Maekawa; J. Phys. Soc. Jpn. **55** 1814 (1986)
6. S. Hikami, A. I. Larkin and Y. Nagaoka; Prog. Theor. Phys. **63** 707 (1980)
7. J. Seto and T. Van Duzer; Low Temperature Physics LT-13 **3** 328 (1972)
8. H. Sakaki, L. L. Chang, G. A. Sai-Halasz, C. A. Chang and L. Esaki; Solid State Comm. **26** 589 (1978)

Part VI

Theory of Weak Localization Effects

The Temperature of the Superconducting Transition in Amorphous Films

A.M. Finkel'stein

L.D. Landau Institute for Theoretical Physics of the USSR Academy of Sciences, SU-117334 Moscow, ul. Kosygina 2, USSR

Abstract

The reduction of the superconducting transition temperature in homogeneous amorphous films has been studied theoretically. It is shown that in order to give a correct description of this effect it is not sufficient just to take into account the first-order corrections in $t=(e^2/2\pi^2\hbar)R_\square$ even at $t\sim 0.01$ and it is necessary to use the renormalization group equations. The dependence of the transition temperature on sheet resistance $R_\square$ found by means of these equations fits the experimental results well.

1. After the paper /1/ it has been established that the Coulomb interaction of electrons in combination with scattering by impurities strongly affects the properties of conduction electrons in normal metals (see the review /2/). These effects manifest themselves in particular in deviations of the superconductivity critical temperature from the predictions of BCS theory. It has been predicted in /3,4/ that the presence of disorder lowers the transition temperature T_c in the following manner:

$$\ln\left(\frac{T_c}{T_{c0}}\right)\approx -\frac{e^2}{6\pi^2\hbar}\, g_1 R_\square \ln^3\left(\frac{1}{T_c\tau}\right) , \qquad (1)$$

where T_{c0} is the transition temperature in a bulk system, $R_\square$ is the sheet resistance, τ is the mean free path time, and g_1 is the constant describing the screened Coulomb interaction. This formula was first derived in /5/, but the sign of the effect was incorrect.

To exclude the influence of other effects when studying this phenomenon, it is necessary that the bulk properties of the electron gas be the same in films with different T_c. In homogeneous films of different thickness but of the same composition, bare values of the Fermi liquid parameters like the density of states, coupling constants and so on, can be regarded as invariable (we neglect the effect of transverse quantization on T_c, which effect has been discussed in /5/, for instance, and which is probably irrelevant for the compounds being studied). Therefore, measurements of T_c versus $R_\square$ on a number of such films make it possible straightforwardly to find the effect of the interplay of the Coulomb interaction and impurity scattering /1,2/ on superconductivity in disordered two-dimensional systems.

For this purpose, a few years ago measurements of T_c were performed in amorphous W-Re /6/, Mo-Ge /7/ and Nb_3Ge /8/ (the effect of disorder on T_c had previously been studied in /9/; see also the review /10/). The choice of these materials was governed by two reasons. First, in these materials $\varepsilon_F\tau\sim 1$ and accordingly $\ln(1/T_{c0}\tau)\sim 5-10$. Therefore, although T_c in amorphous films is influenced by a diversity of factors /10/, the mechanism of interest, thanks to a large value of $\ln^3(1/T_{c0}\tau)$, must be dominant. Second, to distinguish the discussed effect from "geometric" phenomena like percolation or formation granules, it is necessary that films of different thickness be homogeneous. To ensure homogeneity of films both the technology of production and the selection of composition components are important. In particular,

Mo-Ge films were produced in such a way that they were homogeneous down to a thickness of 10 Å. Recently, the effect of the reduction of T_c has been studied in very thin amorphous Pb /11/. In this work to control the homogeneity of films, the ratio Δ/T_c has been neasured. The fact that this ratio remains constant for films of different thickness testifies to the absence of granule formation.

It has been found in /6 - 8,11/ that in comparison with the value T_{c0} in the bulk, the transition temperature in these films is already halved at $R_\square \sim 1\,k\Omega$ and in extremely thin but still homogeneous films /7/ it has been found that at $R_\square \sim 2-3\,k\Omega$ the superconducting transition is completely suppressed. In the experiments discussed, $T_c(R_\square)$ at first falls linearly with increasing $R_\square$, which accords with the predictions of (1). However, at $R_\square \gtrsim 1\,k\Omega$, T_c decreases more slowly, and in /7/ it has been pointed out that (1) should be modified in some way.

2. Before discussing how (1) should be modified, let us touch upon a very important issue. This formula has been derived by calculating diagrams a and b of Fig.1, which presents the main corrections to the superconductivity response function. Note that here the wavy line is the screened Coulomb interaction $V(q,\Omega)$, Γ_c is the amplitude of the interaction in the Cooper channel and all impurity lines are dropped. In deriving (1) it was assumed that as usual the main contribution comes from the "diffusion" region of momenta $q \sim (\Omega/D)^{1/2}$ (D is the electron diffusion coefficient). However, the region of small momenta

$$\left(\frac{\Omega}{D}\right)^{1/2} \gtrsim q \gtrsim \frac{\Omega}{D\kappa} \tag{2}$$

(κ is the inverse screening radius) could in principle also be important, since, due to the singular behaviour of the Coulomb interaction on small momenta, each of the diagrams a - d in Fig.1 adds to T_c a correction of the same order of magnitude as in (1). By their origin all these corrections are akin to the double logarithmic correction to the one-particle density of states which also stems from the region (2). Yet, the sum contribution of all four diagrams in Fig.1 calculated in the small momenta region (2) vanishes, and thus the interaction which transfers momenta smaller than "diffusion" momenta does not affect the superconducting transition temperature. This occurs for the following reason. Integration over small momenta $q \lesssim (\Omega/D)^{1/2}$ does not involve diffusion propagators and does not involve the Coulomb interaction $V(q,\Omega)$. In the result of the integration this long-range part of the Coulomb interaction proves to be dependent effectively only on the frequency and does not actually transfer momentum. This kind of interaction can be excluded by means of a gauge transformation and therefore it does not manifest itself in thermodynamic quantities. (Note that the one-particle density of states does not belong to such quantities.) In particular, it cannot affect the transition temperature. A contrary conclusion (see,e.g., /12/) is accounted for by the fact that the diagrams c and d of Fig.1 have been disregarded.

3. Below we shall demonstrate that to find $T_c(R_\square)$ it is not sufficient to confine ourselves to the first-order correction even at $R_\square \sim 1\,k\Omega$, although the parameter $t = (e^2/2\pi^2\hbar)R_\square$ in which the expansion is performed is still extremely small: $t \sim 0.01$. Despite this it appears that to give a correct description of the effect of the reduction in transition temperature it is necessary to employ the renormalization group equations describing a disordered system of interacting electrons /13/. Since

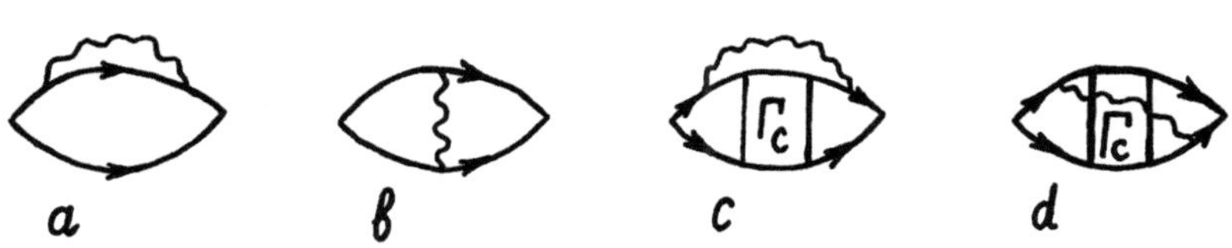

Fig.1. Diagrams used to calculate (1)

momenta that are smaller then the diffusion momenta are irrelevant, T_c can be found from the moment when the amplitude Γ_c becomes infinite.

In the first order in t a renormalization group equation for Γ_c is /13,14/

$$\frac{d\Gamma_c}{dy} = \frac{1}{2}t(Z + m\Gamma_2) - \frac{\Gamma_c^2}{Z} \quad , \tag{3}$$

where $y = \ln 1/\omega\tau$; Γ_2 is the amplitude of electron interaction in the spin density channel; Z describes interaction of long wavelength fluctuations of electron density, and is also a parameter which describes renormalization of the frequency coefficients at the propagator of diffusion and cooperon modes /13/; m is the number of the effective triplet components of a cooperon: m = 3 in the absence of the spin-orbit interaction and m = 0 if the spin relaxation time $\tau_{so} < T_c^{-1}$ (Eq.(3) with m = 0 was recently used in discussing the interplay of the metal-insulator transition with superconductivity /15/). The first term in (3) emerges from renormalization of the amplitude Γ_c by impurity-scattering charge and spin density fluctuations of the electron gas. As long as the temperature decrease is relatively minor, the value of $T_c(R_\square)$ obtained from (3) corresponds to (1), where the parameter g_1 should be replaced by $(Z + m\Gamma_2)/2$.

The author has no concrete data on the value of Γ_2 and the spin-orbit interaction in the systems under study. However, bear in mind that the slope of the critical field dH_{c2}/dT in a bulk sample permits one to determine the diffusion coefficient. In conjunction with this circumstance, in /6 - 8/ the value of the parameter g_1 has been estimated by processing the experimental data on $T_c(R_\square)$ according to (1). In all three cases g_1 appeared to be rather close to 0.5. Since $(Z + m\Gamma_2)/2$ in (3) plays the role of the parameter g_1 and the bare value of Z is unity, the closeness of g_1 to 0.5 means that $m\Gamma_2$ is not large. We shall assume that m = 0 since the presence of rather heavy elements points to the fact that the spin-orbit interaction must be of importance. Then, to first order in t, the renormalization group equations describing the behaviour of the system are /14/

$$\frac{d\Gamma_c}{dy} = \frac{1}{2}tZ - \frac{\Gamma_c^2}{Z} \; ; \qquad \frac{dZ}{dy} = t\left(-\frac{Z}{2} + \Gamma_c\right) \tag{4}$$

$$\frac{dt}{dy} = t^2\left(\frac{1}{2} - \frac{\Gamma_c}{Z}\right) . \tag{5}$$

To find T_c, renormalization of t can be ignored, since (5) starts to be relevant only at $T_c/T_{c0} < 0.1$. Then, integrating (4) we get

$$\frac{T_c}{T_{c0}} = \exp\left(-\frac{1}{\gamma}\right) \left[\left(1 + \frac{(t/2)^{1/2}}{\gamma - t/4}\right) \Big/ \left(1 - \frac{(t/2)^{1/2}}{\gamma - t/4}\right)\right]^{1/\sqrt{2t}} \tag{6}$$

where $\gamma = 1/\ln T_{c0}\tau$; $\gamma < 0$.

It follows from (6) that there are two characteristic scales of t describing the superconducting transition temperature decrease with increasing resistance R : first, T_c falls considerably at $t \sim |\gamma|^3$ in conformity with (1), then the decrease slows down and superconductivity vanishes at larger $t \sim \gamma^2$. Thus we conclude that to describe the suppression of superconductivity in amorphous films correctly it is not sufficient just to take into account the correction of the first order in t, despite the fact that $t \sim 0.01$. In this connection note that the reentrant behaviour of $T_c(R_\square)$ obtained from (1) is not found in reality.

In Fig.2 it is shown that (6) describes the superconductivity suppression observed in homogeneous amorphous $Mo_{79}Ge_{21}$ films very well. Processing the experimental data of /7/ in accordance with (6) we assumed that $\ln 1/T_{c0}\tau = 8.2$, which is rather close to the value employed in /7/. However, it should be noted that (6) is based on a

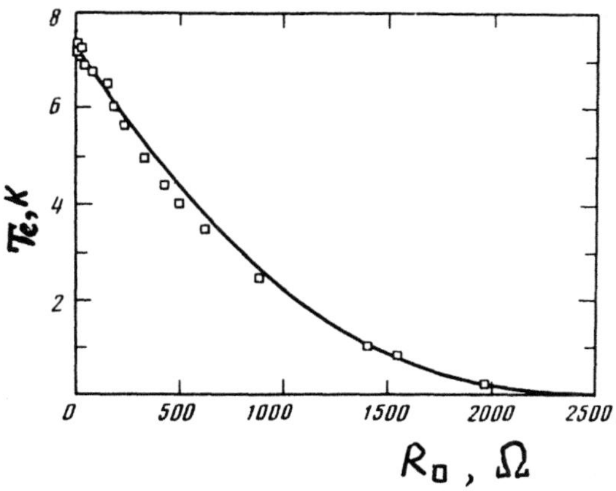

Fig.2. Suppression of superconductivity in amorphous $Mo_{79}Ge_{21}$ films: (□) data from /7/; (-) result of experimental data processing in accordance with (6)

certain specific circumstance. The thing is that in the compounds considered $\tau^{-1} > \omega_D$ (ω_D is the Debye frequency), therefore the equation for the amplitude Γ_c should be integrated separately in each of the regions $\tau^{-1} > \omega > \omega_D$ and $\omega_D > \omega$. Yet at reasonable values of the parameters, renormalization of the value of Γ_c, occurring due to the interplay of the interaction and of the scattering of two-dimensional electrons, practically does not depend on the value of Γ_c in the region $\tau^{-1} > \omega > \omega_D$ since the term containing t in (3) does not depend on Γ_c. This is the reason why in deriving (6) we have not considered separately frequencies higher or lower than ω_D, although, strictly speaking, this is not always correct.

4. In conclusion, let us point out that we have focused on a case, of interest for the theory of disordered systems, where, on one hand, to describe the effect it is necessary to employ the renormalization group equations, and, on the other, the applicability conditions for these equations in a low order in t are fulfilled. Superconductivity suppression in the films discussed occurs by virtue of the joint effect of electron density fluctuations and of impurity scattering. It is quite probable that this mechanism (plus a similar action of spin density fluctuations) is responsible for the fact that in three-dimensional homogeneous materials superconductivity vanishes on the metallic side of the metal-insulator transition.

References

1. B.L. Altshuler, A.G. Aronov: Sov. Phys. - JETP 77,2028(1979)
   B.L. Altshuler, A.G. Aronov, P.A. Lee: Phys. Rev. Lett. 44,1288(1980)
2. P.A. Lee, T.V. Ramakrishnan: Rev. Mod. Phys. 57,287(1985)
3. S. Maekawa, H. Fukuyama: J. Phys. Soc. Jpn. 51,1380(1982)
   S. Maekawa, H. Ebisawa, H. Fukuyama: J. Phys. Soc. Jpn. 52,1352(1983)
4. H. Takagi, Y. Kuroda: Solid State Commun. 41,643(1982)
5. Yu.N.Ovchinnikov: Sov. Phys. - JETP 64,719(1973)
6. H.R. Raffy, R.B. Laibowitz, P. Chaudhari, S. Maekawa: Phys. Rev. B 26,6607(1983)
7. J.M. Graybeal, M.R. Beasley: Phys. Rev. B 29,4167(1984)
   J.M; Graybeal, M.R. Beasley, R.L. Greene: Proc. 17th Int. Conf. Low Temp. Phys. (North-Holland, Amsterdam 1984) p.731
8. P.H. Kes, C.C. Chi, C.C. Tsuei: Proc. Int. Conf. Localization Interaction and Transport Phenomena in Impure Metals, ed. by L. Schweitzer, B. Kramer. PT-8 (Braunschweig 1984) p.181
9. M. Strongin, R.S; Thompson, O.F. Kammerer, J.E. Crow: Phys. Rev. B 2,1078(1971)
10. Yu.F. Komnik: FNT 8,115(1982)
11. R.C; Dynes, A.E. White, J.M. Graybeal, J.P. Garno: Phys. Rev. Lett. 57,2195(1986)
12. H. Ebisawa, S. Meakawa, H. Fukuyama: J. Phys. Soc. Jpn. 54,2257(1985)
13. A.M. Finkel'stein: Z. Phys. B 56,189(1984)
14. C. Castellani, C. DiCastro, G. Forqacs, S. Sorella: Solid State Commun. 52,261 (1984)
15. M. Ma, E. Fradkin: Phys. Rev. Lett. 56,1416(1986)

Magnetic Exchange in Disordered Metals

E. Abrahams, A. Jagannathan, and M.J. Stephen

Serin Physics Laboratory, Rutgers University,
Piscataway, NJ 08855-0849, USA

Abstract

The indirect RKKY exchange interaction in the presence of weak non-magnetic disorder is found to have a broad probability distribution whose odd moments are exponentially short ranged and whose even moments are long ranged as in the pure metal. This is in agreement with an old result on the first moment (de Gennes) and recent results on the second moment (Zyuzin and Spivak and Bulaevskii and Panyukov). The influence of spin-orbit scattering and the relevance to experiments will be discussed.

The conduction electron mediated indirect exchange interaction between localized magnetic moments was originally derived by Ruderman and Kittel [1] in 1954. In the next few years, Kasuya [2] and Yosida [3] gave the detailed theory of this "RKKY" interaction when the local moments were due to ion cores in a metallic environment. Since then, the RKKY interaction has been invoked in a variety of physical phenomena, particularly spin glass behavior in metals containing a random distribution of local magnetic moments.

The RKKY interaction arises from the propagation of spin information from one local moment to the other by means of conduction-electron spin-density fluctuations. The effective local moment-local moment interaction term can be found in second order perturbation theory. It depends on the zero frequency nonlocal conduction electron susceptibility $\chi(R)$:

$$H_{int} = -J^2 S_1 S_2 \chi(R), \tag{1}$$

where J is the local spin-conduction electron spin coupling and R the distance between the magnetic impurities. In the pure metal, the familiar RKKY result is, for $k_F R \gg 1$,

$$\chi_o(R) = -(N_o/2\pi R^3)\cos 2k_F R\,, \tag{2}$$

where k_F is the Fermi wave number and N_o the one spin density of states at the Fermi level. We note the long range (R^{-d} in d-dimensions) R dependence and the presence of characteristic Friedel oscillations which are responsible for spin glass behavior in metals containing a random distribution of local magnetic moments.

In an early paper, de Gennes (4) addressed the question of the behavior of the RKKY interaction in a weakly disordered metal. He presented a calculation of the <u>average</u> (over configurations of non-magnetic impurities) interaction $\bar{\chi}(R)$. While his result for general $R > k_F^{-1}$ is complicated, it has a simple form for $R \gg \ell$, where ℓ is the mean free path for conduction electron scattering from the non-magnetic disorder. Using modern methods of impurity averaging [5,6], we may easily obtain the de Gennes result at $R \gg \ell$ since the zero frequency susceptibility $\chi(R)$ involves only retarded-retarded or advanced-advanced pairs of electron propagators. Thus, no "diffuson" corrections [6] enter $\bar{\chi}(R) \propto \sum_\omega \overline{G_\omega(R)G_\omega(-R)}$ which is then controlled by the exponential fall-off of the average propagator $\bar{G}(R) \propto \exp(-R/2\ell)$.

Therefore,

$$\bar{\chi}(R) = \chi_0(R)e^{-R/\ell} \quad . \tag{3}$$

This result of de Gennes has been widely misinterpreted in the literature since its appearance. That it does not obey the sum rule

$$\int dR\bar{\chi}(R) = 2N_0 \tag{4}$$

has often been mistakenly invoked to argue that it is incorrect. The failure of the sum rule is due only to the fact that Eq. (3) is valid only when $R \gg \ell$. When the correct de Gennes form [4] is used, the sum rule is obeyed.

A more serious misapplication of the de Gennes result has been in the analysis of experiments on metallic spin glasses where the exponential decay of the average RKKY interaction has been (mis)used to argue for a strong mean free path dependence of the spin glass transition temperature, T_g. In fact, the average interaction in a spin glass is not relevant. It is the strength of <u>typical</u> interactions which determine T_g. Thus, if the $\chi(R)$ are gaussian distributed for fixed R, then $\overline{\chi^2}(R)$ determines T_g [7] and not $\bar{\chi}(R)$. We can understand that the range of typical interactions does not have a mean free path dependence, as follows: in a given configuration of impurities, $\chi(R)$ acquires a configuration dependent phase shift $\phi(R)$ so that

$$\chi(\vec{R}) \propto \frac{1}{R^3}\cos[2k_F + \phi(\vec{R})] \tag{5}$$

which remains long range. After averaging, since the $\phi(R)$ are random over the configurations, an exponential decay, as in Eq. (3) results.

These remarks on the relevance of $\bar{\chi}(R)$ have already been made by Zyuzin and Spivak [8] and Bulaevskii and Panyukov [9]. Even earlier, de Châtel [10] pointed out the importance of investigating the nature of the distribution of $\chi(R)$ before drawing conclusions from the first moment $\bar{\chi}(R)$. We can show that the distribution of $\chi(R)$ in the disordered system is broad, that it is not self-averaging and, as in Refs. 8,9, that typical interactions fall off as R^{-d} in d-dimensions.

The calculations are carried out using the usual impurity averaging technique. For example, for the second moment of the distribution of $\chi(R)$, we have

$$\overline{\chi^2}(R) \propto \sum_{\omega_1\omega_2} \overline{G_{\omega_1}(R)G_{\omega_1}(-R)G_{\omega_2}(R)G_{\omega_2}(-R)} \quad . \tag{6}$$

The leading contributions at $R > \ell$ come when $\mathrm{sgn}(\omega_1) = -\mathrm{sgn}(\omega_2)$ so that two of the four G's in Eq. (6) are retarded and two advanced. Then diffusion corrections enter and the exponential decay disappears. The situation resembles that encountered in the calculation of conductivity fluctuations [11]. We give here the expression for the leading contribution to the 2n-th moment of the distribution (the odd moments decay exponentially) for d=3:

$$\overline{\chi^{2n}}(R) = (2n-1)!(N_0/4\pi)^{2n} I_{2n}/R^{6n} \quad , \tag{7}$$

where, with $x_{2n+1} = x_1$,

$$I_{2n} = \int_0^\infty \cdots \int_0^\infty dx_1 \ldots dx_{2n} \prod_{i=1}^{2n} \exp(-\sqrt{x_i + x_{i+1}}) \tag{8}$$

is an integral less than one which decreases slowly with n. For example, $I_2 = .75$, $I_4 \approx 0.4$. The square root of the fourth moment does not differ substantially from the second moment, indicating that the distribution is fairly broad. We conclude, in support of the remarks in Refs. 8-10, that changes of the disorder configuration phase shift the Friedel oscillations which arise from the sharp energy surface [12] and lead to a broad distribution of $\chi(R)$ with a width proportional to R^{-3} (at d=3). Similarly, at d=2, the R^{-2} variation of typical RKKY interactions is obtained.

We mention here, the effect of finite temperature and of the presence of impurity spin-orbit scattering. As pointed out first by Zyuzin and Spivak [8], when $R > L_T = \sqrt{hD/k_BT}$ (D is the diffusion constant), then the coherent diffusion corrections disappear and exponential decay of typical interactions reappears but with a decay length L_T.

As is to be expected, the presence of spin-orbit scattering destroys, at $R > L_{so} = \sqrt{\ell\ell_{so}}$, the spin information carried by the conduction electrons. Here, ℓ_{so} is the spin-orbit scattering mean free path and $\ell_{so} > \ell$ is assumed. Then the RKKY exchange interactions typically decay according to $\exp(-R/L_{so})$. However, as pointed out in [8], anisotropic spin interactions of Dzyaloshinskii-Moriya type remain long ranged. These may be of the form $\vec{S}_1 x \vec{S}_2 \cdot \hat{A}$ and $(S_1 \cdot \hat{B})(S_2 \cdot \hat{B})$ where $\hat{A}, \hat{B}$ are random vectors depending upon the precise configuration of spin-orbit scattering impurities.

Although the physical arguments we have given apply above the mobility edge even in the case of strong disorder, application of the formulas of Ref. 8 and those given here depend upon the conditions

$$k_F R >> k_F \ell >> 1 \tag{9}$$

being satisfied. For example, in a metallic spin glass, where in mean field theory the transition temperature T_g is determined by the second moment of the RKKY distribution (if one assumes that it is gaussian), we can use Eq. (7) and show that T_g is independent of ℓ and scales with concentration of magnetic impurities. Most experiments which investigate the effect of non-magnetic disorder on spin glass properties do not satisfy the conditions of Eq. (9). At the same time, as remarked earlier, the original de Gennes result of Eq. (3) has often been used to interpret the data although it is neither relevant to the problem nor valid outside the regions of Eq. (9). There are however experiments which do satisfy the dilution conditions of Eq. (9) [13] and which show that T_g is independent of mean free path and that it scales with concentration.

Nuclear magnetic resonance in a spin glass system should give information on both $\bar{\chi}$ (Knight shift) and $\overline{\chi^2}$ (inhomogeneous linewidth). Unfortunately, the only data [14] of which we are aware is on samples with average moment separations $\bar{R}$ less than ℓ. Thus Eq. (9) is violated and the present results are inapplicable.

The details of these calculations will be published elsewhere. The research reported here was supported in part by the National Science Foundation under Grant Nos. DMR-85-20190 and DMR-84-05619.

References

1. M.A. Ruderman and C. Kittel, Phys. Rev. 96, 99 (1954)
2. T. Kasuya, Phys. Rev. 100, 893 (1957)
3. K. Yosida, Prog. Theor. Phys. 16, 45 (1956)

4. P.G. deGennes, J. Phys. Rad. 23, 630 (1962)
5. T. Tsuneto, Phys. Rev. 121, 402 (1960)
6. P.A. Lee and T.V. Ramakrishnan, Rev. Modern Phys. 57, 287 (1985)
7. K.H. Fischer, Phys. Rev. Lett. 34, 1438 (1975)
8. A.Yu. Zyuzin and B.Z. Spivak, Pis'ma Zh. Eksp. Teor. Fiz. 43, 185 (1985) [JETP Lett. 43, 234 (1986)]
9. L.N. Bulaevskii and S.V. Panyukov, Pis'ma Zh. Eksp. Teor. Fiz. 43, 190 (1985) [JETP Lett. 43, 240 (1986)]
10. P.F. de Châtel, J. Mag. Mag. Mat. 23, 28 (1981).
11. P.A. Lee and A.D. Stone, Phys. Rev. Lett. 55, 1622 (1985)
12. J.H. Samson, J. Phys. F 13, 1173 (1983)
13. D.C. Vier and S. Schultz, Phys. Rev. Lett. 54, 150 (1985)
14. A.J. Heeger, A.P. Klein and P. Tu, Phys. Rev. Lett. 17, 803 (1966)

The Kondo Effect in the Weakly Localized Regime

S. Suga, H. Kasai, and A. Okiji

Department of Applied Physics, Osaka University,
Suita, Osaka 565, Japan

1. INTRODUCTION

The effects of the randomness on the s-d exchange interaction have been investigated recently. It has been shown by the perturbational calculations that the characteristic temperature dependence appears in the third order conductivity and the second order susceptibility with respect to the coupling constant of the s-d exchange interaction in the weakly localized regime (abbreviated as WLR) [1,2]. In order to investigate the effects of the randomness on the s-d exchange interaction in detail, it is necessary to examine the higher order contributions to the physical quantities mentioned above. For that purpose we perform the calculations of the conductivity both analytically and numerically in the WLR with the use of the diagrammatic method.

2. QUANTUM CORRECTIONS IN THE WLR

The system considered here contains randomly distributed non-magnetic impurities and a single magnetic one. The Hamiltonian may be written as

$$H = \sum_{k\sigma} \varepsilon_k a^+_{k\sigma} a_{k\sigma} + u \sum_{r_\ell} \sum_{kk'} \sum_{\sigma} \exp[-i(k-k')r_\ell] a^+_{k\sigma} a_{k'\sigma} - J \sum_{kk'} \sum_{\sigma\sigma'} a^+_{k\sigma} a_{k'\sigma'} d^+_{\sigma'} d_\sigma \quad , \tag{1}$$

where the third term describes the Coqblin-Schrieffer exchange interaction. This term is equivalent to the s-d exchange interaction with respect to the spin-flip processes.

Introducing the chemical potential, $-\mu$, in the localized state of a magnetic impurity [3], we calculate the self-energy of the conduction electrons, $\Sigma^{(\mu)}(i\varepsilon)$, and the number of the electrons in the localized orbital of a magnetic impurity, $n^{(\mu)}$. The self-energy, $\Sigma(i\varepsilon)$, in this treatment is given by the relation, $\Sigma(i\varepsilon)=\lim_{\mu\to\infty} \Sigma^{(\mu)}(i\varepsilon)/n^{(\mu)}$ [4]. Since the quantum corrections to the conductivity, σ', come out from the diagrams of the self-energy type [2,5], σ' can be written in terms of the self-energy, $\sigma'=2\sigma_0\tau_0 \mathrm{Im}\Sigma(0)$, where τ_0 is the life time of the conduction electrons due to non-magnetic impurities and $\sigma_0=n_e e^2\tau_0/m$, n_e and m being the density of the conduction electrons and the mass of them.

The diagrams of the self-energy, $\Sigma^{(\mu)}(i\varepsilon)$, which give rise to the most singular behaviors with respect to the temperature in the order of $J^n\lambda^{d-1}$ are shown in Fig.1, where d is the dimensionality and λ the degree of disorder given by $1/2\pi\varepsilon_F\tau_0$, ε_F being the Fermi energy. The solid lines represent the thermal Green functions of the conduction electrons given by $G(iz,k)=[iz-\varepsilon_k+i\mathrm{sgn}(z)/2\tau_0]^{-1}$, the broken lines those of the electrons in the localized orbital of a magnetic impurity given by $g(iz)=[iz-\mu]^{-1}$. The double solid lines and the wavy lines are the particle-hole and the particle-particle pair diffusion propagators. The diagrams of Fig.1 represent the quantum corrections to the parquet diagrams.

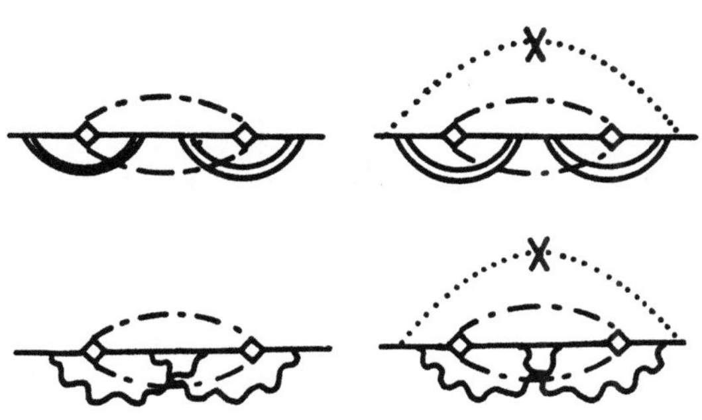

Fig.1 The quantum corrections to the self-energy. Dotted lines with a cross represent scatterings by normal impurities. Square vertices correspond to Abrikosov's T-matrix [3]

We obtain $\Sigma^{(\mu)}(i\varepsilon\to 0)$ in the order of $J^n\lambda^{d-1}$ from the calculations of the diagrams shown in Fig.1. Substituting $\Sigma^{(\mu)}(0)$ and $n^{(\mu)}$ given by $2f(\mu)$ into the relation among $\Sigma(0)$, $\Sigma^{(\mu)}(0)$ and $n^{(\mu)}$ mentioned above, we obtain $\Sigma^n(0)$ in the order of $J^n\lambda^{d-1}$, where $f(x)$ is the Fermi distribution function. Summing up $\Sigma^n(0)$ over n and using the relation between σ' and $\Sigma(0)$, we can derive the quantum corrections to the conductivity as [5]

$$\sigma' = 8\sigma_0(\tau_0/\tau_S)\rho_0 J\lambda^{d-1}(2\pi T\tau_0)^{d/2-2}\,F_d(1)(1-2\rho_0 J\ln T/\varepsilon_F)^{-3} \quad , \tag{2}$$

where $F_2(x)=\sum_{\ell=0}^{\infty}(x+\ell)^{-2}$ and $F_3(x)=(3^{3/2}\pi^2/4)\sum_{\ell=0}^{\infty}(x+\ell)^{-3/2}$. Here τ_S is the life time of the conduction electrons in the Born approximation due to a magnetic impurity, ρ_0 the density of states of them and T the temperature. From eq.(2) it is shown that the conductivity in the present system decreases with more singular temperature dependence than that in the usual Kondo systems.

The results obtained within the framework of the most divergent approximation show that the quantum corrections, σ', except for the characteristic temperature-dependent part, $F_d(1)(2\pi T\tau_0)^{d/2-2}$, are scaled by the Kondo temperature T_K defined by $\varepsilon_F\exp(1/2\rho_0 J)$. From this fact, we consider that the self-consistent ladder approximation [4] can be used in order to investigate the behaviors of the quantum corrections to the conductivity in the low temperature region of the WLR [6].

The diagrams of $\Sigma^{(\mu)}(i\varepsilon)$ which give rise to the most singular behaviors with respect to the temperature in the self-consistent ladder approximation are shown in Fig.2(a). The double dash-dotted lines represent the renormalized Green functions, $G_d(iz)$, of the electron in the localized orbital in the self-consistent ladder approximation, the Dyson equation for which is shown in Fig.2(b). Following the procedure used in the calculations within the most divergent approximation, we can derive the quantum corrections to the conductivity, σ', in the self-consistent ladder approximation. The numerical results for σ' are shown in Fig.3(a) for d=2, where T_L=42.3[K] and T_K=76.1[K], T_L being the localization energy defined by $(1/\tau_0)\exp(-1/2\lambda)$ [1]. The results for d=3 are shown in Fig.3(b).

From the numerical results we can see that the conductivity in the present system decreases more rapidly through T_K than that in the usual Kondo systems as the temperature is decreased.

As for the quantum corrections to the susceptibility, the analytic calculations are performed in a similar way. It is shown that the quantum corrections to the susceptibility of a magnetic impurity, χ_M', is proportional to $J^2\lambda^{d-1}A_d(T)(1-2\rho_0 J\ln T/\varepsilon_F)^{-2}$, where $A_2(T)=-\ln^2(2\pi T\tau_0)$ and $A_3(T)=6\sqrt{3}\pi[\ln(2\pi T\tau_0)+\pi(1-\sqrt{2\pi T\tau_0})]$. The quantum corrections to the spin polarization of the conduction electrons caused by the localized spin on a

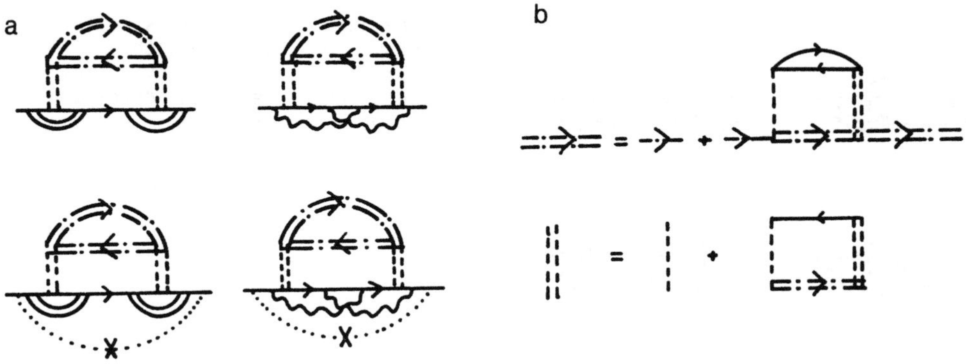

Fig.2 The quantum corrections to the self-energy in the self-consistent ladder approximation (a). The Dyson equation for $G_d(iz)$ in the self-consistent ladder approximation (b). The broken and the double broken lines are the bare and the renormalized exchange interaction, respectively

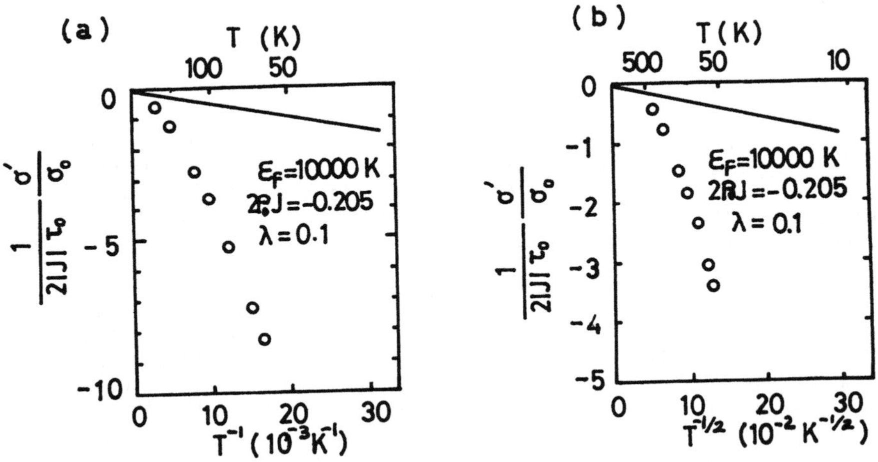

Fig.3 Plots of σ' as a function of T^{-1} for d=2 (a) and $T^{-1/2}$ for d=3 (b). The open circles represent the results in the self-consistent ladder approximation. The solid lines are the results in the order of $J^3\lambda^{d-1}$

magnetic impurity are calculated as well. There are two terms in their expressions. One is expressed by $\rho_0 J X_M'$ as has been expected [7] and the other proportional to $J^2\lambda^{d-1}B_d(T)(1-2\rho_0 J\ln T/\varepsilon_F)^{-2}$, where $B_2(T)=\ln(2\pi T\tau_0)$ and $B_3(T)=-(3\sqrt{3}/2)\pi^2(1-\sqrt{2\pi T\tau_0})$. Those results are obtained within the most divergent approximation. The detailed study of the quantum corrections to the susceptibility will be published elsewhere [8].

3. SUMMARY AND DISCUSSIONS

In this paper we have investigated the interplay between the s-d exchange interaction and the randomness in the WLR. From the calculations performed in the previous section, it is shown that the conductivity and the susceptibility in the present system decrease with more singular temperature dependence in the WLR than those in the usual Kondo systems. This tendency is conspicuous for d=2. It can be said that as compared with the usual Kondo systems, the spin of the conduction electrons in the present system may couple tightly to the localized spin even at high temperatures due to the diffusive motion of the conduction electrons.

REFERENCES

1. F.J. Ohkawa, H. Fukuyama and K. Yosida: J. Phys. Soc. Jpn. 52, 1701 (1983)
2. F.J. Ohkawa and H. Fukuyama: J. Phys. Soc. Jpn. 53, 2640 (1984)
3. A.A. Abrikosov: Physics 2, 5 (1965)
4. S. Kashiba, S. Maekawa, S. Takahashi and M. Tachiki: J. Phys. Soc. Jpn. 55, 1341 (1986)
5. S. Suga, H. Kasai and A. Okiji: J. Phys. Soc. Jpn. 55, 2515 (1986)
6. S. Suga, H. Kasai and A. Okiji: J. Phys. Soc. Jpn. 56, 863 (1987)
7. K. Yosida and A. Okiji: Prog. Theor. Phys. 34, 505 (1965)
8. S. Suga, H. Kasai and A. Okiji: in preparation

Part VII

The Quantum Hall Effect and Localization in High Magnetic Fields

Recent Experiments on the Fractional Quantum Hall Effect

D.C. Tsui[1], *H.L. Störmer*[2], *and R. Willett*[3]

[1]Princeton University, Princeton, NJ 08544, USA
[2]AT&T Bell Laboratories, Murray Hill, NJ 07974, USA
[3]Physics Department, MIT, Cambridge, MA 02129, USA

Abstract

A review is given of recent experiments on the fractional quantum Hall effect concerning the odd denominator rule and the quasi-particle excitation gap.

1. Introduction

This is a review of more recent experiments on the fractional quantum Hall effect (FQHE). The discussions will be based primarily on the magneto-transport measurements made at the National Magnet Laboratory in Cambridge on the two-dimensional (2D) electron system in $GaAs/Al_xGa_{1-x}As$ heterostructures [1,2]. These more recent experiments differ from the earlier ones [3,4,5] in that they are carried out at lower temperatures (to T = 60 mK) and higher magnetic fields (to B = 31 T), to smaller fractions of the Landau level filling factor and, most significantly, on much higher mobility samples [6]. While the earlier experiments focused more on identifying the unique qualitative characteristics of the FQHE and its phenomenology, these recent experiments are more systematic studies intended to discern theoretical concepts and to make direct contact with theory.

The organization of the review is as follows: Sec. 2 introduces the symbols and terminologies. It also gives an overview of the FQHE and provides the general background material to facilitate our discussion of the experiments. The discussion is divided into two parts. Sec. 3 discusses the odd denominator rule, which, as an experimental fact, is conceptually an enigma. Sec. 4 discusses the energy gap for creating quasi-particles and quasi-holes from the electron fluid. References listed at the end of this article are given to facilitate the reading, not for completeness. We refer the reader to Ref. 5 for a more complete list of references prior to 1986.

2. 2D charge carriers in a B field and the FQHE

Charge carriers confined to move along the interface between two different semiconductors are a unique 2D electron system. Their perpendicular motion is quantized by the electric field binding them to the interface. The application of a perpendicular B further quantizes their in-plane motion and renders the system to pseudo-zero-dimensional. The energy spectrum of the single particle states, as a result, is a series of highly degenerate Landau levels, each having a degeneracy $N_d = B/\phi_o$ (where $\phi_o = h/e = 4.14 \times 10^{-7}$ G–cm^2 is the flux quantum).

Disorder and electron-electron interaction, however, are expected to lift degeneracies and broaden each Landau level into a band. In the high B limit, when the cyclotron energy is larger than the broadening due to disorder and electron-electron interaction and the Landau levels remain sufficiently well defined, the electron population of the system, at low T, is characterized by $\nu \equiv n_s/N_d = hn_s/eB$. It is intuitively obvious that charge transport in the system must be strongly dependent on ν and the magneto-transport coefficients are expected to show singular structures at ν=integers, corresponding to complete Landau level fillings. Indeed, the value of the transport coefficients at ν=integer is observed in a remarkably wide range of $\Delta\nu$ about ν=integer, giving rise to quantized plateaus in the Hall resistance and vanishing resistivity ρ_{xx} along the excitation current. This is the integral quantum Hall effect (IQHE) and it has its physical origin in the unusual single particle energy structure of the 2D system. It is explained by introducing localization, due to disorder, of these pseudo-zero-dimensional states.

The FQHE is superficially similar to the IQHE. The Hall conductance plateaus, quantized in units of e^2/h to odd denominator fractions f (i.e. $G_H = fe^2/h$), are observed in finite $\Delta\nu$ about ν=f, while ρ_{xx}, vanishes in the same ranges of $\Delta\nu$ at $T \rightarrow 0$. However, it involves fundamentally different physics. It is the manifestation of a series of a new type of correlated ground states of the 2D electrons and the following conclusions are drawn from the early experiments [7-9]:

1. The states are formed when an electron, or hole, Landau level is partially filled to an odd denominator, fractional filling factor $\nu = p/q$ independent of the host material or its band structure.
2. They are sensitive to disorder in the 2D system and are not existent in sufficiently disordered systems.
3. Their Hall conductance is quantized to $G_H = fe^2/h$ and the quantum number is a rational fraction of odd denominator given by $f = \nu = p/q$.
4. The states can carry electrical current with no dissipation at $T = 0$. At finite T, dissipation, as measured by ρ_{xx}, is thermally activated with well-defined activation energies.

Laughlin [10] explains the IQHE in terms of the gauge invariance. He argues that the Hall conductance quantization results directly from the existence of energy gaps in the excitation spectrum of the single particles that can carry currents and the quantization of the electronic charge. The quantized G_H, as a topological invariant, is the ratio to the flux quantum of the charge quantum carried by the single particle excitation in the 2D system. This same argument can be used for the FQHE and it suggests that an energy gap must exist at E_F and the excitations must carry fractions of the electronic charge. Since no gap is expected in the single particle energy spectrum for $\nu < 1$, the energy gap must be many-body in origin.

Central to our current understanding of the FQHE is the assumption that the 2D charge carriers condense into a series of specially stable ground states at $\nu = p/q$. The validity of this assumption and the prevailing theoretical picture of these new quantum states are based primarily on exact numerical calculations on systems of a few particles and on calculations using trial wave functions. For the fundamental states at $\nu = 1/q$, Laughlin's trial wave function for a system of N electrons is [11]

$$\psi(z_1,z_2,...z_N) = C\prod_{i<j}^{N} (z_i-z_j)^m \exp\left[\frac{-1}{4l_o^2}\sum_{k}^{N} |z_k|^2\right],$$

where $z = x\text{-}iy$, $l_o^2=\hbar/eB$, and C is a normalization constant. The parameter m is an integer and is related to the electron density of the system through $\frac{1}{m}=2\pi l_o^2 n = \nu$ to yield a stable state of the system. By making the wave function antisymmetric, m must be odd and it is identified as our q. Moreover, the fundamental state at $\nu = 1/q$, as described by Laughlin's wave function, has the following properties:

1. Its pair correlation function indicates that it is a novel quantum fluid rather than a Wigner solid for $q < 10$.
2. Its elementary excitations are separated from the ground state by a finite energy gap.
3. These quasi-particle excitations are fractionally charged with $e^* = e/q$.
4. The quantum fluid is incompressible and has no low-lying excitations. Hence, it flows with no dissipation at $T = 0$.
5. For $q > 10$, the quantum fluid is expected to crystallize into a Wigner solid.

The FQHE at $\nu = 1/q$ can be explained by considering the case for $\nu = 1/3$. When B is at the position corresponding to $\nu = 1/3$, the 2D electrons are in the highly correlated fluid-like state, ψ_3, which can carry electrical current without dissipation at $T = 0$. For ν slightly above 1/3, extra electrons not accommodated in the ground state go into fractionally charged quasi-particles. Likewise, for ν slightly below 1/3, the system will be made up of some quasi-holes on top of the ground state. The Hall plateau around $\nu = 1/3$ is explained by localization of these quasi-particles and quasi-holes at low concentrations. It can be caused either by disorder in the 2D system via Anderson localization or by mutual interactions through their residual Coulomb repulsion to form Wigner solids. In either case, ρ_{xx} remains constant in a finite range of B as ν deviates from 1/3, giving rise to a Hall plateau. Since the elementary charge carried by the quasi-particle is $e^* = e/3$, the plateau, by the gauge invariance argument, is quantized to $G_H=1/3(e^2/h)$. By invoking electron-hole symmetry in the Landau level, similar fluid states are expected at $\nu = 1\text{-}(1/q)$ to explain the FQHE with $f = (q\text{-}1)/q$. The FQHE observed at other filling factors in the lowest Landau level is explained by the existence of higher order fluid states using a hierarchical model [12,13]. More specifically, when ν deviates sufficiently from 1/m, the density of the quasi-particles, or quasi-holes, can be sufficiently high to overcome localization. These quasi-particles, or quasi-holes, will interact via their Coulomb repulsion and, under appropriate conditions, form new fluid states. Likewise, the quasi-particles, or quasi-holes, from these new ground states will, in turn, under appropriate conditions, form new ground states. The condition that a higher order state is favored is given by:

$$\nu = \cfrac{1}{q+\cfrac{\alpha_1}{p_1+\cfrac{\alpha_2}{p_2+\cdots}}}$$

where q is odd, p_i's are even, and α_i's are either 0 or ±1.

3. Odd denominator rule

The so-called odd denominator rule is a generalization of results from early experiments that only rational fractions with odd denominators have been observed. Clearly, it is of fundamental importance to examine with improved experiments the validity of this generalization and to find out whether such a selection rule indeed exists. In this section, we summarize all the $\nu < 1$ fractions that have been identified to date and assess the data concerning even fractions $\nu = 1/2$, $1/4$, and $3/4$. Our conclusion is that there is no evidence for $\nu < 1$ even fractions and the odd denominator rule is valid in this regime.

The plot shown as Fig. 1 is a simple scheme to visualize the various $\nu = p/q$ fractions in the $\nu < 1$ regime, accessible to experiments. The y-axis is labeled for odd integers q and the x-axis is B, which in units of $n_s h/e$ is also ν^{-1}. The fraction $\nu = p/q$ in this scheme is given by the intercept of the $y = p{\cdot}x$ line (i.e., slope = p) with the y = q horizontal line. The crosses indicate those odd denominator fractions that have been observed repeatably in a number of experimental runs on several samples. In case of the encircled crosses, quantized Hall plateaus have been observed and the fractional quantum number determined from the experi-

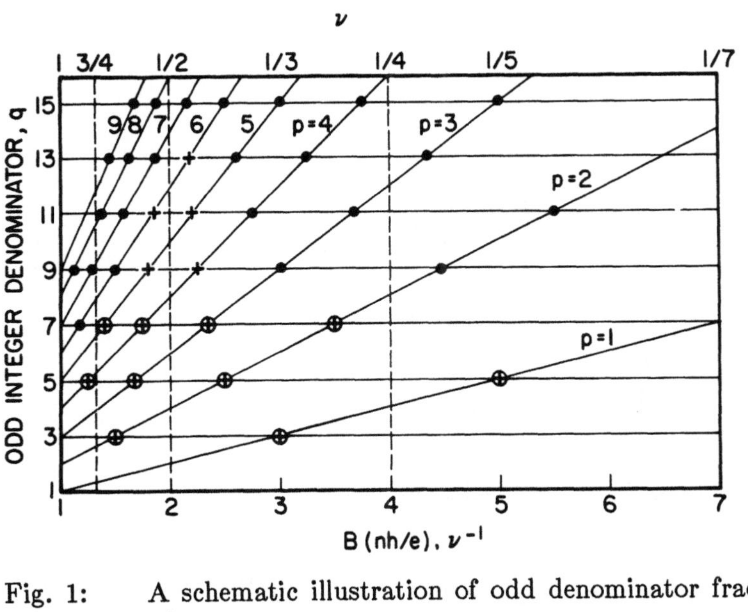

Fig. 1: A schematic illustration of odd denominator fractions given by $\nu = p/q$.

ment is f = ν to better than 1% accuracy. These fractions include f = 1/3, 2/3; f = 1/5, 2/5, 3/5, 4/5; and f = 2/7, 3/7, 4/7, 5/7.

The three vertical lines in Fig. 1 show the expected B positions for even fraction states ν = 3/4, 1/2, and 1/4. It is clear from the plot that, in the range of B between the ν = 2/3 state and the ν = 1/3 state, there are two sequences of closely spaced higher order states. They are given by ν = (m+1) / (2m+1) and m/(2m+1), respectively, where m is a positive integer. Clearly, as m $\rightarrow\infty$, $\nu\rightarrow$ 1/2. Similarly, two sequences given by ν = (3m+1) / (4m+1) and (3m-1) / (4m+1) can be identified in the range of B between the ν = 4/5 and the ν = 5/7 states, and two sequences given by ν = m / (4m-1) and m / (4m+1) between the ν = 2/7 state and the ν = 2/9 states. The former two sequences converge to ν = 3/4 as n $\rightarrow\infty$ and the latter to ν = 1/4 as n $\rightarrow\infty$.

The development of the two sequences of higher order states between ν = 2/3 and ν = 1/3 has been observed in several experiments [4,14,15]. Fig. 2 shows an example, which is an experimental trace of ρ_{xx} vs. B taken at T = 100 mK from a GaAs/Al_xGa_{1-x}As sample with n_s=1.6×10^{11}/cm^2 and a mobility μ=5×$10^6$$cm^2$/Vs. It is apparent that for the progressively higher order states, the separation in B between two neighboring states becomes progressively smaller. Since the quasi-particle excitation gap is also smaller for higher order states, the FQHE is progressively weaker and more difficult to resolve. For this extremely high mobility sample, the 2/3 sequence is observed up to ν = 6/11 and the 1/3 sequence to ν = 6/13. The Hall resistance as a function of B shows quantized plateaus up to ν = 4/9 and then follows the classical straight lines through ν = 1/2. Thus, it is clear that the broad ρ_{xx} minimum around ν = 1/2 results from two unresolved sequences of higher order odd-denominator states and there is no evidence for a ν = 1/2 state. Similar arguments can also be made for ν = 3/4 and ν = 1/4. In both cases, the Hall resistance vs. B is observed to follow the classical line through

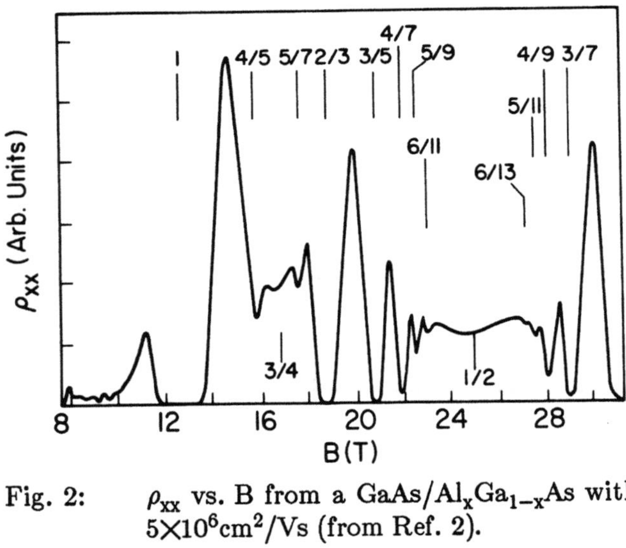

Fig. 2: ρ_{xx} vs. B from a GaAs/Al_xGa_{1-x}As with n_s = 1.6×10^{11}/cm^2 and μ = 5×$10^6$$cm^2$/Vs (from Ref. 2).

$\nu = 3/4$ and $1/4$, while quantized plateaus are observed for neighboring $4/5$ and $5/7$ states and also the $2/7$ state [16].

Two additional remarks should be made. First, the data on higher order states as shown in Fig. 2 are consistent with the hierarchical model discussed in Sec. 2. In particular, the "parent" state is always stronger than the "daughter" state and the two electron-hole symmetric sequences are comparable in strength. Second, our discussion thus far is limited entirely to the extreme quantum limit, when only the (N = 0, spin ↑) Landau level is fractionally occupied, but it is valid for the (N = 0, spin ↓) case as well [16]. On the other hand, for the N=1 Landau level, recent calculations [17] have shown that the stability of the 1/3 state is marginal and the 1/5 is the more stable state. Experiments have thus far been surprising in that ρ_{xx} minima have been observed at $\nu = 19/7$, $5/2$, and $7/3$, and they are accompanied by developing quantum Hall plateaus [15,18].

4. Activation energies

The dissipation in the FQHE is thermally activated and the activation energy is a direct measure of the quasi-particle energy gap, which the theory should be able to predict. Fig. 3 summarizes the results on the 1/3 and 2/3 states from the systematic investigation carried out at the National Magnet Laboratory. The crosses are from Ref. 1 on samples with electron mobility $0.4\times10^6 cm^2/Vs \lesssim \mu \lesssim 1\times10^6 cm^2/Vs$ and the four solid circles are from Ref. 2 on two samples with $\mu = 5\times10^6 cm^2/Vs$. The energy gap Δ is obtained from fitting the T dependence of the ρ_{xx} minima at $\nu = 1/3$ and $\nu = 2/3$ to $\rho_{xx} = \rho_o \exp(-\Delta/2T)$. From theory $\Delta = Ce_2/\epsilon l_o$, where ϵ is the dielectric constant. The proportionality constant C

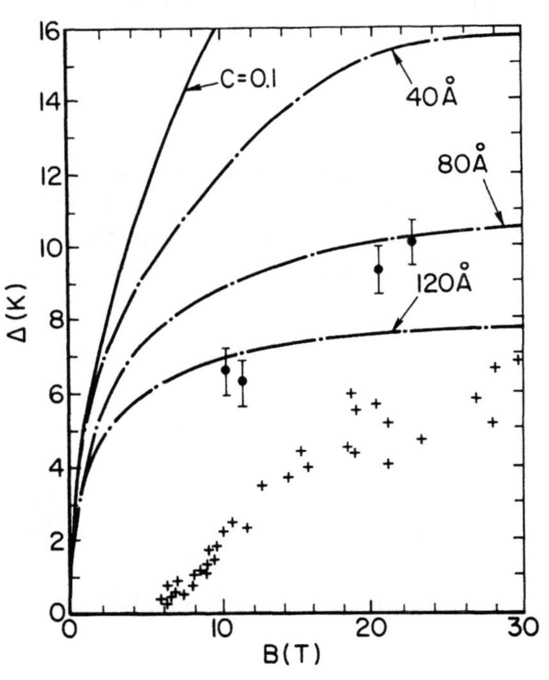

Fig. 3: Δ vs B for 1/3 and 2/3 states (from Ref. 2).

from different calculations varies from 0.04 to 0.11, but is converging to ~0.10 [19]. The full line in Fig. 3 is for $\Delta = 0.1e^2/\epsilon l_o$, and the three dash-dot lines are after the corrections, due to the wavefunction thickness, are made following Zhang and Das Sarma [20].

The difference between theory and experiment is most noteworthy in Fig. 3. The data from all the lower mobility samples, in spite of a significant amount of scatter, show a threshold B of ~6T, below which Δ vanishes. Above threshold, Δ shows an almost linear dependence over an appreciable range of B and then saturation at still higher B. The simple $\sqrt{B}$ dependence is not observed. Even more worth noting is the large difference in the magnitude of Δ. The observed value is almost ten times smaller than that expected from the ideal theory.

This disagreement between theory and experiment has been attributed to the fact that the theory has left out (1) mixing of higher Landau levels [21], (2) the finite thickness of the electron wave function [20,21], and (3) the effect of disorder [22]. Yoshioka [21] has performed numerical calculations on a finite system and found that Δ is reduced by ~20% when mixing of the next Landau level is taken into account. Corrections due to the finite extent of the electron wave function have been treated in several recent publications. Its importance is illustrated in Fig. 3 by the three dash-dot lines which have the wave function thickness taken into account following Zhang and Das Sarma. The effect of disorder is most clearly demonstrated in the data from the high mobility samples shown as solid circles in Fig. 3. These samples have approximately six times higher mobility ($\sim 5\times10^6 cm^2/Vs$); their Δ is close to that predicted by theory if the other two corrections are properly taken into account. Boebinger et al. [1] also investigated lower mobility samples with $\mu \sim 1.5\times10^5 cm^2/Vs$. In that case Δ is lower and shows a higher threshold B ~ 8T. Theoretically, MacDonald et al. [22] have done a self-consistent treatment of the effect of remote ionized impurities. They obtained the qualitative behavior of a threshold as observed in the experiment.

Acknowledgments

The high magnetic field experiments are carried out at the Francis Bitter National Magnet Laboratory, Cambridge, MA 02139, which is supported by the NSF through its Division of Materials Research. The work at Princeton University is supported by the NSF through Grant No. DMR-8212167 and by the ONR through contract No. 00014-82-K-0450.

References

1. G.S. Boebinger, A.M. Chang, H.L. Störmer, and D.C. Tsui, Phys. Rev. Lett. **55**, 1606 (1985) and Phys. Rev. B (to be published).
2. R. Willett, G.S. Boebinger, D.C. Tsui, H.L. Störmer, J.M. English, and G. Weimann, Bull. Am. Phys. Soc. **32**, 891 (1987) and R. Willett, H.L. Störmer, and D.C. Tsui, Conf. EP2D-VII, Santa Fe, July, 1987.
3. E.E. Mendez, M. Heiblum, L.L. Chang, and L. Esaki, Phys. Rev. **B28**, 4886 (1983).

4. A.M. Chang, P. Berglund, D.C. Tsui, H.L. Störmer, and J.C.M. Hwang, Phys. Rev. Lett. **53**, 997 (1984).

5. D.C. Tsui and H.L. Störmer, IEEE J. Quan. Elec. QE-22, 1711 (1986).

6. J.H. English, A.C. Gossard, H.L. Störmer, and K.W. Baldwin, Appl. Phys. Lett. **50**, 1826 (1987).

7. D.C. Tsui, H.L. Störmer, and A.C. Gossard, Phys. Rev. Lett. **48**, 1559 (1982).

8. H.L. Störmer, A. Chang, D.C. Tsui, J.C.M. Hwang, and A.C. Gossard, Phys. Rev. Lett. **50**, 1953 (1983).

9. D.C. Tsui, H.L. Störmer, J.C.M. Hwang, J.S. Brooks, and M.J. Naughton, Phys. Rev. **B28**, 2274 (1983).

10. R.B. Laughlin, Phys. Rev. **B23**, 5623 (1981).

11. R.B. Laughlin, Phys. Rev. Lett. **50**, 1395 (1983).

12. F.D.M. Haldane, Phys. Rev. Lett. **51**, 605 (1983).

13. B.I. Halperin, Phys. Rev. Lett. **52**, 1583 (1984).

14. J. Wakabayashi, S. Kawaji, J. Yoshiro, and H. Sakaki, J. Phys. Soc. Jpn. **55**, 1357 (1986).

15. R.G. Clark, R.J. Nicholas, A. Usher, C.T. Foxon, and J.J. Harris, Surface Sci. **170**; 141 (1986) and also Proc. 18th Intl. Conf. on Physics of Semiconductors, ed. O. Engström, (World Scientific, 1987) p. 393.

16. G.S. Boebinger, A.M. Chang, H.L. Störmer, and D.C. Tsui, Phys. Rev. **B32**, 4268 (1985).

17. A.H. MacDonald and S.M. Girvin, Phys. Rev. **B33**, 4009 (1986).

18. R. Willett et al. (unpublished).

19. See, for example, B.I. Halperin, Surface Sci. **170**, 115 (1986).

20. See, for example, Zhang and S. Das Sarma, Phys. Rev. **B33**, 2903 (1986).

21. D. Yoshioka, J. Phys. Soc. Jpn. **55**, 885 (1986).

22. A.H. MacDonald, K.L. Liu, S.M. Girvin, and P.M. Platzman, Phys. Rev. **B33**, 4014 (1986).

Numerical Study of Localization in 2D Systems: Effects of Magnetic Field and Spin-Orbit Interaction

T. Ando

Institute for Solid State Physics, University of Tokyo,
7-22-1 Roppongi, Minato-ku, Tokyo 106, Japan

A review is given on the numerical methods, the Thouless-number and the finite-size scaling, for the study of localization effects in two-dimensional systems. Both methods give the result that states are localized exponentially except just at the center of each Landau level in high magnetic fields. The inverse localization length calculated by the two methods turn out to be in agreement with each other. The Thouless-number method is applied also for the localization problem in square lattices, where effects of magnetic field and spin-orbit interaction can be studied. There is a critical randomness above which extended states in Landau levels merge and disappear in high magnetic fields.

1. INTRODUCTION

The localization in two dimensions is interesting because of the marginal dimensionality. In strong magnetic fields, there exists a breakdown of the single-parameter scaling argument /1/ that states are all localized in two dimensions. The existence of current-carrying extended states has been shown in connection with the exact theorem on the quantum Hall effect /2/ and has been confirmed by both numerical study /3,4/ and perturbation calculations /5,6/. The purpose of this paper is to give a brief review on the numerical study of localization problems in two dimensions.

One of the exciting phenomena is the quantum Hall effect in which the Hall conductivity is quantized into the integer multiple of $-e^2/h$ when the Fermi level lies in localized states existing between different Landau levels /7,8/. The quantum Hall effect requires the presence of both localized and extended states. Consequently, the localization problem in strong magnetic fields and the quantum Hall effect are closely related to each other. A direct way to study the localization problem is numerical computer simulations. Although the size of systems is limited, appropriate criteria of localization or delocalization enable correct extrapolations to infinite systems.

This paper is organized as follows: In Sect. 2 the two numerical methods, the Thouless-number method and the finite-size scaling method, are introduced and their merits and demerits are discussed together with examples of results. Section 3 describes the localization problem in two-dimensional lattice systems, where symmetry effects, i.e., those of magnetic fields and spin-orbit interactions, can be studied by the numerical methods.

2. NUMERICAL METHODS: LOCALIZATION IN HIGH MAGNETIC FIELDS

One of the criteria or methods which have turned out to be most successful is the Thouless number method, in which the system-size dependence of the Thouless number is studied. The Thouless number, $g(L)$, is proportional to the conductance of the system with size L and is also a scaling variable in the single-parameter scaling theory. Another powerful method is the finite-size scaling, in which we consider systems having a form of a long strip. If the length L_x is much larger than the width L_y, the system is essentially one-dimensional and states are all localized exponentially in the x direction. We can determine the inverse localization length $\alpha(E)$ in two dimensions by a study of the L_y dependence of the inverse localization length $\alpha(L_y,E)$ in long strips.

Take a two-dimensional system in high magnetic fields for example. In the Thouless number method, we consider a system with a finite size LxL in which scatterers are distributed at random /3/. The Thouless number is determined by $\langle\Delta E\rangle L^2 D(E)$, where D(E) is the density of states per unit area and $\langle\Delta E\rangle$ is the geometric average of the shifts of the individual energy levels due to a change in boundary conditions from periodic to antiperiodic in one direction. In order to get reliable answers we have to take averages of quite a large number of sample systems with careful analysis on possible errors in addition to making system sizes as large as possible. The inverse localization length $\alpha(E)$ is determined by $g(L)=g_0\exp(-\alpha L)$, if g(L) decreases with L exponentially. This method is very handy, powerful, and above all requires no essential assumption in order to determine the inverse localization length. Note, however, the inverse localization length may possibly be not so accurate in the weak-localization case, especially when the localization length becomes much larger than the maximum size of sample systems.

Figure 1 shows an example of calculated Thouless number as a function of energy for the lowest Landau level N=0 in high magnetic fields /3/. Within each energy interval, g(L) with increasing L ($15.0<L/\ell<50.1$) is shown, where ℓ is the radius of the cyclotron orbit of the lowest Landau level N=0 defined by $\ell^2=c\hbar/eH$ with H the magnetic-field strength. The energy is normalized by the broadening of the Landau level calculated in the self-consistent Born approximation. The figure demonstrates a clear exponential localization except just in the vicinity of the center of the Landau level E=0. Figure 2 gives the inverse localization length for the lowest two Landau levels as a function of energy measured from the center. The critical exponent s defined by $\alpha(E)\propto|E|^s$ is very close to 2 for N=0.

In the finite-size scaling method, the inverse localization length $\alpha(L_y,E)$ can be calculated recursively for strip lengths as long as desired. Usually, we plot $\alpha(L_y,E)L_y$ as a function of L_y in logarithmic scales and seek the following scaling relation:

$$\alpha(L_y,E)L_y = f(\alpha(E)L_y), \tag{2.1}$$

where $\alpha(E)$ is the inverse localization length in two dimensions. In square lattices in the absence of a magnetic field, MACKINNON and KRAMER have convincingly demonstrated that this single-parameter scaling is valid at the band center /9/. They have successfully determined both inverse localization length and scaling function in two and three dimensions and shown that the results are consistent with those of the single parameter scaling theory.

One should be warned, however, that the scaling hypothesis (2.1) can sometimes lead to wrong conclusions. It is absolutely essential to check first that such a hypothesis is correct, although it is not an easy task since it requires calculations of $\alpha(L_y,E)$ for a large range of L_y.

Figure 3 gives the inverse localization length of the lowest Landau level N=0 for systems in a long strip as a function of the width L_y in high magnetic fields /4/. It demonstrates clearly that the hypothesis (2.1) is not valid. As a matter of fact, $\alpha(L_y,E)L_y$ first decreases with L_y for small L_y when $|E|/\Gamma<0.64$, but it eventually starts to increase for sufficiently large L_y except at E=0 in agreement with the conclusion obtained in the Thouless-number method that states are all localized (i.e. $\alpha(E)>0$) except at the Landau-level center E=0. However, if we use the results only for a narrow range of L_y, between 1.4 and 5.6 in Fig. 3 for example, we may be easily misled to the wrong conclusion that (2.1) is valid and there is a mobility edge at about $|E|/\Gamma=0.15$. As a matter of fact such a conclusion has been reached in a similar study using a square lattice system in high magnetic fields (the L_y range mentioned above corresponds to that used for this study) /10,11/. Since (2.1) is not valid, we have to seek two parameter scaling $\alpha(L_y,E)L_y=h(\alpha(E)L_y,L_y/\ell)$. The resulting inverse localization length is plotted in Fig. 2. The finite-size scaling method gives more accurate results especially for small $\alpha(E)$ in comparison with the Thouless number method.

Fig. 1

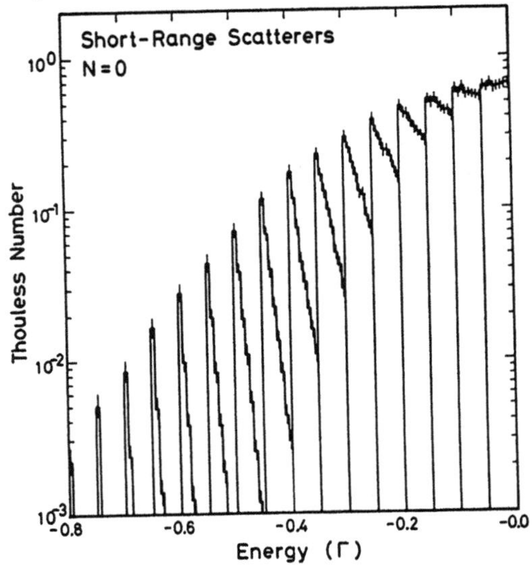

Fig. 2

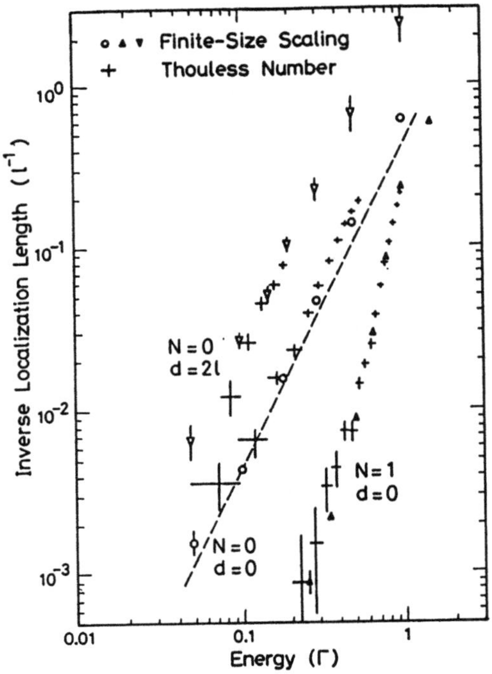

Fig. 1 Calculated Thouless number as a function of energy for the lowest Landau level (N=0). The energy is measured from the center of the Landau level and normalized by the broadening calculated in the self-consistent Born approximation. Within each energy interval, the Thouless number with increasing system size (L/ℓ = 15.0, 20.0, 25.0, 30.1, 35.1, 40.1, 45.1, 50.1) is plotted. Scatterers with short-range potentials are distributed densely in the system (white-noise random potential). After Ref. 3

Fig. 2 Log-log plot of the inverse localization length $\alpha(E)$ in units of the inverse of the cyclotron radius of the lowest Landau level for N=0 and 1 versus energy. The energy is measured from the center of each Landau level and normalized by the broadening calculated in the self-consistent Born approximation. Scatterers with a Gaussian potential having range d are distributed densely in the system. The crossed error bars are the result of the Thouless number study and the squares that of the finite-size scaling study. The white-noise random potential corresponds to d=0 and correlated long-range random potential corresponds to $d=2\ell$. After Ref. 4

Numerical methods can be applied for the study of various kinds of properties on localization in two-dimensional systems in high magnetic fields. In the following, some of the representative results are summarized (see Ref. 12 for more details): (1) States are all localized exponentially except just at the center of each Landau level in high magnetic fields. The inverse localization length exhibits a power-law dependence on energy in the vicinity of the center and the critical exponent is very close to 2 for systems containing high concentrations of scatterers with short-range potentials (white-noise random potentials). (2) The same is applicable to scatterers with long-range potentials although the localization effect becomes stronger. Figure 2 contains the inverse localization length for scatterers with a Gaussian potential having range $d=2\ell$. (3) When the field decreases and effects of mixings between different Landau levels become important, the energy of extended states is shifted to the higher energy side in comparison with the energy of peaks of density of states /3/. (4) There is a scaling relation

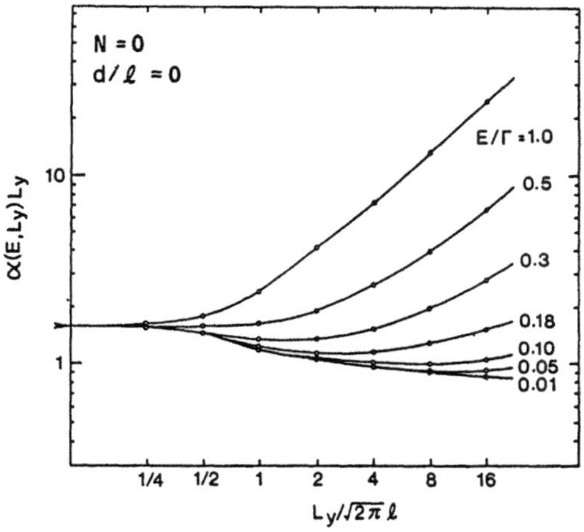

Fig. 3 Log-log plot of numerically obtained $\alpha(L_y,E)L_y$ vs. L_y for the lowest Landau level N=0. The horizontal arrow indicates the asymptotic value of $\alpha L_y=\pi/2$ for $L_y \to 0$ irrespective of the energy E. Scatterers with short-range δ potentials are distributed densely in the system (white-noise random potential). After Ref. 4. The system width considered in a lattice model /10,11/ is in the range between 1.4 and 5.6. Using the results in such a narrow range of L_y, we may be misled to the conclusion that the single-parameter scaling is valid and may place the mobility edge around $|E|/\Gamma=0.15$

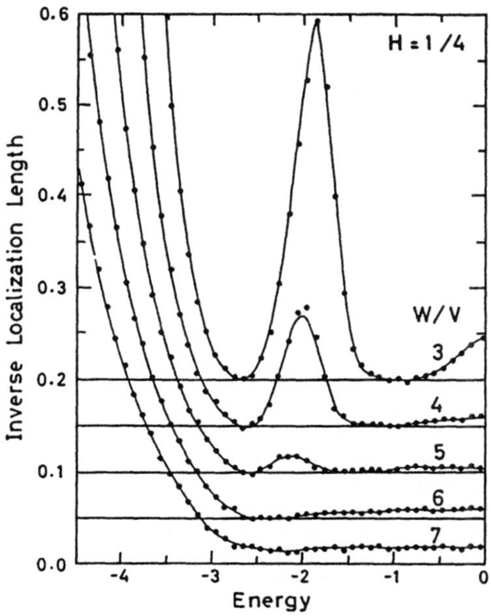

Fig. 4 Calculated inverse localization length of a square lattice in the magnetic field $Ha^2/\Phi_0=1/4$. Only the low energy half is shown because of the symmetry about the band center E=0. The numbers appearing in the figure show the randomness W/V

between the diagonal and off-diagonal element of the conductivity tensor in system with a finite size /13/, i.e., $\sigma_{xx}(L,E_F)$ and $\sigma_{xy}(L,E_F)$ fall on a single curve for different E_F's and L's where E_F is the Fermi energy. The scaling curve depends on the Landau level reflecting that the diagonal conductivity depends on the level index N, i.e., $\sigma_{xx}=(N+1/2)e^2/h$ in the self-consistent Born approximation. This relation is strongly modified by level mixing effects when the magnetic field becomes weaker.

A nonlinear σ model with σ_{xy} as a θ term in the Yang-Mills theory has been proposed to explain the localization and quantum Hall effect /14/. In this theory σ_{xx} and σ_{xy} are independent scaling variables and a flow diagram determines the localization or delocalization. The predicted flow lines are periodic in the σ_{xy} axis with the periodicity e^2/h and possess two kinds of fixed points, unstable fixed points on the line $\sigma_{xy}=-e^2(N+1/2)/h$ and stable fixed points at $\sigma_{xx}=0$ and $\sigma_{xy}=-Ne^2/h$. Although this field theory has several attractive features, its main predictions contradict results of the numerical study mentioned above.

The major difficulty underlying the localization problem in high magnetic fields is the absence of expansion parameters. As a matter of fact, each term in perturbation expansion contributes almost equally, which forces the perturbation to be up to the infinite order. The present author was the first that calculated the density of states by summing up all the terms of such perturbation series although in an approximate way /3/. Later, WEGNER has succeeded in summing up the series exactly for the lowest Landau level /15/. The resulting density of states has turned out to almost coincide numerically with the approximate result. Unfortunately, similar exact summation is not possible for two-particle Green's function which contains the information on the localization. HIKAMI calculated perturbation series up to ninth order and summed the asymptotic series by the Borel-Pade extrapolation /5/. He has concluded that σ_{xx} has a nonzero value close to that in the self-consistent Born approximation. He estimated also the inverse localization length and obtained the critical exponent s=1.9 for the lowest Landau level, in good agreement with that of the numerical study. ONO developed a self-consistent perturbation method, in which a part of perturbation series can be summed up, and concluded that all the states except those just at the center of Landau levels are localized /6/. However, obtained $\alpha(E)$ does not show the power law dependence on energy but instead $\alpha(E) \propto \exp(-\Gamma^2/E^2)$.

3. TWO-DIMENSIONAL SYSTEM ON LATTICE

It is widely accepted that the conclusion of the single-parameter scaling theory is valid in the absence of a magnetic field, i.e., that states are all localized in two dimensions however weak randomness may be. There exist current-carrying extended states in high magnetic fields as has been discussed in the previous section. It is quite interesting, therefore, whether there exists a critical magnetic field below which states become all localized. The numerical method in usual two-dimensional systems on a plane cannot give a definite answer to this question, although it shows that the energy of extended states is shifted to the higher energy side with decreasing magnetic field. One way to overcome such difficulties is to consider systems on lattices, where the number of basis set is always finite for systems with a finite size. As a matter of fact, lattice two-dimensional systems have been used by various different authors for the localization study.

The symmetry of the system plays an important role in localization effects. The usual system in the absence of a field belongs to the universality class called orthogonal. When a magnetic field is applied, the time-reversal symmetry is broken and the universality class is now unitary. We have another symmetry called symplectic in the presence of strong spin-orbit interactions. It is quite important theoretically to elucidate roles of such different symmetries. Two-dimensional systems on lattices can be ideal also for such purposes.

We consider the tight-binding model for a square lattice described by the following Hamiltonian:

$$\mathcal{H} = \sum_{i} \varepsilon_i c_i^{\dagger} c_i + \sum_{ij} V c_i^{\dagger} c_j \ . \tag{3.1}$$

where V is the transfer integral between the nearest-neighbor sites, and the site energy ε_i varies at random with distribution width W. The degree of randomness is characterized by the ratio W/V. We use the Thouless number method in determining the inverse localization length. The system size is in the range 5 L/a 32 with a the lattice constant. (More details will be published elsewhere /16/.)

The magnetic field H is included in the form of Peierls' phase factor in the nearest-neighbor transfer integral. Its strength is characterized by the ratio of the flux Ha^2 passing through a unit cell to the unit flux Φ_0=ch/e. The energy spectrum in the absence of disorder, W=0, has been obtained by Hofstadter and comprises p different bands (Landau levels) when $Ha^2/\Phi_0=q/p$ with integers p and q /17/. This model has explicitly been used in studying localization effects on the Hall conductivity including the demonstration of its quantization in terms of a topological quantity called the winding number /18/.

To study effects of spin-orbit interactions, we have to consider a slightly more complex system. The simplest model will be a square lattice consisting of s-like atoms and the transfer between them is mediated by a p-like atom lying in between. The p orbitals are located on a plane parallel but displaced from the plane of the square lattice formed by s orbitals. We assume a strong spin-orbit interaction in p-like atoms and consequently introduce spin off-diagonal transfer integral between neighboring s orbitals. The transfer becomes dependent on the direction of the bond, i.e., $V(i\sigma;j\sigma')=V_x$ or V_y depending on the direction of the nearest-neighbor sites in x or y direction. We have in the matrix form,

$$V_x=\begin{bmatrix} V_1 & V_2 \\ -V_2 & V_1 \end{bmatrix},\quad V_y=\begin{bmatrix} V_1 & iV_2 \\ iV_2 & V_1 \end{bmatrix},\quad |\sigma=+1>=\begin{bmatrix} 1 \\ 0 \end{bmatrix},\quad |\sigma=-1>=\begin{bmatrix} 0 \\ 1 \end{bmatrix}. \tag{3.2}$$

The strength of the spin-orbit interaction is characterized by $S=V_2/V$ where $V=(V_1{}^2+V_2{}^2)^{1/2}$ is the total transfer. This model simulates actual two-dimensional systems in n-channel inversion layers associated with the conduction band of III-V compounds semiconductors /19/.

Figure 4 gives the inverse localization length in the magnetic field $Ha^2/\Phi_0=1/4$ as a function of energy for different degrees of disorder. In this magnetic field, the band splits into four Landau levels (symmetric about the band center chosen as the energy origin) and the result clearly shows the presence of extended states in each Landau level for small randomness. The two extended states below the band center (the two above the band center using the symmetry) merge together with increasing randomness and disappear above a certain randomness of the order of $(W/V)_c\sim 6$. Similar results are obtained for $Ha^2/\Phi_0=1/8$, where the band splits into eight Landau levels. Again, the four extended states below the band center merge together and disappear above a similar critical randomness.

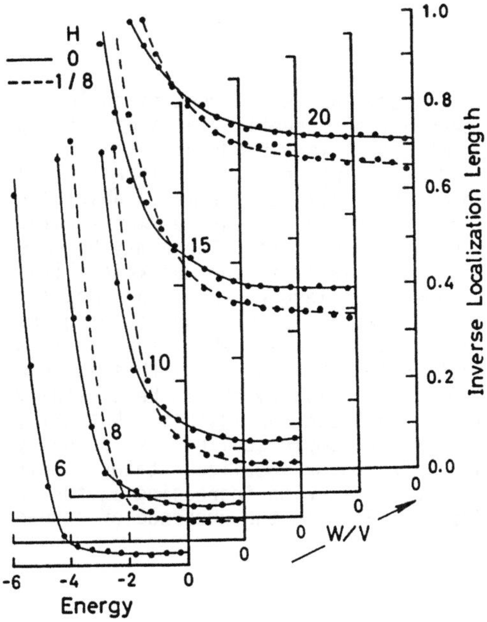

Fig. 5

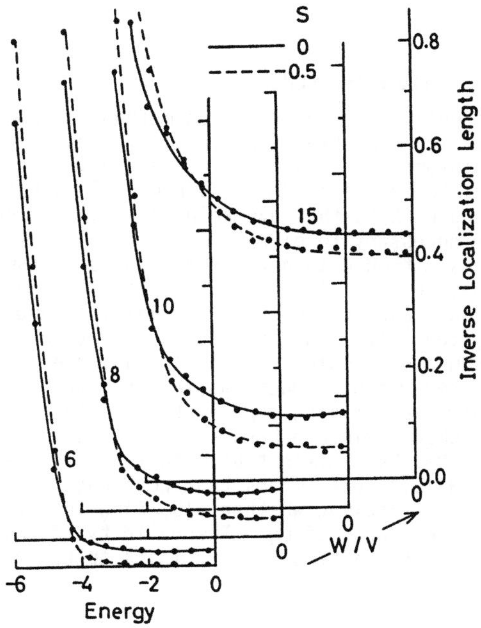

Fig. 6

Fig. 5 Calculated inverse localization length of a square lattice. The solid lines represent results in the absence of a field and the dashed lines those in the magnetic field $Ha^2/\Phi_0=1/8$. The numbers appearing in the figure show the randomness W/V

Fig. 6 Calculated inverse localization length of a square lattice. The solid lines represent results in the absence of a spin-orbit interaction (S=0) and the dashed lines those in its presence (S=0.5). The numbers appearing in the figure show the randomness W/V

When the randomness is sufficiently large, i.e., under weak-magnetic-field conditions in the conventional sense $\omega_c\tau<<1$ where ω_c is the cyclotron frequency and τ is the relaxation time, states are always localized exponentially. Figure 5 compares the inverse localization length in the presence and absence of a magnetic field. The magnetic field tends to reduce the localization for states close to the band center, i.e., the negative magnetoresistance, while it enhances the localization for states in the region of the band tails, i.e., the positive magnetoresistance. Note here that the density of states for such large randomness is not influenced by a magnetic field at all.

Figure 6 shows the inverse localization length as a function of energy for strong spin-orbit interactions S=0.5. The spin-orbit interaction reduces the localization effect for states near the band center, while it enhances the localization effect in the band tail region. This is quite analogous to the effect of weak magnetic fields. For W/V=6 the inverse localization length at the center almost vanishes in the presence of spin-orbit interaction, while states already exhibit clear exponential localization in its absence. It is tempting, therefore, to suggest the presence of a metal-insulator transition. From the present results alone, however, it is not possible to conclude definitely the presence of extended states and the mobility edge for sufficiently small randomness.

The decrease of the localization effect due to the presence of a weak magnetic field and strong spin-orbit interaction in the region of the band center is consistent with the results obtained by perturbation calculations in weak-localization theory /20/. However, the enhancement of localization in band tail regions, which may be easily understood by an intuitive picture, has not been predicted by the weak localization theory.

4. SUMMARY AND CONCLUSION

A review has been given on methods of numerical study and recent results on localization in two-dimensional systems in the presence of magnetic fields or strong spin-orbit interactions. The Thouless number method is very handy but has turned out to be one of the most powerful methods in determining the inverse localization length. Note that averaging over very large numbers of sample systems is absolutely necessary to give reliable results. Inaccuracies present in earlier works using this method originate from inadequate number of sample systems. There may be some inaccuracies in the inverse localization length especially in the weak-localization case where the localization length becomes much larger than the maximum system size considered. On the other hand, the finite-size scaling method gives accurate predictions of the inverse localization length even when the maximum system width is much smaller than the localization length. However, this is true only when a single-parameter scaling is valid. Its validity should be carefully checked, since otherwise we can be easily misled to wrong conclusions. Actually, it has been demonstrated in magnetic fields that the simple scaling is actually invalid. This means that we cannot determine extended or localized states simply by just looking whether $\alpha(L_y,E)L_y$ increases or decreases with increasing L_y for a narrow range of L_y. The validity of the simple scaling in the presence of strong spin-orbit interactions remains to be seen.

REFERENCES

1. E. Abrahams, P.W. Anderson, D.C. Licciardello and T.V. Ramakrishnan: Phys. Rev. Lett. **40**, 471 (1979)
2. H. Aoki and T. Ando: Solid State Commun. **18**, 1079 (1981)
3. T. Ando: J. Phys. Soc. Jpn. **52**, 1893 (1983); **53**, 3101 (1983); **53** 3126 (1983)
4. T. Ando and H. Aoki: J. Phys. Soc. Jpn. **54**, 2238 (1985)
5. Y. Ono: J. Phys. Soc. Jpn. **51**, 2055 (1982); **51**, 3544 (1982); **52**, 2493 (1983); **53**, 2342 (1984)
6. S. Hikami: Prog. Theor. Phys. **72**, 722 (1984); **76**, 1210 (1986); **77**, 602 (1987)
7. K. von Klitzing, G. Dorda and M. Pepper: Phys. Rev. Lett. **45**, 494 (1980)

8. S. Kawaji and J. Wakabayashi: Physics in High Magnetic Fields, edited by S. Chikazumi and N. Miura (Springer, Berlin, 1981), p. 284
9. A. MacKinnon and B. Kramer: Phys. Rev. Lett. **47**, 1546 (1981)
10. B. Kramer and A. MacKinnon: Proc. Int. Seminar on Localization in Disordered Systems (Leipzig, Teubner).
11. L. Schweitzer, B. Kramer and A. MacKinnon: J. Phys. C **17**, 4111 (1984).
12. T. Ando: Prog. Theor. Phys. Suppl. **84**, 69 (1985)
13. T. Ando: Surf. Sci. **170**, 243 (1986); J. Phys. Soc. Jpn. **55**, 3199 (1986)
14. H. Levine, S.B. Libby and A.M.M. Pruisken: Phys. Rev. Lett. **51**, 1915 (1983); Nucl. Phys. B **240** FS12 , 30; 49; 71 (1984)
15. F. Wegner: Z. Phys. B **51**, 279 (1983)
16. T. Ando: to be published
17. D.R. Hofstadter: Phys. Rev. B **14**, 2239 (1976)
18. H. Aoki and T. Ando: Phys. Rev. Lett. **57**, 3093 (1986)
19. T. Ando, A.B. Fowler and F. Stern: Rev. Mod. Phys. **54**, 436 (1982)
20. S. Hikami, A.I. Larkin and Y. Nagaoka: Prog. Theor. Phys. **63**, 707 (1980).

Edge and Bulk Extended States in Two-Dimensional Disordered Electronic Systems in Strong Magnetic Fields

T. Ohtsuki and Y. Ono

Institut für Theoretische Physik, Universität zu Köln,
D-5000 Köln, Fed. Rep. of Germany and
Department of Physics, Faculty of Science, University of Tokyo,
Hongo 7-3-1, Bunkyo-ku, Tokyo 113, Japan

1. Introduction

To understand the (integer) quantized Hall effect (QHE), the existence of both localized and extended states seems inevitable [1]. Therefore the investigation of the electronic states in 2D disordered systems with perpendicular magnetic fields is important. Recently many theoretical works [2-6] have pointed out the convenience of using the model system with confinement (cylinder geometry) instead of the system with periodic boundary condition in both directions (torus geometry). One of the advantages of the former is that the Hall conductivity can be expressed in terms of the quantities at the Fermi level [7]. Microscopic investigation of the electronic states in the cylinder geometry has not yet been done so extensively as in the torus geometry [8-10].

In this paper, we report the results of numerical calculations on the electronic states in the cylinder geometry. In this geometry, the system has no isotropy, and therefore we regard a state as extended if it is extended at least in one direction, since such a state can carry the Hall current. Particularly we consider the properties of the extended states in connection to the QHE. For disorder, we take a sum of randomly distributed δ potentials, the coefficients of which are also random variables.

2. Model and Formulation

The model treated here is described by the following Hamiltonian,

$$H = (\mathbf{p} + e\mathbf{A}/c)^2/2m + \sum_i u_i \delta(\mathbf{r}-\mathbf{r}_i) \quad , \tag{1}$$

where $\mathbf{A} = (0,Bx)$ giving rise to a uniform magnetic field B perpendicular to the system, and u_i and $\mathbf{r}_i = (x_i,y_i)$ are random variables distributed uniformly in the regions, $u_i \in [-u/2,u/2]$, $x_i \in [-L_x/2,L_x/2]$ and $y_i \in [0,L_y]$. The strength of impurity scatterings is characterized by the following dimensionless quantity,

$$\gamma = (\hbar\omega_c)^{-1}\sqrt{n_i u^2/\pi l^2} \quad , \tag{2}$$

where l is the magnetic length ($=\sqrt{c\hbar/eB}$), n_i the impurity concentration and ω_c the cyclotron frequency. The confinement is taken into account by imposing the following boundary condition (b.c.) for the wave function $\psi(x,y)$,

$$\psi(\pm L_x/2,y) = 0 \ . \tag{3}$$

For the y-direction we take the following b.c. which is more general than the simple periodic one,

$$\psi(x,y+L_y) = e^{-i\vartheta}\psi(x,y) \quad . \tag{4}$$

The parameter ϑ is introduced to investigate the sensitivity of the wave functions to the boundary condition. By this sensitivity we can characterize the degree of extendedness of a wave function (at least in y-direction).

The eigenfunction of H is expanded in terms of the following basis functions satisfying (3) and (4),

$$\varphi_{p,k(\vartheta)}(\mathbf{r}) = \sqrt{2/L_xL_y}\; e^{ik(\vartheta)}\sin[p(x+L_x/2)] \tag{5}$$

with $p = (\pi/L_x)m$ and $k(\vartheta) = (2\pi n - \vartheta)/L_y$ (m,n; integers);

$$\psi_{\nu(\vartheta)}(\mathbf{r}) = \sum_{p,k(\vartheta)} C_{p,k(\vartheta)}{}^{\nu}\varphi_{p,k(\vartheta)}(\mathbf{r}) \quad . \tag{6}$$

Then the Schrödinger equation reduces to an eigenvalue problem of a matrix. The dimension of the Hamiltonian matrix can be made finite by introducing appropriate cutoffs for p and k, which will not influence the low energy eigenstates. Details of the numerical method are discussed in [11].

In the following we consider the expectation value of the "center coordinate", $X_\nu(\vartheta)$, as one of the quantities characterizing the electronic states,

$$X_\nu(\vartheta) = l^2 \langle\nu(\vartheta)|i\partial/\partial y|\nu(\vartheta)\rangle \quad . \tag{7}$$

Using (5), (6) and the completeness of the set of functions $\{\psi_{\nu(\vartheta)}\}$, we can immediately derive the following sum rule which is independent of disorder.

$$N_T^{-1}\sum_\nu X_\nu(\vartheta) = l^2\vartheta/L_y \equiv \Delta_\vartheta \quad , \tag{8}$$

where N_T is the total number of states in the system and the integer n for $k(\vartheta)$ is taken symmetrically around zero. This sum rule is trivial in the pure case, since $X_\nu(\vartheta)$ is a good quantum number and each state suffers the same shift Δ_ϑ in changing ϑ.

3. Results

In Fig. 1 we show the energy spectrum for a system with $L_x = 50\Delta X$ and $L_y = 120\Delta X$ where ΔX ($\equiv 2\pi l^2/L_y$) is the interval between neighbouring "center coordinates" in the pure case. The plot is given as the eigenenergy E_ν vs. X_ν. The results for four different samples with the same impurity parameter $\gamma = 0.4$ are plotted together; the broken lines represent the dispersion in the pure case with $L_y = \infty$. The parameter ϑ is fixed at zero.

In contrast to the pure case, there is no state with $E_\nu = 3/2$ and $X_\nu = \pm L_x/2$. This is because the degenerate edge and bulk extended states are strongly mixed by the impurity scatterings. On the other hand, the edge states in the gap of the bulk subbands seem not to be significantly influenced by disorder, as is usually believed [3,10]. In the bulk region, the spectrum is broadened and most of the states are localized. Precise investigation of the wave functions shows that there is at least one state extended in both directions near the center of each broadened bulk subband.

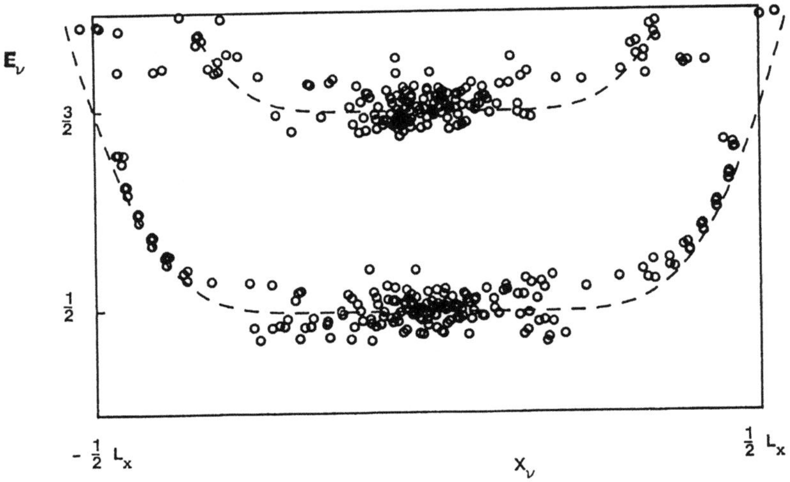

Fig.1: Eigenenergy E_ν(by $\hbar\omega_c$) vs. X_ν; γ=0.4 and ϑ=0

It is quite interesting to see the effect of changing ϑ on the electronic states. For example, a slight change of ϑ can remove the strong degeneracy between the edge and bulk states at E_ν = 3/2 and therefore we find a continuous transition from a mixed state to a purely edge-like state [11]. We have analyzed the effect of ϑ on the electronic states in terms of the ϑ-dependence of $X_\nu(\vartheta)$ for a fixed impurity configuration. As expected, the localized states are not affected by ϑ, namely $X_\nu(\vartheta)$ is essentially independent of ϑ, while $X_\nu(\vartheta)$ for the bulk extended states depends strongly on ϑ. For the edge states, the shift of $X_\nu(\vartheta)$ due to finite ϑ is almost equal to Δ_ϑ (defined in (8)); this should be compared with the pure case. In order to check the sum rule (8), we have averaged $X_\nu(\vartheta)$ over the states in the lowest subband, which consist of the edge states and the bulk extended and localized states, for different values of ϑ with fixed impurity configuration (L_x = 22ΔX, L_y = 52ΔX and γ = 0.2). The result is given in Table 1, where the value of Δ_ϑ is used to specify ϑ. Quite astonishingly, the sum rule (8) is satified with a good accuracy even within the lowest subband. Furthermore if we note the fact that the shifts of $X_\nu(\vartheta)$ for the edge states are essentially equal to Δ_ϑ, the sum rule is understood to be valid *within the lowest bulk subband*. In other words, the shifts of $X_\nu(\vartheta)$ for the bulk extended states compensate almost exactly the loss of the shifts for the localized states to recover the shift in the pure case when the shifts are averaged over all states of the lowest subband in the bulk region.

Table 1: The value of $X_\nu(\vartheta)$ averaged within the lowest subband for different Δ_ϑ (both $X_\nu(\vartheta)$ and Δ_ϑ are scaled by ΔX). The impurity configuration is fixed and the system parameters are L_x=22ΔX, L_y=52ΔX and γ=0.2.

Δ_ϑ	0.0	0.15	0.20	0.25	0.30
$\overline{X_\vartheta(\vartheta)}$	0.00071	0.14958	0.19916	0.24871	0.29822

4. Discussion

We have performed a numerical calculation on the electronic states in 2D disordered systems with perpendicular magnetic fields under the confinement in the x-direction. For the y-direction we have taken the boundary condition (4) characterized by a parameter ϑ. The ϑ-dependence of $X_\nu(\vartheta)$ (the expectation value of the "center coordinate") has been discussed and a disorder-independent sum rule for $X_\nu(\vartheta)$ has been found. It will be worthwhile to discuss the relation between the sum rule and the QHE.

According to the theory by Hajdu, Janßen and Viehweger[5], the Hall conductivity at zero temperature is expressed in the form,

$$\sigma_{yx} = \frac{e^2}{h}\sum_\nu{}' \frac{1}{L_x}\int d\vartheta \, \frac{\partial x_{\nu\nu}(\vartheta)}{\partial\vartheta} \quad , \quad [\ x_{\nu\nu}(\vartheta) \equiv \langle\nu(\vartheta)|x|\nu(\vartheta)\rangle\ ], \tag{9}$$

where the sum is over the states below the Fermi level. Although in their theory the parameter ϑ is introduced to parametrize inequivalent representations of translation on the cylinder surface, it is practically the same as that used in this paper. From this formula we find at least in the thermodynamic limit ($L_x, L_y \to \infty$) that σ_{yx} takes the quantized values when the filling factor μ is an integer, if the sum of $x_{\nu\nu}(\vartheta)$ over each subband is stable against (or in other words, independent of) disorder. Since $x_{\nu\nu}(\vartheta)$ is essentially equal to $X_\nu(\vartheta)$ in the bulk region, the sum rule as shown in Table 1 indicates the stability of the sum of $x_{\nu\nu}(\vartheta)$ over the lowest subband. Once we understand the quantization of σ_{yx} at μ = integers, it is evident that σ_{yx} as a function of the filling factor μ has plateaus at quantized values, because the localized states which exist at least in the tails of subbands do not contribute to σ_{yx} or to the Hall current. In this sense the sum rule (8) (or Table 1) expresses in an indirect way the fact that the bulk extended states compensate the loss of the Hall current by the localized states.

As a preliminary trial we have calculated σ_{yx} for the present system by using the numerical data for eigenfunctions and the original Kubo formula in the form of the current-current correlation function. The result shows that, when μ is increased from zero, σ_{yx} remains essentially zero for $\mu < 0.3$ and takes almost the same values as in the pure case [12] for $0.7 < \mu < 1$ and that it has large fluctuations for $0.3 < \mu < 0.7$. More details will be presented elsewhere.

References

1. H. Aoki, T. Ando: Solid State Commun. **38**, 1079 (1981)
2. R.B. Laughlin: Phys. Rev. B**23**, 5632 (1981)
3. B. Halperin: Phys. Rev. B**25**, 2185 (1982)
4. M. Jonson: Phys. Scr. **32**, 435 (1985)
5. J. Hajdu, M. Janßen, O. Viehweger: Z. Phys. B**66**, 433 (1987)
6. Y. Ono, B. Kramer: Z. Phys. B**67** (1987) in print
7. P. Streda: J. Phys. C**10**, 2153 (1977), C**15**, L717 (1982)
8. T. Ando: J. Phys. Soc. Jpn. **52**, 1740 (1983), **53**, 3101 (1984)
9. H. Aoki: J. Phys. C**16**, 1893 (1983)
10. L. Schweitzer, B. Kramer, A. MacKinnon: J. Phys. C**17**, 4111(1984)
11. T. Ohtsuki, Y. Ono: to be submitted to Solid State Commun.
12. Y. Ono, T. Ohtsuki: Z. Phys. B in print

On the Absence of a Quantum Electrodynamical Correction to the Quantum Hall Effect

K. Ishikawa[1] *and T. Matsuyama*[2]

[1]Department of Physics, Hokkaido University, Hokkaido, Japan
[2]Research Institute for Fundamental Physics, Kyoto University, Kyoto, Japan

We demonstrate the absence of a quantum electrodynamical correction [1] in the quantized value of the linear Hall conductance, and the absence of non-linear current in the quantum Hall regime where the transverse conductance σ_{xx} vanishes.

Since the σ_{xx} vanishes, there is no energy dissipation and the process is reversible. The fluctuation-dissipation theorem is not needed and a field theoretical calculation, which also fits well with our purpose of investigating the effect of higher-order quantum electrodynamical interactions, is suitable. Hence we apply the method of quantum field theory. In fact the σ_{xx} may not vanish at the edge of the sample due to the source-drain voltage. These regions are excluded from the electron system which we study. We investigate the electrons in the region where the σ_{xx} vanishes.

If the conductance σ_{xy} on the plateau is equal to e^2/h x integer, this effect is used as (1) a standard of electric resistance and (2) a method for a precise determination of the fine structure constant. The first is very important in electronics and the second is very important in fundamental physics since it can supply a test of quantum electrodynamics (QED). QED is the theory which determines the properties of almost all macroscopic matter. The electron is a fundamental object in QED and is considered usually as point-like. However, the electron in fact may have some internal structure, like the proton. If so, some signal would be seen. The electron's anomalous magnetic moment is the most probable quantity to show that signal. The precise value of α is needed in order to see that possible signal.

The Lagrangian density of pure QED in terms of bare quantities is given by

$$\mathcal{L} = \bar{\psi}_b \gamma_\mu (i\hbar\partial^\mu + e_0 A_b{}^\mu)\psi_b - m_b \bar{\psi}_b \psi_b - \frac{1}{4} F_{b\mu\nu} F_b{}^{\mu\nu} \quad . \tag{1}$$

One way to carry out the renormalization program is to write the above fields and Lagrangian density in terms of the renormalized quantities. By introducing three constants Z_i ($i = 1\sim3$), which are actually infinite in the perturbative calculation, the infinites of the S-matrix can be absorbed. Let us define renormalized quantities as

$$\psi_b = \sqrt{Z_2}\,\psi \,, \quad A_b{}^\mu = \sqrt{Z_3}\, A^\mu \,, \quad e_0 = \frac{Z_1}{Z_2\sqrt{Z_3}}\, e = Z_3^{-\frac{1}{2}} e \quad . \tag{2}$$

The constants Z_i satisfy $Z_1=Z_2$ from the Ward-Takahashi identity and are fixed by the normalization of the renormalized propagators and the vertex function in the following manner :

$$S_F(p)\Big|_{\not p\to m}=\frac{1}{\not p-m}\ ,\quad D_F(q)_{\mu\nu}\Big|_{q^2\to 0}=-\frac{g_{\mu\nu}}{q^2}+\text{gauge term},\quad \Gamma_\mu(p,p)=e\gamma_\mu\ . \tag{3}$$

The electron's anomaly (g-2) is the one physical quantity that has been computed and has been measured most seriously. Theoretical calculation using the renormalization theory is given by

$$g-2=\frac{1}{2}\frac{\alpha}{\pi}-0.328478966\left(\frac{\alpha}{\pi}\right)^2+\cdots\cdot \tag{4}$$

The fine structure constant α is the only parameter that must be determined experimentally.

The determination of α is made by a solid state physics method. Inside matter, there are electrons, photons, and many nuclei. It may be best to use field operators for all freedoms and to start from the QED Lagrangian density, which describes their dynamics. However, the effect of heavy objects or the effect of a particle state whose energy is far from the corresponding chemical potential is neglibible in the low energy region, except for the renormalization effect. The renormalization in matter may be different from that in the vacuum. We must pay attention to the possible difference of the renormalization in studying the QED correction.

The total Lagrangian we study is written as

$$\mathcal{L}=\mathcal{L}_{QED}+\mathcal{L}_2\ (\text{nucleus},\ A_\mu)$$

$$\mathcal{L}_2(\text{nucleus},\ A_\mu)=\psi_N^+(i\hbar\frac{\partial}{\partial x_0}+eZA_0)\psi_N-\psi_N^+\left(\frac{(\vec{P}+eZ\vec{A})^2}{2m}\right)\psi_N+\cdots\ . \tag{5}$$

In the Lagrangian, the nucleus field and electron field are used. The nucleus and the electrons are bound by the Coulomb force to become an atom or ion. These bound states have lower energy than scattering states and play the role of basic units in many materials. Particularly in metals, the ions are bound and located periodically. Electrons near the Fermi surface are in the scattering states and others which have much lower energy are bound in the nucleus to constitute ions. For the problem of ordinary solid state physics it may be sufficient to use the bound ion fields or something similar. For our purpose of studying the QED correction in the material, it is better to use (5), since $\mathcal{L}_{QED}$ in (5) is the same as that of pure QED. We use (5) and study the effect of loops of photons and electrons.

In a metal, the nuclei are located periodically. Their pair annihilation effect is negligible in the low energy region. Hence it is good to approximate their freedom by a c-number term and a q-number phonon term. We write the $\mathcal{L}_2$ (nucleus, A_μ) as

$$\mathcal{L}_2=eZA_0\rho(x)+\tilde{\mathcal{L}}_2(\Phi(x),eA_\mu)\quad , \tag{6}$$

where $\rho(x)$ is the c-number and $\Phi(x)$ is the q-number phonon field. The electromagnetic field also has c-number part which is determined from

$$\partial_\mu F^{\mu 0}_{cl.}=eZ\rho(x)\quad . \tag{7}$$

The phonon field $\Phi(x)$ has an ultraviolet cutoff, and ultraviolet divergence is not generated from this freedom.

From the Lagrangian, we see that the electron couples with only the photon. It might be almost trivial that all the ultraviolet divergences of the theory can be subtracted by the counter term of pure QED. We use the pure QED counter term.

The total current integrated in the z-direction, $\tilde{J}_\alpha$ ($\alpha \neq z$), is conserved in three-dimensional space time. Since the vertex function for $\tilde{J}_\alpha$ is normalized by $\Gamma_\alpha(p,p) = \frac{\partial s^{-1}}{\partial p_\alpha}(p)$ from the Ward-Takahashi identity. The antisymmetric part of the conductance which is defined by the ratio between the current $\tilde{J}_\alpha$ and the integrated voltage difference $\tilde{V}_\alpha$ is written as

$$\sigma_{xy} = \frac{e^2}{h} \frac{1}{24\pi^2} \int d^3p \, \varepsilon^{\alpha\beta\gamma} \, \mathrm{Tr} \, \{\frac{\partial s^{-1}}{\partial p_\alpha} s \frac{\partial s^{-1}}{\partial p_\beta} s \frac{\partial s^{-1}}{\partial p_\gamma} s\} \quad , \tag{8}$$

where the s(p) is the electron propagator under the magnetic field and is a large matrix with a Landau level index. A detailed account of the derivation of the above formula is given in [2]. This integration has a particular topological meaning and is equal to e^2/h x integer if the integration region is compact and the singularity of s(p) is not in the integration region. Actually the integration region is that of a pinched torus and s(p) has no singularity in the spatial component of the momentum variable in the localized state region and in the gap region. Thus σ_{xy} is equal to e^2/h x integer in the localized state region and in the gap region.

The charge e in (8) and the propagator s(p) are defined by a minimal counter term, namely by the counter term of pure QED, and are not quantities on the mass shell. The charge on the mass shell in the matter may depend on the material and on the definition. What will happen if these on-mass-shell quantities are used? Let e_M and e_M' stand for the charge defined through the zeroth component $A_{0,M}$ and spatial component $A_{i,M}$ respectively. The e^2 in (8) is replaced by $e_M e_{M'}$, and the propagator s(p) is replaced by the one defined through the on-shell condition. The Ward-Takahashi identity leads, however, $e_M A_{0,M}$ to agree with $e_b A_{0,b}$ and $e_{M'} A_{i,M}$ to agree with $e_b A_{i,b}$. Therefore, the voltage V_M and current I_M in the matter are measured as $\frac{e_M}{e} V_M(V)$ and $e/e_{M'} I_M(I)$ in the ordinary method [3] in which the charge e is used as the unit. The ratio between I and V is given by the same formula as (8). Thus the current and voltage difference normalized by the charge e lead the quantized Hall conductance to be exactly e^2/h x integer.

Let us turn to the second problem, non-existence of non-linear current when the σ_{xx} vanishes. We use an effective action in order to see it. A general form of gauge-invariant effective action in three-dimensional space time is given by

$$S = \int d^3x [\frac{c_0}{2}] \varepsilon_{\alpha\beta\gamma} A^\alpha \partial^\beta A^\gamma + \tilde{S}[\partial_\alpha A_\beta - \partial_\beta A_\alpha] \quad . \tag{9}$$

The second part $\tilde{S}$ is an arbitrary functional of $F_{\alpha\beta}$.

The current density is obtained from the above action S as

$$J_\alpha(x) = c_0 \varepsilon_{\alpha\beta\gamma}\partial^\beta A^\gamma - (\partial_\beta g_{\alpha\gamma} - \partial_\gamma g_{\alpha\beta}) \frac{\delta\tilde{S}}{\delta F_{\beta\gamma}(x)} \quad . \tag{10}$$

The observed current is actually the one integrated in one of two spatial directions, the x-direction for instance, and becomes

$$I_\alpha = c_0 \, V_\beta \quad . \tag{11}$$

From the derivatives to the term which depends on only $F_{\beta\gamma}$, the second term drops out of (11) and the linearity between the current and voltage difference in the dissipationless situation is shown.

Acknowledgement

The present work is partially supported by a Grand-in-Aid for Scientific Research, The Ministry of Education, Science and Culture, Japan (61540189).

References

[1] T. Kinoshita, in Proc. Int. Symp. on Foundations of Quantum Mechanics, ed. S. Kamefuchi et al. (Phys. Soc. Japan 1983), p.357.

[2] K. Ishikawa and T. Matsuyama; Nucl. Phys. B280 (1987) 523: Zeit. Phys. C33 (1986) 41, Hokkaido University preprint (1987).
K. Ishikawa, in Proc. Int. Symp. on Fundations of Quantum Mechanics, ed. M. Namiki et al. (Phys. Soc. Japan 1986), p.165.

[3] D.N. Langenberg and J.R. Schrieffer; Phys. Rev. B3 (1971) 1776: J.B. Hartle et al., Phys. Rev. B3 (1971) 1778.

Degenerate Ground State for the Fractional Quantum Hall Effect

K. Takano[1] *and A. Isihara*[2]

[1]Department of Physics, Nagoya University, Chikusa-ku, Nagoya 464, Japan

[2]Department of Physics, State University of New York at Buffalo, Buffalo, NY 14260, USA

A new variational wave function is shown to yield a ground state energy at $\nu=1/m$ of the fractional quantum Hall effect slightly below that of Laughlin's wave function. There are indications that the new ground state is degenerate.

Since Tsui et al.[1] discovered the fractional quantum Hall effect at one-third filling of the lowest Landau level in GaAs/GaAlAs heterostructures, quantized Hall plateaus have been observed at series of odd denominator fractions of the filling factor of the lowest Landau level[2]. At these plateaus of the Hall conductivity, the magnetoconductivity vanishes or at least dips markedly and shows activation-type temperature dependences. In the case of the one-third quantized Hall state, the quantization has been found[3] to be better than 1 part in 10^4. To explain the basic $\nu=1/m$ case, where ν is the filling factor and m is an odd integer, Laughlin introduced a wave function describing an incompressible fluid state[4]:

$$\Psi_L = \Pi_{i<j}(z_i-z_j)^m \exp(-\Sigma_k|z_k|^2/4), \tag{1}$$

where $z_j=x_j+iy_j$ $(j=1,\cdots,N)$ is the coordinate of the *jth* electron in units of the magnetic length $a_0=\sqrt{(c\hbar/eB)}$, N being the number of electrons in the system. This wave function yields a ground state energy which is lower than that of the charge density wave state[4,5]. It is constructed on electron states of the lowest Landau level such that the total angular momentum is conserved and is given by $L=mN(N-1)/2$. Its first factor represents electrons' mutual exclusion very strongly with m-fold zeros.

Laughlin's wave function is not the exact eigenstate but a Jastrow-type trial function. Based on numerical analysis of small systems, Laughlin[4] and Yoshioka et al.[6] showed that it is very close to the exact ground state. However, the latter authors noticed a possible deviation from the m-fold zero property although its cause was unclear[7]. Also, effects of many-electron correlations on the wave function may be studied because the Jastrow-type function is constructed only on electron pairs independently of the details of electron-electron interactions.

The necessity of the m-fold degeneracy in the ground state for the 1/m effect has been pointed out by several authors[8]. On the other hand, Trugman and Kivelson[9] showed that Laughlin's function is nondegenerate and exact for short range repulsive potentials. Combining these two lines of work, the m-fold degeneracy might be sought with attention focused on the long range Coulomb potential.

In the present article, we try to approach the true ground state by using a variational method. We shall introduce a new trial wave function with a variational parameter, modifying Laughlin's wave function. We shall show that such a modification of Laughlin's function lowers the ground state energy.

Our search for a new trial wave function for the $\nu=1/m$ quantum Hall effect is based on the following criteria: i) It is constructed on electron states of the lowest Landau level only. ii) It modifies Laughlin's wave function with new terms representing many-body correlations. iii) It conserves the total angular momentum, $L=mN(N-1)/2$. Under these conditions, we have found a new wave function which is given by

$$\Psi = \Pi_{i<j}(z_i-z_j)(f_{ij})^p(f_{ji})^p \exp(-\Sigma_k |z_k|^2/4), \tag{2}$$

$$f_{ij} = z_i-z_j + c \sum_{k\neq i,j}' z_k, \tag{3}$$

where p is the integer determined by $m=2p+1$ and c is a variational parameter. This wave function is unchanged even if c is replaced by -c, antisymmetric with respect to electron exchanges, and is reduced to Laughlin's function when c vanishes. On the other hand, it weakens the strong m-fold zero character of Laughlin's wave function and vanishes not only when electron pairs overlap but also when f_{ij} or f_{ji} is zero. In qualitative agreement with what the work of Yoshioka et al. [7] indicates, the m-fold zeros are shifted from each other. Ψ is not a Jastrow-type function of electron pairs only, and yet conserves the total angular momentum. The form of L suggests that a wave function can be composed of $N(N-1)/2$ factors. The antisymmetry of the wave function determines the form of each factor almost uniquely.

While it is rather difficult to explain the origin of the c term in f_{ij}, we were based on the following gedanken experiment. Suppose the electron at z_i approaches the one at z_j in the presence of all other electrons. If the center of mass of these other electrons coincides with the origin, the disk geometry of the back-ground positive charges, which is assumed in the evaluation of the ground state energy, would cancel their combined effect on the pair. If the center is shifted from the origin, the charge cloud of these electrons would act on the i*th* electron as if the charge cloud at z_j were shifted. Such a shift is due to intermediate or long range Coulomb repulsion so that the parameter c may depend on the Coulomb potential.

The role played by the c term in the new trial wave function may be found by using the OCP (one component plasma) analogue and comparing it to Laughlin's result[4]. Let us assume that the c term is small, because otherwise the function is not close to Laughlin's state, and that the electrons do not overlap each other so that none of z_i-z_j vanishes. We find then to first order in c the OCP potential energy given by

$$\Phi = -m^2 \sum_{i<j} \ln|z_i-z_j| + \frac{m}{4}\sum_i |z_i|^2 + \frac{m-1}{2m} \sum_{i<j} |c|^2 |\Sigma_k' z_k|^2/|z_i-z_j|^2, \tag{4}$$

Thus the c term adds a repulsive potential to Laughlin's OCP potential. This additional term decreases rather fast with electron distances, depends on the location of the center of gravity of the OCP except for a chosen pair i and j, and also on the density $1-\nu$ of the background positive charges. Equation (4) has been derived on the assumption that z_i and z_j do not overlap. The probability for such overlaps may be considered to be small because the trial wave function vanishes due to its first factor.

Using the new trial wave function, we performed Monte Carlo calculations of the ground state energy of the $\nu=1/3$ state as a function of the parameter c. This parameter can be complex, but we investigated only real cases. For the background charges, we assumed the standard disk geometry. For this case, we could use the exact analytical formulae for the electron-background and background-background energies which have been derived in Eqs. (16), (17) and (18) of our article[10]. The use of these formulae reduces the computation time. We adopted the standard Metropolis algorithm, and performed 20 Monte Carlo simulations, each consisting of 250,000 Monte Carlo steps, for a given value of the parameter c. For improved reliability of several points, we repeated the 20 Monte Carlo simulations a few times with different initial random numbers. In order to avoid accumulation of computing errors, calculations were made in double precision whenever possible. In the case of c=0 (Laughlin case) Morf and Halperin[11] have performed similar calculations.

Figure 1(a) represents our result for N=10. Each point corresponds to the average taken from 20 Monte Carlo simulations. The error bars indicate possible variations of the average values. In the limit of $c \to 0$, the ground state energy ε_g is -0.395057(63) in the unit of $e^2/\kappa a_0$ (κ: the dielectric constant), which corresponds to Laughlin's wave function. As c increases from 0, the energy decreases. In the region $0.05<c<0.09$, it appears that there might be some structures in the energy curve, but due to fluctuations it is difficult to identify them. Thus, we have introduced just one arrow to indicate a possible minimum point at c=0.085, where the minimum energy is -0.395548(77).

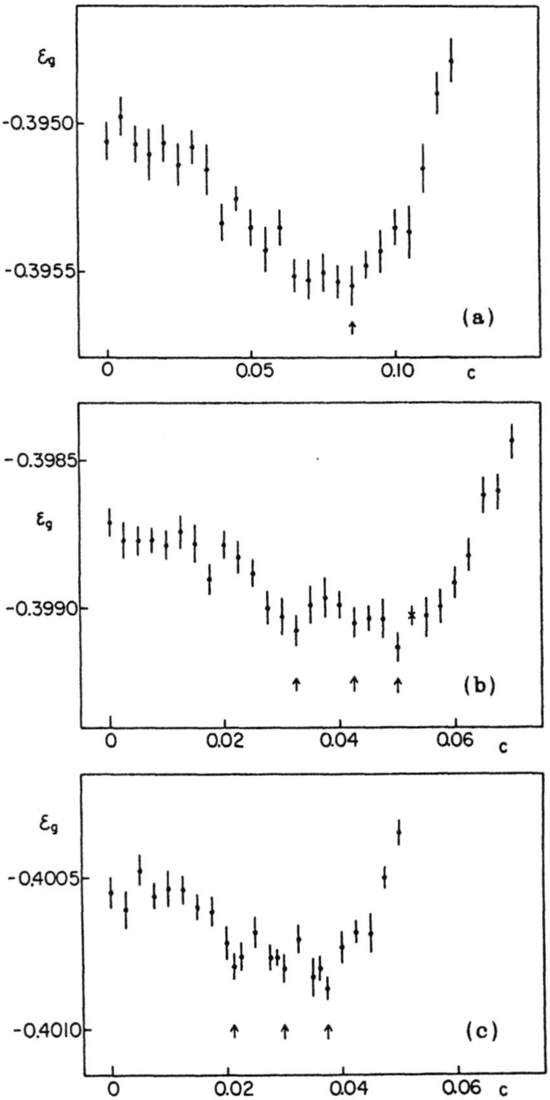

Fig.1. The ground state energies at ν=1/3 for (a) N=10, (b) 20, and (c) 30.

Figure 1(b) represents the case of N=20. The energy decreases again as c increases from zero where ε_g= -0.398717(52). It appears that there are minima at c= 0.0325, 0.0425 and 0.05 with minimum energies -0.399075(56), -0.399050(53) and -0.399132(51) respectively. These points are indicated by arrows. For some unknown reasons, our initial Monte Carlo run at c= 0.0525 showed rather large fluctuations. Hence, we performed 3 sets of 20 Monte Carlo simulations for this particular point which is indicated by a cross. As a result, the error bar at this point became relatively short. At c= 0.0175, the Monte Carlo average came out rather low. However, since the ground state energy decreased as c was further increased, we did not repeat the Monte Carlo calculations at this point. We were more interested in the minima around the arrows.

Figure 1(c) is the case of N=30. In spite of the large computation time (about 7000 seconds for a single Monte Carlo simulation), we tried to confirm the three minima near the arrows by increasing the density of c values. The ground state energies are -0.400795(47), -0.400798(51) and -0.400865(41) at c= 0.02125, 0.03 and 0.0375 respectively. Note that the limit c→0 corresponds to Laughlin's state which has a higher energy of -0.400549(57) than any of the above three lowest energies. Figure 1(c) shows some initial fluctuations but, for the same reason as before, we did not try to do more Monte Carlo simulations for these relatively small c cases. On the other hand, the structure between c=0.02 and 0.04 is now more clearly identifiable than the previous cases.

Comparing all three cases, we find that the three minima become clearer as N increases and the corresponding c values decrease. This decrease is approximately proportional to $1/\sqrt{N}$. On the other hand, the c term in Eq.(3) includes a sum of the coordinates of N-2 electrons and we might expect a contribution $\langle|\Sigma_k' z_k|^2\rangle^{1/2} \sim \sqrt{N}$ from the last term in Eq.(4). Thus, we expect that there will be a net effect from the c term even in the thermodynamic limit. We remarked that the energies at c=0 for N=20 and 30 agree with those reported by Morf and Halperin[11]. Combining the present result with that for N = 5 and 6 in our previous paper[12], we extrapolate the difference between the energies of c=0 (Laughlin case) and of the deepest minimum to the limit of N → ∞ as

$$\Delta\varepsilon_g = \varepsilon_g(c=0) - \varepsilon_g(\text{minimum}) = 0.000350 + 0.044578\ N^{-2.6} \qquad (5)$$

(Fig.2). The limiting value is 0.000350(119).

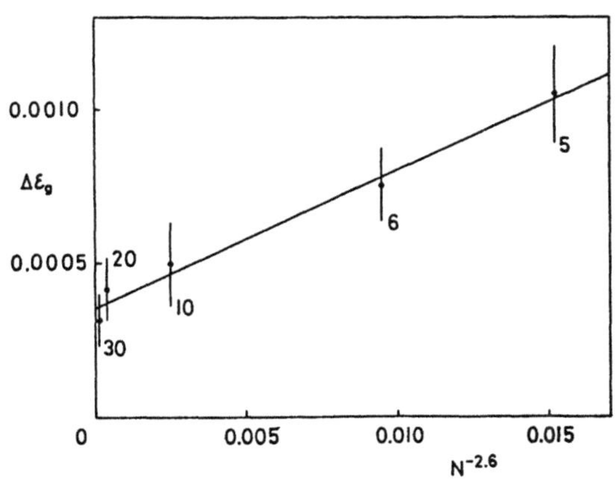

Fig.2. The difference between the energies of c=0 and of the minimum for N=5, 6, 10, 20 and 30.

References

1. D.C. Tsui, H.L. Störmer, A. Gossard: Phys. Rev. Lett. 48, 1559 (1982)
2. H.L. Störmer, A. Chang, D.C. Tsui, J.C.M. Hwang, A.C. Gossard, W. Wiegmann: Phys. Rev. Lett. 50, 1953 (1983)
3. D.C. Tsui, H.L. Störmer, J.C.M. Hwang, J.S. Brooks, M.J. Naughton: Phys. Rev. B28, 2274 (1983)
4. R.B. Laughlin: Phys. Rev. Lett. 50, 1395 (1983)
5. D. Yoshioka, H. Fukuyama: J. Phys. Soc. Jpn. 47,394 (1979); D. Yoshioka, P.A. Lee: Phys. Rev. B27, 4986 (1983)
6. D. Yoshioka, B.I. Halperin, P.A. Lee: Phys. Rev. Lett. 50, 1219 (1983); D. Yoshioka: Phys. Rev. B39, 6833 (1984)
7. D. Yoshioka, B.I. Halperin, P.A. Lee: Surf. Sci. 142, 155 (1984)
8. J.E. Avron, R. Seiler: Phys. Rev. Lett. 54, 259 (1985); Q. Niu, D.J. Thouless, Y.S. Wu: Phys. Rev. B31, 3372 (1985); R. Tao, F.D.Haldane: Phys. Rev. B33, 3844 (1986)
9. S.A. Trugman, S. Kivelson: Phys. Rev. B31, 5280 (1985)
10. K. Takano, A. Isihara: Phys. Rev. B34, 1399 (1986)
11. R. Morf, B.I. Halperin: Phys. Rev. B33, 2221 (1986)
12. K. Takano, A. Isihara: J. Phys. C20, L281 (1987)

Conductivity of Band Electrons in a Magnetic Field

I. Sadakata

Department of General Education, Higashi Nippon Gakuen University, Toubetu, Hokkaido 061-02, Japan

1. Introduction

Recently the motion of electrons in a disordered system has been intensively investigated [1,2,3]; it has been shown that among diagrams which contribute to the conductivity the maximally crossed diagrams are dominant in two-dimensional systems and decrease the conductivity to zero. Magnetic field supresses the effect of the maximally crossed diagrams and increases the conductivity [4].

For electrons in a narrow band, on the basis of the coherent potential approximation (CPA), we have studied the conductivity by taking into account the maximally crossed diagrams [5] and have shown that the conductivity vanishes in two-dimensional systems. If a magnetic field is in existence, the motion of band electrons is more complex than that of free electrons. In this study we investigate the conductivity of two-dimensional band electrons in a magnetic field by taking into account the maximally crossed diagrams.

2. Magnetic Field

The Hamiltonian of electrons on a two-dimensional square lattice in a magnetic field [6,7] is given as

$$\hat{H} = \sum_{i,j} t_{ij} \exp[-i(e/\hbar c)(A_i+A_j)(R_i-R_j)/2]\,|i\rangle\langle j|, \tag{1}$$

where t_{ij} is $-(1/2)$ if the site i and j are nearest neighbours and the vector potential $A_i = (\,0, (R_i)_x H\,)$. An eigenfunction of $\hat{H}$ is factorised as $\exp[ik_y i_y a]f(i_x)$. $f(m)$ satisfies the equation [8] as

$$Ef(m) = -(1/2)\{\, f(m+1)+f(m-1)+2\cos(2\pi gm+k_y a)f(m)\,\}, \tag{2}$$

where $g=a^2 eH/(2\pi\hbar c)$. If g is rational and expressed as p/q (p and q are relatively prime), the site diagonal potential in (2) has the superlattice symmetry of period qa.

We will study the special case where g=1/2. This corresponds to a magnetic field of the order 20 Tesla for $a \sim 100$ A. Such a value of a will be realized in a superlattice structure. In this case the potential in (2) is alternating. The eigenfunctions and eigenvalues f_{ks}, E_{ks} ($s=1,2;\ -\pi/2a<k_x<\pi/2a;\ -\pi/a<k_y<\pi/a$) are easily obtained. Two subbands E_{k1} $(=-E_k; E_k=(\cos^2 k_x a+\cos^2 k_y a)^{1/2}$) and

E_{k2} ($=E_k$) touch at points such as ($\pi/2a$, $\pi/2a$); the density of states $D(E)$ ($|E|< \sqrt{2}$) tends to zero as $E\to 0$ and diverges logarithmically as $E\to 1$. The eigenvector $|ks>$ is given as

$$<i|ks> = \exp[ik_y i_y a] f_{ks}(i_x). \tag{3}$$

The current operator J_r ($r=x,y$) is given as follows

$$J_r = ie/\hbar \sum_{i,j} t_{ij}(R_i-R_j)_r \exp[-i(e/\hbar c)(A_i+A_j)(R_i-R_j)/2]|i><j|. \tag{4}$$

The matrix element $<ks|J_r|k's'>$ is obtained by using (3); it is obvious that $<ks|J_r|k's'>$ is zero if $k\neq k'$. By using the Kubo formula, we get the frequency-dependent conductivity as

$$\sigma_{xx}(\omega) = -\frac{i}{\omega}(ae/\hbar)^2 \sum_k \frac{\partial^2 E_k}{\partial(k_x a)^2} (F(E_{k1})-(F(E_{k2})) \tag{5}$$

$$\sigma_{xy}(\omega) = 0, \tag{6}$$

where $F(E)$ is the Fermi function. As to the Green function $G(z)=(z-H)^{-1}$, we note that $<n|G|n>$ is independent of the site n and given as

$$<n|G|n> = \int dE\, D(E)/(z-E).$$

3. Impurities

Next we will add impurities to the system; the impurity potential is assumed to be given by $H_i = \sum_i e_i|i><i|$, where the probability density function $p(e_i)$ is such as $p(e_i)=1/W$ if $|e_i|<W/2$; $=0$ otherwise. Using CPA, we get the averaged Green function $\bar{G}=<G>$ as

$$\bar{G}(z) = (z-\hat{H}- \Sigma(z))^{-1}, \tag{7}$$

where the self-energy $\Sigma(z)$ satisfies the equations as

$$< (e_n- \Sigma)/[1-(e_n-\Sigma)<n|\bar{G}|n>] > = 0 \tag{8}$$

$$<n|\bar{G}|n> = \int dE\, D(E)/(z-E- \Sigma). \tag{9}$$

Solving (8) and (9), we get $<n|\bar{G}|n>$. The graph of $<n|\bar{G}(E-i0)|n>$ vs. E is shown in Fig.1 for W=0.5 and 4.0.

To obtain the conductivity, we need the averaged two-particle Green function $<<m|GJ_rG'|n>>$, where $G'=G(z')$. In CPA, we have

$$<<m|GJ_rG'|n>> = <m|\bar{G}J_r\bar{G}'|n> + \sum_l <m|\bar{G}|l>D_l<l|\bar{G}'|n> \tag{10}$$

where D_l is the summation of the particle-hole ladder-type diagrams and satisfies the equation as

$$D_l = L[<l|\bar{G}J_r\bar{G}'|l> + \sum_m <l|\bar{G}|m>D_m<m|\bar{G}'|l>]$$

and L is the effective particle-hole interaction given as

$$L = (\Sigma - \Sigma')/(\bar{G} - \bar{G}'). \tag{11}$$

In our system we find that $D_1=0$ and the second term of (10) does not contribute to the conductivity; we get the conductivity as

$$\sigma_{xx}(E) = \hbar/\pi \sum_m \langle m|J_x \mathrm{Im}\bar{G}(E-i0) J_x \mathrm{Im}\bar{G}(E-i0)|m\rangle \tag{12}$$

$$\sigma_{xy}(E) = 0.$$

Here E denotes the Fermi energy. The graph of $\sigma_{xx}(E)$ vs. E is shown in Fig.2 for W=0.5, 2.0, 4.0 and 8.0. We note that (a) when $W \to 0$, $\sigma_{xx}(E)$ is not peaked at E=0 and decreases to $(e^2/\hbar)\pi^{-2}$ as $E \to 0$; (b) when W is large, $\sigma_{xx}(E) \to 0.5(e^2/\hbar)\pi W^{-2}$.

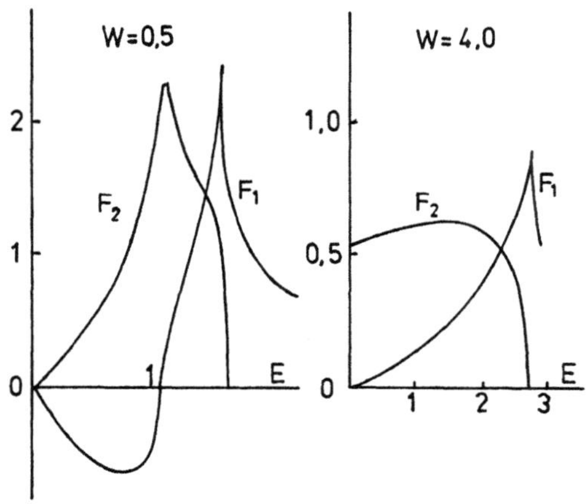

Fig.1 The graph of $\langle n|\bar{G}(E-i0)|n\rangle$. The real and imaginary partsof $\langle n|\bar{G}(E-i0)|n\rangle$ are denoted by F_1 and F_2, respectively

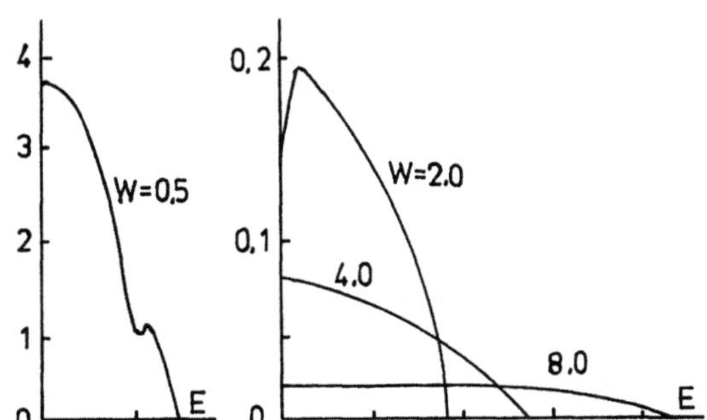

Fig.2 $\sigma_{xx}(E)$. The unit of the vertical axis is $e^2/\hbar$. $\sigma_{xx}(0)$ is about 0.1 for W=0.5 and 2.0

4. Maximally Crossed Diagrams

In CPA, the effective particle-particle interaction is identical with the effective particle-hole interaction L in (11). In maximally crossed diagrams two particles interact with L in the form of ladder. The contribution of the maximally crossed diagrams to the conductivity is given as

$$\delta\sigma_{xx}(E) = \hbar/2\pi \sum_{m,n,r,s} \langle m|J_x|n\rangle\langle r|J_x|s\rangle \sum_{l,l'} K_{l,l'}$$

$$\times\langle n|\bar{G}|l\rangle\langle l'|\bar{G}|r\rangle\langle s|\bar{G}'|l\rangle\langle l'|\bar{G}|m\rangle , \tag{13}$$

where $z=E-i0$ and $z'=E+i0$; and $K_{l,l'}$ is the summation of the particle-particle ladder-type interactions and given as follows

$$K_{l,l'} = L[\ \delta_{l,l'} + L\langle l|\bar{G}|l'\rangle\langle l|\bar{G}'|l'\rangle$$
$$+ L^2 \sum_m \langle l|\bar{G}|m\rangle\langle m|\bar{G}|l'\rangle\langle l|\bar{G}'|m\rangle\langle m|\bar{G}'|l'\rangle + \dots\ ]. \tag{14}$$

It is difficult to sum up the series in $K_{l,l'}$. We note, however, that the " average " of the second term of $K_{l,l'}$ is reformed as

$$L \sum_{l'} \langle l|\bar{G}|l'\rangle\langle l|\bar{G}'|l'\rangle = -(\Sigma - \Sigma')/(\ z-\Sigma - z'+\Sigma'). \tag{15}$$

Here we used the symmetry relation

$$\langle l|\bar{G}|l'\rangle = \langle l'|\bar{G}|l\rangle\ . \tag{16}$$

In the case where $z=E-i0$ and $z'=E+i0$, the right hand side of (15) is equal to +1. This fact suggests that $\delta\sigma_{xx}(E)$ is infinite and so the conductivity in our system vanishes. However, a careful treatment of $K_{l,l'}$ is necessary to confirm this suggestion.

We have studied the case where $g=1/2$. This case is a special case, because (6) and (16) are satisfied. To state more general statements about the conductivity, we must investigate another case, for example, $g=1/3$.

References

1. E.Abrahams, P. W. Anderson, D. C. Licciardero and T. V. Ramakrishnan: Phys. Rev. Lett. 42, 672 (1979)
2. L. P. Gorkov, A. I. Larkin and D. E. Khmel'nitskii: Sov. Phys.-JETP Lett. 30, 228 (1979)
3. D. Vollhardt and P. Wölfle: Phys. Rev. B22, 4666 (1980)
4. A. Kawabata: J. Phys. Soc. Jpn. 53, 3540 (1984)
5. I. Sadakata: J. Phys. Soc. Jpn. 54, 736 (1985); 55, 3991 (1986)
6. D. J. Thouless, M. Kohmoto, M. P. Nightingale and M. den Nijs: Phys. Rev. Lett. 49, 405 (1982)
7. H. Aoki: J. Phys. C18, L67 (1985)
8. D. R. Hofstadter: Phys. Rev. B14, 2239 (1976)

Magnetic Field and Electronic States in Modulation-Doped Heterostructures

Shi-jie Xiong

Department of Physics, Nanjing University, Nanjing, The People's Republic of China

Since the discovery of the quantum Hall effect, a number of models have been proposed to explain the experimental results /1,2/. Most of the models are based on the features of the 2-dimensional electronic gas subject to an external field. A realistic system in which the quantum Hall effect is observed is usually of the form of modulation-doped heterostructures. The modulation-doping technique is always used to enhance the mobility of carriers /3/. For these systems we should consider the motion of electrons in a film consisting of several layers with different potentials and different disorder extents. In this work we attempt to propose a simple model to describe the feature of electrons in modulation-doped heterostructure subject to an external magnetic field, and to calculate the densities of states and the localisation characteristics of electrons based on this model.

We suppose that the electrons are moving in a film of s.c. lattice with lattice constant a, which is of length La, width Ma and thickness Na, in the x,y,z directions, respectively. The external magnetic field $\vec{B}$ is applied in the z direction. The one-electron tight-binding Hamiltonian is given by

$$H = \sum_{lmn} E_{lmn} |lmn\rangle \langle lmn| + \sum_{lmn} \sum_{l'm'n'} |lmn\rangle \langle l'm'n'| \, V_{\ell mn,\ell'm'n'} \tag{1}$$

where l, m, n are the site indices in the x, y, z directions, respectively, E_{lmn} is the potential energy in the corresponding site, $V_{lmn,l'm'n'}$ being the hopping integral. In modulation-doped heterostructure, the extents of randomness and the average potentials are different for different atomic planes. The average potential in an atomic plane with large randomness is usually higher than that in the atomic plane with less randomness. We introduce randomness parameters W_1, W_2,..., W_N and average potentials V_1, V_2,..., V_N to express the probability distributions of E_{lmn},

$$P(E_{lmn}) = P_n(E) = \begin{cases} 1/W_n, & |E-V_n| \leq W_n/2, \\ 0, & |E-V_n| > W_n/2. \end{cases} \tag{2}$$

Here we consider the nearest neighbour hoppings of electrons only. In absence of the external magnetic field the N-N hopping integrals may be set to unit by choosing the energy unit,

$$V_{lmn,l'm'n'} = \begin{cases} 1, & \text{for } (lmn),\ (l'm'n') \text{ being nearest neighbours,} \\ 0, & \text{otherwise.} \end{cases} \tag{3}$$

A homogeneous magnetic field $\vec{B} = (0, 0, B)$ leads to complex hopping integrals /4/. In Landau gauge we have

$$V_{lmn,l'm'n'} = \begin{cases} 1, & \text{for } m'=m,\ l'=l \text{ and } n'=n\pm 1 \text{ or } m'=m,\ n'=n,\ l'=l\pm 1, \\ \exp(\pm 2\pi i\alpha), & \alpha=eBa^2/h, \quad \text{for } l'=l,\ n'=n \text{ and } m'=m\pm 1. \end{cases} \tag{4}$$

For the boundary conditions we use the strip geometry by taking

$$V_{11n,1Mn} = V_{1Mn,11n} = 0. \tag{5}$$

In order to calculate the densities of states and the localisation lengths we use the iterative method proposed by MACKINNON /5/. By this method we can evaluate the quantities corresponding to a sample of extension ((L+1)a, Ma, Na) from the quantities of the sample of extension (La, Ma, Na):

$$A^{(L+1)} = f(A^{(L)}). \tag{6}$$

The Hamiltonian $H^{(L+1)}$ is related to $H^{(L)}$ by the following expression:

$$H^{(L+1)} = H^{(L)} + H^{0}_{L+1} + H', \tag{7}$$

where

$$H^{0}_{L+1} = \sum_{m,n} E_{L+1,m,n} \left| L+1,m,n\right\rangle \left\langle L+1,m,n\right| + \sum_{m',n';m,n}\left|L+1,m',n'\right\rangle \left\langle L+1, m,n\right| \tag{8}$$

$$H' = \sum_{m,n}\left(\left|L,m,n\right\rangle \left\langle L+1,m,n\right| + \left|L+1,m,n\right\rangle \left\langle L,m,n\right|\right). \tag{9}$$

The Green's functions, defined as

$$G^{(L)}_{1,1'} = \left\langle 1 \left| (z-H^{(L)}+i\gamma)^{-1} \right| 1'\right\rangle \tag{10}$$

are MN×MN matrices. They satisfy the following iterative relations:

$$G^{(L+1)}_{L+1,L+1} = (z - H^{0}_{L+1} - G^{(L)}_{L,L} + i\gamma)^{-1} \tag{11}$$

$$G^{(L+1)}_{1,1'} = G^{(L)}_{1,1'} + G^{(L)}_{1,L} \, G^{(L+1)}_{L+1,L+1} \, G^{(L)}_{L,1'}. \tag{12}$$

The local density of states in an atomic plane is given by

$$\rho_n(z) = -1/\pi \lim_{\substack{L,M\to\infty \\ \gamma\to 0}} (1/LM) \sum_{1,m} \left[G^{(L)}_{1,1}(z)\right]_{mn,mn}. \tag{13}$$

The localisation length of the electronic waves in the x direction may be calculated from the Green's functions:

$$\lambda(z) = -\lim 2(L-1)a/\ln \sum_{m,n,m',n'} \left|\left[G_{1,L}(z)\right]_{mn,m'n'}\right|. \tag{14}$$

In order to investigate the influence of modulation-doping on the electronic states in a film subject to a magnetic field, we calculate ρ and λ for a sample film with extension of L=36,M=8,N=2. The results are shown in Figs.1-2. Figure 1 shows the local densities of states in a modulation-doped film. It can be seen that the magnetic sub-bands are considerably broadened as compared with the case of the purely periodic potential in strip geometry. The broadening also appears in the atomic plane without disorder. This reflects the influence of the disorder in the doped layer on the electronic states in undoped layer. The calculations for other modulation potential parameters show that this influence is reduced as the potential in the doped layers is enhanced. The localisation lengths for different

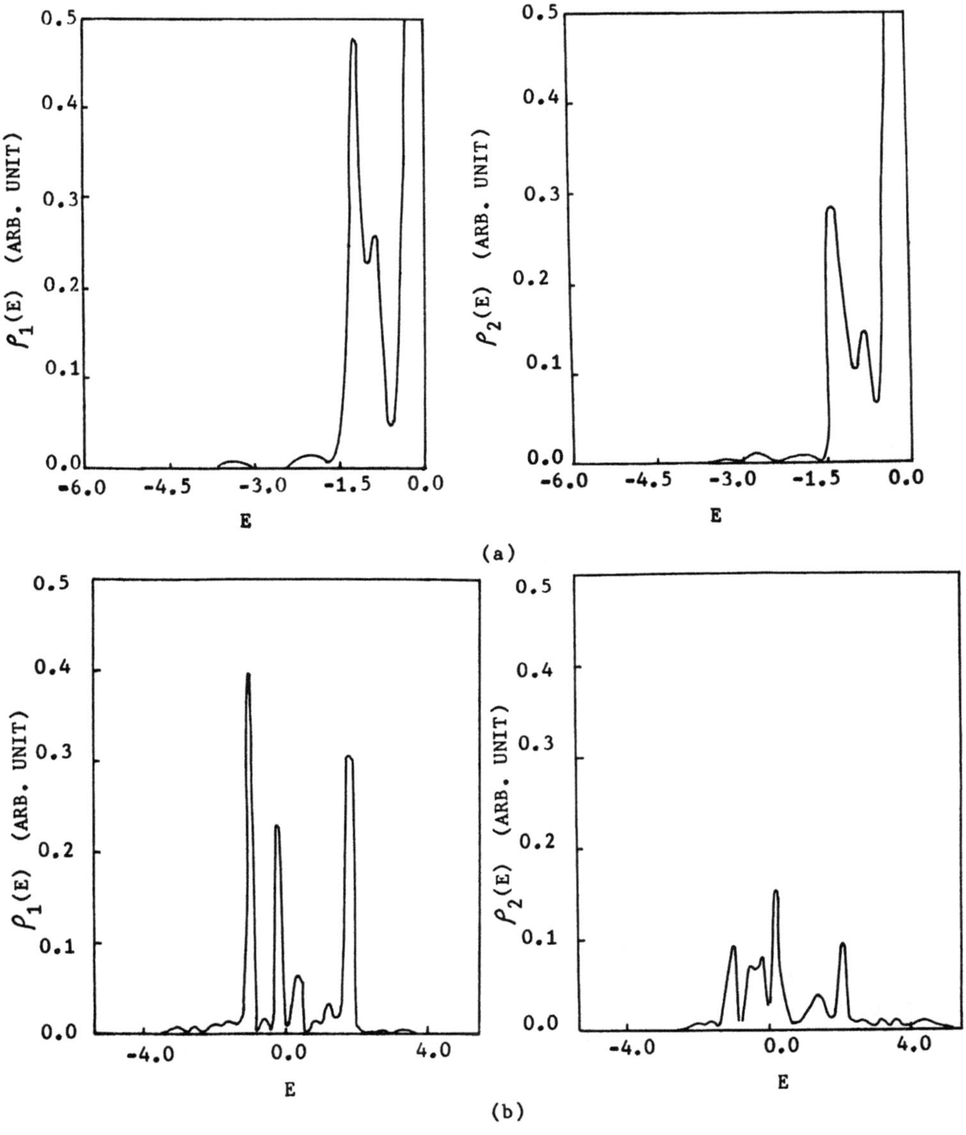

Fig.1 The local densities of states as functions of energy E. The magnetic field strength corresponds to α=0.125. The modulation-doping parameters are (a) W_1=0, W_2=1, V_1=V_2=0; (b) W_1=0, W_2=1, V_1=0, V_2=1 . L=36, M=8, N=2.

modulation-doping parameters are shown in Fig.2. The qualitative characteristics of localisation at the centers of sub-bands are not much effected by the disorder of doped layer, provided that there exists layer with very small randomness. This layer provides the channel for transmission of electronic waves, which is almost independent of the disorder parameters in the doped layer. As concerns the quantum Hall effect, it seems to be true that if the modulation-doping parameters are such that the electronic states are less effected by the disorder in the doped layers, the magnetic sub-bands are still separated, and there should exist the channel for electronic transmission free of the disorder. These features pro-

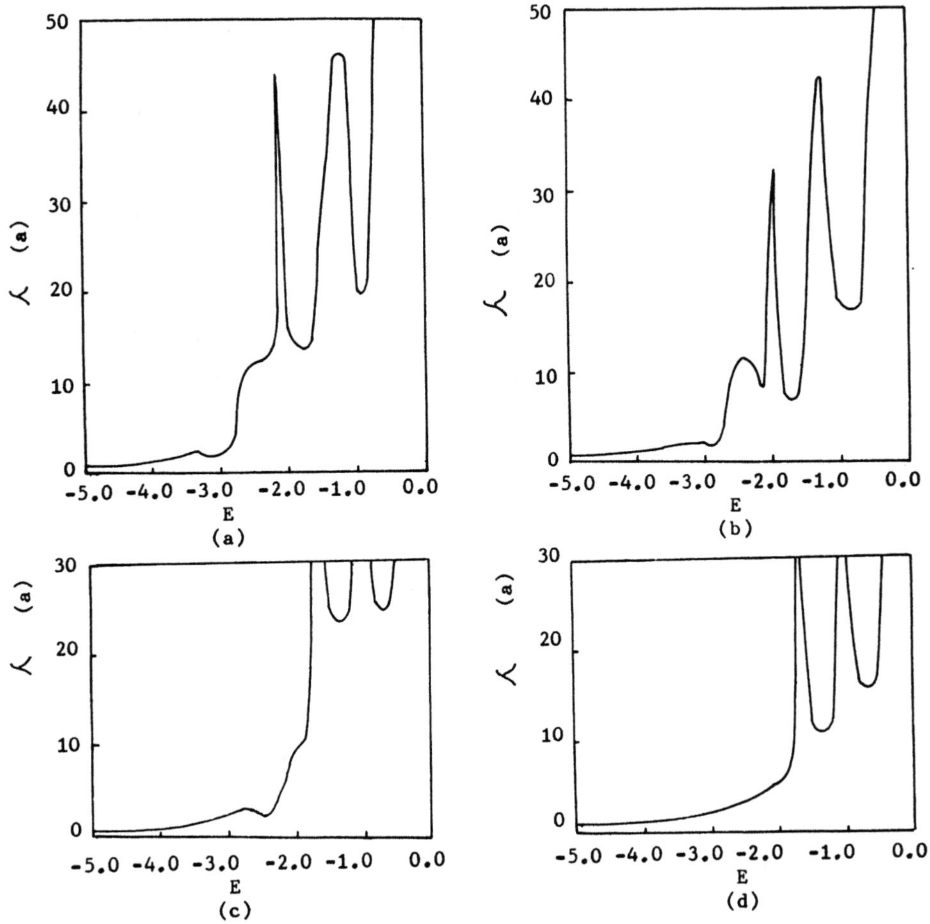

Fig. 2 The localisation lengths as functions of energy E. The magnetic field is such that α=0.125, and the modulation-doping parameters are (a) W_1=0.2, W_2=1, V_1=V_2=0; (b) W_1=0.2, W_2=3, V_1=V_2=0; (c) W_1=0.2, W_2=1, V_1=0, V_2=2; (d) W_1=0.2, W_2=3, V_1=0, V_2=2 . The extension of the system is L=36,M=8,N=2.

vide the conditions for the appearance of the plateaus of voltage in the y direction due to the pinning of the Fermi level, and the voltage drop in the x direction

References

1. H. Aoki and T. Ando, Solid State Commun. 38, 1079 (1981)
2. R. B. Laughlin, Phys. Rev. B23, 5632 (1981)
3. D. C. Tsui and A. C. Gossard, Appl. Phys. Lett. 38, 550 (1981)
4. J. M. Luttinger, Phys. Rev. 84, 814 (1951)
5. A. MacKinnon, J. Phys. C: Solid State Phys. 13, L1031 (1980)

Part VIII

Quasicrystals

The Localization Problem in Low-Dimensional Quasicrystals

M. Kohmoto

Department of Physics, University of Utah,
Salt Lake City, UT 84112, USA

Abstract

Quasiperiodic systems can be regarded as being intermediate between ordered and disordered systems, and novel physical properties are expected. We discuss the electronic states of a one-dimensional quasicrystal model (Fibonacci lattice) and a two-dimensional model (Penrose lattice).

1. Introduction

There is much current interest in quasiperiodic structures. These systems are intermediate between the completely periodic perfect crystals, and the random or disordered amorphous solids. Undoubtedly, a major push for understanding was given by the experiments of SHECHTMAN et al., [1] which seem to show some evidence for a quasicrystal in the material Al Mn. [2] The theoretical understanding of these structures is based on the non-periodic tiling of the two-dimensional plane first introduced by Penrose, and described in an article of GARDNER.[3] A particularly simple quasicrystal structure is obtained by projecting a higher-dimensional lattice.

A one-dimensional version of quasicrystals can be obtained by projecting a two-dimensional square lattice on to a line. For a choice of angle between the line and an edge of the square lattice whose tangent is an inverse of the golden mean $\phi^{-1} = (\sqrt{5}-1)/2$, we obtain a Fibonacci lattice which is quasiperiodic or, more precisely, almost periodic.

In this article, we review the properties of the electronic tight-binding models on the Fibonacci lattice and the Penrose lattice. For the Fibonacci lattice, the exact renormalization-group (RG) procedure is known.[4,5] In fact, original work was done before the discovery of the quasicrystals, and the motivation was to construct a model with an exact RG structure. Based on the RG, detailed investigations of the electronic states are possible. A very striking feature of this model is that the states are always critical for all the energy levels and the coupling constant. This means that the states are at the metal-insulator transition point. The energy spectrum and the wave functions have rich scaling structures.

Electronic properties of the two-dimensional Penrose lattice were studied by several groups.[6-10] KOHMOTO and SUTHERLAND [9] found localized states which are infinitely degenerate at the center of the spectrum. These states are localized due to the lattice topology and form rings of various dimensions (ring states). Although numerical studies suggest the existence of critical states for the remaining parts of the spectrum, they are not yet completely understood.

2. Fibonacci Lattice

A natural and the simplest model for electronic properties of the one-dimensional quasicrystal is a tight-binding equation,

$$t_{n+1}\psi_{n+1} + t_n\psi_{n-1} = E\psi_n \quad , \tag{2.1}$$

where ψ_n denotes the wave function at the nth site and $\{t_i\}$ is the Fibonacci sequence with two kinds of hopping matrix elements t_A and t_B. This is an off-diagonal version of an almost periodic Schrödinger equation,

$$\psi_{n+1} + \psi_{n-1} + V_n \psi_n = E\psi_n \quad , \tag{2.2}$$

where the potential V_n takes two values arranged in the Fibonacci way. Although it seems that there is no simple transformation between (2.1) and (1.2). KOHMOTO and BANAVAR [11] obtained a RG equation for (2.1) which is the same as the one originally obtained by KOHMOTO et al.[4] and OSTLUND, et al.[5] for (1.2). Therefore we would expect the same properties in the localization problem (energy spectrum, wave function etc.) for the two models. The tight-binding model (2.1) was also studied numerically by many authors [12]. These models are examples of the almost periodic Schrödinger equation which has been of considerable interest among mathematicians [13] as well as physicists.[14]

In order to study the quasiperiodic problems, it is useful to consider systematic approximations by periodic systems. For the Fibonacci sequence, this can be done as follows: Construct recursively a series of finite sequences S_ℓ as $S_{\ell+1} = (S_\ell , S_{\ell-1})$ for $\ell \geq 1$ with $S_o = B$ and $S_1 = A$. We obtain the following S_ℓ's:

$$S_1 = A \quad , \qquad S_2 = AB \quad , \qquad S_3 = ABA \quad ,$$

$$S_4 = ABAAB \quad , \qquad S_5 = ABAABABA \quad , \qquad \cdots\cdots\cdots\cdots .$$

A symbol sequence which is obtained by an infinite application of the above procedure is the Fibonacci sequence.

We are interested in the properties of the energy spectrum and the wave functions. The transfer matrix method is standard for investigating the one-dimensional problems. Rewrite (2.2) as

$$\vec{\psi}_{n+1} = M_n \vec{\psi}_n \quad , \tag{2.3}$$

where $\vec{\psi}_n$ is a column vector $(\psi_n, \psi_{n-1})^t$, M_n is a 2×2 matrix

$$M(n) = \begin{bmatrix} E - V_n & -1 \\ 1 & 0 \end{bmatrix} \quad , \tag{2.4}$$

which is called a transfer matrix. It depends on V_n and the energy E, but its determinant is always unity. The wave function at an arbitrary site is represented by products of the transfer matrices as

$$\vec{\psi}_{N+1} = M^{(N)} \vec{\psi}_1 \quad , \tag{2.5}$$

with

$$M^{(N)} = M(N)M(N-1)\cdots M(1) \quad . \tag{2.6}$$

Since the energy spectrum is a set in which the wave functions do not diverge exponentially, it can be determined by the behavior of $M^{(N)}$ in the limit $N\to\infty$. Thus we understand the electronic properties completely once the product of the transfer matrices are calculated. However, so far we just rewrote the original problem in a matrix form. Therefore, the calculation of the matrix product is as difficult as the original problem in its present form.

Guided by the recursive construction of the Fibonacci sequence, we now consider matrix products of F_ℓ transfer matrices (F_ℓ is a Fibonacci number defined recursively as $F_{\ell+1} = F_{\ell-1} + F_\ell$):

$$M_\ell \equiv M^{(F_\ell)} = M(F_\ell)M(F_\ell-1)\cdots M(2)M(1) \quad . \tag{2.7}$$

This matrix satisfies the recursion relation

$$M_{\ell+1} = M_{\ell-1} M_\ell \quad . \tag{2.8}$$

This is the basic equation of this method. This recursive scheme is a far better calculational scheme than straightforward matrix multiplications or a diagonalization of the original hamiltonian. However, the importance of (2.8) is much more profound than being merely a good calculational scheme.

Let us consider (2.8) as a nonlinear dynamical mapping problem. The 2×2 real matrix M_ℓ has a restriction that the determinant is one, hence it is parametrized by three real numbers. Since the map (2.8) transforms $(M_{\ell-1}, M_\ell)$ to $(M_\ell, M_{\ell+1})$, it is regarded as a six-dimensional map. In dynamical system studies, it is usual to study low-(one- or two-) dimensional systems. The six-dimensional problem seems very complicated.

Now, let us focus on obtaining the energy spectrum from (2.8). Firstly, consider a periodic system in which one period consists of F_ℓ sites, namely S_ℓ. The transfer matrix corresponding to a unit cell S_ℓ is given by M_ℓ and this is multiplied (infinitely) many times. Therefore the condition that a wave function does not diverge exponentially requires the absolute value of the eigenvalues of M_ℓ to be unity. This is nothing but Floquet theorem (or Bloch theorem). Since the determinant of M_ℓ is unity, the condition is written as

$$|x_\ell| \le 1 \quad , \tag{2.9}$$

where

$$x_\ell = (\text{Trace } M_\ell)/2 \quad .$$

The quasiperiodic system is obtained by the limit $\ell \to \infty$, so the spectrum is obtained as a set of energy value E which satisfies (2.9) in the limit. It can be shown from (2.8) that x_ℓ obeys the following remarkable recursion equation:[4]

$$x_{\ell+1} = 2\, x_\ell x_{\ell-1} - x_{\ell-2} \quad . \tag{2.10a}$$

This equation can also be written as

$$r_{\ell+1} = T\, r_\ell = (2\, x_\ell y_\ell - z_\ell,\ x_\ell,\ y_\ell) \quad , \tag{2.10b}$$

where $r_\ell = (x_\ell, y_\ell, z_\ell) = (x_\ell, x_{\ell-1}, x_{\ell-2})$ is a three-dimensional vector. The nonlinear map T represents a three-dimensional dynamical system and it is a sub-dynamical system of the original six-dimensional map. Namely, a three-dimensional orbit which is a projection from a six-dimensional space is solely determined by the sub-dynamical system (2.10). Thus, we need to study only the three-dimensional map to obtain the energy spectrum. This sub-dynamical system (2.10) is called the trace map below. Moreover, there is an important constant of motion.[4] Consider a quantity,

$$I = x_\ell^2 + y_\ell^2 + z_\ell^2 - 2x_\ell y_\ell z_\ell - 1 \quad . \tag{2.11}$$

It can be checked from (2.10) that I is independent of ℓ. The value of I is determined from the initial condition

$$r_1 = ((E-A)/2,\ (E-B)/2,\ 1) \tag{2.12}$$

to be

$$I = (A-B)^2/4 \quad . \tag{2.13}$$

Here, note that I represents the strength of the potential and is independent of the energy E. The special case $I = 0$ corresponds to the simple tight-binding model without the potential term. The wave function is given by $\psi_n = \exp(ikn)$ and is extended. On the other hand, the first two terms in (2.2) which represent hopping of electrons can be neglected in the limit $I \to \infty$. The wave function in this case is simply localized at each site. Equation (2.8) or (2.10) can be thought of as a RG transformation. The fact that I is invariant under this transformation--namely, it does not flow to either $I = 0$ or $I = \infty$--suggests that the wave functions are neither extended nor localized in a standard manner.

An example of the energy spectrum obtained by the RG method is shown in Fig. 1. Sets of E which satisfy (2.9) are shown for $\ell = 3,4,5$ and 6. Each of these represent the energy bands of the periodic system with a period of F_ℓ. Note that each spectrum consists of F_ℓ bands. This figure already shows clearly the features of the quasiperiodic system.

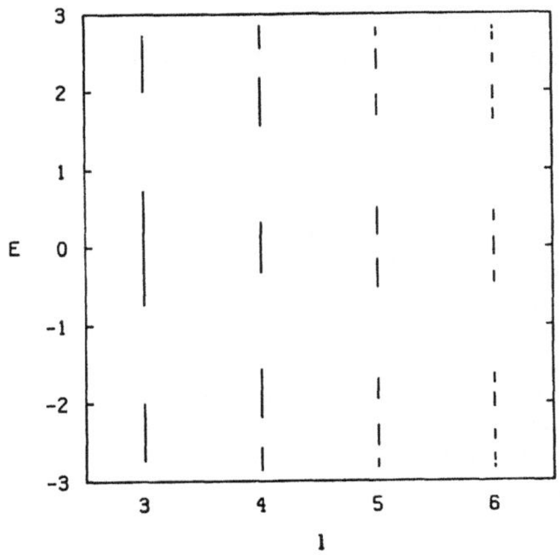

Fig. 1. Band structure of the periodic systems of periods F_ℓ with $\ell = 3,4,5$, and 6.

First of all, as ℓ gets larger, more energy gaps appear. The gaps for the limit $\ell \to \infty$ correspond to the escaping orbits of the trace map. In fact, most of the orbits are escaping and the non-escaping orbits are very rare. (They exist only with probability zero.) This implies that the Lebesgue measure of the spectrum (roughly speaking, the sum of the the band-widths) is zero. This excludes an absolutely continuous spectrum, and the wave functions are not extended in a usual sense.

Another notable feature is that the spectrum is divided into three parts and each part looks as a rescaled version of the entire spectrum. Each part then is divided into three parts and there exists a corresponding scale change. This self-similarity of the spectrum can be explained by the Smale horseshoe structure of the trace map.[15] The self-similarity and the densely populated gaps mean that the spectrum is a Cantor set. An quantitative analysis of scaling of the spectrum is given in [16].

It is enough, as explained above, to study the trace map to obtain the energy spectrum. However, we have to consider orbits of the full six-dimensional map (2.8) in order to study the wave functions. This is a very complicated problem. Fortunately, the orbits of the full map corresponding to cycles of the trace map can be understood well. Here, we only describe main results. For details, see [16].

It can be shown that the six-dimensional orbits which correspond to cycles of the trace map have a rather simple structure. In fact wave functions for these orbits have self-similarity and obey following scaling properties:

$$|\psi_n| \sim n^{\beta} . \qquad (2.14)$$

Here n is taken for the increasing peaks which are encountered sequentially as n is increased:

$$\sum_{m \le |n-n_o|} |\psi_m|^{\lambda} \sim n^{\alpha(\lambda)} \quad , \qquad (2.15)$$

where n_o is a site of the local maximum of the wavefunction and λ is a parameter. When $\lambda = 2$, (2.15) represents the behavior of the electronic probability density.

In order to specify an energy in the Cantor set, denote the top, middle and bottom subclusters as 1,0, and $\bar{1}$ respectively. By repeating this procedure infinitely many times, one can get a one-to-one correspondence between a point in the Cantor set and an infinite sequence of the symbols 1,0, and $\bar{1}$.

The center of the spectrum where the scaling index takes the maximum value, is given by a code $\{0,0,0,\cdots\}$. This point is represented by the six-cycle of the trace map. It turns out that the corresponding orbit in the full six-dimensional map also is a six-cycle. This special property enables one to calculate the scaling indices for the wave function exactly.[11,16] The wave function is shown in Fig. 2. The self-similarity is clearly seen.

The edge of the spectrum which is most concentrated has a code {1,1,1,···}. This point is represented by a two-cycle of the trace map. Unlike the six-cycle, the corresponding orbit of the full map is not a two-cycle. In the remaining directions, the orbit is divergent. In general, the orbit of the full map diverges even if a projection to the trace space is periodic. Nonetheless it can be shown that wave functions corresponding to those orbits are still self-similar.[16] The wave function corresponding to the two-cycle is shown in Fig. 3.

So far, the wave functions corresponding to cycles have been discussed. The cycles, however, are very rare (although there are infinitely many of them) and most of the non-escaping orbits are chaotic. An example of a wave function for a chaotic orbit is shown in Fig. 4.

Wave functions of this type do not have the scaling properties of (2.14) and (2.15). Although the trace orbit is completely chaotic, the wave function does not seem totally random. Understanding of this type of wave functions in relation to quantum chaos is an important unsolved problem.

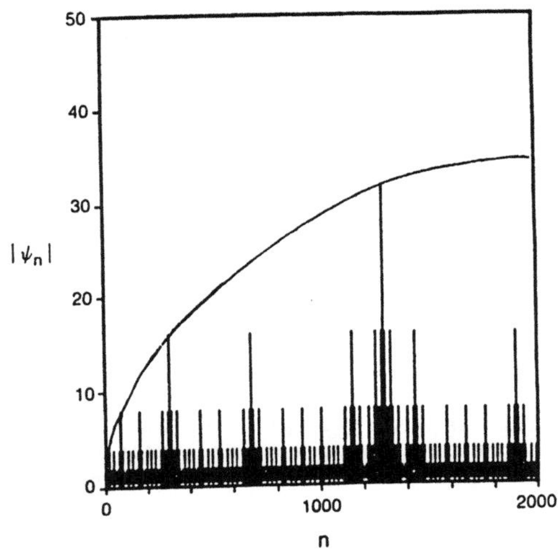

Fig. 2

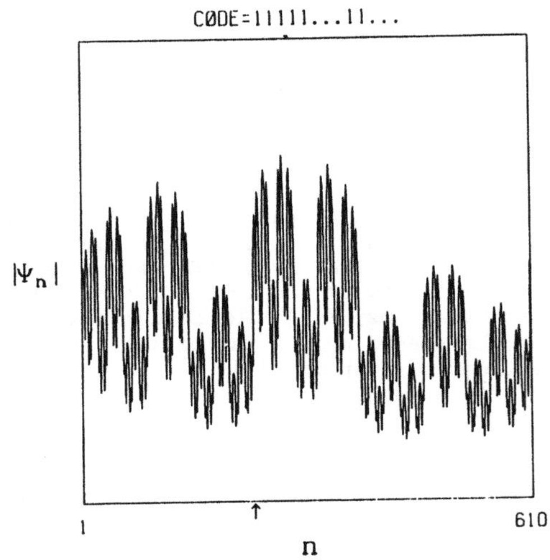

Fig. 3

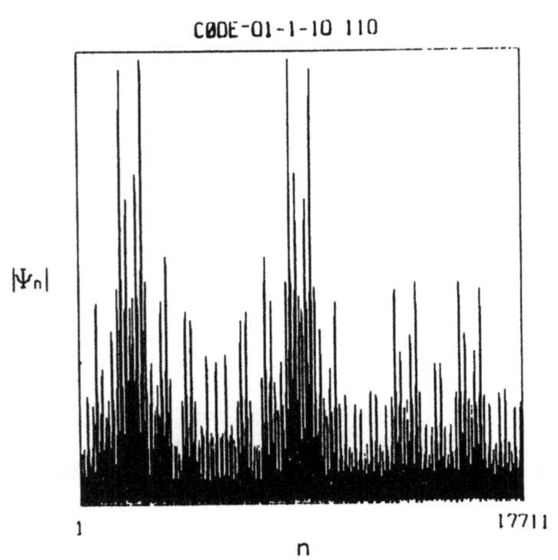

Fig. 4

Fig. 2 The wave function at the center of the spectrum.

Fig. 3. The wave function at the edge of the spectrum.

Fig. 4. A wave function corresponding to a code $\{0,1,\bar{1},\bar{1},0,\bar{1},1,0\cdots\}$

3. Electronic States of the Penrose Lattice

An example of the Penrose lattice is shown in Fig. 5.

It consists of two types of rhombuses and the ratio of these is the golden mean $(\sqrt{5} + 1)/2$. Since the ratio of the two types of the elements is an irrational number, the Penrose lattice cannot be periodic. Also, this pattern has a pentagonal symmetry which is prohibited crystographically. (A periodic pattern cannot have a pentagonal symmetry.) The Penrose lattice is invariant under a certain transformation of lattices and has a self-similarity which is analogous to that of the Fibonacci lattice.[3,17]

Electronic states of the Penrose lattice have been investigated mainly numerically by several groups.[6-10] Denote a lattice point of the Penrose lattice by n, then

$$-\sum_{n'} \psi_{n'} + V_n \psi_n = E\psi_n \qquad (3.1)$$

gives an electronic tight-binding model, where the summation is over the sites which are connected to the site n by one edge. The integrated density of states for the model with $V_n = 0$ is shown in Fig. 6.

The main difference from the one-dimensional case is that energy gaps (which would appear as horizontal parts of the graph) are not clearly seen. One cannot conclude from the data that the spectrum is a Cantor set. Another feature is that the integrated density of states jumps vertically at $E = 0$ and has small horizontal portions around it. This means the density of states has a δ-function-like singularity at $E = 0$ and energy gaps exist around the singularity. This singularity resembles a Van Hove singularity if seen from low accuracy numerical data. However, it turns out that this δ-function represents infinitely degenerate localized states and is not related to a Van Hove singularity of periodic systems.[6,9] In Fig. 5, let the values of the wave function be +1 and -1 alternately at those sites marked by dots. If the wave function vanishes on all the other sites, it satisfies (3.1) with $V_n = 0$ and the eigenvalue is $E = 0$. The existence of this type of ring states is due to the local lattice structure of the Penrose lattice rather than the global quasiperiodicity. The Conway theorem [3]

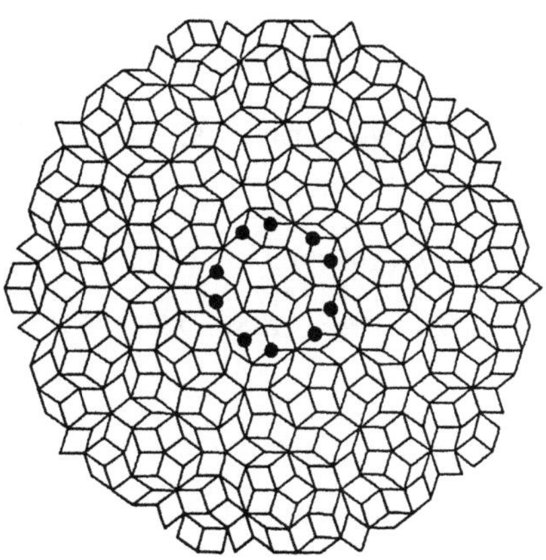

Fig. 5

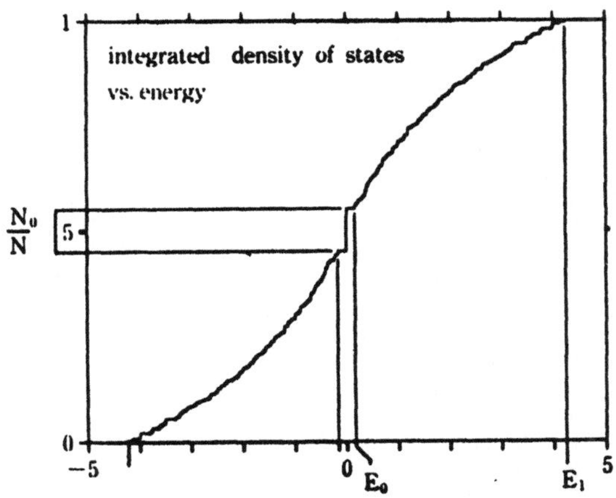

Fig. 6

Fig. 5. An example of the Penrose lattice. The lattice points marked by • represent a ring state.

Fig. 6. The integrated density of states of the electronic states of the Penrose lattice.

tells us that the same local pattern can be found everywhere and the separation between them is at most the size of the pattern. This implies that the ring state of Fig. 5 can exist everywhere in the lattice. Moreover, the self-similarity of the lattice allows ring states with larger radii.

These ring states are apparently localized and they are all degenerate at $E = 0$. It is possible to superpose these wave functions to produce an extended state but it does not carry electronic current. As mentioned previously, the existence of those ring states is due to the local lattice structure and is not necessarily a feature of quasiperiodic systems. It is known that similar localized states occur even in periodic systems.

Except for these special localized states, the localization problem of the Penrose lattice is not well understood numerically or theoretically. In general, a higher dimension tends to delocalize states. Since it is known that the one-dimensional Fibonacci lattice does not have localized states,[18] it is natural to conjecture that the two-dimensional Penrose lattice does not have localized states except for the special ring states. The remaining problem is how critical states and extended states would exist in the spectrum. SUTHERLAND [19] obtained an exact lowest energy state by appropriately choosing V_n in (3.1) depending on the types of the vertex. (There are eight types of vertex due to coordination number and other factors.) The wave function has similar properties to the wave function of the Fibonacci lattice. (See Fig. 2). The scaling indices $\alpha(\lambda)$, β of (2.14) and (2.15) can also be defined for this wave function.

4. Concluding Remarks

The structure of the quasicrystal in real materials is not completely understood yet. In any event, the structure will not have pure quasiperiodicity represented by three-dimensional extensions of the Penrose lattice. Therefore in the study of real quasicrystals much care must be taken in using the theory of the perfect quasiperiodic systems described here. In addition, real materials are, of course, three-dimensional. It is possible to make low-dimensional quasiperiodic systems artificially , for example, as semiconductor superlattices [18] superconducting networks [19] and optical multilayers [20].

Acknowledgement

This work was in part supported by a grant from the Alfred P. Sloan Foundation through a fellowship.

References

1. D. Schechtman, I. Blech, D. Gratias and J.W. Cahn, Phys. Rev. Lett. **53**, 1951 (1984).
2. A.L. Mackay, Physica **114A**, 609 (1982);
   A.L. Mackay, Sov. Phys. Crystallogr. **26**, 517 (1981);
   D. Levine and P.J. Steinhardt, Phys. Rev. Lett. **53** 2477 (1984);
   P. Kramer and R. Neri, Acta Crystallogr. Sect. **A 40**, 580 (1984);
   P.A. Kalugin, A. Yu Kitaev, and L.C. Levitov, Pis'ma Zh. Eksp. Teor. Fiz, **41**, 119 (1985) [JTEP Lett **41**, 119 (1985)];
   V. Elser, Phys. Rev. **B32**, 4892 (1985);
   M. Duneau and A. Katz, Phys. Rev. Lett. **54**, 2688 (1985);
   R.K.P. Zia and W.J. Dallas, J. Phys. **A18**, L341 (1985).
3. M. Gardner, Sci. Am. **236**, 110 (1977).
4. M. Kohmoto, L.P. Kadanoff and C. Tang, Phys. Rev. Lett. **50**, 1870 (1983).
5. S. Ostlund, R. Pandit, D. Rand, H.J. Schellnhuber, and E. Siggia, Phys. Rev. Lett. **50**, 1873 (1983).
6. K. Semba and T. Ninomiya, (unpublished).
7. T. Odagaki and D. Nguyen, Phys. Rev. **B33**, 2184 (1986).
8. T.C. Choy, Phys. Rev. Lett. **55**, 2915 (1985).
9. M. Kohmoto and B. Sutherland, Phys. Rev. Lett. **56**, 2740 (1986);
   M. Kohmoto and B. Sutherland, Phys. Rev. **B 34**, 3849 (1986).
10. H. Tsunetsugu, T. Fujiwara, K. Ueda, and T. Tokihiro, J. Phys. Soc. Jpn. **55**, 1420 (1986).

11. M. Kohmoto and J.R. Banavar, Phys. Rev. B **34**, 563 (1986).
12. See, e.g. J.P. Lu, T. Oadgaki and J.L. Birman, Phys. Rev. **B33**, 4809 (1986); F. Nori and J.P. Rodriguez, Phys. Rev. **B34** 2207 (1986); M. Fujita and K. Machida, Solid State Commun. **59**, 61 (1986).
13. B. Simon, Adv. Appl. Math. **3**, 463 (1982);
14. J.B. Sokoloff, Phys. Rep. **126**, 189 (1985).
15. M. Kohmoto and Y. Oono, Phys. Lett. **102A**, 145 (1984).
16. M. Kohmoto, B. Sutherland, and C. Tang, Phys. Rev. B **35**, 1020 (1987).
17. N.G. de Brujin, Ned. Akad. Weten, Proc. Ser. **A43**, 39 (1981).
18. F. Delyon and Petritis, Commun. Math. Phys. **103**, 441 (1986).
19. B. Sutherland, Phys. Rev. **B 34**, 3904 (1986).
20. R. Merlin, K. Bajema, R. Clarke, F.-Y. Juany, and P.K. Bjattacharya, Phys. Rev. Lett. **55**, 1768 (1985); P. Hawrylak and J. Quinn, Phys. Rev. Lett. **57**, 300 (1986).
21. A. Behrooz, M.J. Burns, H. Deckman, D. Levine, B. Whitehead, and P.M. Chaikin, Phys. Rev. Lett. **57**, 368 (1968).
22. M. Kohmoto, B. Sutherland, and K. Iguchi, Phys. Rev. Lett. **58**, 2436 (1987); D. Wurtz, T. Schneider, and M.P. Soerensen (preprint).

Power-Law Conductance of a Fibonacci Lattice

M. Goda

Faculty of Engineering, Niigata University, Niigata 950-21, Japan

Abstract

The one-electron problem in a Fibonacci lattice of size N=135000 is numerically studied and the following power law with respect to length n is observed in the off-diagonal term of the Green function $G_{\ell,\ell\pm n}(E)$ (ℓ:lattice site, E:energy) and the conductance g(n,E) averaged over subsystems (denoted by < >) of size n:

$$\langle|G_{\ell,\ell\pm n}(E)|\rangle \propto n^{-\alpha(E)} \exp\{-A(E)n\}, \qquad \langle g(n,E)\rangle \propto n^{-\beta(E)} \exp\{-B(E)n\}.$$

The exponents of the power $\alpha(E)$ and $\beta(E)$ seem to have, in a sense, gradual energy dependence. Near the ground state energy 2α and $\beta>1$ corresponds to the power-law localization of the eigenfunctions, while near the energy 0.4 2α and $\beta<1$ corresponds to the power-law delocalization. The exponent A(E) is a Lyapunov exponent with an energy-sensitive and self-similar structure.

As a result power law conductivity and resistivity with respect to temperature T are expected to be observed in a suitable range of temperature.

1. Introduction

The quasi-crystal discovered recently[1] lies just between periodic and random systems and is thus expected to reveal some critical feature between delocalization and localization. Two characteristics, a singular continuous spectrum and critical eigenfunctions, have been observed numerically and theoretically, for the one-electron problem in a Fibonacci lattice[2~6] and Penrose lattice[7~9]. To understand what is deduced from the characteristics we discuss in this paper the one-particle Green function and conductance of electrons in a Fibonacci lattice.

2. Green Function, Transfer Matrix and Conductance

A tightly bound one-electron system described by the Hamiltonian

$$H = \sum_{\ell=1}^{N} |\ell\rangle\varepsilon_\ell\langle\ell| + \sum_{\ell=1}^{N-1}(|\ell\rangle\langle\ell+1| + |\ell+1\rangle\langle\ell|) \qquad (1)$$

is considered. The site energy ε_ℓ has been taken to be 1 (for τ) or -1 (for 1) following the Fibonacci sequence $\tau 1 \tau\tau 1 \cdots\cdots$. The matrix element of the Green function G(E) is constructed by two independent solutions, an increasing solution $U^+(E)=\{U^+_\ell(E)\}$ and a decreasing solution $U^-(E)=\{U^-_\ell(E)\}$, of the homogeneous equation $(H-E)U(E)=0$ as follows:

$$G_{\ell,\ell'}(E) = C\, U^+_\ell(E)\, U^-_{\ell'}(E) \quad \text{for } \ell \leqq \ell' ,$$
$$= C\, U^-_\ell(E)\, U^+_{\ell'}(E) \quad \text{for } \ell' \leqq \ell , \tag{2}$$

where C is a constant. The homogeneous equation is rewritten by using the transfer matrix $T_\ell(E)$. The eigenvalues $\lambda^+_{(n)}(E)$ and $\lambda^-_{(n)}(E)$ ($|\lambda^+_{(n)}| \geqq |\lambda^-_{(n)}|$, $\lambda^+_{(n)}\, \lambda^-_{(n)} = \det T_{(n)} = 1$) of the product of the transfer matrices $T_{(n)}(E)=T_{\ell+n-1} \ldots T_{\ell+1}\, T_\ell$ determine the asymptotic behaviour of the $U^+_{\ell+n}(E)$ for large n. If the eigenvalue $|\lambda^+_{(n)}(E)|$ increases exponentially (by power law) with respect to n the same is true for $|U^+_{\ell+n}(E)|$ and the Green function $|G_{\ell,\, \ell\pm n}(E)|$ decreases exponentially (by power law) with respect to the distance n.

We can also calculate the conductance g and the registance $r=1/g$ with the transfer matrix by adopting an approximate expression for dimensionless conductance $g(n,E)$ (in e^2/h units) [10-12]

$$g(n,E)=\mathrm{tr}\{2/[T_{(n)}T^+_{(n)} + (T_{(n)}T^+_{(n)})^{-1} + 2]\}. \tag{3}$$

3. Numerical Results[13]

As the distribution of the energy spectrum is known to have a self-similar (fractal) structure seen in the Cantor set, it follows that the Green function and the conductance should decay exponentially, in infinit system, with respect to the distance n, for almost all energies.

A part of the Lyapunov exponent $A(E) \equiv (1/n)\ln|\lambda^+_{(n)}(E)| \equiv 1/n^*(E)$ for large n averaged over $M=2^{17}/n$ subsystems of the size $n=2^{13}$ of the system (1) with $N=135000$ (> 2^{17}) is shown for each energy in Fig. 1. In the main band gaps the function A(E) forms "hills". When we enlarge one of the "valleys", where the spectrum concentrates, in Fig. 1 the same structure, consisting of a new class of

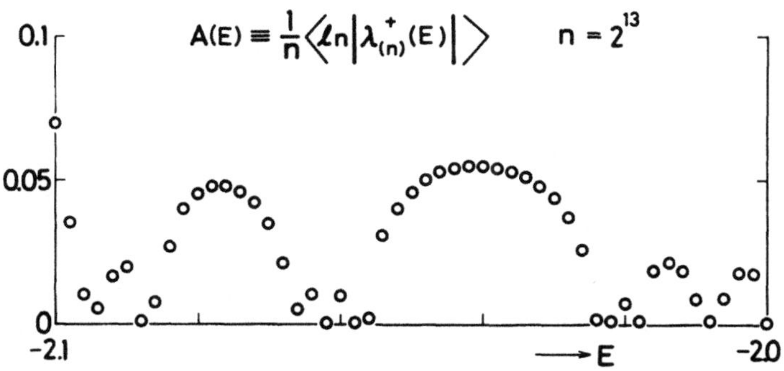

Fig. 1 A part of the energy (E) dependence of the Lyapunov exponent A(E) averaged over $2^4=16$ subsystems of the size $n=2^{13}=8192$ of the system (1) with N=135000.

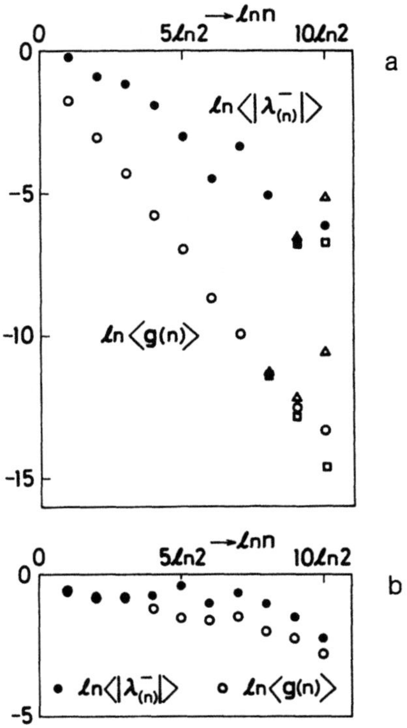

Fig. 2 Ln-ln plots of the size (n) dependence of the eigenvalues $|\lambda^-|$, and the conductance g, averaged over $2^{17}/n$ subsystems of size n for E=-2.097633 (a) and E= 0.397023 (b).
Those quantities for slightly different energies $E_+=E_0+\Delta E$ (denoted by △) and $E_-=E_0-\Delta E$ (denoted by □) for $\Delta E=10^{-6}$ are also plotted for reference.

"hills" and "valleys", appears again just corresponding to the self-similar structure of the spectrum. Thus by enlarging successively the hierachy of the "valleys" we can find many energy regions in which the Lyapunov exponent has a given (arbitrarily small) upper bound, A_M.

Selecting two energy values E=-2.097633 and 0.397023, as examples, in the energy regions where $A_M<0.0005$ and $n^*>2000$, we calculate the n dependence of the two mean values, the conductance g(n) and the minimum eigenvalue $|\lambda^-{}_{(n)}|$, averaged over the $2^{17}/n$ subsystems of length n of the system (1). Figure 2 shows that, in the interval of $n=2^1$ to $2^{10}=1024$, these obey a power law, although the exponents of the power are distributed forming "multi-power". To see how the power is distributed, the averaged eigenvalues and the conductances for slightly different energies $E=E_0 \pm \Delta E$, $\Delta E=10^{-6}$ from the above-mentioned two particular energies (denoted by E_0) are also plotted in Fig. 2. A remarkable feature, seen in the figure, is that the power is insensitive to the change of energy for small size n, but is sensitive to it when the size exceeds a critical value n_c. The critical value n_c depends on the energy E and the energy difference ΔE. The quantity $n_c(E,\Delta E)$ seems to be a decreasing function of the energy difference ΔE. The variance and the mean value of each quantity are of the same order for each n and the fluctuations of the mean values around the power law depend on the way of averaging. The power law which governs the two mean values crosses over an exponential law roughly at size $n^*(E)$.

4. Power-Law Conductivity

In the case where the exponent $B(E\tilde{=}E_F)$ (E_F: Fermi energy) is sufficiently small or zero we can expect a power law conductance which does not obey the scaling relation of ABRAHAMS et al.[14] which insists $\beta=+1$. At finite temperature $T(>0)$ a phenomenological argument suggests the following energy-dependent DC conductivity:

$$\sigma(T,E) = n_\varepsilon(T)g(n_\varepsilon(T),E) \propto D^{-(\beta-1)/2}\, T^{p(\beta-1)/2} \exp\{-BD^{1/2}/T^{p/2}\}, \qquad (4)$$

where the mean free path n_ε of an electron has been assumed to be equal to $(D(E)\,\tau_\varepsilon(T))^{1/2}$ ($D(E)$: diffusion constant, $\tau_\varepsilon(T)$: relaxation time of an electron due to inelastic scattering which has been assumed to be proportional to T^{-p}). The DC conductivity $\sigma(T)$ must be a linear combination of the $\sigma(T,E)$ in a small energy interval $[E_F-k_BT,\ E_F+k_BT]$. Irrespective of the details of the linear combination we can guess the following. When the quantities $p(E)$ and $\beta(E)$ can be treated roughly as constants for the necessary energy values in the energy interval we can draw the term $T^{p(\beta-1)/2}$ out of the linear combination. The residual quantity further contains a power of T due to the fractal structure of the distribution of the spectrum. Power law conductivity with respect to temperature T is thus expected to be observed for a suitable Fermi energy in a suitable range of temperature.

References

1 D. Shechtman, I. Blech, D. Gratias and J. W. Cahn: Phys. Rev. Lett. **53** (1984) 1951.
2 M. Kohmoto, L. P. Kadanoff and C. Tang: Phys. Rev. Lett. **50** (1983) 1870.
3 S. Ostlund, R. Pandit, D. Rand, H. J. Schellnhuber and E. D. Sigga: Phys. Rev. Lett. **50** (1983) 1873.
4 J. P. Lu, T. Odagaki and J. L. Birman: Phys. Rev. B**33** (1986) 4809.
5 M. Fujita and K. Machida: Solid State Commun. **59** (1986) 61.
6 T. Ninomiya: J. Phys. Soc. Jpn. **55** (1986) 3709.
7 H. Tsunetsugu, T. Fujiwara, K. Ueda and T. Tokihiro: J. Phys. Soc. Jpn. **55** (1986) 1420.
8 M. Kohmoto and B. Sutherland: Phys. Rev. Lett. **56** (1986) 2740.
9 B. Sutherland: Phys. Rev. B**34** (1986) 3904.
10 P. W. Anderson, D. J. Thouless, E. Abrahams and D. S. Fisher: Phys. Rev. B**22** (1980) 3519.
11 Y. Imry: Europhys. Lett. **1** (1986) 249.
12 K. Ueda and T. Tsunetsugu: Phys. Rev. Lett. **58** (1987) 1272.
13 M. Goda: J. Phys. Soc. Jpn. **56** (1987) 1294.
14 E. Abrahams, P. W. Anderson, D. C. Licciardello and T. V. Ramakrishnan: Phys. Rev. Lett. **42** (1979) 673.

Presence of Diffusion in Certain One-Dimensional Random Lattices

K. Kawamura, M. Hibino, K. Kawase, and K. Goto

Department of Physics, Faculty of Sciences and Technology, Keio University, Hiyoshi 3-chome, Kohoku-ku, Yokohama 223, Japan

1. Introduction

According to the scaling theory on Anderson localization/1/, the electronic states in one-dimensional disordered conductors are localized always. Before this theory appeared, the localization of the electronic and vibrational waves in one-dimensional random lattices had been proved for various models/2/. Nowadays, many people believe that electronic states in one-dimensional random lattices are always localized, although not all the types of models have been studied.

The disorder of an electron system may be classified into quantitative and topological disorders/3/. In the former, the topological structure of a disordered lattice is the same as the ideal lattice but the atomic energy and/or the hopping parameter take random values. The disorder which has been proved to localize electron wave functions belongs to this category. But the influence of the topological disorder has not been studied yet from the point of view of the Anderson localization.

The typical example of the topological disorder is found in dislocated lattices/4/. While the quantitative disorder can exist in a linear chain, the dislocated lattice is extended more or less in three-dimensional space. However, we can construct a quasi-one-dimensional electron system in dislocated lattices in the following manner.

Figure 1a represents a screw dislocation in a slab of otherwise perfect crystal. When this dislocation line is straight and extends to infinity the wave vector in the direction of the dislocation line (the z-axis) is good quantum number. In this case, non-trivial motion takes place only in the (x, y)-plane. If the slab is thin, the electron motion in the direction of y-axis (the axis perpendicular to the slab surface) is quantized and, for electrons whose energy levels lie near the bottom of the lowest subband, the degree of freedom in that direction is frozen and the system of electrons in this slab can be regarded as one-dimensional.

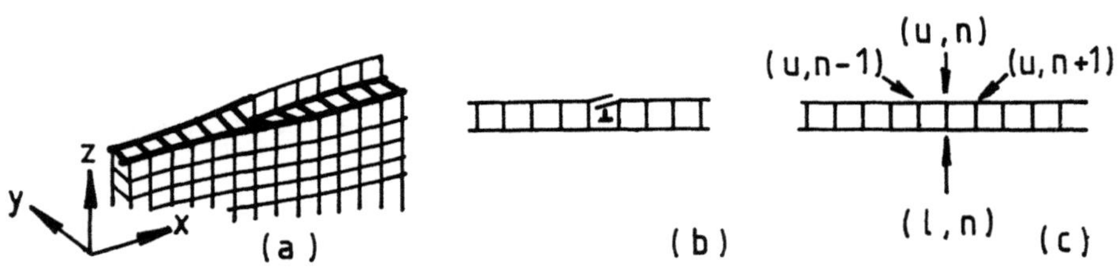

Fig.1 Screw dislocation in a slab (a), topological structure of cross sections on x-y plane in the dislocated (b) and perfect (c) slabs

2. The Tight-Binding Equation

The network which is represented with the bold lines in Fig. 1a has the same topological structure as the network represented in Fig.1b. In the absence of the dislocation, the corresponding network is the perfect ladder represented by Fig. 1c. The central plaquette in Fig.1b, which breaks the translational invariance in the horizontal direction, will be called the dislocation plaquette.

The tight-binding equation for the electron in the perfect ladder is

$$E\, C_u(n) = T\,(C_u(n+1) + C_u(n-1) + C_l(n)) \tag{2.1}$$

$$E\, C_l(n) = T\,(C_l(n+1) + C_l(n-1) + C_u(n)) \tag{2.2}$$

where n labels the step of the ladder and u (l) denotes the atom on the upper (lower) end of each step (see Fig.1c). The above two equations have solutions which are symmetric and antisymmetric combinations of C_u and C_l. They will be called, respectively, the symmetric and antisymmetric modes, hereafter. The value of E/T of symmetric (antisymmetric) modes is extended between 3 and -1 (1 and -3).

In the dislocated network represented by Fig.1b, the equation (2.2) is not altered but (2.1) is modified in the following manner/5/;

when the n-th step is the left-hand side of a dislocated plaquette;
$C_u(n+1) \rightarrow \exp(ip_z)C_u(n+1)$,
when the n-th step is the right-hand side of a dislocated plaquette;
$C_u(n-1) \rightarrow \exp(-ip_z)C_u(n-1)$.

The transmission and reflection amplitudes are defined as usual. Since there are two channels of electron propagation, each matrix has 2*2 elements.

Now we consider a slab which includes mutually parallel N dislocations. Then the network consists of N dislocation plaquettes which are combined by the ladder in Fig. 1c. The transmission and reflection amplitude matrices for the whole system are expressed in terms of those matrices for a single dislocation and propagators for perfect ladders. Some computational results will be described in the following section.

3. Numerical Studies of the Transmission Coefficent

Coupling of an electron with a screw dislocation is specified by p_z, the wave vector in the direction of the dislocation axis through the factor $(\exp(ip_z) - 1)$/5/. When $p_z = 0$, each dislocation has no influence, while, when $p_z = \pi$, electron is reflected completely by a single dislocation. In the numerical calculation, the value of p_z has been varied between 0 and 1.5.

The value of E/T was fixed at a certain value between 3 and 1 so that the symmetric mode can exclusively propagate through the perfect region; its wave vector is denoted by p_s. In this case, the transmission amplitude has physical meaning only when both of the incident and transmitted waves are symmetric mode. The transmission coefficient T is defined by the square of the absolute value of this amplitude. This depends on the number N of dislocations as well as p_z and p_s of the incident wave.

The semilogarithmic plot of the transmission coefficient against the number of dislocations for a given sample of the dislocated network is presented in Fig. 2 for various values of p_z. The wave number p_s of the incident wave of a symmetric mode is 0.05. The separation between neighboring dislocated plaquettes is a random variable which follows the Poisson distribution. In our present study, the average separation is five and the total length of the disordered network is

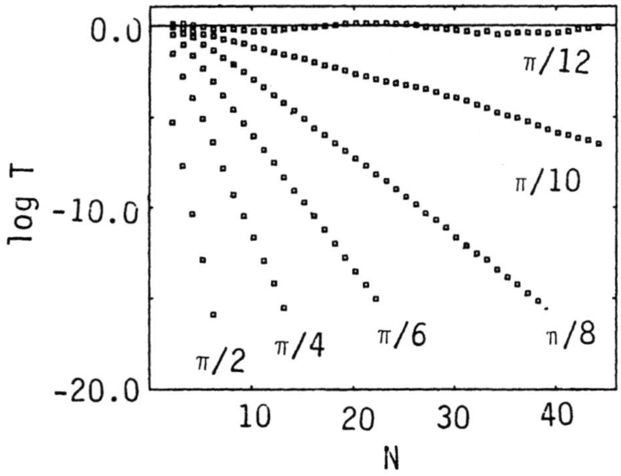

Fig. 2 Transmission coefficient against number of dislocations for various values of p_z

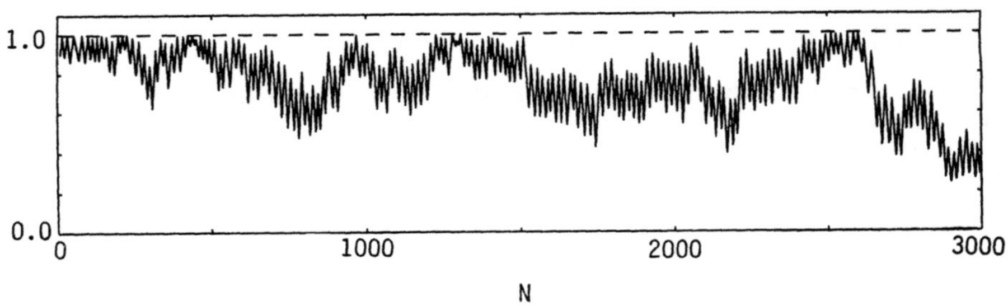

Fig.3 Oscillation of transmission coefficient with number of dislocations

roughly 5N. When p_z is not vanishingly small, T(N) decays exponentially with N. The decay constant decreases with decreasing value of p_z. It was confirmed that the sample dependence is very weak.

When the value of p_z is lowered below $\pi/12$, the behavior of T(N) changes qualitatively; T(N) exhibits oscillation instead of the exponential decay. To study this new feature, T(N) was computed in a larger region of N for $p_z = \pi/18$ and $p_s = 0.04$ for a network in which mean separation of neighboring dislocations is six. The result is presented in Fig. 3. In this case, T(N) oscillates on rather high level in the region $N \lesssim 2600$. Quite remarkable thing is that it takes the value 0.99935 for N = 2602; more than 99% of incident electron can be transmitted through such a highly dislocated ladder. For $N \gtrsim 2600$, T(N) still oscillates but the center of oscillation decreases gradually.

The transition from the exponential decay to oscillation takes place also with the change of p_s. In the region $0<p_s<0.1$ and $0<p_z<1.0$, T(N) exhibits oscillation in a certain region expressed by $p_s > f(p_z)$. The value of $f(p_z)$ increases with p_z but its functional form is not known. Further, for given value of p_s and p_z, the oscillation disappears when the mean separation of the dislocated plaquettes is decreased.

4. Discussion

Since the symmetric mode propagates exclusively, our electron system is no doubt of one dimension and we can invoke the general theory on one-dimensional disordered conductors. The exponential decay of T(N) with N described in the preceding section is then attributed to the exponential localization of the wave function/2/.

The transmission coefficient T(N) oscillates with N when p_z is small and it occasionally takes values near unity even for large N. By the title of the present paper, we do not assert that electron wave function is extended over a one-dimensional conductor of infinite length. What is implied is that the electron can diffuse through a certain linear chain on a macroscopic scale. The fact that the transmission coefficient of the ladder with 2602 dislocations is 99.935% implies that the conductance of this sample a few microns long is almost the same/6/ as the perfect conductor.

This behavior is seemingly inconsistent with the exponential localization of the wave function. The nature of electronic states which contribute to such a behavior of the transmission coefficient is not clear. Since T(N) vanishes ultimately when N is sufficiently large, wave functions of those electrons are not extended as in the perfect lattice. It is possible/7/ that the transmission coefficient becomes unity even in the localization regime, if the center of a localized wave function is located at the center of the sample and if its tails reach the two ends of the sample. For large p_z, the decay rate of this wave function is so large that it has a practically zero amplitude at the ends of the sample. But if the disorder is weak, the amplitude of the tail at the boundary is sufficiently large and the electron incident to the sample can be transmitted over this state.

However, since the oscillating decay is evidently different in quality from the simple exponential decay, the electronic wave functions are possibly not localized in the usual sense; such wave functions may be weakly localized. If so, the present numerical result implies that the weak localization takes place even in one-dimensional lattice when the disturbance due to each scatterer is weak.

Although there is discrepancy between this idea and the usual expectation on one-dimensional conductors, let us apply this idea to the electronic states in the grain boundary. Uchida and Landwehr invoked the weak localization theory on disordered two-dimensional conductors/8/, when they studied the magnetoresistance of the grain boundary of Si. But where is the two-dimensional system at the grain boundary? Although the grain boundary is geometrically two-dimensional space, the potential well for the holes at the boundary is some hundred angstrom thick. Consequently, the positively charged region at the boundary is three-dimensional space. However, if the energy levels of the hole lie near the bottom of the lowest subband, and if the dislocations at the boundary are straight and mutually parallel, the system of the hole is quasi-one-dimensional as the electron system considered in the present model. If this is the case, the electronic wavefunctions at the grain boundary are weakly localized, not because they are in a two-dimensional space but because they are in a weakly disordered one-dimensional space.

References

1. E. Abrahams, P. W. Anderson, D. C. Liccirdello, and T. V. Ramakrishnan: Phys. Rev. Letters 42, 673 (1979)
2. K. Ishii: Prog. Theor. Phys. Supplement No.53 77 (1973) and references therein.
3. M. H. Cohen: Topological Disorder in Condensed Matter, ed. by F. Yonezawa and T. Ninomiya,Springer Ser.Solid-State Sci.,Vol.46(Springer,Berlin,Heidelberg 1983)
4. K. Kawamura: Z. Physik B 29, 101 (1978)
5. K. Kawamura: Z. Physik B 30, 1 (1978)
6. R. Landauer: Phil. Mag. 21, 863 (1970)
7. M. Kastner: Lecture at Conference on Condensed Matter on Mesoscopic Scale, UCSB (June, 1987)
8. G. Landwehr and S. Uchida: Surf. Sci. 170, 719 (1986)

Part IX

Mesoscopic Systems: Theory

Statistics of Mesoscopic Fluctuations and Electric Current Relaxation in Disordered Conductors

B.L. Altshuler[1], V.E. Kravtsov[2], and I.V. Lerner[2]

[1]Leningrad Nuclear Physics Institute, Academy of Sciences, Gatchina, Leningrad 188350, USSR
[2]Institute of Spectroscopy, Academy of Sciences, Troitsk, Moscow 142092, USSR

The distribution function of the mesoscopic fluctuations as well as the time dependence of the relaxation current are shown to have logarithmically normal asymptotics. These asymptotics could be obtained in the framework of the single formalism based on the nonlinear σ model. Mesoscopic fluctuations of the relaxation time scales in disordered conductors are considered.

1. Introduction

Up to very recently the quantum interference effect on electron transport in weakly disordered conductors seemed basically to be well comprehended. In the framework of the weak localization theory (see for reviews [1,2]) one was able to describe the conductance G dependence on the sample size L, temperature T, frequency ω, etc. The scaling hypothesis based on the assumption that the dimensionless conductance $g = hG/e^2$ was the only scaling parameter has enabled *Abrahams* et al. [3] to formulate a statement on the absence of extended states in $d = 2$ dimensions and to construct a qualitative picture of the metal-insulator transition at $d > 2$. The scaling considerations have been supported with rigorous calculations in the framework of the effective field theory (nonlinear σ model) proposed for the localization problem by *Wegner* [4] and developed by a number of authors [5-8]. The σ model treatment has produced quantitative results for $g(\omega)$ or $g(L)$ provided only $g > 1$, i.e. even in the case of quantum interference corrections to conductance being not small in comparison with its classical value.

However, as has been recently realized, there are serious limitations on the validity of the results mentioned above. The point is that perturbative (or σ model) results for $g(L)$ apply naturally to the conductance averaged over the ensemble of samples with macroscopically identical characteristics which differ from one another only by realization of the random impurity potential. But conductance fluctuations from sample to sample (which have received the name mesoscopic fluctuations) prove to be of the same order of magnitude as quantum corrections to g and, which is of great importance, remain finite for $L \to \infty$, as has been shown by *Altshuler* [9] and *Lee* and *Stone* [10]. They found the dispersion of the mesoscopic fluctuations in the region of weak localization as follows:

$$\langle(\delta g)^2\rangle = (h/e^2)^2 (\langle G^2\rangle - \langle G\rangle^2 \simeq 1 \ . \qquad (1)$$

Here and elsewhere the angle brackets denote ensemble averaging.

As the scatter of the physical quantities from sample to sample turns out to be relevant, the treatment of the averaged quantities does not provide sufficient information to describe completely the properties of disordered conductors. That is why the study of mesoscopic fluctuations is now becoming a subject of considerable interest. The mesoscopic fluctuations lead to unusual properties of disordered conductors which reveal themselves, for example, as aperiodic reproducible oscillations of G as a function of magnetic field or Fermi energy. The discussion of various "mesoscopic" properties may be found in recent reviews [11,12].

In this paper we will be interested in statistical properties of the mesoscopic fluctuations. We will consider in the framework of the broadened σ model the irreducible fluctuation moments (cumulants) $\langle\langle(\delta g)^K\rangle\rangle$ (double brackets stand for the irreducible averages) in order to determine the distribution function of mesoscopic fluctuations $f(\delta g)$. We will show this function to be Gaussian (and universal) only in the region of weak localization with δg being not too large. Even in this region the probability of large fluctuations ($\delta g > g_0{}^{1/2}$) exceeds drastically the Gaussian one: the asymptotics of the distribution function prove to be logarithmically normal. These long tails are not described by the one-parameter scaling and thus turn out to be nonuniversal, i.e. they depend on both the renormalized value of the dimensionless conductance $g = g(L)$ and its bare value $g_0 = g(L=l)$, where l stands for the electron mean free path [13].

Similar nonexponential long tails manifest themselves in the current relaxation processes in disordered conductors. For example, the long-time asymptotic behaviour of relaxation current after a step change in voltage proves to be logarithmically normal [$\sim\exp(-\ln^2 t)$] rather than exponential. The analogy between the asymptotics of the fluctuation distribution and electric current relaxation is not accidental. The long-time-tail behaviour of relaxation current which will be treated in the next section is closely connected with mesoscopic fluctuations of phase relaxation times τ_ϕ.

2. Current Relaxation in Disordered Conductors

The electric current density j(t) is governed for arbitrary time dependence of external electric field E(t) by frequency-dependent conductivity $\sigma(\omega) = e^2 g(\omega) L^{2-d}/h$:

$$\langle j(t)\rangle = \int E(t - t') \langle\sigma(t')\rangle\, dt' , \tag{2}$$

where $\sigma(t)$ is the Fourier transform of $\sigma(\omega)$:

$$\langle\sigma(t)\rangle = \int \langle\sigma(\omega)\rangle \exp(-i\omega t) d\omega/2\pi . \tag{3}$$

The quantity $\sigma(t)$ will be referred to as the response function, for it corresponds to the current response to the $\delta(t)$-shaped pulse of electric field. The classical expression for the averaged response function

$$< \sigma_0(t)> = (\sigma_0/\tau) \exp(-t/\tau) \tag{4}$$

results according to (3) from the well-known Drude formula $\sigma(\omega) = \sigma_0(1-i\omega\tau)^{-1}$. Here σ_0 is the classical value of the conductivity related to the electron density of states per one spin ν and diffusion constant $D = l^2/d\tau$ by the Einstein relation $\sigma_0 = 2e^2 D\nu$ with l and τ being, respectively, the mean free path and the time of free path.

The response function is determined by (4) for the very short time $t < \tau$ only. For $t > \tau$ the quantity $\sigma(t)$ is governed totally by quantum effects. The usual cooperon correction to $\langle\sigma(\omega)\rangle$ is given by the expression [14,1,2]

$$\langle\sigma_1(\omega)\rangle = -(4e^2/h) \int (q^2 - i\omega/D + L_{\phi 0}^{-2})^{-1} d^d q/(2\pi)^d , \tag{5}$$

where $L_{\phi 0} = D\tau_{\tau 0}$ with $\tau_{\phi 0}$ being the phase breaking time due to inelastic processes. The appropriate contribution to the response function is found as follows:

$$\langle\sigma_1(t)\rangle = -(e^2/\pi h) t^{-d/2} (4\pi D)^{1-d/2} \exp(-t/t_0) \tag{6}$$

where $t_0 = \tau_{\phi 0}$ for the isolated sample and

$$t_0 = \min\{\tau_{\phi 0}, L^2/D\} \tag{7}$$

for a cube of volume L^d with massive contacts. Therefore one finds from (6) the attenuation of averaged relaxation current changing from a power law at $\tau < t < t_0$ to the exponential law for $t > t_0$. With decreasing temperature T the phase breaking time $\tau_{\phi 0} \simeq ?\ T^{-p}$ increases so that the attenuation time t_0 becomes equal to the time L^2/D of electron diffusion through the sample.

However, the quantum correction (5) proves insufficient to obtain the response function for very long times t. An allowance for the higher order quantum correction to $\sigma(\omega)$ leads to attenuation which is not nearly so rapid as the exponential one, although it exceeds the power-law attenuation [15]:

$$\langle\sigma(t)\rangle = -(\sigma/t)\exp\left\{-\left[\ln(t/\tau) - 2u\right]^2/8u\right\}. \tag{8}$$

Here $\sigma = e^2 g(L)L^{2-d}/h$ is the renormalized (observable) value of static conductivity while u is given by

$$u = \ln(\sigma_0/\sigma)\ , \tag{9}$$

where $\sigma_0 = e^2 g_0 l^{2-d}/h$ is the bare (classical) conductivity, $g_0 = g(L{=}l)$.

The very fact that $\sigma(t)$ depends on both σ_0 and σ reveals an insufficiency of the one-parameter scaling in deriving the asymptotics (8) of the response function. Indeed, the renormalization group procedure of deriving this asymptotic behaviour presented in Sect.5 looks much more complicated than a one-parameter procedure. However, the static conductance g(L) on which the quantity u depends is to be taken from the usual Gell-Mann-Low equation [1-6] with the β-function given in the one-loop approximation by $\beta(g) = (d-2)g-1$. It is important that the quantity u may be large enough even in the region of applicability of the one-loop results. In two dimensions it reaches the value of $\ln g_0 >> 1$ for $g \simeq 1$. At $d > 1$, (9) and the Gell-Mann-Low equation for g result in $u = -\ln(l/L)^{d-2}+(g_0-g_c)/g_c]$ in metal region $g > g_c = (d-2)^{-1}$. Herefore, u grows rapidly with g_0 approaching the critical value g_c and equals $(d-2)\ln(L/l)$ in the critical region $g = g_c$.

Let us consider qualitatively the relaxation current j(t) after stepwise switching on the electric field $E(t) = E\theta(t)$. In a short time $t \simeq \tau$ the current rises up to its classical value $j_0 = \sigma_0 E$. After that it decreases at first by the power (or logarithmic at d=2) law then by the exponential law (6) and finally by the logarithmically normal law (8) down to its steady-state value $j = \sigma E$. With increasing u the relaxation slows down so that the steady-state regime is reached only after a rather long time.

Now we will discuss the reason for the time-delayed relaxation (8). The nonexponential decay of $\sigma(t)$ is due to the faint singularity in the frequency dependence of the ac conductivity $\sigma(\omega)$ at $\omega \to 0$. Such a singularity was first discovered by *Kravtsov* and *Lerner* [16] on investigating the stability of the one-parameter scaling. To reveal the singularity it is convenient to expand $\sigma(\omega)$ as

$$\sigma(\omega) = \sigma_0 \sum_{n=0}^{\infty} C_n (i\omega\tau)^n + (e^2/h) \sum_{n=0}^{\infty} S_n (i\omega t_0)^n . \tag{10}$$

On neglecting the quantum corrections one finds $C_n = 1$ and $S_n = 0$, thus obtaining the Drude formula from (10). With the quantum corrections taken into account, the coefficients C_n result (in the framework of the impurity diagram technique) from expanding the electron Green's function in ω while S_n result from expanding diffusons (or cooperons) in ω. The crossover terms proportional to $t_0^n\tau^m$ are omitted in (10) because they are of no importance for further analysis. In spite of τ and t_0 obeying the inequality $\tau \ll t_0$, it is the first sum in (10) which governs the asymptotics (8). It is caused by the extremely rapid increase of coefficients C_n with n:

$$C_n \sim \exp[2u(n^2 + n - 1/2)]. \tag{11}$$

A similar increase law has been found for various $(\omega\tau)^n$-dependent contributions to the ω-dependent diffusion coefficient $D(\omega)$ [16] and rigorously proved [17] by considering the high moments of mesoscopic fluctuations of conductance and density of states (see next section). An analogous expression has been previously obtained by *Wegner* [18] by treating the moments of inverse participation ratio. In Sect.5 we will outline the technique used for obtaining (11) which is based on the renormalization group treatment of the broadened nonlinear σ model.

In order to sum the first asymptotic series in (10) we will use a trick similar to the Borel summation which is based on the following identity:

$$e^{2u(n^2 + n)} = \frac{1}{\sqrt{8\pi u}} \int_0^{\infty} (t_\phi/\tau)^n \, e^{-[\ln(t_\phi/\tau) - 2u]^2/8u} dt_\phi/t_\phi . \tag{12}$$

On substituting (11), (12) into (10), changing the order of summation and integration and performing the Fourier transformation (3) one obtains the asymptotics of the response function as

$$\langle\sigma(t)\rangle = \sigma \int \exp(-t/t_\phi) f(t_\phi) dt_\phi/t_\phi \tag{13}$$

where

$$f(t_\phi) \sim t_\phi^{-1} \, e^{-[\ln(t_\phi/\tau) - 2u]^2/8u} \, . \tag{14}$$

The expression (8) comes out after performing the integration in (13). On the other hand, (13,14) allow a meaningful interpretation of the results obtained. One can suspect that the nonaveraged response function has the exponential decay only. The long-time tail (8) arises from averaging (13) over the wide distribution $f(t_\phi)$ of relaxation times t_ϕ with the asymptotics (14). In this way the relaxation (6) at intermediate t_ϕ can also be obtained from (13) if the distribution function $f(t_\phi)$ is assumed to have a more or less sharp maximum at $t_\phi = t_0$. This assumption is verified by direct calculations.

The same type of distribution function will be obtained in the next section for the mesoscopic fluctuations of the static conductance and density of states. This coincidence is not accidental and will be discussed in Sect.4.

3. Statistics of Mesoscopic Fluctuations

As the dispersion of the conductance (1) proves to be large, one faces the situation where the knowledge of only the averaged conductance is insufficient. In this situation knowledge of the dispersion may also be insufficient and it is interesting to determine the whole distribution function of the fluctuations. Such a function is governed by the irreducible fluctuation moments (cumulants) $\langle\langle(\delta g)^k\rangle\rangle$. The technique of calculating these cumulants which is similar to that for expansion coefficients C_n and S_n in (10) will be outlined in Sect.5. Here we will restrict ourselves to describing the principal results.

As has been proved in [17], the cumulants $\langle\langle(\delta g)^k\rangle\rangle$ are governed for k being not too large by the one-parameter scaling. This means that the summation of all the size-dependent perturbation corrections (which are proportional for $d = 2$ to powers of $g_0^{-1}\ln L/l$) reduces to a substitution of the renormalized conductance g for its bare value g_0 into all nonlogarithmic perturbation results. As the dispersion $\langle(\delta g)^2\rangle$ is independent of g_0 to the lowest perturbation order [9,10], it therefore remains universal with all quantum corrections being incorporated. However, there are nonlogarithmic contributions to the dispersion which are proportional to g_0^{-1} [17]. Then the scaling procedure (equivalent to summation of all the quantum corrections) replaces them by g^{-1}-proportional contributions. Therefore, in the region $g \lesssim 1$ the universality of the dispersion may be broken down but it remains valid in the metal region ($g \gg 1$) with an accuracy to g^{-1}.

The higher cumulants $\langle\langle(\delta g)^k\rangle\rangle$ have been found [17] within the lowest order perturbation analysis combined with the scaling considerations as follows:

$$\langle\langle(\delta g)^k\rangle\rangle \sim g^{2-k} . \tag{15}$$

This expression is valid for k obeying the inequality

$$k \lesssim \Delta = u^{-1}\ln(L/l) . \tag{16}$$

As for $g \gg 1$ the cumulants (15) are small compared to the dispersion (1), in this region the distribution function of the mesoscopic fluctuations is close to the Gaussian one $f(\delta g) \sim \exp[-(\delta g)^2]$ for the fluctuations being comparatively small: $\delta g \lesssim \Delta^{1/2}$. But for $g \sim 1$ all of the cumulants (15) are found to be of the order of unity, thus resulting in a markedly non-Gaussian distribution.

The one-parameter scaling approach and, therefore (15) for $\langle\langle(\delta g)^k\rangle\rangle$ breaks, down at $k > \Delta$ (16). In this region the cumulants increase with k by a law similar to (11):

$$\langle\langle(\delta g)^k\rangle\rangle \sim g(l/L)^{2k-2} \exp[u(2k^2 - 3k + 1)] . \tag{17}$$

Making use of the familiar expression for the distribution function in terms of cumulants

$$f(\delta g) = \int_{-\infty}^{\infty} \exp\left[i\varsigma\delta g + \sum_{k\gtrsim 2} (-i\varsigma)^k \langle\langle(\delta g)^k\rangle\rangle/n!\right] d\varsigma/2\pi ,$$

one finds the asymptotics of $f(\delta g)$ at $\delta g > \Delta^{1/2}$ to be analogous to the one of the distribution of relaxation times:

$$f(\delta g) \sim (\delta g)^{-1} e^{-[\ln(\delta g t_0/\tau) + 2u]^2} . \tag{18}$$

Mesoscopic fluctuations $\delta\nu$ of the density of states [19] are described with a similar distribution function having in the metal region a close-to-Gaussian maximum and a logarithmically normal tail. All the expressions for $f(\delta\nu)$ may be obtained from those for $f(\delta g)$ by substituting $(g/\nu)\delta\nu$ for δg.

We may try to extend the results presented to the insulator region $g \ll 1$, which is naturally far beyond the region of formal applicability. Nevertheless, if (9) for u is used with $\sigma = \sigma_0 \exp(-L/l_{loc})$, then the asympototic behaviour (18) checks well with the distribution function of resistance fluctuations obtained exactly for one-dimensional disordered chains by *Abrikosov* [20] and *Mel'nikov* [21]. Moreover, the results for $\langle\sigma(t)\rangle$ and $f(\delta\nu)$ obtained for the one-dimensional case by such a continuation (with l_{loc}=4l) have been confirmed, including even the numerical factor before the $\ln^2$ term, by recent exact calculations performed directly for a one-dimensional chain [22].

A coincidence of the mesoscopic distribution asymptotics for the insulator and metal cases seems very natural. The reason is that extremely untypical realizations of the random potential which form the tail should be less sensitive to the metal-insulator transition than the typical ones determining the mean value of conductance.

The distribution functions of fluctuations in the vicinity of their maxima are quite different for the metal and insulator: the distribution in this region proves to be a normal one with a unit dispersion for the metal and a logarithmically normal one for the insulator. The metal-insulator transition is thus accompanied by a reduction of the Gaussian region and speading of the logarithmically normal distribution to the whole range of fluctuations. In the critical region ($g_0 \simeq g_c$) for $d > 2$ as well as in the whole range of parameters at $d = 2$, the distribution function appears to be nonuniversal (as the tail depends on the bare conductivity σ_0 playing the role of a disorder parameter) and it is not described by the one-parameter scaling.

4. The Mesoscopic Fluctuation of the Response Function

Now we shall discuss the problem of response function fluctuations from sample to sample in order to answer the natural question: is the long-time tail of the averaged response function (8) defined by the processes in a particular sample or does it result simply from averaging over the whole ensemble?

We begin with the case of temperature $T = 0$. In the region of validity of the expression (6) for the averaged $\sigma(t)$, the analysis within the impurity diagram technique yields the response function dispersion within the accuracy of a numerical factor as

$$\langle[\delta\sigma(t)]^2\rangle/\langle\sigma(t)\rangle^2 \sim (t/t_0)^{d/2} = [(Dt)^{1/2}/L]^d \, . \tag{19}$$

Therefore for $t > t_0$ the dispersion exceeds the averaged value although the distribution of $\sigma(t)$ remains nearly Gaussian in the metal region g

>> 1 because the cumulants are small as compared to the dispersion, for t not too large:

$$\langle\langle[\delta\sigma(t)]^k\rangle\rangle^2/\langle[\delta\sigma(t)]^2\rangle^{\,k} \sim \left[(t_0/t)g^{4/(d-4)}\right]^{k/2-1} . \qquad (20)$$

In the region of large t when $\langle\sigma(t)\rangle$ is given by the asymptotics (8), the mesoscopic fluctuations of $\sigma(t)$ turn out to exceed drastically the averaged value:

$$\langle\langle\left[\delta\sigma(t)\right]^k\rangle\rangle \,/\, \langle\sigma(t)\rangle^{\,k} = A^{1-k} \exp\{(k-1)\,[\ln(t/\tau) - 2n]^2/8n\} , \qquad (21)$$

where $A \sim g\, t_0/t \sim h/(tS)$ with $S = (\nu L^d)^{-1}$ being the mean spacing between energy levels. In spite of A being small the fluctuations are large in the whole region of applicability of the asymptotics (8). Thus the averaged response function (8) has nothing to do with the relaxation processes in a particular sample. The asmyptotic behaviour arises from the ensemble averaging, the main contributions to it being made with the untypical, low probability realizations of the random potential.

However, the situation is changed considerably in the case of $T \neq 0$. Let us first consider the temperature effects on neglecting electron-electron (or electron-phonon) inelastic collisions. The inclusion of temperature has naturally no influence on the mean values. But the fluctuations are reduced. The σ model analysis which will be presented elsewhere [23] shows the logarithmically normal factor for the dispersion $\langle\delta\sigma(t_1)\delta\sigma(t_2)\rangle$ to be multiplied by the damping factor proportional to $\exp[-T(t_1-t_2)/h]$. On evaluating the response function fluctuations from sample to sample one should perform averaging over a time interval $\Delta t > h/T$ as the simultaneous dispersion (at $t_1=t_2$) becomes meaningless. Such a time averaging results in the dispersion differing from that in (21) (for k=2) only by reduction of A with the additional small factors $h/(T\Delta t)$. As a result, a time interval arises in which the averaged-over Δt response function of a particular sample differs only slightly from the ensemble-averaged asymptotics (8). The non-averaged-over-time response function is characterized by aperiodic reproducible (for a given sample) time oscillations against the logarithmically normal background, the mean period of oscillations being equal to h/T. But for any fixed temperature T, the fluctuations at $t \to \infty$ become large so that the relaxation processes in a particular sample appear to have an exponential decay.

Finally, let us briefly consider the effects of real inelastic collisions on $\sigma(t)$. The asymptotic behaviour of $\langle\sigma(t)\rangle$ remains, in the case

$\tau_{\phi 0} < L^2/D$, logarithmically normal rather than exponential. This is due to the existence of mesoscopic fluctuations of phase breaking times τ_ϕ. The direct calculations which have been performed for electron-phonon inelastic scattering [23] (considered as a particular example of any inelastic process) shows the distribution function of τ_ϕ^{-1} to have the same shape as that of t_ϕ for T = 0. Namely, it has a maximum at $\tau_\phi = \tau_{\phi 0}$ and the logarithmically normal asymptotics (14) for τ_ϕ large. The calculations have been carried out for $\tau_{\phi 0} \gtrsim L^2/D$ as it is the case where one may leave out of account the higher order inelastic interactions as well as the self-consistency procedure for τ_ϕ. But it is physically clear that for $L \to \infty$ one has to discuss the fluctuations of τ_ϕ averaged over the volume $L_{\phi 0}{}^d = (D\tau_{\phi 0})^{d/2}$ rather than over the whole volume L^d. All the results obtained deal with the fluctuations of τ_ϕ averaged in such a way. In this case the fluctuations of the response function are reduced by the additional factor proportional to $(L_{\phi 0}/L)^d$. Thus there exists a fairly wide region of applicability of the asymptotics (8) to the relaxation in a particular sample. Formally, even in this case the $\sigma(t)$-fluctuations become large for $t \to \infty$. We do not give estimates here but should like to emphasize that by choosing parameters one would make the time when the fluctuations become large arbitrarily long. Therefore, the logarithmically normal long-time tails may be observable in relaxation processes in real conductors.

5. The Nonlinear σ Model Technique

The nonlinear σ model has usually been used in localization theory to evaluate the density-density correlator and thus obtain the renormalized diffusion coefficient. We will use the previously developed [24,17] technique (which is slightly modified here) presenting directly the conductance and its fluctuation moments. The moments of static conductance may be written as

$$\langle(\delta g)^k\rangle = (-8N^2L^2)^{-k} Z^{-1} \int DQ \prod_{i=1}^{k} \mathrm{tr}\{(\partial^2/\partial A_i^2) \exp(-F)\}|_{N=0,\ A=0} , \tag{22}$$

where $Z = \int DQ\exp(-F)$ is the normalization factor. The replica trick is used here in the framework of [5]. The functional F consists of two parts, $F = F_0 + F_{add}$, where the "normal" part F_0 has the form of a conventional σ model functional

$$F_0 = (\pi\nu D/8) \int \mathrm{Tr}(\nabla_\alpha Q)^2 d^d r \tag{23}$$

while the "additional" part F_{add} (which has been derived along with F_n from the initial microscopic model of noninteracting electrons in a random potential) is given by

$$F_{add} = \sum_{m=2}^{k} F_m; \qquad F_m = X_m \int \mathrm{Tr}(\nabla Q)^{2m} d^d r\ , \tag{24}$$

where for simplicity the vector indices are omitted. A bare value of the charge X_m is proportional to $\nu D l^{2(m-1)}$. The matrix fields Q(r) obey the usual geometrical constraints

$$Q^2 = I, \quad \mathrm{Tr}\{Q\} = 0\ . \tag{25}$$

There are two principal distinctive features of the model considered as opposed to the conventional σ model. Firstly, the symbol ∇_α signifies here a covariant derivative differing from the usual gradient ∂_α by a commutator:

$$\nabla_\alpha Q \equiv \partial_\alpha Q - i[A_\alpha, Q]\ . \tag{26}$$

Here A_α is a background source field which must be put equal to zero after differentiating in (22). Second, some new (as compared to the usual case [5]) indices are added to all the fields. The matrix field Q, e.g., has the structure

$$Q = Q^\mu \tau_\mu; \quad Q^\mu = Q^{pp';\mu}_{ab;ij}; \quad (Q^\mu)^* = Q^\mu\ , \tag{27}$$

where τ_μ are the quaternion units: $\tau_0 = I$, $\tau_{1,2,3} = i\sigma_{x,y,z}$ (with σ being the Pauli matrices), a, b are the replica indices running from 1 to N (with N=0 for final results), p,p′ originate from retarded (p=1) or advanced (p=2) Green's functions in a nonaveraged current loop, and the new indices i and j numerate different current loops in an initial nonaveraged expression for g^k. The source field A is chosen to be diagonal in the new indices, $A_{ij} \equiv A_i \delta_{ij}$. The symbol Tr in (23-25) denotes the trace over all the indices while tr in (22) denotes the trace over all the indices but i,j. Then the structure of differentiation in (22) ensures all the unphysical diagrams (which include loops containing more than two external vertices) annihilate.

The renormalization of the functional (24) containing no gradient terms (i.e. including only the powers of AQ) has been performed in [17]. It is worth noting that the global symmetry of the σ model which is broken in (24) is restored by generating under renormalization the additional vertices which are obtained from the vertex F_m (24) by partitioning $Tr(AQ)^{2m}$ into all possible products of traces. Then the renormalization problem for the charges X_m reduces to some problem of the theory of representations of symmetric groups yielding the renormalized value of X_m as [17] $X_m \sim \exp(2um^2-3um)$ with u defined by (9). A recent analysis [25] of the full functional (24) containing both gradients and the source terms has confirmed this result.

The renormalized values of the fluctuation moments are obtained now by differentiating in (22) where the renormalized functional is substituted for the bare one, the functional integral being calculated to the lowest nonvanishing perturbation order. Then the "normal" contribution (15) to the irreducible fluctuation moment is obtained after k-fold differentiation of the "normal" functional (23) while the "additional" contribution (17) proves to be proportional to the renormalized value of the charge $X_{k(24)}$. Naturally, there exist also "crossover" contributions which are relevant only in the intermediate fluctuation region left out of consideration here.

Equation (22) will determine the ac conductivity $\langle\sigma(\omega)\rangle$ (with $\langle\sigma(0)\rangle$ differing from g only by a factor L^{2-d}) as well as the fluctuation moments $\langle\langle\sigma(\omega_1)...\sigma(\omega_k)\rangle\rangle$ if the functional F is complemented by the following fequency-dependent vertices which have also been derived from the initial microscopic model:

$$F_\omega = \sum_{n=1}^{\infty}\sum_{m=0}^{\infty} F_n^m ; \qquad F_n^m = X_n^m \int Tr\{(\nabla Q)^{2m}(\omega\Lambda Q)^n\}d^d r \ . \quad (28)$$

The bare value of the charge X_n^k is proportional to $\nu D l^{2(k-1)}\tau^n$ and depends on the permutations of the noncommutating matrices ∇Q and $\omega\Lambda Q$ where the matrix $\omega\Lambda$ is given by

$$\omega\Lambda = \begin{pmatrix}\omega_1 & & \\ & \ddots & \\ & & \omega_k\end{pmatrix}_{ij} \otimes \begin{pmatrix}1 & 0\\ 0 & -1\end{pmatrix}_{pp'} \otimes \delta_{ab} \otimes \tau_0 \ . \quad (29)$$

When the mean value $\langle\sigma(\omega)\rangle$ is calculated, the matrix ω degenerates into a single frequency ω. The first sum in (10) is produced by F_n^1, the coefficients C_n being proportional to the renormalized charges X_n^1. The renormalization of this vertex by itself leads to an increase law similar to that of (11) but with the exponent $u(n^2-n)$ instead of $2u(n^2+n)$. The correct exponent appears due to a contribution of the vertex F_0^{n-1} to the renormalization process according to the following scheme:

$$\begin{array}{ccccccccc} F_0^{n+1} & \longrightarrow & g^{-1}F_1^{n} & \longrightarrow & g^{-2}F_2^{n-1} & \longrightarrow & \ldots g^{-n+1}F_{n-1}^{2} & \longrightarrow & g^{-n}F_n^{1}\ . \\ & \nearrow & & \nearrow & & \nearrow & & \nearrow & \\ F_1^0 & & F_1^0 & & F_1^0 & & F_1^0 & & \end{array} \tag{30}$$

It means that the maximum exponent is determined by forced rather than fundamental solution of the renormalization group equation. As X_0^n is proportional to $\exp[u(2u^2-3n)]$ and g is proportional to $\exp(u)$ (for $d=2$), one comes to the increase law (11).

The fluctuation moments $\langle\langle\sigma(\omega_1)\ldots\sigma(\omega_k)\rangle\rangle$ may be represented by analogy to (10) as follows:

$$\langle\langle\sigma(\omega_1)\ldots\sigma(\omega_k)\rangle\rangle \sum_n \left\{ \sum_{n_1+n_2+\ldots+n_k=n} (i\omega_1\tau)^{n_1}\ldots(i\omega_k\tau)^{n_k}\, C_n^k \right\}. \tag{31}$$

The renormalized values of C_n^k are proportional to the renormalized charges X_n^k attached to the vertices F_n^k. The renormalization scheme which is similar to that of (30) results in an increase law $C_n^k \sim \exp[2u(n+k)^2-2un-3uk]$. Then the Fourier transform of the series (31) leads to the expression (21) for the mesoscopic fluctuation moments of the response function.

Note in conclusion that a technique similar to that described here has been applied [23] to obtain the results mentioned above on the temperature dependence of fluctuations as well as to consider the effects of inelastic collisions on the relaxation.

6. Concluding Remarks

We have considered two problems which at first sight are not closely connected to each other: the statistics of mesoscopic fluctuations of static conductance and the electric current relaxation in disordered conductors. The solution of both problems requires the limits of conventional one-parameter scaling theory to be exceeded. Namely, in order to obtain large-fluctuation asymptotics of the mesoscopic distribution function as well as a long-time tail of the current relaxation one should complement the nonlinear σ model with an infinite number of additional vertices, these vertices having been derived from the initial microscopic model of noninteracting electrons in a random potential. In spite of such a complication, the theory remains renormalizable. The renormalization group treatment yields the logarithmically normal asymptotics for both the distribution function of the mesoscopic fluctuations and the time dependence of the electric current relaxation. On approaching the Anderson transition the region of applicability of the asymptotics is extended. Thus a description of the vicinity of the transition becomes more complicated and seems to be nonuniversal (i.e. dependent on the strength of disorder). It appears that the logarithmically normal law for the asymptotics is universal in the sense that it is inherent in various relaxation processes.

Acknowledgements We are grateful for valuable discussions to A.G. Aronov, P.B. Wiegmann, D.E. Khmel'nitskii, V.N. Prigodin and V.I. Yudson.

References

1 P.A. Lee, T.V. Ramakrishnan: Rev. Mod. Phys. **57**, 287 (1985)
2 B.L. Altshuler, A.G. Aronov: In *Electron-Electron Interactions in Disordered Systems*, ed. by A.L. Efros, M. Pollak (Elsevier, New York 1985) p.1
3 E. Abrahams, P.W. Anderson, D.C. Licciardello, T.V. Ramakrishnan: Phys. Rev. Lett. **42**, 673 (1979)
4 F. Wegner: Z. Phys. B **35**, 207 (1979)
5 K.B. Efetov, A.I. Larkin, D.E. Khmel'nitskii: Zh. Exp. Teor. Fiz. 79, 1120 (1980) [Sov. Phys.-JETP **52**, 568]
6 Z. Schäfer, F. Wegner: Z. Phys. B **38**, 113 (1980)
7 A. Houghton, A. Jevicki, R.D. Konway, A.M.M. Pruisken: Phys. Rev. Lett. **45**, 394 (1980)
8 S. Hikami: Phys. Rev. B **24**, 2671 (1981)
9 B.L. Altshuler: Pis'ma Zh. Exp. Teor. Fiz. **41**, 530 (1985) [Sov. Phys.-JETP Lett. **41**, 649]
10 P.A. Lee, A.D. Stone: Phys. Rev. Lett. **55**, 1622 (1985)
11 Y. Imry: "Physics of Mesoscopic Systems" in *Directions in Condensed Matter Physics*, ed. by G.Grinstein, E. Mazenko (World Publishing, Singapore 1986) pp.101-163
12 S. Washburn, R.A. Webb: Adv. Phys. **35**, 375 (1986)
13 Here and elsewhere we use a notation g to mean conductance, while δg is used for nonaveraged dispersion. In the case of a weak disorder considered here the inverse classical conductance $g_0 \sim (h/pl)^{d-1} \ll 1$ plays the part of the weak-disorder parameter
14 L.P. Gor'kov, A.I. Larkin, D.E. Khmel'nitskii: Pis'ma Zh. Exp. Teor. Fiz. 30, 248 (1979) [Sov. Phys.-JETP Lett. **30**, 228]

15 B.L. Altshuler, V.E. Kravtsov, I.V. Lerner: Pis'ma Zh. Exp. Teor. Fiz. **45**, 160 (1987)
16 V.E. Kravtsov, I.V. Lerner: Zh. Exp. Teor. Fiz. **88**, 1281 (1985); Solid State Commun. **52**, 593 (1984)
17 B.L. Altshuler, V.E. Kravtsov, I.V. Lerner: Pis'ma Zh. Exp. Teor. Fiz. **43**, 342 (1986) [Sov. Phys. - JETP Lett. **43**, 441]; Zh. Exp. Teor. Fiz. **91**, 2276 (1986)
18 F. Wegner: Z. Phys. B **36**, 209 (1980)
19 B.L. Altshuler, B.I. Shklovskii: Zh. Exp. Teor. Fiz. **91**, 220 (1986) [Sov. Phys.-JETP **91**, 220]
20 A.A. Abrikosov: Solid State Commun. **37**, 997 (1981)
21 V.I. Mel'nikov: Fiz. Tverd. Tela **23**, 782 (1981)
22 B.L. Altshuler, V.N. Prigodin: Pis'ma Zh. Exp. Teor. Fiz. **45**, 538 (1987)
23 B.L. Altshuler, V.E. Kravtsov, I.V. Lerner: Zh. Exp. Teor. Fiz., to be published
24 V.E. Kravtsov, I.V. Lerner: Fiz. Tverd. Tela **29**, 456 (1987)
25 V.E. Kravtsov, I.V. Lerner, V.I. Yudson: to be published

Universal Conductance Fluctuations, Random Matrix Theory and the Scaling Theory of Localization

A.D. Stone[1], *K.A. Muttalib*[2], *and J.-L. Pichard*[3]

[1]Applied Physics, Yale University, P.O. Box 2157, New Haven, CT 06520, USA
[2]Address as of 9/87, University of Florida, Gainesville, USA
[3]SPSRM, CEA-Saclay, F-91191 Gif-Sur-Yvette, France

A framework is proposed for reconciling the one-parameter scaling theory of localization with the non-self-averaging behavior characteristic of the conductance in both the localized and metallic regime. The joint probability distribution of the eigenvalues of the transfer matrix which determines the conductance is obtained analytically from symmetry arguments and a maximum entropy hypothesis. It is shown that this probability distribution naturally factorizes into a universal term describing the local fluctuations of the eigenvalues, and a term describing the mean eigenvalue density, which contains all information about the system parameters. The first term gives statistical behavior identical to the well-known random matrix ensembles of Wigner and Dyson, and leads naturally to the universal conductance fluctuations in the metallic regime. This result is verified by independent numerical calculations. The eigenvalue density is found numerically to bear a simple relationship to the limiting density of Liapunov exponents of the transfer matrix. This density is known to satisfy one-parameter scaling, thus these results strongly support the one-parameter scaling hypothesis.

1. FLUCTUATION EFFECTS AND SCALING THEORY

The importance of fluctuation effects in the study of localization and quantum transport phenomena has been appreciated for some time [1,2], however attempts to reconcile these effects with the one-parameter scaling hypothesis of the scaling theory of localization [3] have not been entirely successful. Clearly the goal is to show that not only does the average of the conductance (or some other suitable variable) satisfy one-parameter scaling, but the entire probability distribution also is determined by one parameter for each combination of system parameters [4] (sample dimensions, disorder, fermi energy, etc.). Until recently almost all theoretical studies of the statistics of the transport coefficients focused on the one-dimensional case, which is special because the localization length, ξ, and the elastic mean free path, ℓ, are the same [5]. Thus these studies focused on the strongly localized regime, in which there are certainly enormous fluctuations in the conductance or resistance, but it is clear that this behavior is determined by one parameter, ξ. Moreover, these theoretical ideas are not so amenable to experimental test, because in real systems at finite temperature, inelastic hopping dominates the conduction in the localized regime [6]; and this process has somewhat different statistical behavior, from that expected at T = 0.

Recently the presence of important fluctuation phenomena in the conductance of metals has been widely observed experimentally [7,8] and understood theoretically [9-14]. By a metal, we mean any conductor whose samples dimensions, L_x, L_y, L_z, are greater than ℓ and less than ξ, so that the transport is diffusive. The most striking result of these investigations is that the statistical variance of the conductance, averaged over an ensemble of metal samples with the same macroscopic characteristics satisfies

$$\mathrm{Var}(g) \cong 1 \qquad (1)$$

independent of its size (at T = 0), or of its average conductance (g is measured in units of e^2/h). Equation (1) predicts much larger fluctuations in g than would be expected by naive classical reasoning [11], and has been confirmed experimentally for a wide range of materials whose dimensions are of order the inelastic diffusion length, L_{in}, ("mesoscopic" samples). For larger samples the classically expected self-averaging behavior is restored by inelastic effects [11,13], $Var(g)/\langle g\rangle^2 \propto (L_{in})^d/(L_xL_yL_z)$; and this explains why these fluctuations were not seen in the many earlier experiments which observed weak localization effects in the average conductance of metals.

Although (1) is not itself in conflict with one-parameter scaling (Var(g) depends on no parameters), recently Altshuler et al [15] have calculated higher moments of g in perturbation theory and found that the statistical behavior of g is not determined by one parameter. Moreover (1) itself has only been established theoretically by numerical calculations and perturbation theory in $(k_f\ell)^{-1}$, where k_f is the fermi wavevector. Here we present a more fundamental explanation for (1), based on the general theory of random matrices developed by Wigner, Dyson and others [16-19], thus confirming a conjecture by Imry [20]. We show that this approach naturally leads to a new formulation of scaling theory, which relates it to a "maximum entropy" hypothesis; and present preliminary numerical results that strongly support one-parameter scaling.

2. UNIVERSAL CONDUCTANCE FLUCTUATIONS AND SPECTRAL RIGIDITY OF THE TRANSFER MATRIX

We consider an ensemble of disordered conductors of length L_x and cross-sectional area $(L_y)^{d-1}$ attached to infinite perfectly ordered "leads." The scattering properties of each conductor are completely characterized by a unitary scattering matrix, $S = \begin{pmatrix} r & t' \\ t & r' \end{pmatrix}$, (where r, t and r', t' are the reflection and transmission matrices from right to left and vice-versa), which relates the incident fluxes I, I' to the outgoing fluxes O, O'; or by a transfer matrix, T, which relates the fluxes on one side to those on the other.

$$S\begin{pmatrix} I \\ I' \end{pmatrix} = \begin{pmatrix} O \\ O' \end{pmatrix} \qquad T\begin{pmatrix} I \\ O \end{pmatrix} = \begin{pmatrix} O' \\ I' \end{pmatrix}. \tag{2}$$

The transfer matrix is often a simpler object to calculate for a given model of a random conductor because it is multiplicative, the total transfer matrix, $T(L_x)$ satisfies

$$T(L_x) = \prod_{i=1}^{L_x} T^{(i)} , \tag{3}$$

where $T^{(i)}$, the transfer matrix across the ith cross-section of the conductor, is typically determined very simply from the random Hamiltonian. For a fixed energy, E, there will be N propagating channels in each direction ($N \propto (Ly)^{d-1}$, for N >> 1), so that S and T are 2N x 2N matrices, and t, t', r, r' are N x N matrices. The conductance of the disordered region, for N >> 1 and $\ell/L_x \ll 1$, is given by [21] $g = Tr(t^+t)$. From the definition (2) and the unitary of S, it follows that [22]:

$$g = Tr(t^+t) = 2Tr\{[T^+T + (T^+T)^{-1} + 2]^{-1}\} = \sum_{i=1}^{N} \frac{1}{1 + X_i} , \tag{4}$$

where we have defined $X = (1/4)\,[T^+T + (T^+T)^{-1} - 2]$, $\{X_i\}$ are the N non-degenerate eigenvalues of this 2N x 2N matrix, and the form of X implies $X_i > 0$. Equation (4) holds for each particular conductor, and we thus see that the ensemble-averaged conductance, $\langle g\rangle$, just measures the average number of eigenvalues of X in a small interval near zero, e.g., [0,1]. For a metallic conductor $\langle g\rangle \gg 1$, thus there must be many X_i in this interval, and this number must decrease with increasing disorder, approaching unity at the metal-insulator transition. Var(g) measures the fluctuations in the number of X_i in the interval [0,1] from conductor to conductor; so that if the X_i were uncorrelated, one would expect $Var(g) \cong \langle g\rangle$, i.e., nonuniversal fluctuations, typically much <u>larger</u> than those predicted by (1). The validity of (1) thus requires a high degree of correlation in the spectrum of X.

This point of view strongly suggests an analogy between the result of (1), and the phenomenon of spectral rigidity observed in the spectra of complex nuclei [19]. Briefly, in many complex nuclei it is found that the energy level sequences are very regular, they deviate very little from uniform spacing, and the deviation only grows logarithmically with the numbers of levels considered. Note that this is a stronger statement than simple level-repulsion and implies a long-range correlation of energy levels. This behavior has been quantitatively understood by hypothesizing that the nuclei exhibit statistical fluctuations characteristic of the well-known random matrix ensembles introduced by Wigner and Dyson [16-19], the Gaussian orthogonal ensemble (GOE) describing systems with time-reversal symmetry, the Gaussian unitary ensemble (GUE) for systems without time-reversal symmetry, and the related circular ensembles. These ensembles can be shown to exhibit such long-range eigenvalue correlations [18, 19].

In particular, Dyson and Mehta [18] have shown for these ensembles that in an interval L of the spectrum where the density is constant, any linear statistic $Z = \sum_i f(X_i)$ has a variance which is independent of the average number of eigenvalues in L, and typically of order unity (f is any smooth function on L, and the sum is only over X_i in L). As Imry [20] pointed out, <u>if</u> this result is valid for the matrix X, it immediately yields (1), since g is essentially such a statistic (if one introduces an irrelevant cut-off for large X).

3. MAXIMUM ENTROPY HYPOTHESIS AND THE PROBABILITY DISTRIBUTION $(\{X_i\})$

It is not obvious that the matrix X in (2) <u>should</u> have eigenvalues with statistical properties described by the standard ensembles. The standard ensembles are defined by relatively simple statistical postulates [16] (e.g., statistical independence of matrix elements for the GOE); whereas X is determined from the random Hamiltonian of the metal in the complicated manner indicated by (3) and (4), and in general we find that this procedure does not generate one of the standard ensembles. Nonetheless it is now known that a very wide class of random matrix ensembles exhibit the eigenvalue fluctuations characteristic of the GOE (or GUE). The reason for this is that the spectral <u>correlations</u> are typically determined solely by the symmetries of the matrices in the ensemble, whereas only the average spectral <u>density</u> is sensitive to the specifics of a given model. This idea may be quantified for our problem by introducing a "maximum entropy" hypothesis: the ensemble of X matrices is assumed to be "as random as possible" given a certain spectral density [23]. This assumption can be given a precise meaning in terms of minimizing an information functional on the matrix space of X. This immediately implies [23] that the joint probability distribution of the eigenvalues X_i has the form

$$\rho(\{X_i\}) = J(\{X_i\}) \prod_{i=1}^{N} \exp[- f(X_i)], \tag{5}$$

where f(X) is determined from the spectral density $\sigma(X)$ by the relation $f(X) = \int dX' \sigma(X') \ln|X-X'|$. Since the term involving $\sigma(X)$ factorizes, all the eigenvalue correlations are contained in the Jacobean, $J(\{X_i\})$, which is universal, i.e., determined solely by the symmetries of X. If X were e.g., a real symmetric matrix, this maximum entropy hypothesis would immediately yield GOE statistics for the local fluctuations of the X_i, however we emphasize that the symmetries of X do not correspond to any of the standard ensembles and $J(\{X_i\})$ was only very recently calculated by the authors (24), and independently by Mello et al. [25]. Mello et al. [26] have earlier introduced a very interesting form of the maximum entropy hypothesis to treat the 1d case, and recently have generalized the approach to the many-channel case (without, however, making the connection to (1), or verifying the correctness of the ansatz by microscopic calculations).

To derive $J(\{X_i\})$ it is convenient to consider the matrix $Q = T^+T$ appearing in (4) whose eigenvalues, $\{\lambda_i\}$, are simply-related to $\{X_i\}$ by $X_i = 1/4[\lambda_i + \lambda_i^{-1} - 2]$. T^+T is itself of interest because the matrix $(T^+T)^{1/2L_x}$ is a self-averaging quantity as $L_x \to \infty$ (Osledec's Theorem). In this limit the logarithm of its eigenvalues yields the Liapunov spectrum of the random map defined by the matrix product $T(L_x)$ [27], a fact which will be of crucial importance below. The matrix T has U(N,N) symmetry, i.e., it preserves the norm $|I|^2 - |O|^2$. This symmetry represents current conservation, as can be seem from (2) since the unitarity of S implies $|I|^2 + |I'|^2 = |O|^2 + |O'|^2$. If the conductor is not in a magnetic field, T will be further constrained by time-reversal symmetry. In [24,25] this symmetry was implemented by mapping the ensemble of T matrices to the real symplectic ensemble. Here we show how this somewhat artificial step can be avoided.

Time-reversal symmetry implies that $S = S^T$ in (2), therefore, from (2), KTK = T^*, with $K = \binom{01}{10}$ (here and below this notation denotes N X N sub-blocks of a 2N X 2N matrix). Therefore Q is an SU(N,N), Hermitian positive matrix satisfying $KQK = Q^*$; this means the 2N λ_i are positive real numbers which occur in inverse pairs (we choose the ordering $\lambda_i = \lambda_{i-n}$, i = N + 1,...2N). We need to define a volume element in the matrix space of Q, so that we can change variables in the probability distribution, P(Q), and then integrate over the eigenvectors of each Q, leaving the appropriate Jacobean factor for $\rho(\{X_i\})$ in (5). We follow closely Dyson's approach in [17]. Define

$$Q + dQ = T^+(1 + dM)T. \tag{6}$$

P(Q) and dM define a class of measures on Q, where by our ansatz (5), P(Q) only depends on the eigenvalues, λ_i, in the combination $X_i = 1/4[\lambda_i + \lambda_i^{-1} - 2]$, and these measures are specified by the <u>independent</u> elements of dM. We can show that this definition of dM does not depend on the choice of T in the factorization $Q = T^+T$, so we have only to calculate explicitly the independent elements of dM. Q is Hermitian, so $Q = U^+DU$, where D is a real diagonal matrix with the further symmetries mentioned above. The U(N, N) symmetry of Q can be written $Q^+JQ = J$, with $J = \begin{pmatrix}1 & 0\\ 0 & -1\end{pmatrix}$. Using this symmetry, one can show that U can be chosen to satisfy $UJU^+ = K$. Time-reversal symmetry (if present) is satisfied by the choice $UK = -U^*$. We differentiate the diagonalizing equation of Q to get

$$T^+dMT = Q = U^+[DdA - dAD + dD]U, \tag{6}$$

where we have defined $dU = dAU$, and since $UU^+ = 1$, $dA = -dA^+$.

The crucial step is that since dM is independent of our choice of T, we can choose a convenient T to simplify (6). We choose $T = RD^{1/2}U$, where the orthogonal matrix $R = 1/\sqrt{2}\begin{pmatrix}1 & 1\\ -1 & 1\end{pmatrix}$. One can check that this T has U(N,N) symmetry. Substituting into (6) yields

$$RdMR^T = [D^{1/2}dAD^{-1/2} - D^{-1/2}dAD^{1/2} + D^{-1}dD]. \tag{7}$$

Since the measure derived from dM is unchanged by orthogonal transformations [17], we may ignore the R matrices in (7), and write

$$dM_{ij} = \begin{cases} \lambda_i^{-1} d\lambda_i & i = j = 1,2,\ldots,2N \\ [(\lambda_i/\lambda_j)^{\frac{1}{2}} - (\lambda_j/\lambda_i)^{\frac{1}{2}}]dA_{ij} & i \neq j = 1,2,\ldots,2N \end{cases}. \tag{8}$$

The symmetries of dA and dD determine the independent dM_{ij} in (8). The conditions $U^+U = 1$ and $UJU^+ = K$ imply the symmetries $dA = -dA^+$ and $[K,dA] = 0$, and time-reversal symmetry implies $dA = -dA^*$. These conditions reduce the numbers of independent elements of dA to N^2 in the presence of time-reversal symmetry and $2N^2$

in its absence. If we make a choice of the independent dA_{ij} consistent with these symmetries, and observe the eigenvalue symmetry of $D, (\lambda_{i-n} = \lambda_i^{-1})$, (8) yields the measure

$$\mu(dQ) = \prod_{i=1}^{N} d\lambda_i (1 - \lambda_i^{-2}) \exp[-f(X_i)] \prod_{i<j}^{N} [(\lambda_i + \lambda_i^{-1}) - (\lambda_j + \lambda_j^{-1})]^{\beta} dA_{ij} . \quad (9)$$

Changing variables to X_i and integrating over the dA_{ij} gives $J(\{X_i\})$ in (4),

$$\rho(\{X_i\}) = \prod_{i<j}^{N} |X_i - X_j|^{\beta} \prod_{i}^{N} \exp[-f(X_i)] , \quad (10)$$

where $\beta = 1$ for the time-reversal symmetric case, and $\beta = 2$ otherwise.

The crucial result of (10) is that the "interaction" term which induces eigenvalue correlations is of exactly the same form as for the (N X N) GOE ($\beta = 1$) and GUE ($\beta = 2$). Thus our symmetry analysis, combined with the maximum entropy hypothesis supports Imry's conjecture that random matrix theory may provide a more fundamental explanation of the universal conductance fluctuations, (1).

4. TEST OF THE MAXIMUM ENTROPY HYPOTHESIS

Equation (9) represents only an ansatz for the statistical behavior of a class of random matrix ensembles generated by the standard microscopic models describing quantum transport. It is only of great interest if in fact those models generate distributions of this form. We therefore tested the applicability of the ansatz by comparing its statistical predictions with the behavior of an ensemble of X matrices generated numerically for the Anderson random tight-binding model [9,24].

Equation (10) predicts that after correcting for variation in the average spectral density, $\sigma(X)$, the $\{X_i\}$ should exhibit GOE fluctuations in the absence of a magnetic field, B, and GUE fluctuations in a field. There are two commonly studied fluctuation measures for these distributions. The probability density of the nearest neighbor spacing, $P(S) = a_\beta S^\beta \exp[-b_\beta S^2]$, ($a_\beta$, b_β are known constants), which measures the local eigenvalue repulsion; and the statistic $\Delta_3(L)$ which measures the deviation of the spectrum from uniform spacing as a function of the average number of eigenvalues, L, in the interval [18,19]. $\Delta_3(L)$ increases logarithmically with L [18], as opposed to the linear increase expected for uncorrelated levels, thus $\Delta_3(L)$ measures the spectral rigidity crucial for (1). In Fig. 1 we have corrected for the variation in $\sigma(X)$ by rescaling X to a variable with uniform density by the method described in [24]. We find essentially perfect agreement with the GOE and GUE fluctuations predicted by (10); both for P(S) and $\langle\Delta_3(L)\rangle$ as long as $\langle g\rangle \gg 1$. This strongly supports the maximum entropy hypothesis leading to (10), and the use of random matrix theory, in the manner proposed by Imry, in order to provide a fundamental justification for (1). From this point of view, the universal conductance fluctuations follow simply from the symmetries of X and the fact that g is a linear statistic on the eigenvalues of X. Note that the resistance, 1/g, is not a linear statistic on X so this approach picks out the conductance as the object which is expected to have particularly simple statistical properties in the metallic regime. Also, g, is _not_ a linear statistic on the eigenvalues of the Hamiltonian, H, so the spectral rigidity of H, recently studied by Altshuler and Shklovskii [14], is not so directly related to (1) as that of X.

For $\langle g\rangle \cong 1$, we find significant departures from the GOE result for $\Delta_3(L)$. This raises two possibilities. In [24] we argued that the procedure for correcting for the variation of $\sigma(X)$ may break down for $\langle g\rangle \cong 1$. Alternatively, the maximum entropy hypothesis leading to (1) may be breaking down as the system passes from the metallic to the localized regime. Note, this may not be important for the statistics of g, since for $\langle g\rangle \ll 1$, g is determined only by the smallest X_i.

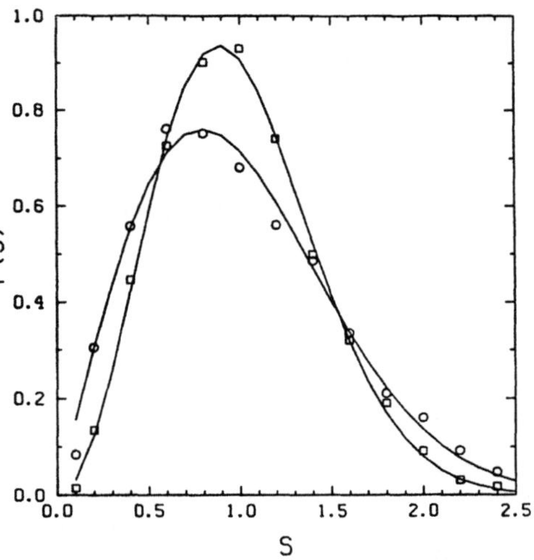

Fig. 1a: P(S) for rescaled X_i. Circles (squares) are numerical results for $B = 0$ ($B \neq 0$), solid lines are analytic results for GOE (GUE). Parameters are as given in ref. [24].

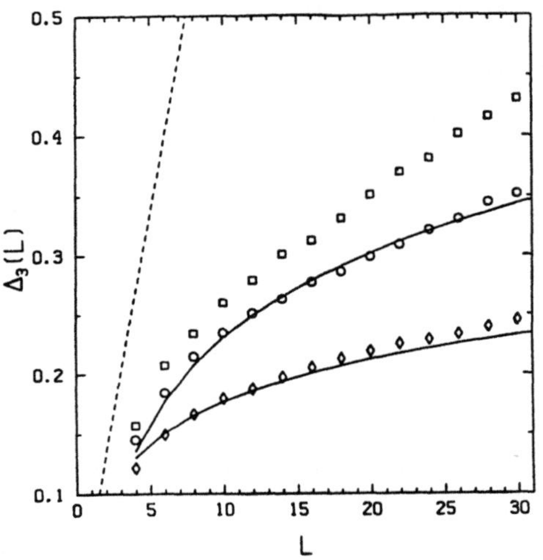

Fig. 1b: $\langle\Delta_3(L)\rangle$ numerical results, circles, (diamonds), are for $B = 0$ ($B \neq 0$) and $\langle g\rangle \cong 10$. Solid lines are GOE (GUE) results. Squares are for $\langle g\rangle \cong 1$. Dashed line is Poisson result.

5. THE MAXIMUM ENTROPY HYPOTHESIS AND SCALING THEORY

It is worth emphasizing the affinity between the maximum entropy hypothesis (5) and the scaling hypothesis. Scaling theory requires that both the average and the fluctuation behavior of the conductance become universal for samples much larger than ℓ. Thus, whatever the physical origin of the microscopic disorder (or whatever the details of the microscopic models used to describe the system on atomic scales), at length scales much larger than the elastic mean free path, the distribution of the conductance has "forgotten" its microscopic origin. However, this distribution is not self-averaging for $L_x \gg \ell$, the fluctuations in g have important physical effects. What then determines this universal distribution? In our approach, the distribution is determined simply by the fact that, for $L_x \gg \ell$, the matrix X, which determines g, becomes as random as possible, consistent with its symmetries, and with some limited information about the system (the spectral density, $\sigma(X)$). Thus the maximum entropy formulation and the result (10) provide a natural framework for the scaling theory of localization insofar as the theory postulates a universal, model-independent distribution for g, and that distribution must be consistent with the universal conductance fluctuations. However, the one-parameter scaling hypothesis imposes a stronger condition on $\rho(\{X_i\})$ than the maximum entropy hypothesis; it requires that $\sigma(X)$ be a function of only one parameter for each combination of parameters specifying the system (degree of disorder, sample dimensions, fermi energy). _If_ this is so, then $\rho(\{X_i\})$ and hence the distribution of g, is a universal, one-parameter distribution, and the scaling hypothesis is valid. We present below preliminary, but striking numerical evidence that this strong constraint on $\sigma(X)$ is satisfied.

6. EVIDENCE FOR THE VALIDITY OF ONE-PARAMETER SCALING

In order to test the validity of one-parameter scaling for $\sigma(X)$, we establish a connection to the finite-size scaling studies of the localization length by Pichard et al. [22,27,28]. Consider a 2N X 2N transfer matrix $T(L_x)$ describing a disordered conductor of cross-section $(L_y)^{d-1}$. Osledec's theorem [27] says that the matrix $O(L_x) = (T^+(L_x)T(L_x))^{1/2L_x}$ is self-averaging. i.e., $O_\infty = O(L_x \to \infty)$ is a unique matrix for each combination of system parameters, independent of the particular realization of the random matrix product, T. The logarithms of the eigenvalues of O_∞, $\{\alpha_i\}$, $i = 1, \ldots, N$ define N numbers characterizing the asymptotic behavior of the model, which are called Liapunov exponents in the context of dynamical systems, but in this context should be thought of as a set of inverse localization lengths. (The symmetry of T implies the $\alpha_i = -\alpha_{i-N}$, $i = N + 1, \ldots, 2N$). The localization length $\xi = \alpha_1^{-1}$, (where α_1 is the smallest α_i) is the longest length scale and Osledec's theorem combined with (4), roughly speaking, mean that the conductance decays exponentially, $g \propto \exp[-L_x/\xi]$, for $L_x \gg \xi$. Note that the theorem only applies for a system of fixed width, L_y, whose length, $L_x \to \infty$; thus it essentially provides a rigorous statement of the well-known fact that in any dimension a system which is sufficiently long compared to its width behaves one-dimensionally, and becomes localized. However the great insight of the finite-size scaling idea is that the dependence of ξ on the width, L_y, will indicate towards which phase the system will tends as the system size goes to infinity, for each particular value of the system parameters. The one-parameter scaling hypothesis then takes the form $\xi(L_y)/\xi = f(L_y/\xi)$, where ξ is the localization (or correlation) length of the infinite system. References [22,28] numerically verified this hypothesis for $\xi^{-1} = \alpha_1$. Reference [27] then showed numerically that the other α_i are essentially determined by α_1; that is to say there exists a limiting density $\rho(\alpha)$ towards which the $\{\alpha_i\}$ coverage rapidly as the $L_y \to \infty$, (i.e., the number N of $\alpha_i \to \infty$). Thus the eigenvalue density of the asymptotic, self-averaging matrix O_∞ is determined by only one parameter. However, this is not sufficient to confirm the scaling hypothesis, since, in general, g does not depend on this self-averaging density, which describes the strongly-localized regime; instead (from (4) and (10)) g depends on the average density of eigenvalues of the non-self-averaged matrix $T^+(L_x)T(L_x)$ considered (in the metallic regime) for $L_x < \xi$. Although a priori there is no simple mathematical relationship between these two densities, we have studied numerically the relevant spectral density, $\sigma(X)$, and initial results indicate that the simplest possible relationship holds under rather general conditions. The quantities $\bar{\alpha}_i = \langle (1/2L_x)\cosh^{-1}(X_i+1)\rangle = \langle (1/2L_x)\ln(\lambda_i)\rangle$, are the analogues of the α_i for the matrix $T^+(L_x)T(L_x)$. In Fig. 2. We show the striking new result that $\bar{\alpha}_i \cong \alpha_i$, $i = 1, \ldots, N$, i.e., the average density $\rho(\bar{\alpha})$ (which is simply-related to $\sigma(X)$) approximately equals the self-averaging, asymptotic density, $\rho(\alpha)$.

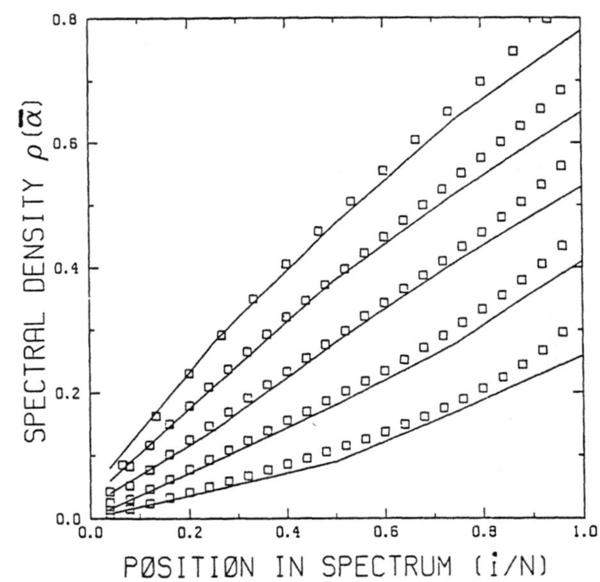

Fig. 2: Numerical results for the average spectral density, $\rho(\bar{\alpha})$(squares) for a 2d, 25x25 site Anderson model. We plot $\bar{\alpha}_i$, i = 1,...,N vs. relative position in the spectrum (i/N); the derivative of the curve gives $\rho(\bar{\alpha})$. Solid lines are the density of Liapunov exponents, $\rho(\alpha)$ taken from [27]. The curves are, in ascending order, for disorder W = 2 - 6. The smaller $\bar{\alpha}_i$ determine g, and there the agreement with α_i is best. It improves even further for $L_x > L_y$.

The equality of the two densities is found both for lengths $(\bar{\alpha}_i)^{-1}$ longer than and shorter than L_x, however it apparently only holds for samples with $L_x \geq L_y$. We expect that the two densities are still related for $L_y > L_x$, but the relationship involves the aspect ratio, L_y/L_x. Results have only been obtained in two dimensions and further careful study is in progress [29]. However, these results strongly suggest that the spectral density, $\sigma(X)$ which by our earlier arguments determines the parameter-dependence of the distribution of the conductance, is simply related to the distribution of inverse localization lengths, $\rho(\alpha)$. Since $\rho(\alpha)$ is a one-parameter distribution this relationship then establishes the validity of the one-parameter scaling hypothesis for the distribution of the conductance.

We acknowledge helpful discussions with A. Auerbach, O. Bohigas, B. Derrida, Y. Gefen, Y. Imry, P. Lee, and S. Kivelson. A.D. Stone thanks CEN Saclay for support during a visit, J.-L. Pichard thanks Yale University for the same. A.D. Stone thanks A. Bjorkli for her careful and timely preparation of the manuscript.

References

1. R. Landauer, Philos. Mag. 21, 863 (1970)
2. P.W. Anderson, D.J. Thouless, E. Abrahams and D.S. Fisher, Phys. Rev. B22, 3519 (1986)
3. E. Abrahams, P.W. Anderson, D.C. Licciardello, and T.V. Ramakrishnan, Phys. Rev. Lett. 42, 673 (1979)
4. B. Shapiro, Phys. Rev. B34, 4394 (1986)
5. D.J. Thouless, J. Phys. C 6, L49 (1973)
6. P.A. Lee, Phys. Rev. Lett. 52, 1641 (1984)
7. C.P. Umbach, S. Washburn, R.B. Laibowitz and R.A. Webb, Phys. Rev. B30, 4048 (1984); S. Washburn and R.A. Webb, Adv. Phys. 35, 375 (1986), and references therein
8. W.J. Skocpol, P.M. Mankiewich, R.E. Howard, L.D. Jackel, D.M. Tennant and A. Douglas Stone, Phys. Rev. Lett. 56, 2865 (1986)
9. A.D. Stone, Phys. Rev. Lett. 54, 2692 (1985);
10. P.A. Lee and A.D. Stone, Phys. Rev. Lett. 55, 1622 (1985)
11. P.A. Lee, A.D. Stone and H. Fukuyama, Phys. Rev. B 35, 1039 (1987)
12. B.L. Al'tshuler, Pis'ma ZH. Eksp. Teor. Fiz. 41, 530 (1985) [JETP Lett., 41, 648 (1985)];
13. B.L. Al'tshuler and D.E. Khmel'nikskii, Pis'ma Zh. Eksp. Teor. Fiz. 42, 291 (1985) [JETP Lett. 42, 359 (1986)];
14. B.L. Al'tshuler and B.I. Shklovskii, Zh. Eksp. Teor. Fiz. 91, 220 (1986) [Sov. Phys. JETP 64, 127 (1986)]
15. B.L. Al'tshuler, V.E. Kravtsov and I.V. Lerner, Pis'ma Zh. Eksp. Teor. Fiz. 43, 342 (1986) [JETP Lett. 43, 441 (1986)]
16. see M.L. Mehta, Random Matrices, Academic Press, NY (1967); and C.E. Porter, Ed., Statistical Theory of Spectra: Fluctuations, Academic Press, N.Y. (1965)
17. F.J. Dyson, J. Math, Phys. 3, 140 (1962);
18. F.J. Dyson and M.L. Mehta, J. Math. Phys. 4, 701 (1963)
19. T.A. Brody, J. Flores, J.B. French, P.A. Mello, A. Pandey and S.S.M. Wong, Rev. Mod. Phys. 53, 385 (1981)
20. Y. Imry, Europhys. Lett. 1, 249 (1986)
21. D. S. Fisher and P.A. Lee, Phys. Rev. B23, 6851 (1981)
22. J.-L. Pichard, Thesis, Université de Paris-Sud (1984)
23. R. Balian, Il Nuevo Cim. 57, 183 (1968)
24. K. Muttalib, J.-L. Pichard, and A.D. Stone, preprint
25. P.A. Mello, P. Pereyra and N. Kumar, preprint
26. N. Kumar and P.A. Mello, Phys. Rev. B31, 3109 (1985); P.A. Mello, Phys. Rev. B35, 1082 (1987).
27. J.-L. Pichard and G. Andre, Europhys. Lett. 2, 477 (1986)
28. J.-L. Pichard and G. Sarma, J. Phys. C 14, L617 (1981)
29. J.-L. Pichard, A.D. Stone, and K. Muttalib, unpublished

Nonlinear Effects in Mesoscopic Kinetics of Microjunctions

D.E. Khmelnitskii

Institute of Solid State Physics, Chernogolovka,
Moscow District, 142432, USSR

Fluctuations of the current-voltage (IV) characteristic and nonlinear thermopower of a mesoscopic sample are discussed. Such a sample can be realized as a junction of two normal metals. A diagrammatic approach to these problems based on the Keldysh method is developed. It is shown that the voltage dependence of current is a random function, and the IV-characteristic has a form of "grass-blades" on a usual ohmic background. The scale of each "grass-blade" in voltage is of order $V_c \sim \hbar/\tau_f e$, where τ_f is the time of flight through the junction and e is the electron charge. The scale in current δI depends on the voltage V and temperature T, and of order $(e^2/\hbar)(VV_c)^{1/2}$, if $V \gg V_c$, T/e. As a result, there appear regions of negative differential resistance provided the voltage is sufficiently large. Nonlinear thermopower versus temperature difference behaves in a similar way. Application of this approach to properties of bridges between two superconductors is discussed.

Current-voltage characteristic (IV-characteristic) of a junction of two normal metals obtained via the kinetic equation is ohmic. Its deviations from the linear form due to the non-equilibrium distribution function in the junction and energy dependence of the rate of inelastic processes arise at large voltage and will not be considered here. Another source of non-linearity is related to quantum interference effects called now mesoscopic fluctuations. In what follows we discuss the case when the junction resistance $R_0 \ll \hbar/e^2$ and interference effects lead only to small corrections to the Ohm's law (LARKIN and KHMELNITSKII /1/). Quantum corrections are maximal when interference occurs during a long time interval, larger than the time of flight τ_f of an electron through the junction. Such interference is possible only for electrons with close energy values $|\varepsilon_1-\varepsilon_2| \lesssim eV_c = \hbar/\tau_f$. If the mean free path ℓ is shorter than the junction length L, then $\tau_f \sim L^2/D$, where D is the diffusion coefficient. Interference effects depend on a given impurity distribution, leading to non-self-averaging of the IV-characteristic. If the junction voltage, $V \ll V_c$, the IV-characteristic is ohmic and conductance g=1/R has a non-self-averaged correction δg. In this case the theory predicts irregular dependence of $\delta g(\mu,H)$ on the magnetic field H and the Fermi energy μ of electrons (see reviews by LANDAUER /2/, AZBEL /3/, IMRY /4/, WASHBURN and WEBB /5/, and original papers by ALTSHULER /6/, LEE and STONE /7/, ALTSHULER and KHMELNITSKII /8/, ALTSHULER and SHKLOVSKY /9/).

The current through the junction is formed by electrons of energies ε near the Fermi level, where $|\varepsilon| \sim V$. If $V \gg V_c$, this range is divided into V/V_c intervals of values of ε, for which interference is important. Each such interval gives an independent random contribution of order $e^2V_c/\hbar$ to the total current. Therefore the current I is a random function of the voltage V with a characteristic scale $\Delta V \sim V_c$ and an amplitude $\Delta I \sim (e^2/\hbar)(VV_c)^{1/2}\Phi(V/V_c)$. In each given junction the IV-characteristic is an irregular curve with above-mentioned irregularity scale, which, however, is reproduced during a long time interval. Fluctuation contribution to the differential conductance g(V)=dI/dV is of the order $\delta g \sim (e^2/\hbar)(V/V_c)^{1/2}$. If $V > (\hbar/e^2R_0)V_c$, the IV-characteristic has regions with negative differential resistance.

A random function I(V) is specified by a correlator

$$K(V_1, V_2)=\overline{I(V_1)I(V_2)}-\overline{I(V_1)}\,\overline{I(V_2)}. \tag{1}$$

In the present paper the dependence of K on $V=(V_1+V_2)/2$, $\Delta V=V_1-V_2$, and temperature T is found. We suppose that the junction has the form of a bridge of cross-sectional area S and length $L\gg S^{1/2}$ between bulk electrodes. The mean free path ℓ is believed to be shorter than L.

In what follows it is convenient to use the Keldysh diagrammatic approach (KELDYSH /10/). In this method the Green function has a matrix form

$$\hat{G}=\begin{pmatrix} G^R & G^K \\ 0 & G^A \end{pmatrix}.$$

Here

$$G^{R(A)}(1,2)=\mp i\theta(\pm t_1\mp t_1)\langle\psi(1)\psi^+(2)+\psi^+(2)\psi(1)\rangle,$$
$$G^K(1,2)=-i\langle\psi(1)\psi^+(2)-\psi^+(2)\psi(1)\rangle,$$

and ψ^+ and ψ are creation and annihilation operators in the Heisenberg representation. In thermodynamic equilibrium

$$G_\varepsilon^K=(G_\varepsilon^R-G_\varepsilon^A)[1-2n(\varepsilon)], \tag{2}$$

where $n(\varepsilon)$ is the Fermi distribution function.

Inelastic relaxation processes usually are not important in microjunctions (they are assumed to occur in the bulk metals), and therefore G satisfies the equation

$$[\varepsilon+(\hbar^2/2m)\nabla^2-U(\mathbf{r})-e\varphi(\mathbf{r})]\hat{G}_\varepsilon(\mathbf{r},\mathbf{r}')=\delta(\mathbf{r}-\mathbf{r}'), \tag{3}$$

where $\phi(r)$ is the electrostatic potential, and U(r) is the random impurity potential with the Gaussian distribution, for which

$$\overline{U(\mathbf{r})}=0, \quad \overline{U(\mathbf{r})U(\mathbf{r}')}=2\pi\nu\tau\delta(\mathbf{r}-\mathbf{r}').$$

Here ν is the density of one-electron levels, and τ is the mean free time. Calculating averaged Green functions in the $\mu\tau/\hbar\gg 1$ approximation, we get

$$\overline{G}_\varepsilon^{R(A)}(\mathbf{r},\mathbf{r}')=\int(d\mathbf{p})\exp[i\mathbf{p}(\mathbf{r}-\mathbf{r}')]\{\varepsilon-p^2/2m\pm i/2\tau\}^{-1}, \tag{4}$$

and $\overline{G}_\varepsilon^K(r,r')$ satisfies the diffusion equation

$$D\nabla^2\overline{G}^K(\mathbf{r},\mathbf{r})=0, \quad D=v_F^2\tau/3\,. \tag{5}$$

Boundary conditions are

$$\overline{G}_\varepsilon^K(\mathbf{r},\mathbf{r})|_{x=0,L}=2\pi i\nu[1-2n(\varepsilon\pm eV/2)], \tag{6}$$

where $V=\phi(L)-\phi(0)$ is the voltage.

The current I is given by the relation

$$I=-i\frac{e\hbar}{2m}\int d\varepsilon\int dS(\nabla-\nabla')G_\varepsilon^K(\mathbf{r},\mathbf{r}')|_{\mathbf{r}=\mathbf{r}'}. \tag{7}$$

Averaging (7) over distribution of random potential, and using (5) and (6), we obtain a usual expression for the total current $\overline{I}$ through the junction:

$$\bar{I}=DS\nabla\int\bar{G}_{\varepsilon}^{K}(\mathbf{r},\mathbf{r})\,d\varepsilon=(e\nu DS/L)\int_{-\infty}^{+\infty}\{n(\varepsilon-eV/2)-n(\varepsilon+eV/2)\}\,d\varepsilon. \tag{8}$$

The correlation function $K(V_1,V_2)$ of currents is given by diagrams shown in Fig.1 a-d.

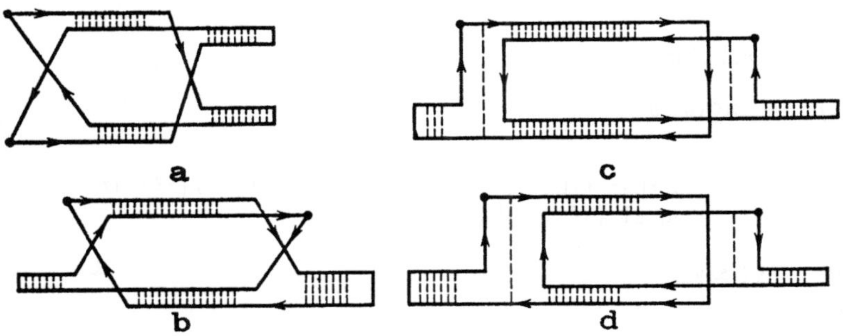

Fig. 1.

The "appendices" in these diagrams represent the functions $\overline{G}^K(r,r)$ given by (5) and (6), and other ladders - two-particle Green functions

$$P_{\varepsilon_1-\varepsilon_2}(\mathbf{r},\mathbf{r}')=\overline{G_{\varepsilon_1}^{R}(\mathbf{r},\mathbf{r}')G_{\varepsilon_2}^{A}(\mathbf{r},\mathbf{r}')}. \tag{9}$$

One-particle Green functions in (9) describe electron motion in different electrostatic potentials ϕ_1 and ϕ_2. In diffusion approximation $P_\omega(r,r')$ satisfies the equation

$$\{D\nabla^2+i\omega/\hbar+i(e/\hbar)[\varphi_1(\mathbf{r})-\varphi_2(\mathbf{r})]\}P_\omega(\mathbf{r},\mathbf{r}')=-2\pi\nu\delta(\mathbf{r}-\mathbf{r}'). \tag{10}$$

At boundaries with bulk electrodes we have

$$P_\omega(\mathbf{r},\mathbf{r}')|_{x=0,\,L}=0. \tag{11}$$

The sum of diagrams in Fig.1 a-d is given by

$$K(V_1,V_2)=\frac{2}{(2\pi)^4}\left(\frac{eSD}{\nu\hbar L^2}\right)^2\iint_0^L dx_1\,dx_2\iint_{-\infty}^{+\infty}d\varepsilon_1\,d\varepsilon_2\cdot$$
$$\cdot\left\{|P_{\varepsilon_1-\varepsilon_2}(x_1,x_2)|^2+\frac{1}{2}\operatorname{Re}P^2_{\varepsilon_1-\varepsilon_2}(x_1,x_2)\right\}\cdot$$
$$\cdot\left[n\left(\varepsilon_1-\frac{eV_1}{2}\right)-n\left(\varepsilon_1+\frac{eV_1}{2}\right)\right]\left[n\left(\varepsilon_2-\frac{eV_2}{2}\right)-n\left(\varepsilon_2+\frac{eV_2}{2}\right)\right]. \tag{12}$$

At $V_{1,2}\ll V_c$ we can neglect potential ϕ in (10), and (12) reduces to the formula for the correlation function of currents in the ohmic region. If $V_{1,2}\gg V_c$, ΔV, then the main contribution in (12) is given by the term in $|P_{\varepsilon_1-\varepsilon_2}(x_1,x_2)|^2$, and the energy range $|\varepsilon_1-\varepsilon_2|<<eV$ becomes important. Therefore V, T and ΔV-dependences of K are factorized:

$$K(V,\Delta V,T)=\frac{4}{(2\pi)^4}\left(\frac{e^2}{\hbar}\right)^2 VV_c\left[\operatorname{cth}\left(\frac{eV}{T}\right)-\frac{T}{eV}\right]f\left(\frac{\Delta V}{V_c}\right). \tag{13}$$

Here the dimensionless function $f(\alpha^3)$ is given by

$$f(\alpha^3)=\alpha^{-2}\iint\limits_{-\alpha/2}^{\alpha/2} dy_1\,dy_2\int\limits_{-\infty}^{+\infty} dz\,|\Pi_z(y_1,y_2)|^2, \tag{14}$$

where $\Pi_z(y,y')$ satisfies the equation

$$\{\partial^2/\partial y^2-i(y-z)\}\Pi_z(y,\ y')=-2\pi\delta(y-y'). \tag{15}$$

The coefficient in (14) is chosen so that $f(0)=1$. At $\alpha>>1$ the region $|y_1-z|\sim|y_2-z|\sim1<<\alpha$ is of importance in the integral (14). Therefore

$$f(\Delta V/V_c)=C(V_c/\Delta V)^{1/3},\qquad \Delta V\gg V_c; \tag{16}$$

$$C=\frac{V}{V_c}\iint\limits_{-\infty}^{+\infty} dy_1\,dy_2|\Pi_z(y_1,y_2)|^2, \tag{17}$$

where the function $\Pi_z(y_1,y_2)$ has a weak z-dependence in the interval $-\alpha/2<z<\alpha/2$.

At high temperature $T>>V_c$, V, the factorization may be performed at arbitrary relations between V and ΔV. In this case the term in $\mathrm{Re}P^2$ in (12) can also be ignored, and the relation is rewritten in the form

$$K(V_1,V_2,T)=\frac{1}{(2\pi)^4}\left(\frac{e^2}{\hbar}\right)^2 V_1V_2\frac{eV_c}{T}f\left(\frac{\Delta V}{V_c}\right). \tag{18}$$

The correlator of differential conductances K_g is related to $K(V_1,V_2)$ by the expression

$$K_g(V_1,V_2)=\overline{g(V_1)g(V_2)}-\overline{g(V_1)}\,\overline{g(V_2)}=\frac{\partial^2}{\partial V_1\partial V_2}K(V_1,V_2). \tag{19}$$

At $V>>V_c$ the main contribution to K_g is obtained by differentiating the function $f(\Delta V/V_c)$:

$$K_g=-\frac{4}{(2\pi)^4}\left(\frac{e^2}{\hbar}\right)^2\frac{V}{V_c}\left[\operatorname{cth}\frac{eV}{T}-\frac{T}{eV}\right]f''\left(\frac{\Delta V}{V_c}\right). \tag{20}$$

The correlator K_g grows with increasing V, and at

$$V[\operatorname{cth}(eV/T)-T/eV]>V_c(\hbar/e^2R_0)^2 \tag{21}$$

fluctuations of g are larger than $\bar{g}$. This means that if relation (21) holds, the IV-characteristic has regions of width $\Delta V\sim V_c$ with negative differential conductance.

Quantitatively the form of the IV-characteristic does not depend on the explicit form of the function $K(V_1,V_2)$, but it would be interesting to obtain this correlation function experimentally and, in particular, to verify the asymptotic expression (17). We have found $K(V_1,V_2)$ averaging over all realizations of random potential. The correlation function can also be obtained by measuring the IV-characteristic of a junction in a wide range of voltage V and averaging over V with fixed ΔV.

The shape of the junction is not important. For arbitrary shape instead of L one should use the specific dimension of the region defining the junction resistance, e.g., for an orifice in a thin dielectric layer between normal metals this dimension is of the order of the orifice diameter. The condition $\ell \ll L$ is also not crucial. If $\ell \gg L$, an irregular IV-characteristic should also be observed, but the time of flight $\tau_f \sim L/v_F$ in this case is shorter, and the specific fluctuation scale $V_c \sim \hbar v_F/L$ is less the for an impure junction.

The diagrammatic approach developed above is applicable to many nonlinear mesoscopic problems. As an example, we give an expression for a correlator K_{QI} of thermal (Q) and electric (I) currents through a bridge, connecting two bulk electrodes of temperatures T_1 and T_2 (T_1' and T_2'), the voltage between electrodes being equal to $V_1(V_2)$ (LESOVIK and KHMELNITSKII /11/):

$$K_{QI}(T_1,T_2,V_1;T_1',T_2',V_2)=\frac{2}{(2\pi)^4}\left(\frac{eSD}{\nu\hbar L^2}\right)^2\iint_0^L dx_1\,dx_2\iint_{-\infty}^{+\infty} d\varepsilon_1\,d\varepsilon_2\cdot$$
$$\cdot\left\{|P_{\varepsilon_1-\varepsilon_2}(x_1,x_2)|^2+\frac{1}{2}\operatorname{Re}P^2_{\varepsilon_1-\varepsilon_2}(x_1,x_2)\right\}\cdot$$
$$\cdot\left[n\left(\varepsilon_1-\frac{eV_1}{2},T_1\right)-n\left(\varepsilon_1+\frac{eV_1}{2},T_2\right)\right]\left[n\left(\varepsilon_2-\frac{eV_2}{2},T_1'\right)-\right. \tag{22}$$
$$\left.-n\left(\varepsilon_2+\frac{eV_2}{2},T_2'\right)\right].$$

This expression is obtained by substituting the thermal current operator $|\varepsilon|p$ instead of the electric current operator ep into the expressions for the diagrams a-d. It can be used, when studying linear mesoscopic thermoelectric effects. The differences of occupation numbers should be expanded in V and T_1-T_2. It can also be used to obtain the dependence of thermopower on temperature difference in nonlinear regime. In this way the mesoscopic fluctuations of thermopower versus temperature difference have been predicted (LESOVIK and KHMELNITSKII /11/).

To study mesoscopics in a bridge between two superconductors, it is necessary to use the Green function in a double matrix form (a product of the Gor'kov and Keldysh matrices), introducing the anomalous Gor'kov F-function. Here it is also possible to predict a lot of new mesoscopic effects.

I am grateful to A.I. Larkin, G.B. Lesovik and A.A. Golubov for fruitful collaboration.

References

1. A.I. Larkin and D.E. Khmelnitskii: ZhETF 91, 1815 (1986)
2. R. Landauer: In Localization, Interaction, and Transport Phenomena (LIPTIM), ed. by B. Kramer, G. Bergmann, and Y. Bruinsraede, Springer Ser. Solid-State Phys., Vol.61 (Springer, Berlin, Heidelberg, New York 1984) p.38
3. M. Ya. Azbel: ibid., p.162
4. Y. Imry: In Directions in Condensed Matter Physics, ed. by G. Grinstein and G. Mazenko, World Sci. Ser. Condensed Matter Phys., Vol. 1, p.101

5. S. Washburn and R.A. Webb: Adv. Phys. **35**, 375 (1986)
6. B.L. Altshuler: Pisma v ZhETF **41**, 530 (1985)
7. P.A. Lee and A.D. Stone: Phys. Rev. Lett. **55**, 1622 (1985)
8. B.L. Altshuler and D.E. Khmelnitskii: Pisma v ZhETF **42**, 291 (1985).
9. B.L. Altshuler and B.I. Shklovsky: ZhETF **91**, 220 (1986)
10. L.V. Keldysh: ZhETF **47**, 1515 (1964)
11. G.B. Lesovik and D.E. Khmelnitskii: ZhETF **93** (1987)

Asymmetry of Magnetoresistance in Microstructures

Y. Isawa[1], *H. Ebisawa*[2], *and S. Maekawa*[3]

[1]Research Institute of Electrical Communication, Tohoku University, Sendai 980, Japan

[2]Department of Engineering Science, Tohoku University, Sendai 980, Japan

[3]Institute for Materials Research, Tohoku University, Sendai 980, Japan

1. INTRODUCTION

Quantum transport phenomena in microstructures, which are attributed to the wave property of the electrons, are the matter of great interest. A typical one is the asymmetry of the magnetoresistance upon reversal of the magnetic field H. Since the pioneering work of Webb et al.[1-3] on the Aharonov-Bohm effect, the asymmetry has been observed in the microstructures, constructed from several branches of the small metallic fine wires, which correspond to 2-port in circuit theory. However its origin was not made clear. The 2-port means any 4-terminal (the terminals are denoted by 1,1',2 and 2') network in which the four terminals have been paired into ports 1 1' and 2 2' (see for example Fig.1). The condition for 2-port is that the current entering the network by terminal 1(2) is equal to that leaving the network by terminal 1'(2'). Recently Benoit et al.[4] have confirmed that the asymmetry is consistent with the Onsager-Kubo's reciprocity theorem by 4-terminal measurements of the matrix elements of 2×2 impedance. Buttiker[5] has also pointed out the importance of the many-terminal effects on the asymmetry. Ma and and Lee[6] have shown that the transverse conductance G_{xy} in rectangular sample is asymmetric in the magnetic field.

In microstructures, electrons are coherent so that the current density at a point r is directly influenced by the electric field not only in the immediate neighbourhood of r but also within the coherence length L_ϕ around r . Hence if the so-called "sample" is smaller than L_ϕ, the current density becomes highly non-local, which means that the sample and leads should be treated as an inseparable system[7-9]. We show in this paper that the asymmetry of the magneto-resistance in microstructures resulting from this non-locality is in agreement with experimental results.

2. MODEL

We investigate a simple 2-port constructed by fine metallic wires as shown in Fig.1. The circuit equations for voltages $V_1(H)$ and $V_2(H)$ in the input and output terminals, respectively, are written as the linear combination of the currents I_1 and I_2 :

$$V_1(H)=Z_{11}(H)I_1+Z_{12}(H)I_2, \tag{1}$$
$$V_2(H)=Z_{21}(H)I_1+Z_{22}(H)I_2. \tag{2}$$

We represent the voltage $V_{1,2}(H)$ as the sum of two terms: (i) the classical part of the voltage $V^0_{1,2}(H)$ determined by the Ohm's law and (ii) the quantum fluctuating part $\delta V_{1,2}(H)$. Similar representation for the impedance leads to

$$Z_{ij}(H)=Z^0_{ij}(H)+\delta Z_{ij}(H) \quad (i,j=1,2). \tag{3}$$

The voltage fluctuation is written as follows

$$\delta V_i(H) = \Sigma\, \delta Z_{ij}(H) I_j. \tag{4}$$

When the input port is open(I_1=0) and the constant current flows in the output port(I_2=constant), the voltage fluctuation, $\delta V_1(H)$, averaged over the cross section of the wire is given by the integral of the electric field fluctuation $\delta E(r)$ along the branches 1,2 and 3 :

$$\begin{aligned}\delta V_1(H) &= \delta Z_{12}(H) I_2 \\ &\equiv \int_{r\in 1,2,3} \delta E(r) dr.\end{aligned} \tag{5}$$

When the fine wire is much longer than its width, the average field fluctuation $\delta E(x)$ over the cross section S_X at x(x is the component of r along the wire) is determined by

$$S\sigma_0 \delta E(x) + \int_{r'\in 2,4,5} \delta\sigma(r,r') E_0(r') dr' dS_X = 0, \tag{6}$$

where σ_0 is the Drude formula for the conductivity, $E_0(r') = I_2/S\sigma_0$ is the average electric field in the absence of the quantum correction, and $\delta\sigma(r,r')$ is the quantum correction to the conductivity which causes the current fluctuation at r by the electric field imposed on r'. We note that $\delta\sigma(r,r')$ is not a scalar but a diagonal component of a tensor with directions along the length of the wire at r and r'. The symmetrical and the anti-symmetrical parts of the impedance fluctuations:

$$\delta Zs(H) = (\delta Z_{21}(H) + \delta Z_{12}(H))/2, \tag{7.a}$$
$$\delta Za(H) = (\delta Z_{21}(H) - \delta Z_{12}(H))/2, \tag{7.b}$$

are given by

$$\left.\begin{matrix}\delta Zs(H) \\ \\ \delta Za(H)\end{matrix}\right\} = \iint_{r\in 1,2,3\; r'\in 2,4,5} dr dr' \quad (\delta\sigma(r',r,H) \pm \delta\sigma(r,r',H))/2S^2\sigma_0^2. \tag{8}$$

There is no contribution to $\delta Za(H)$ from the integral region where both r and r' are in 2-branch. This means that the non-local correlation in the conductivity extending over the different branches is the origin of the asymmetry in the impedance fluctuation. When the length of 2-branch is much smaller than $L\phi$, the asymmetry associated with 2-branch is negligible. Then the non-local coupling between 1,3-branches and 4,5-branches dominates the magnitude of $\delta Za(H)$.

The microscopic reciprocity theorem derived by Onsager and Kubo for the conductivity leads to

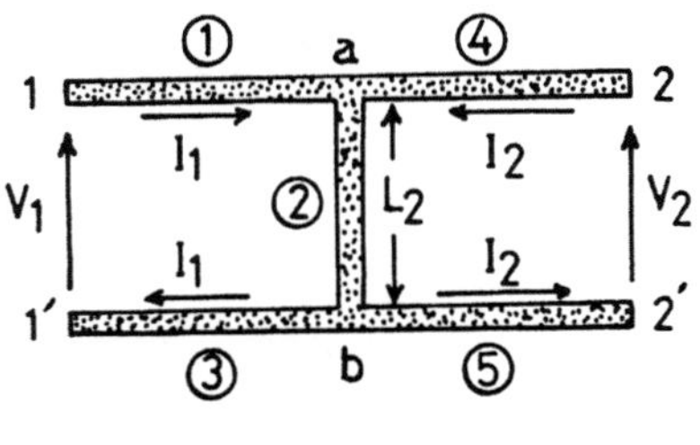

Fig.1 A simple 2-port. Each branch is made of fine metallic wire.

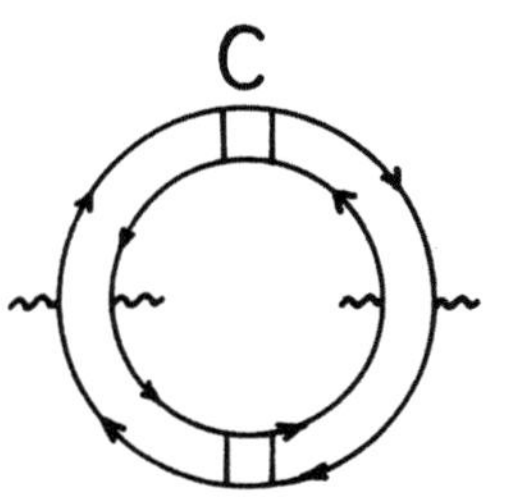

Fig.2 The simplest scattering process for the conductance fluctuation. C is the particle-hole diffusion propagator.

$$\sigma(r,r',H)=\sigma(r',r,-H), \tag{9}$$

which implies that the conductivity fluctuation satisfies the relation

$$\delta\sigma(r,r',H)=\delta\sigma(r',r,-H). \tag{10}$$

Here we neglect the asymmetry of σ from the classical Hall effect since it is too small to explain the observed asymmetry. Thus we obtain

$$\delta Za(H)=-\delta Za(-H) \text{ and } \delta Zs(H)=\delta Zs(-H). \tag{11}$$

This is consistent with the experiments by Benoit et al.[4]. In order to determine the magnitude of $\delta Za(H)$, we evaluate the second moment of the impedance fluctuation which is denoted by $\langle\delta Za(H)^2\rangle$ and $\langle\delta Zs(H)^2\rangle$ with ensemble average <-->. By use of the microscopic theory in which the simplest scattering process shown in Fig.2 is taken into account, we obtain

$$\langle\delta Za(H)^2\rangle = 8e^4/\{h^2(\sigma_0 S)^4\}\times S^2\Big(\int\int_{r\in 1,2,3\ r'\in 2,4,5} drdr' - \iint_{r\in 2\ r'\in 2} drdr'\Big)|C(r,r')|^2 \tag{12}$$

and

$$\langle\delta Zs(H)^2\rangle=8e^4/\{h^2(\sigma_0 S)\}^4\times S^2\Big(\iint_{r\in 1,2,3\ r'\in 2,4,5} drdr' + \iint_{r\in 2\ r'\in 2} drdr'\Big)$$
$$\times[\mathrm{Re}\{C(r,r')C(r',r)\}+C(r',r)^2], \tag{13}$$

where $C(r,r')$ is the particle-hole diffusion propagator in spatial representation. We restrict ourselves to the case where the field is so strong that the contribution from the particle-particle diffusion propagator is negligible, though it is indispensable in weak field particularly for the theory to satisfy the exact relation $\langle\delta Za(H=0)^2\rangle=0$[6,10]. In Fig.3 we show numerical results of the symmetric and anti-symmetric voltage fluctuations, ΔVs and ΔVa, normalized by ΔV_ϕ as a function of $\sqrt{L_2/L_\phi}$. Here ΔV_ϕ is the voltage fluctuation in a fine wire with length L_ϕ and width w when the conductance fluctuation is assumed to be universal

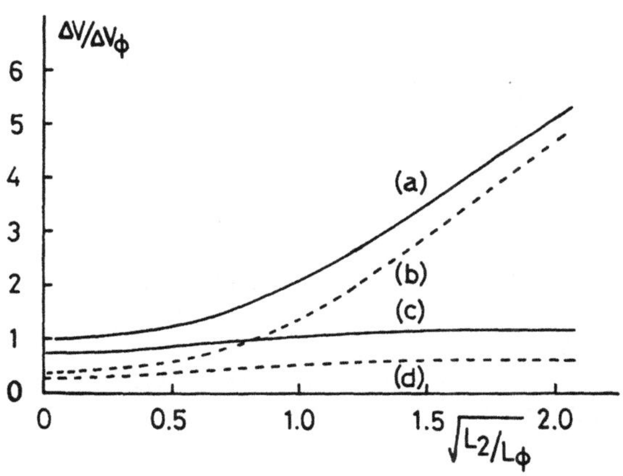

Fig.3 The voltage fluctuations as a function of $\sqrt{L_2/L_\phi}$. (a) and (b) are symmetric voltage fluctuations, $\Delta Vs/\Delta V_\phi$, while (c) and (d) are anti-symmetric ones, $\Delta Va/\Delta V_\phi$. The values of parameters are $L_1+L_2+L_3=10\times L_\phi$, $L_1=L_3$ for (a)~(d). The length $L_4=L_5$ is $3\times L_\phi$ for (a) and (c), and $0.5\times L_\phi$ for (b) and (d).

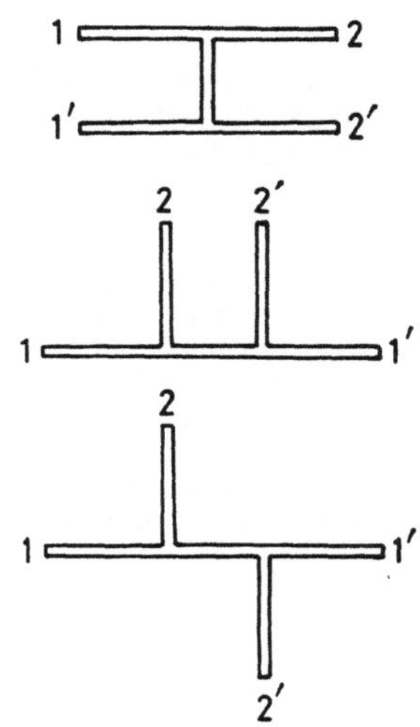

Fig.4 Three microstructures which give rise to the same voltage fluctuations.

$$\Delta V\phi = IR_0{}^2 L\phi^2 e^2/(hw^2), \qquad (14)$$

where I is the current, R_0 is the sheet resistance. As is seen from Fig.3, both ΔVs and ΔVa are nearly independent of L_2 when it is shorter than the coherence length L_ϕ. When L_2 is larger than L_ϕ, we can see the clear difference between ΔVs and ΔVa: ΔVs increases linearly as a function of $\sqrt{L_2/L_\phi}$, while ΔVa is almost independent of L_2. These features are in good agreement with the experimental results by Webb et al.[11], which can be explained as follows. The open voltage fluctuation ΔV at the output port 2 2' is the sum of the voltage fluctuations in 2,4 and 5-branches: $\Delta V=\Delta V_2+\Delta V_4+\Delta V_5$. When $L_2 \ll L_\phi, L_4 > L_\phi$ and $L_5 > L$, the voltage fluctuation in 2-branch ΔV_2 is much smaller than that in 4 and 5-branches and ΔV is only weakly dependent on $\sqrt{L_2}$. The ratio of ΔVs to ΔVa is given by $\Delta Vs/\Delta Va=\sqrt{2}$ for $L_2/L_\phi \rightarrow 0$. When L_2 is larger than L_ϕ, only the regions within the coherence length from the contacts a and b in 2-branch in Fig.1 is effective on the asymmetric part of the magnetoresistance since the asymmetric part comes from the correlation between different branches, while the outside of these regions in 2-th branch is ineffective. Hence for $L_2 \gg L_\phi$, the asymmetric part is independent of L_2. On the other hand, the whole region in 2-branch contributes to the symmetrical part of the magnetoresistance. Thus the symmetric part continues to increase for $L_2 > L_\phi$.

Our theory predicts that ΔVs and ΔVa are independent of the angle between different branches at the contacts and the curvature of wire as long as the length of the fine wire is much larger than the thickness. For example, three microstructures, which are topologically equivalent, as shown in Fig.4 will give rise to the same voltage fluctuations.

3. CONCLUSION

By applying the microscopic theory to a simple microstructure which should be treated as a 2-port, we find that the asymmetry of the magnetoresistance in small metallic system is attributable to the non-local fluctuation of the conductivity extending over the different branches. We also find that our theoretical results are in good agreement with the experimental ones.

4. REFERENCES

1. C.P.Umbach, S.Washburn, R.B.Laibowitz and R.A.Webb:Phys.Rev.30(1984)4048.
2. R.A.Webb,S.Washburn,C.P.Umbach and R.B.Laibowitz:Phys.Rev.Lett.54(1985)2696.
3. S.Washburn, C.P.Umbach, R.B.Laibowitz and R.A.Webb:Phys.Rev.32(1985)4789.
4. A.D.Benoit, S.Washburn C.P.Umbach, R.B.Laibowitz and R.A.Webb: Phys.Rev.Lett.57 (1986)1765.
5. M.Buttiker: Phys.Rev.Lett.57(1986)1761.
6. M.Ma and P.A.Lee:preprint.
7. Y.Isawa,H.Ebisawa and S.Maekawa: J.Phys.Soc.Jpn.55(1986)2523.
8. Y.Isawa,H.Ebisawa and S.Maekawa: Proc.2nd Int.Symp.Foundations of Quantum Mechanics,Tokyo,1986,pp.218-224.
9. S.Maekawa, Y.Isawa and H.Ebisawa: J.Phys.Soc.Jpn.56(1987)25.
10. B.L.Altshuler and D.E.Khmelnitzkii:JETP Lett.42(1985)361.
11. R.A.Webb, S.Washburn, A.D.Benoit, C.P.Umbach and R.B.Laibowitz:Proc.2nd Int. Symp.Foundations of Quantum Mechanics,Tokyo,1986,pp.193-206.

Magnetoconductance of Mesoscopic Systems

A. Sawada[1], *K. Tankei*[2], *and Y. Nagaoka*[1]

[1]Department of Physics, Nagoya University, Chikusa-ku, Nagoya 464, Japan

[2]Laboratory of Physics, Matsusaka University, Matsusaka. Mie 515, Japan

The magnetoconductance of mesoscopic systems has been investigated by many authors [1-3]. In particular the Aharonov-Bohm effect and the conductance fluctuation phenomenon have attracted much attention. In these experiments the observed conductance was found not to be an even function of the magnetic field H. At first glance it seems to contradict the Onsager relation which verifies that the diagonal elements of the conductivity tensor are even functions of H. This contradiction was solved by BÜTTIKER [4]. He pointed out that, since the conductance is measured by the four-terminal method, the observed one is not the diagonal element itself. The diagonal elements derived from the observed data were shown to be even functions of H [5].

In this paper we first present a brief review on the symmetry properties of the magnetoconductance to avoid confusion which appears in some references. Then, by introducing a simplified model, we numerically study the magnetoconductance of mesoscopic systems. It is found that the behavior of the conductance is sensitive to the distribution of impurities, and that the AB oscillation does not disappear even when the number of impurities gets larger.

Before Büttiker's work, there were some theoretical studies on the symmetry of the magnetoconductance of mesoscopic systems. GEFEN, IMRY and AZBEL [6] first investigated theoretically the Aharonov-Bohm effect of a ring composed of one-dimensional wires. Based on the Landauer formula of the conductance, they made a phenomenological study on the propagation of electron wave along the wire, and derived the expression for the conductance which oscillates as a function of the magnetic flux penetrating the ring: i.e.

$$g(\phi) = \frac{a+b\cos\phi+b'\sin\phi}{A+B\cos\phi+C\cos2\phi}, \tag{1}$$

where $\phi=2\pi\varphi/\varphi_0$, φ and φ_0 being the flux penetrating the ring and the flux quantum respectively, and a, b etc. are independent of ϕ. It is noticed that Eq.(1) is not an even function of ϕ.

The symmetry of the magnetoconductance was discussed by Büttiker and Imry[7]. Their argument is based on the symmetry properties of the S-matrix and the Landauer formula of the conductance. Let us consider a general multi-channel case which is schematically illustrated in Fig.1. If one denotes the amplitude of an incoming wave in the i-th channel of the right (left)-hand side by A_{ir} $(A_{i\ell})$, and that of an outgoing wave by $B_{ir}(B_{i\ell})$, they are related by

$$B = S'A , \tag{2}$$

where A and B are vectors whose components are $(A_{1r},\cdots,A_{Mr},A_{1\ell},\cdots,A_{N\ell})$ and $(B_{1r},\cdots,B_{Mr},B_{1\ell},\cdots,B_{N\ell})$, respectively and S' is an $(N+M)\times(N+M)$ matrix. The current conservation law is expressed as

$$\sum_{i=1}^{M} v_{ir}|A_{ir}|^2 + \sum_{j=1}^{N} v_{j\ell}|A_{j\ell}|^2 = \sum_{i=1}^{M} v_{ir}|B_{ir}|^2 + \sum_{j=1}^{N} v_{j\ell}|B_{j\ell}|^2 , \qquad (3)$$

where v_α (α=ir,iℓ,jr,jℓ) is the electron velocity in the α-th channel. Therefore, if one introduces the S-matrix defined by [8]

$$S_{\alpha\beta} = \sqrt{\frac{v_\alpha}{v_\beta}}\, S'_{\alpha\beta} \quad , \qquad (4)$$

it is verified that S is unitary:

$$S^+(\phi)S(\phi) = 1 \quad . \qquad (5)$$

It is also proved that due to the time reversal symmetry S satisfies the relation

$$S^*(-\phi)S(\phi) = 1 \; . \qquad (6)$$

The transmission matrices t and t' and the reflection matrices r and r' are introduced as sub-matrices of S by

$$S = \begin{pmatrix} r & t \\ t' & r' \end{pmatrix} \quad . \qquad (7)$$

Then the Landauer formula of the conductance is given in its simplest version by

$$g = \mathrm{tr}\,|t|^2 \; , \qquad (8)$$

which holds when the transmission probability is sufficiently small. Using Eqs. (5) and (6), we can prove [7] that the conductance is an even function of ϕ:

$$g(\phi) = g(-\phi) \; . \qquad (9)$$

In order to study how the conductance depends on the magnetic flux, we consider the ladder-type lattice (M=N=2) as shown in Fig.2, taking the tight-binding model

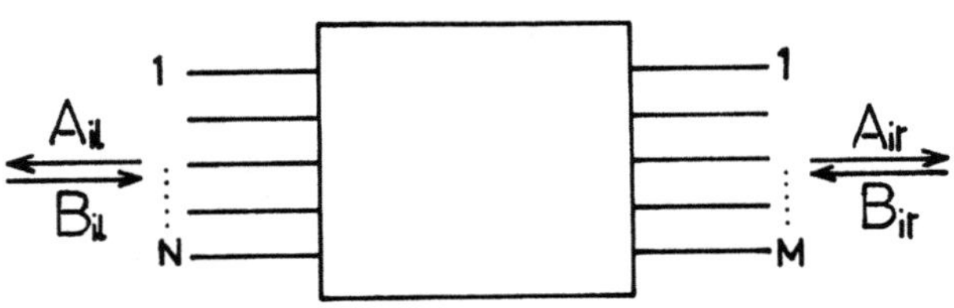

Fig. 1 General multi-channel case

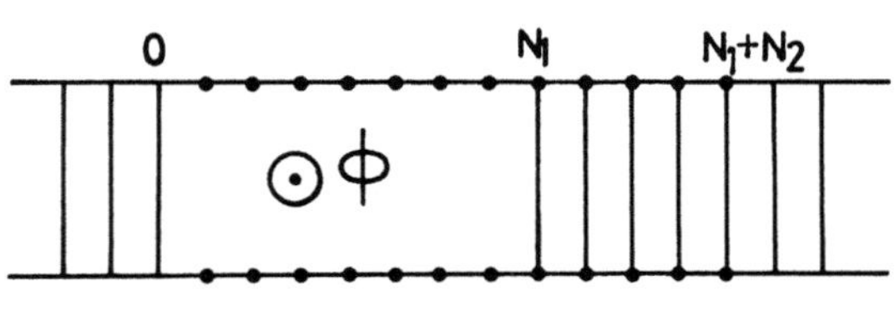

Fig.2 Ladder-type lattice

for electrons. We assume the magnetic flux penetrates an open area between 0 and N_1 sites. Lattice sites between 1 and N_1+N_2 are occupied by impurities, and the lattice is perfect outside this region.

The S-matrix is calculated by using the transfer matrix method. Then it is explicitly shown that S satisfies Eqs.(5) and (6). Therefore the conductance derived from it is an even function of ϕ. Furthermore, it takes the same form as Eq.(1) with b'=0: i.e.

$$g(\phi) = \frac{a + b\cos\phi}{A+B\cos\phi+C\cos2\phi} \quad . \tag{10}$$

In order to study now the ϕ-dependence of $g(\phi)$ in more detail, we have performed numerical calculations. For simplicity, all the nearest neighbor transfer energies are taken to be equal to t, and the impurity potentials are assumed to be distributed uniformly with the width W.

In Fig.3, the ϕ-dependence of $g(\phi)$ is shown for $N_1=50$, $N_2=250$, E/t=-0.6 and W/t=0.5, where E is the electron energy. Solid and broken curves are obtained, respectively, for two distributions of potentials which differ from each other only on one site between 1 and N_1. In this way, the behavior of $g(\phi)$ against ϕ changes drastically with the change in the impurity potential.

Figure 4(a) shows the ratios of the coefficients $|b/a|$ and $|B/A|$ as a function of N_2 with fixed $N_1=300$, E/t=0.6 and W/t=0.35 for five different distributions of potentials. The ratios remain almost constant when N_2 increases. The ratio C/A has the same order of magnitude as B/A, and shows a similar N_2-dependence to B/A.

If N_1 increases with fixed $N_2=300$, the ratio $|b/a|$ has a small variation as shown in Fig.4(b), while the ratios $|B/A|$ and $|C/A|$ decrease with increasing N_1.

As shown in Fig.4, the two ratios b/a and B/A behave quite differently when N_1 increases. It implies that the ϕ-dependences in the numerator and in the denominator of Eq.(10) have different origins. The term with the coefficient b arises from the interference effect between two waves propagating along the separate chains. On the other hand, interference effect between two waves propagating around the flux in opposite directions gives the term $C\cos2\phi$ in the denominator. The term $B\cos\phi$ appears due to the interference effect between a wave propagating around the flux and a wave scattered backward. When N_1 increases, the latter two interference effects disappear more rapidly than the former. Therefore, it is expected that the ratios $|B/A|$ and $|C/A|$ decrease with increasing N_1, while the ratio $|b/a|$ hardly changes.

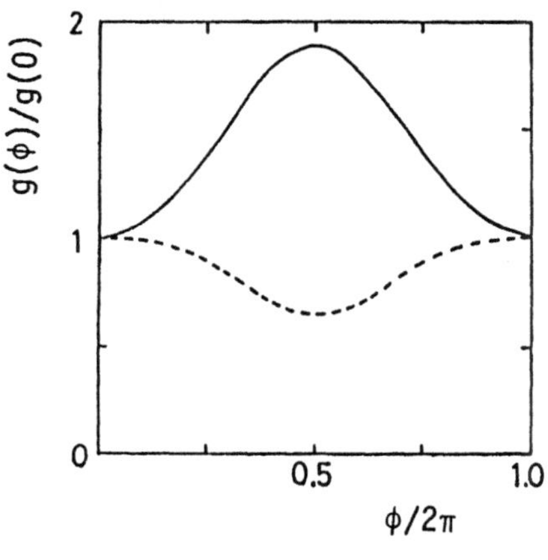

Fig.3 Conductance vs. flux

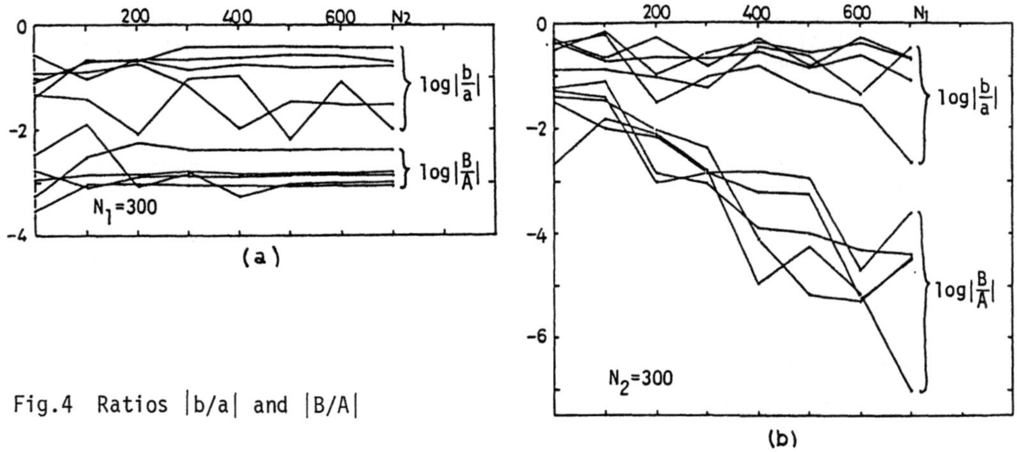

Fig.4 Ratios |b/a| and |B/A|

In this paper we considered a quasi-one-dimensional case with two channels. As Fig.3 shows, the phase difference between waves propagating in each channel is affected sensitively by the variation in the impurity potentials. In a real system which corresponds to the multi-channel case, an inteference between waves with various phase will reduce amplitude of the AB oscillation. This problem will be studied in near future.

References

1. C.P. Umbach, S. Washburn, R.B. Laibowitz and R.A. Webb, Phys. Rev. B30, 4048 (1984).
2. A. Douglas Stone, Phys. Rev. Lett. 54, 2692 (1985).
3. R.A. Webb, S. Washburn, C.P. Umbach, and R.B. Laibowitz, Phys. Rev. Lett. 54, 2696 (1985).
4. M. Büttiker, Phys. Rev. Lett. 57, 1761 (1986).
5. A.D. Benoit, S. Washburn, C.P. Umbach, R.B. Laibowitz, and R.A. Webb, Phys. Rev. Lett. 57, 1765 (1986).
6. Y. Gefen, Y. Imry and M.Y. Azbel, Phys. Rev. Lett. 52, 129 (1984).
7. M. Büttiker and Y. Imry, J. Phys. C18, L467 (1985).
8. M. Büttiker, Y. Imry, R. Landauer and S. Pinhas, Phys. Rev. B31, 6207 (1985). In this reference, the Landauer formula in the multi-channel case is introduced. However, there seems to be some confusion in the definition of the S matrix.

Part X

Mesoscopic Systems: Experiments

Localization and Interference in Small Systems

D.E. Prober, S. Wind, and V. Chandrasekhar

Section of Applied Physics and Dept. of Physics, Yale University, New Haven, CT 06520-2157, USA

1. Introduction

One of the significant achievements of condensed matter physics in the last decade has been the understanding of electron transport in disordered systems, especially in systems of reduced dimensionality. Studies of electron transport at low temperatures in small systems reveal clearly the effects of electron quantum interference. These effects are apparent because the electrons retain phase coherence over relatively long distances. The interference effects include weak localization in narrow wires, Aharonov-Bohm resistance oscillations in rings, and universal conductance fluctuations in short, narrow structures.

This paper reviews our experimental studies of one-dimensional weak localization and Aharonov-Bohm oscillations. Special attention has been given to the understanding of the phase coherence mechanisms, as these determine the overall magnitude of the electron interference effects, and also establish the conditions under which they may be observed.

2. Theoretical Background

2.1 Length Scales and Dimensionality

A sample is considered one-dimensional (1D) if its lateral dimensions are less than the characteristic length scale for a given effect. The effects considered in this work take place in the regime of diffusive electron motion, where elastic scattering by impurities is the dominant momentum relaxation mechanism. Quantum interference occurs between electron partial waves which follow different paths and interfere after several elastic scattering events. The elastic mean free path, ℓ, is of order 1 - 10 nm for the samples discussed here.

Constructive interference can occur between electron partial waves following different trajectories as long as phase coherence is preserved. Elastic scattering does <u>not</u> destroy phase coherence. Instead, phase coherence is destroyed by ***phase-breaking events*** which affect the energy-time evolution of the electron wavefunction. These phase breaking events are due to electron-electron scattering and electron-phonon scattering. At low temperatures, these phase-breaking events are rare, and phase-coherence is maintained over relatively long distances. This phase coherence length is $\ell_\phi = \sqrt{D\tau_\phi}$, where D is the electronic diffusion constant ($D = 1/3\, v_F \ell$ with v_F the Fermi velocity) and τ_ϕ is the phase coherence time. ℓ_ϕ is the characteristic length scale determining dimensionality for the weak localization effect as well as for Maki-Thompson superconducting fluctuations [1]. ℓ_ϕ also sets the size scale for the observation of Aharonov-Bohm oscillations in rings. ℓ_ϕ can be several microns at low temperatures.

Spin-orbit scattering can have an important effect on quantum interference phenomena in metallic systems [2]. The rotation of the electron spin due to the spin-orbit interaction causes the coherent backscattering of electrons to result in destructive interference [3]. The characteristic length scale for spin-orbit scattering is $\ell_{so} = \sqrt{D\tau_{so}}$. In most metals, ℓ_{so} may be comparable to or shorter than ℓ_ϕ.

Interactions between electrons are governed by the electron thermal diffusion length, $\ell_T = \sqrt{\hbar D/k_B T}$. This is the distance over which two electrons that differ by an energy $k_B T$ are correlated [1]. ℓ_T may range from ~ 10 to 100 nm at 1 K for moderately disordered metals. ℓ_T is the length scale determining dimensionality for the electron-electron scattering which gives phase breaking.

2.2 Weak Localization

Weak localization results from the interference of electron partial waves traveling along time-reversed trajectories. The localization contribution to the resistance as a function of temperature and magnetic field for a one-dimensional wire was first calculated for the case of no spin-orbit scattering [5]. This result is

$$\frac{\Delta R^{Loc}}{R} = \frac{R_\square}{W\pi\hbar/e^2}\left[\ell_\phi^{-2} + \ell_H^{-2}\right]^{-1/2} . \tag{1}$$

$R_\square$ is the sheet resistance of the sample and $\ell_H = \sqrt{3}\hbar c/eHW$, with W the sample width. The criterion for the validity of (1) is that $H < 12.5H_W$, where $H_W = \hbar c/4eW^2$ [6].

Equation (1) is the *pure 1D* result, for $W << \ell_\phi$. Evaluation of the leading correction [3] to (1) for ℓ_ϕ of finite size indicates that this correction becomes significant when $\ell_\phi \leq W$ for the field ranges used in typical experiments. Thus, the condition for the applicability of (1) is $\ell_\phi > W$. The claim that the pure 1D result, (1), applies whenever $\ell_\phi > W/\pi$ is only correct at very small magnetic fields, much smaller than are utilized in typical magnetoresistance measurements.

The 1D localization result was extended by SANTHANAM ET AL. [6] to include the effects of spin-orbit scattering. In this case, there are two terms:

$$\frac{\Delta R^{Loc}}{R}(T,H) = \frac{R_\square}{W\pi\hbar/e^2}\left\{\frac{3}{2}\left[\ell_2^{-2} + \ell_H^{-2}\right]^{-1/2} - \frac{1}{2}\left[\ell_\phi^{-2} + \ell_H^{-2}\right]^{-1/2}\right\} . \tag{2}$$

Here, ℓ_2 is a composite length given by

$$\ell_2^{-2} = \ell_\phi^{-2} + (4/3)\ell_{so}^{-2} . \tag{3}$$

The first term in (2) is referred to as the triplet term. The second term is the singlet term; it is not affected by spin-orbit scattering. In the absence of spin-orbit scattering, (2) reduces to (1). The dimensionality of a sample with respect to the triplet term and the singlet term is determined by the characteristic length scales ℓ_2 and ℓ_ϕ, respectively. In the presence of magnetic scattering, $\ell_\phi^{-2} \to \ell_\phi^{-2} + 2\ell_s^{-2}$ and $\ell_2^{-2} \to \ell_\phi^{-2} + (4/3)\ell_{so}^{-2} + (2/3)\ell_s^{-2}$, where ℓ_s is the magnetic scattering length [3].

2.3 Superconducting Fluctuations

For superconductors at temperatures not too close to T_c ($T/T_c \gtrsim 1.3$), the most significant quantum correction is due to Maki-Thompson (MT) fluctuations [7]. The 1D Maki-Thompson contribution is given by [6]

$$\frac{\Delta R^{MT}}{R}(T,H) = -\beta(T/T_c)\frac{R_\square}{W\pi\hbar/e^2}\left[\ell_\phi^{-2} + \ell_H^{-2}\right]^{-1/2} \quad ; \qquad (4)$$

$\beta(T/T_c)$ is the Larkin parameter which characterizes the strength of the interaction between electrons [8]; it is independent of sample dimensionality. Equation (4) is valid for $H << (ck_BT/4eD)\ln(T/T_c)$ and $\hbar/\tau_\phi << (k_BT)\ln(T/T_c)$. ℓ_ϕ is the dimensional length scale for the MT term.

2.4 Aharonov-Bohm Oscillations in Rings

In a multiply connected structure such as a ring, the quantum interference contribution to the resistance may be dominated by those electron trajectories which encircle the ring. Resistance oscillations periodic in the magnetic flux can arise from two effects. The first effect, due to the interference of electron partial waves traversing time-reversed trajectories completely around the ring, leads to oscillations of period h/2e. These oscillations, first predicted by ALTSHULER ARONOV, AND SPIVAK (AAS) [9], are related to the weak localization effect. The second type of oscillatory effect in rings is due to non-time-reversed electron trajectories which interfere at the exit of the ring. These oscillations have a fundamental period of h/e [10]. They arise from the same type of interference as the conductance fluctuations [11]. Observation of either type of oscillation in a ring requires that the size of the ring be not much greater than ℓ_ϕ. In addition, the magnitude of the h/e oscillations is reduced by a factor ℓ_T/L, where L is half the ring circumference [11]. Thus, the ring dimension should also not greatly exceed ℓ_T.

2.5 Phase Breaking Mechanisms

The electron phase breaking rate is given by

$$\tau_\phi^{-1} = \tau_{ee}^{-1} + \tau_{ep}^{-1} \quad ; \qquad (5)$$

where τ_{ee}^{-1} and τ_{ep}^{-1} are the electron-electron and electron-phonon scattering rates, respectively. The electron-phonon scattering rate has a strong temperature dependence, $\simeq T^3$ [3], so that at the lowest temperatures electron-electron scattering makes the dominant contribution to phase breaking.

ALTSHULER ET AL. [12] pointed out that in moderately disordered conductors, there are *two* types of electron-electron scattering which can lead to loss of phase memory. The first type, strongly inelastic scattering, involves collisions with energy transfers on the order of the electron thermal excitation energy ($\Delta E \sim k_BT$). The second type, quasielastic scattering, is due mainly to collisions with small energy transfers ($\Delta E << k_BT$). This latter type is equivalent to the scattering of electrons by thermal electromagnetic fluctuations (Nyquist noise) [13]. For 1D wires, the quasielastic rate is predicted to be much larger than the large-energy-exchange rate. A wire is considered 1D with respect to electron-electron scattering if its lateral dimensions are less than ℓ_T. In this 1D case, the electron-electron scattering rate is given by

$$\tau_N^{-1} = \left\{ \frac{R_\square k_B}{(\hbar/e^2)\hbar} \left[\frac{D}{2} \right]^{1/2} \frac{T}{W} \right\}^{2/3} . \tag{6}$$

For wires of width greater than ℓ_T the electron-electron scattering is two-dimensional. The 2D scattering rates due to quasielastic scattering and due to large energy exchange scattering have the same temperature dependence, $\propto T$, and are similar in magnitude [13,14]. Thus, it is in 1D systems that the predictions regarding the quasielastic electron-electron scattering rate can be most clearly tested.

3. Sample Fabrication and Measurement

The samples studied in this work were narrow wires and small rings defined by electron-beam lithography, and by x-ray lithography for wider wires. Thin films of Al, Ag and Au were deposited onto oxidized Si wafers by thermal evaporation. 2D thin film samples were also deposited along with the wires and rings for comparison of measured quantities. Film thicknesses were ~ 20 nm. Wires were either 50 or 220 μm long with widths ranging from 35 - 600 nm. Rings were 1 or 2 μm in diameter, with linewidths of ~ 60 - 200 nm. Film sheet resistances were ~ 1 - 5 Ω.

Resistance measurements were done using either a three-terminal or a four-terminal ac bridge at 1.2 - 20 K in a perpendicular magnetic field of up to 5 kG. With measuring currents low enough to avoid self-heating, a resolution of $\Delta R/R$ better than 10^{-6} was achieved.

4. Experimental Results

4.1 Narrow Wires

Figure 1 shows the magnetoresistance of a 40 nm wide Al wire at three representative temperatures. The smooth curves are fits by the theory for 1D localization with spin-orbit scattering and 1D Maki-Thompson superconducting fluctuations. Values of ℓ_ϕ and ℓ_{so} inferred from the fits indicate that this wire is clearly in the regime of one-dimensional behavior.

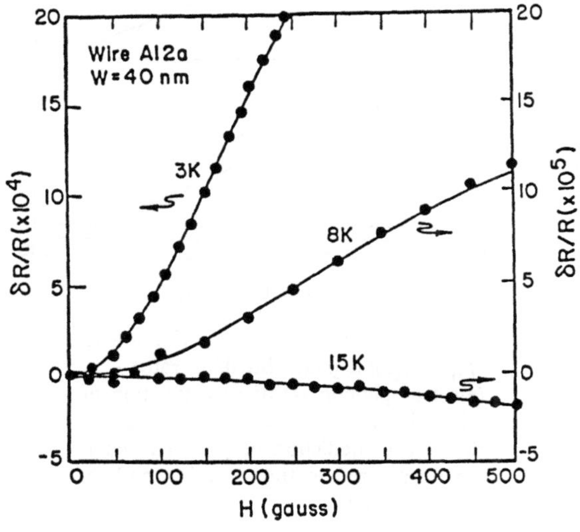

Fig. 1. Normalized magnetoresistance, $\delta R/R$, of a 40 nm wide Al wire at three temperatures; $\delta R(T,H) = \Delta R(T,H) - \Delta R(T,0)$. Fitting parameters for the 1D theory are $\ell_\phi = 1.19$ μm at 3 K, 0.52 μm at 8 K and 0.24 μm at 15 K; $\ell_{so} = 0.55$ μm at all temperatures.

It is instructive to examine the relative importance of the individual contributions to the magnetoresistance. At low temperatures the magnetoresistance is dominated by Maki-Thompson fluctuations. At higher temperatures the triplet term is the largest component of the magnetoresistance, as the prefactor of the triplet term is large, and $\ell_\phi < \ell_{so}$. The Maki-Thompson term remains significant well above T_c.

The phase-breaking rate inferred from the theoretical fits to the magnetoresistance data is plotted in Fig. 2 for several wires of width less than ℓ_T. The solid lines for the wires are fits by the form

$$\tau_\phi^{-1} = A_{2/3}T^{2/3} + A_3T^3 , \tag{7}$$

corresponding to a combination of 1D electron-electron scattering and 3D electron-phonon scattering. The coefficients $A_{2/3}$ and A_3 are in good agreement [15] with theoretical predictions for 1D quasielastic scattering, (6), and with clean limit electron-phonon scattering [16]. The rate found in the codeposited 2D film is also shown in the figure. The solid line for this film is a fit to the form $A_1T + A_3T^3$, with 2D electron-electron scattering. At high temperatures, electron-phonon scattering makes the dominant contribution to τ_ϕ^{-1}. The electron-phonon T^3 contribution is essentially the same for the film and the wires, and is consistent with other studies on Al wires and films [7,17,18]. At low temperatures, electron-electron scattering dominates. This electron-electron rate is seen to increase with decreasing wire width, as predicted by (6); it is significantly greater than the rate found for the 2D film. This is in contrast to the rate found in wires wider than ℓ_T [17], which showed 2D electron-electron scattering rates identical to the rates found in codeposited 2D films. A further test of the 1D electron-electron prediction, (6), is the width dependence of τ_ϕ^{-1} at fixed temperature. This is shown in Fig. 3, where the electron-phonon rate has been

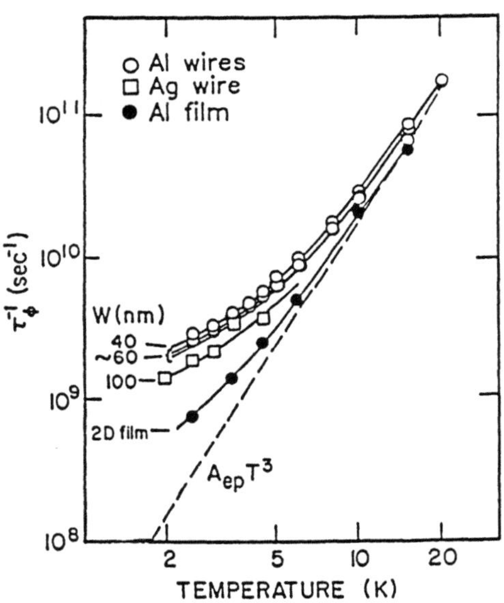

Fig. 2. Phase breaking rate vs. temperature. The solid lines are fits by (5), as explained in the text. The data for the Ag wire (W = 100 nm) from 2 to 4.5 K are normalized to the $R_\square$ and D of the Al wires according to (6). The dashed line plots the electron-phonon rate $\propto T^3$.

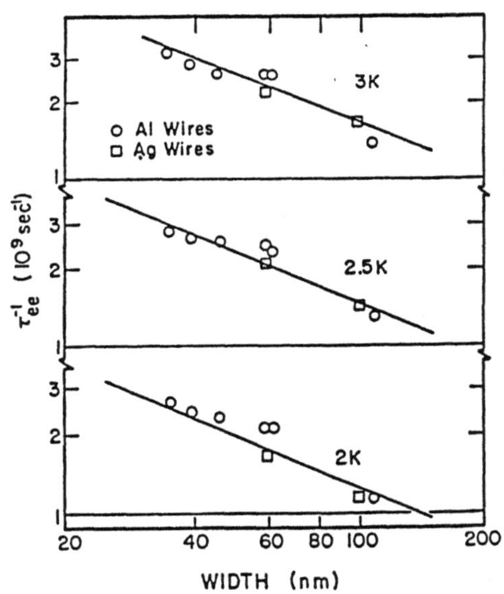

Fig. 3. Electron-electron contribution to τ_ϕ^{-1} = [(total phase breaking rate) - (electron-phonon rate)] as a function of wire width. The solid lines give the theoretical prediction of (6). The data are normalized to the same $R_\square$ and D, according to (6).

subtracted from the total inferred phase breaking rate. The solid lines correspond to the theoretical magnitude and width dependence directly from (6), with no adjustable parameters. The width dependence predicted by the quasielastic theory is seen clearly.

4.2 Aharonov-Bohm Oscillations in Rings

Studies have been carried out of the oscillatory magnetoresistance in rings of Al, Ag, Ag with Co impurities, and Au. For the pure metals, we find AAS h/2e oscillations at low fields, and h/e oscillations at high fields [20]. An example of the h/e oscillations is shown in Fig. 4 for a 1 μm Al ring at 1.5 K. Our earliest measurements had failed to observe these oscillations for Al rings. The present measurement system is an improved version.

A second significant absence was found in experiments on Au rings at IBM [21]. Here, the low-field AAS oscillations, and the low-field background magnetoresistance due to localization were absent. Magnetic impurities in those Au films might explain this absence. We have recently studied Au rings in which the expected AAS h/2e oscillations and localization magnetoresistance are observed. Similarly, AAS oscillations are seen in large arrays of Au rings [22]. The different properties observed at IBM are presumably due to different material parameters.

The absence of AAS oscillations reported for Au rings [21], discussed above, did lead us to study the effects of magnetic scattering on ℓ_ϕ, and whether ℓ_ϕ is the same for time-reversed interference effects (AAS oscillations and localization) as for non-time-reversed interference (h/e oscillations). The effect of magnetic impurities was studied with Ag rings that were undoped, or doped with Co. Magnetoresistance data for two Ag rings are shown in Fig. 5. The rings are of similar material properties, with the exception that the ring in Fig. 5b had ~ 0.01 atomic layer of Co deposited on its surface. As can be seen from the figure, low-field h/2e oscillations are present in the sample with no added magnetic impurities, but are absent in the sample with added magnetic impurities. The value of the phase-breaking length $\ell_\phi^{eff} = (\ell_\phi^{-2} + 2\ell_s^{-2})^{-1/2}$ can be inferred from the weak localization background magnetoresistance. For the Ag ring, this value of ℓ_ϕ^{eff} is consistent with the amplitude of the low-field AAS h/2e oscillations, and with the magnitude of the h/e oscillations at both high and low fields. For the Ag+Co ring, ℓ_ϕ^{eff} as inferred from the localization background is smaller, and this smaller value is consistent with the absence of AAS h/2e oscillations. Using this value of ℓ_ϕ^{eff} for the Ag+Co ring, one would predict that the h/e oscillations should be a factor of ≃10 smaller for this ring than for the Ag ring. However, the measured

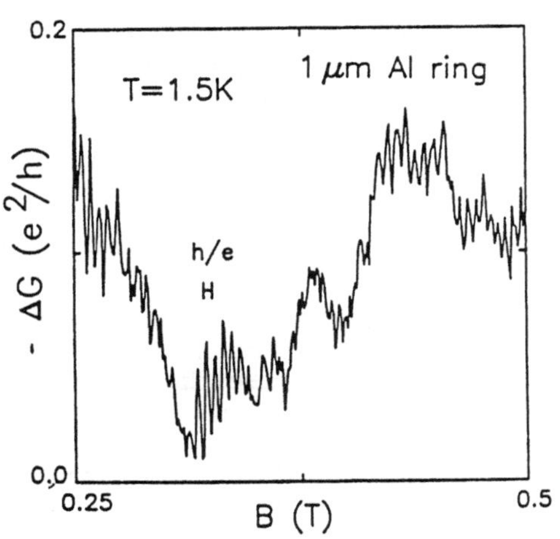

Fig. 4. Magnetoresistance of a 1μm Al ring at 1.5 K showing h/e oscillations at high magnetic fields. Early measurements on Al rings had failed to observe these oscillations.

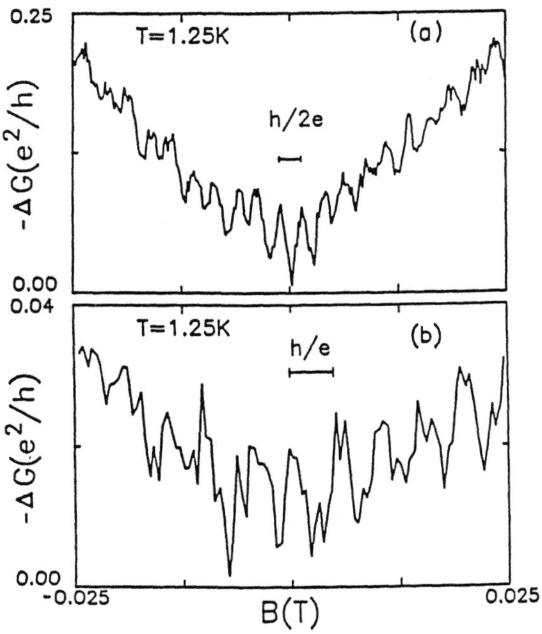

Fig. 5. (a) Magnetoresistance at low fields of a Ag ring with no added magnetic impurities showing clear h/2e oscillations. (b) Low field magnetoresistance of a similar Ag ring with ~ 0.01 atomic layers of Co on its surface. The h/2e oscillations are absent in this ring. The magnitude of the h/e oscillations is nearly the same for both rings.

magnitude of the h/e oscillations is nearly the same for the two rings. This is shown in Fig. 5 for the low-field region. Thus, the roles played by the phase-breaking length for the time-reversed h/2e oscillations and the non-time-reversed h/e oscillations are apparently not the same. Further work will be necessary to determine the relevant length scales for these two effects. We note that our results contrast with the results of similar measurements on Au rings [21], where added magnetic impurities were seen to suppress the low-field h/e oscillations.

We have also compared values of ℓ_ϕ for the time-reversed localization magnetoresistance in long Al wires to the values of ℓ_ϕ for the localization magnetoresistance in rings. The ℓ_ϕ values measured in long wires is shown in Fig. 2. At high temperatures (>6K) the values of ℓ_ϕ for the ring are the same as for long wires. However, at the lowest temperatures, the values of ℓ_ϕ for rings are approximately 0.6 those of long wires. This may be due to the effects of the leads being spaced within ℓ_ϕ of the ring. However, a reasonable estimate of the effect of the leads [23] would give only a small correction of the low-temperature values of ℓ_ϕ inferred for the rings, so that the corrected low-temperature values of ℓ_ϕ for the rings are still only ≈0.7 of those for the long wires. At present it is not clear if ℓ_ϕ is actually reduced in the ring structure, or if the correction formulae used [23] are not fully correct.

5. Conclusions

The effects of electron quantum interference were studied in two types of geometries of reduced dimensionality. In narrow wires of width less than ℓ_ϕ, quantum interference leads to 1D weak localization, which has been utilized as a tool for measuring electron phase breaking rates. For wires which are narrower than ℓ_T, the dominant phase breaking mechanism at low temperatures is one-dimensional electron-electron scattering with small energy transfers. Comparison of these rates with those of wires wider than ℓ_T confirms that ℓ_T is indeed the relevant length scale determining dimensionality for this mechanism. Quantum interference in rings leads to oscillations of period h/2e and h/e. We have observed both types of oscillations in individual rings and have found that the conditions for their observation are apparently not the same.

We thank P. Santhanam, A. D. Stone and R. G. Wheeler for useful discussions and M. J. Rooks for assistance with electron-beam lithography. This work was supported by NSF Grants DMR-8505539 and DMR-8207443. Support of the microlithography facilities at Yale was provided by NSF Grant No. DMR-8213080, IBM Corp., Hewlett-Packard, and Shipley, Inc.

References

1. B.L. Altshuler, A.G. Aronov, D.E. Khmelnitskii and A.I. Larkin, in Quantum Theory of Solids, ed. by I.M. Lifshitz (Mir, Moscow, 1982).
2. S. Hikami, A.I. Larkin and Y. Nagaoka, Prog. Theor. Phys. 63, 707 (1980).
3. G. Bergmann, Phys. Rep. 107, 1 (1984).
4. D.J. Thouless, Phys. Rev. Lett. 39, 1167 (1977).
5. B.L. Altshuler and A. G. Aronov, Pis'ma Zh. Eksp. Teor. Fiz. 33, 515 (1981) [JETP Lett. 33, 499 (1981)].
6. P. Santhanam, S. Wind and D.E. Prober, in Proceedings of the Seventeenth International Conference on Low Temperature Physics, Karlsruhe, Federal Republic of Germany, Part I, p. 495 (North Holland, Amsterdam, 1984).
7. P. Santhanam, S. Wind and D.E. Prober, Phys. Rev. B 35, 3188 (1987).
8. A.I. Larkin, Pis'ma Zh. Eksp. Teor. Fiz. 31, 239 (1980); [JETP Lett. 31, 219 (1980)].
9. B.L. Altshuler, A.G. Aronov and B.Z. Spivak, Pis'ma Zh. Eksp. Teor. Fiz. 33, 101 (1981) [JETP Lett. 33, 94 (1981)].
10. M. Buttiker, Y. Imry, R. Landauer and S. Pinhas, Phys. Rev. B 31, 6207 (1985).
11. P.A. Lee, A.D. Stone and H. Fukuyama, Phys. Rev. B 35, 1039 (1987).
12. B.L. Altshuler, A.G. Aronov and D.E. Khmelnitskii, Sol. St. Commun. 39, 619 (1981).
13. B.L. Altshuler, A.G. Aronov and D.E. Khmelnitskii, J. Phys. C 15, 7367 (1982).
14. E. Abrahams, P.W. Anderson, P.A. Lee and T.V. Ramakrishnan, Phys. Rev. B 24, 6783 (1981); J.M.B. Lopes dos Santos, Phys. Rev. B 28, 1189 (1983).
15. S. Wind, M.J. Rooks, V. Chandrasekhar and D.E. Prober, Phys. Rev. Lett. 57, 633 (1986).
16. W.E. Lawrence and A.B. Meador, Phys. Rev. B 18, 1154 (1978).
17. P. Santhanam, S. Wind and D.E. Prober, Phys. Rev. Lett. 53, 1179 (1984).
18. P. Santhanam and D.E. Prober, Phys. Rev. B 29, 3733 (1984).
19. V. Chandrasekhar, M.J. Rooks, S. Wind and D.E. Prober, Phys. Rev. Lett. 55, 1610 (1985).
20. V. Chandrasekhar and D.E. Prober, Proceedings of LT-18 Conference, Kyoto, Japan.
21. S. Washburn and R.A. Webb, Adv. Phys. 35, 375 (1986).
22. B. Pannetier et al., Phys. Rev. Lett. 53, 718 (1984).
23. P. Santhanam, Phys. Rev. B35, 8737 (1987).

Four-Wire Measurement of Conductance Fluctuations in Small Systems

A. Benoit, S. Washburn†*, C.P. Umbach*†*, R.A. Webb*†*, D. Mailly***, and L. Dumoulin***

Centre de recherche sur les très basses températures CNRS,
25 Av. des Martyrs, F-38042 Grenoble, France

The electrical resistance of small normal-metal rings has been observed to oscillate as a function of magnetic field with period h/e. This is a clear manifestation of the Aharonov-Bohm effect in a disordered electrical conductor. In addition, the magnetic field penetrating the wire results in a randomly fluctuating magnetoresistance. Due to the four-lead experimental configuration, the resistance is not symmetric about H=0. We also show that the measured voltage fluctuation (or oscillation) is independent of the probe position and the measured conductance fluctuation is not universal. This is also true if there is no classical current flow between the voltage probes (non-local effect). Finally, we discuss the effect of magnetic scattering. For paramagnetic impurities, we can separate a low-field region with very short coherence length and a high-field region where the impurities do not affect the oscillations.

It has been shown that in the limit where the phase-coherence length L_ϕ of the electrons is comparable to the sample size L, the conductance of a normal metal wire exhibits sample-specific[1] fluctuations as a function of magnetic field. The same quantum interference behavior induces in a ring, oscillations of period h/e in a magnetic field[2]. Theoretical calculations have shown that the rms value $<\Delta G>$ of the fluctuations (or the oscillations) is "universal"[3,4] in the limit of low temperatures and small sample sizes ($L \approx L_\phi$). In the case of high temperatures or long sample sizes, the experimental results agree with the theory[5-7]. In the small sample limit ($L<L_\phi$), the theory describes a sample connected to two perfect reservoirs. In an experiment, in contrast, we always use a four-wire technique. We discuss first here the new behavior induced by this sample arrangement. The second part of this paper is dedicated to the effect of magnetic impurities on fluctuations and oscillations.

One surprising property observed experimentally is that the magnetoresistance is not symmetric upon reversal of the magnetic field[2,8]. This is apparently in contradiction with fundamental symmetries[9]. By measuring the resistivity for all possible permutations of the four leads on the sample, we can prove that this asymmetry does not violate the

---

† IBM T.J Watson Research Center, P.O. Box 218, Yorktown Hts., NY 10598.

* Laboratoire de microstructures et de microelectronique, CNRS 196 av H.Ravera 92220 Bagneux, FRANCE.

** CSNSM bât. 108 Université Paris-Sud, Orsay, FRANCE

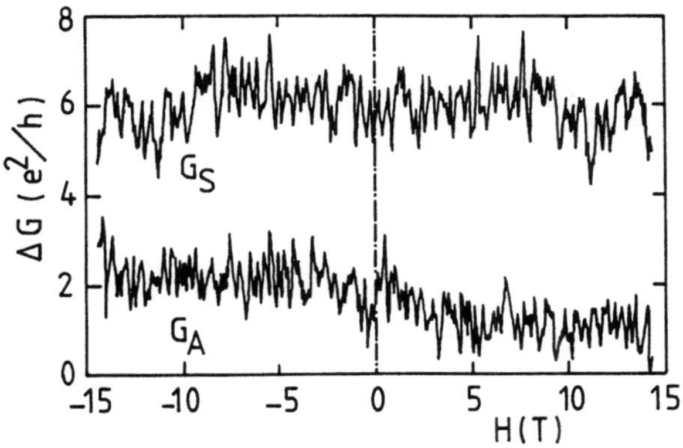

FIG. 1. Magnetoconductance of a gold sample in units of e^2/h: G_S is the sum of two measurements obtained by exchange of current and voltage leads; G_A is the difference of the same two measurements.

fundamental symmetries but is due to the four wire arrangement[10]. The conductance fluctuation G_S in Fig.1 is obtained by summing the two resistances measured after exchange of the current and voltage leads. The curve is perfectly symmetric upon reversing the magnetic field, as expected from the theory[11]. The curve G_A shows the antisymmetrical part, obtained by subtraction of the same two measurements. We observe that the intensity of the fluctuations of the antisymmetric and the symmetric parts of the magnetoconductance are comparable and of order of e^2/h. We deduce that the antisymmetric part is not only produced by the connection of the wires as in the classical Hall effect, but arises from all parts of the sample within one coherence length L_ϕ. The same effect is observed for the oscillations in a ring[10]. It is important to note that there is no hysteresis even after high field reversal. This property rules out an explanation of the asymmetry by magnetic impurities.

The effect of the four-wire measurement becomes particularly important when the dimensions of the sample are smaller than the coherence length L_ϕ. In this case, we can measure[7,12] fluctuations of conductance bigger than the universal value e^2/h. This can be understood by considering the measured fluctuations between two voltage leads with a fixed current into the current leads. In this geometry, we can prove[12] that the antisymmetric part of the fluctuations is independent of the length of the sample (represented by the open symbols in Fig. 2). The symmetric part follows the same behavior for dimensions smaller than the coherence length ($L/L_\phi<1$) but for long samples present a $\sqrt{L}$ variation, explained by classical addition of independent region in the sample. The constant value of the voltage fluctuations for $L<L_\phi$ agree with recent calculations. In the case where there is no current flow between the voltage leads, we do not measure any average voltage but fluctuations and oscillations are the same.

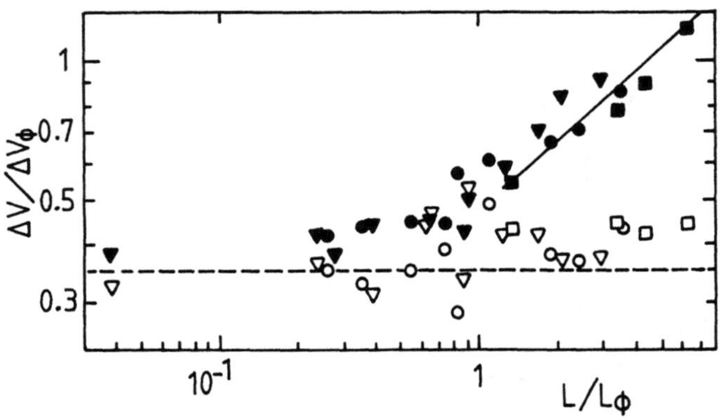

FIG 2. Intensity of measured voltage fluctuations in magnetic field as a function of the distance between voltage probes along a wire: solid and open symbols represent respectively the symmetric and antisymmetric contributions, full and dashed lines represent the corresponding expected behavior.

In contrast with the case of weak-localisation, the magnetic field does not destroy the fluctuations and the oscillations of conductance. The effect of magnetic impurities can be studied in different magnetic fields. We must take into account here that a change in the phase of the electrons does not destroy the interferences but modifies the resulting pattern. However, if the impurity has some degree of freedom, as in the paramagnetic state, the phase of two electrons following the same path at different times can be different. This process causes the destruction of the interferences and in this case, all the conductance fluctuations and oscillations disappear. The corresponding phase-breaking cross-section must be proportional to the probability of an impurity being at two different times in the states up and down. If P_u and P_d are the up and down Boltzmann probabilities for a 1/2 paramagnetic spin, we obtain

$$S = A \times P_u \times P_d = \frac{A}{ch^2(\, g\mu H/2kT\,)} \quad , \tag{1}$$

where A is coupling constant, g the Lande factor, μ the Bohr magneton, H the magnetic field and T the temperature of the impurities. Now, we can compute the equivalent phase breaking time $t_{ph} = t_m/S$ where t_m is the typical time spent by one electron between two magnetic impurities. If v_f is the velocity of the electrons, t_m is given by $t_m = a^3/cd^2v_f = la^3/d^2Dc$, where D is the diffusion constant, l the mean free path, c the impurity concentration, a the atomic spacing and d a characteristic length for magnetic interaction of the order of a few atomic spacings. The phase-breaking length L_ϕ is obtained by

$$L_\phi = \sqrt{D \times t_{ph}} = \sqrt{\frac{l\, a^3}{d^2 S c}} \quad .$$

This expression is valid only when the magnetic effect dominantes the coherence length L_ϕ. This expression allows us to compute the intensity of the conductance fluctuations and h/e oscillations as a function of magnetic field. The intensity of the oscillations is given by the exponential law $\Delta G=B \exp(-L/L_\phi)$, where B is a constant of order e^2/h and L is half the circumference of the ring. We can introduce the effective number N of impurities interacting with an electron during one turn around the loop:

$$\Delta G=B \exp\left(-\sqrt{N}\, ch^{-1}\left(\frac{g\mu H}{2kT}\right)\right) \qquad \text{with} \qquad N=\frac{AcL^2d^2}{l\, a^3} \quad . \quad (3)$$

On increasing the magnetic field, this function presents a step from 0 to its saturated value. The width of this step H_t is simply the field corresponding to the temperature of the impurities ($g\mu H_t=kT$) but the position in the magnetic field is given by $H_c=\mathrm{Log}(N)\times H_t$. The value of N represents a typical number of impurities, and generally, we have $H_c>>H_t$.

If we use a dilute magnetic alloy, in order to respect our condition of free paramagnetic centers, the Kondo temperature T_k and the spin glass transition temperature T_f must be low in comparison to the experimental temperature. For AuMn alloy, the first condition is easily respected as T_k<1mK. The second condition requires a rather dilute sample. We measure down to 160 mK, the magnetoresistance of 40 ppm and 120 ppm AuMn ring of 0.5 μm^2 area. The manganese is introduced by ion implantation at 65 keV on a sample ring previously obtained by electron beam lithography. The impurity concentration was checked by measuring the low-temperature Kondo anomaly in the resistivity of gold film of the same thickness. For the two samples, the oscillations (with a period of h/e) and the fluctuations observed in the high-field region disappear at low field. Figure 3 represents the intensity of the conductance oscillations measured by Fourier analysis, as a function of magnetic field between 0 and 2 tesla. The full and dashed lines represent the calculation, adjusting the total intensity and using the

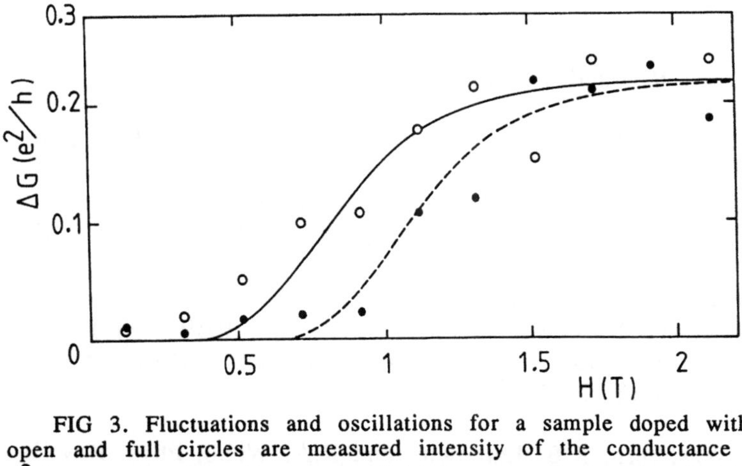

FIG 3. Fluctuations and oscillations for a sample doped with manganese impurities: open and full circles are measured intensity of the conductance oscillations in units of e^2/h as a function of magnetic field for respectively 40 and 120 ppm, full and dashed lines are calculation with N=600 and 6000.

values of 600 and 6000 for the effective number N in equation (3). The calculation is done for a 1/2 spin, assuming that only the two lowest Zeeman levels of the manganese spin 5/2 are occupied. An estimation of N can be obtain from the concentration of manganese, using a mean free path of 10 nm and a characteristic interaction length $d=5\times a$. The result N=500 and 1500 for the respective concentrations 40 and 120 ppm is in relatively good agreement with the data but a quantitative comparison is very difficult without any detailed microscopic description of the interaction. Furthermore, deviations may occur because of the nearness of the spin glass transition.

Another technique to introduce magnetic centers consists in depositing some impurities on the surface of the sample. In this case, using directly a magnetic metal is very dangerous because it can produce clusters at the surface. We use the well-known cerium-magnesium-nitrate paramagnetic salt (magnetic ordering at a few millikelvin) and the deposition is done simply by dipping the sample into an alcoholic dilute solution. In this case, at the measured temperature, the impurities are in a paramagnetic state without interactions. We must take into account the anisotropy of the magnetic properties by an integration over all possible orientations. The measured oscillations[13] are in complete agreement with the model presented here. In this experiment, we cannot easily change the concentration of salt at the surface and it is not possible to check the concentration dependence of the model.

We have here an experimental technique very sensitive to any spin change in the sample. Measurements at lower concentrations are needed to know if one unique reversed magnetic spin is enough to change completely the measured conductance fluctuation. In the case of a spin glass at low temperature, we can have very long spin relaxation times, and with this technique we hope to observe the individual flipping of some groups of spins.

1 M. Büttiker, Y. Imry, R. Landauer, and S. Pinhas, Phys. Rev. B 31, 6207 (1985).

2 A. Webb, S. Washburn, C.P. Umbach, and R.B. Laibowitz, Phys. Rev. Lett. 54, 2696 (1985).

3 P.A. Lee and A.D. Stone, Phys Rev. Lett. 55, 1622 (1985).

4 B.L. Al'tshuler, Pis'ma Zh. Eksp. Teor. Fiz. 41, 530 (1985) ;JETP Lett. 41, 648 (1985).

5 Y. Ymry, Europhys. Lett. 1, 249 (1986).

6 P.A. Lee, A.D.Stone, and H.Fukuyama, Phys. Rev. B 35, 1039 (1987).

7 W.J. Skocpol, P.M. Mankiewich, R.E. Howard, L.D. Jackel, D.M. Tennant and A.D. Stone, Phys. Rev. Lett. 55 2987 (1985).

8 C.P. Umbach, S. Washburn, R.B. Laibowitz and R.A. Webb, Phys. Rev. B 30, 4048 (1984).

9 L. Onsager, Phys. Rev. 38, 2265 (1931); H.B.G. Casimir, Rev. Mod. Phys. 17, 343 (1945).

10 A.Benoit, S. Washburn, C.P. Umbach, R.B. Laibowitz and R.A. Webb, Phys. Rev. Lett. 57, 1765 (1986).

11 M. Büttiker, Phys. Rev. Lett. 57, 1761 (1986).

12 A.Benoit, C.P. Umbach, R.B. Laibowitz and R.A. Webb, Phys. Rev. Lett. 58, 2343 (1986).

13 A.Benoit, D.Mailly, L.Dumoulin, C.P. Umbach, S. Washburn and R.A. Webb, to be published.

Universal Conductance Fluctuations in Mesoscopic n^+-Si Wires

K. Itoh[1], Y. Sasaki[1], S. Tanuma[1], Y. Iye[2], and T. Tamegai[2]

[1]Department of Electronics, Faculty of Technology, Gunma University, Kiryu, Gunma 376, Japan
[2]Institute for Solid-State Physics, University of Tokyo, 7-22-1 Roppongi, Minato-ku, Tokyo 106, Japan

Mesoscopic wires of arsenic doped n^+-Si were prepared by techniques of selective diffusion and anisotropic etching with no use of sophisticated technique of microfabrication. The effective thickness of wires was 0.05 to 0.14 μm. Low temperature magnetoconductance shows aperiodic oscillations superimposed on the overall field dependence due to the electron localization effect. The amplitude of the aperiodic oscillations was in a range of $5\times10^{-4}(e^2/h)$ to $1\times10^{-2}(e^2/h)$. The amplitude of oscillation for wires thinner than inelastic diffusion length ($\sim$ 0.1 μm) shows the length dependence as expected from the theory of the universal conductance fluctuation for a one-dimensional system.

1. Introduction

Since recent experimental observation of the reproducible aperiodic oscillation of magnetoconductance in small metal rings and wires [1], much attention has been paid to the transport properties of such mesoscopic systems that their sizes are between macroscopic scales and atomic ones. It has been clarified theoretically that the observed strange fluctuation of the conductance is due to the quantum interference effect between elastically scattered electrons and has an universal amplitude of e^2/h at zero temperature [2-4]. This quantum interference effect has also been observed in other mesoscopic systems of degenerate semiconductors such as narrow channel Si-MOSFET's [5-7] and n^+-GaAs wires [8-10].

To observe the conductance fluctuation, the wire size should be as small as the phase coherent length L_ϕ of electron waves. Since this length is typically 0.1 to 1 μm at low temperatures, an advanced microfabrication technology is generally required to fabricate mesoscopic systems of such scale.

Heavily doped silicon has been one of the important systems for studying the electron localization effect in the sense that the carrier concentration as well as the disordedness are highly controllable, including various degenerate states. The authors have prepared thin arsenic-doped n^+-Si wires of whose thicknesses were comparable to or less than the inelastic diffusion length L_{in} in order to study the quantum transport properties of heavily doped Si with low dimensions.

For preparing very thin wires of heavily doped n^+-Si, the authors developed a new method with no use of sophisticated microfabrication such as electron beam lithography. This paper presents the fabrication method and the magnetoconductance fluctuations observed in a thus prepared mesoscopic system.

2. Sample Preparation and Experiments

Thin n^+-Si wires were fabricated by techniques of selective diffusion and anisotropic etching. When impurity atoms were diffused into the silicon substrate through the windows cut in a diffusion-preventing mask of SiO_2, they, of course, penetrated only onto the bare region of Si surface, except the edge of SiO_2 mask where the diffusion front should get around under the masking roof almost the same distance as the penetration depth in the bare region. This situation is shown in Fig.1 [11]. Thus, if the diffused layer of uncovered Si surface can be removed leaving the diffused part under the masking roof, we can get thin impurity-diffused "wires" along the edge line of SiO_2 masking roof. As a method to remove the diffused layer in the uncovered region, we employed the anisotropic etching using alkaline etchant. The etching rate of this etchant is much smaller for {111} crystallographic planes of Si than for {100} planes. In addition a very thin SiO_2 film is sufficient to prevent the etching reaction. So if the substrate is chosen to have {100} surfaces and the mask edge line is aligned to <110> directions precisely, the etching does not proceed notably under the masking roof.

The schematic structure of an n^+-Si wire prepared by the above-mentioned procedure is shown in Fig.2. Material for making the sample in this experiment had a three-layer structure of polycrystalline Si(0.15 μm)/SiO_2/p-Si(100). Prior to the deposition of poly-Si film, the (100) Si surface was oxidized by heating at 1000°C in oxygen atmosphere to make SiO_2 coverage and square windows were cut in SiO_2 using a photolithography technique in such a way that the window frames were aligned to <110> directions. Poly-Si was deposited onto the SiO_2 windows in a thickness of about 0.15 μm. Thus prepared poly-Si surface was ion-implanted with arsenic atoms to a dose of 1.5×10^{16} cm^{-2} as the diffusion source. These triple-layered chips were annealed at 1100°C in dry O_2 atmosphere in order to diffuse As atoms into p-type substrate. Thin SiO_2 film was formed on the poly-Si surface during this annealing process. An array of dots of protecting wax was put on this SiO_2 film along one of the frame lines of window of the mask. Then this top surface was removed by buffered HF solution, to leave those dots of heavily doped poly-Si islands which served as electrodes. Alkaline etchant of hydrazine [12] was then applied for the anisotropic etching of diffused layers. A desired thickness of wire was obtainable by choosing appropriate conditions for the diffusion and etching processes. The size of prepared wires ranged from 0.05 to 0.14 μm in thickness and from 1.5 to 10.8 μm in length.

Figure 3 shows a scanning electron micrograph of an n^+-Si wire. There can be seen a well-defined (111) plane (region III) which is inclined to the (100) substrate surface (region II) by the angle of 55°. The conductive channel, called as "wire"

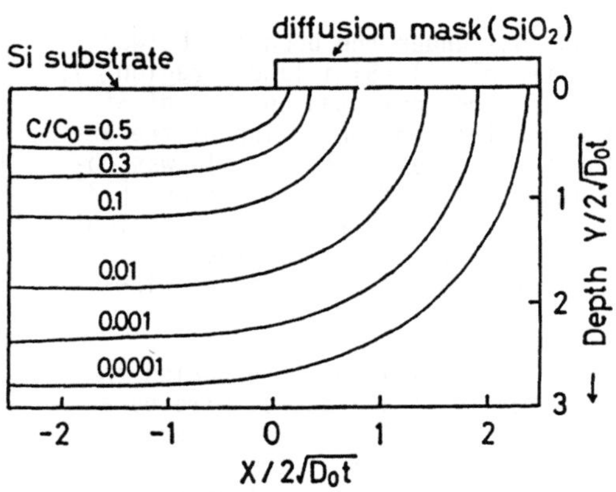

Fig.1 Calculated contours of constant impurity concentration C at the mask edge under the diffusion condition of the constant surface concentration C_0 and constant diffusion coefficient D_0. Distances are normalized with $2\sqrt{D_0t}$.

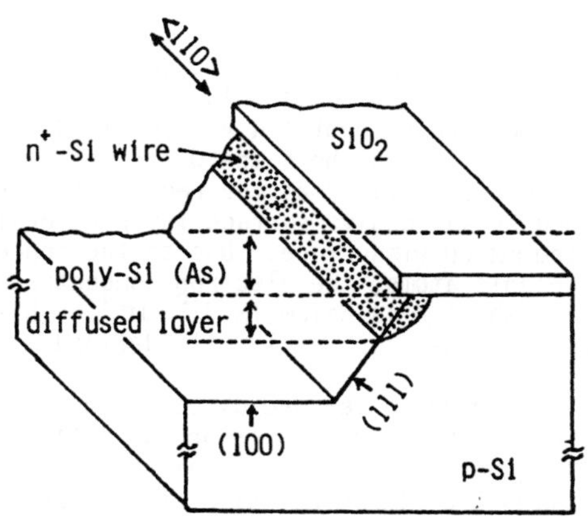

Fig.2 Schematic structure of an n^+-Si wire. Broken lines show the cross sectional structure of the sample before anisotropic etching.

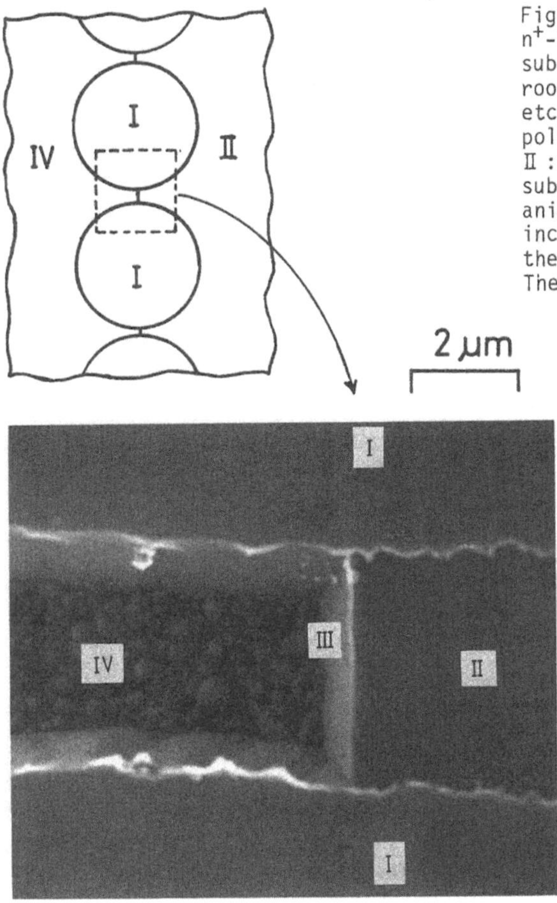

Fig.3 Scanning electron micrograph of an n^+-Si wire viewed from the top of the substrate surface. The SiO_2 masking roof shown schematically in Fig.2 was etched off. Region I: a part of the n^+ poly-Si islands left for electrode, II : unetched (100) surface of p-Si substrate, III: the side wall of the anisotropically etched region - the inclined (111) plane, IV: the bottom of the anisotropically etched region. The sketch shows the surrounding area.

in this paper which cannot be seen directly, runs just under the boundary between region II and III. This thin n^+ wire bridges the two n^+ poly-Si islands (region I). A pair of the current and voltage leads of Al wires were contacted on each n^+-Si island. Figure 4 shows the depth profile of the electron concentration of the arsenic diffused layer at the regions far from the frame edges. The profile was obtained by successive surface layer removal and measurement of the Hall effect [13] on the remaining surface. Due to the concentration-dependent diffusion coefficient of arsenic atoms, the electron concentration n drops very abruptly at the deeper region where n is below 1×10^{19} cm^{-3}. This is advantageous to form thin n^+-wires whose thickness is fairly well defined.

It is hard to measure directly the thickness of such a thin n^+ wire buried in the p-Si substrate. Therefore we estimated the effective thickness of wire by a square root of cross-sectional area $\sqrt{S}$ from its conductance G at room temperature assuming the simple relation of $S=G\cdot L/\bar{\sigma}$, where L is the length of the wire and $\bar{\sigma}$ is the bulk conductivity corresponding to the average electron concentration $\bar{n}$ [14]. The value of $\bar{n}$ was read as 2×10^{20} cm^{-3} or less from Fig.4 but taken to be 1×10^{20} modestly so as to estimate $\sqrt{S}$ not too small.

Although the current and potential leads were common at the n^+ poly-Si island, the quasi-four-terminal conductance of the sample was measured mainly by an ac method with the current in the range of 1×10^{-9} to 5×10^{-8} A to minimize the electron

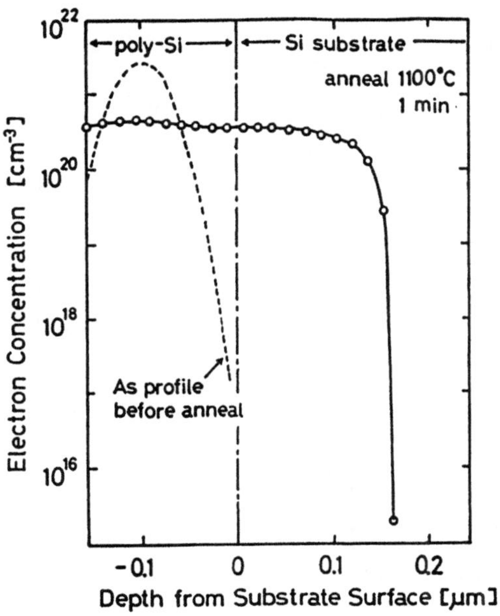

Fig.4 Depth profile of electron concentration of arsenic diffused layer at the region far from the mask edge.

heating. The magnetoconductance of the sample was measured in a ^{3}He cryostat down to the temperature of 0.4 K in the magnetic field range of 0 to 9 T.

3. Results and Discussions

Typical magnetoconductance per unit length along the wire is shown in Fig.5 at several temperatures for a sample whose length L is 7.3 μm and the effective thickness $\sqrt{S}$ is 0.08 μm. The transverse magnetic field is perpendicular to the (100) surface of the Si substrate. The magnetoconductance is positive and increasing up to about 5 T, and then turns to slightly decreasing at higher fields. Superimposed on this back ground magnetoconductance aperiodic oscillations are seen at low temperatures. The amplitude of the aperiodic oscillation increases with decreasing temperature. The trace of the oscillation can be recognized even at 20 K. The pattern of the oscillation is sample dependent but is reproducible unless each sample experiences a trip to room temperature. This phenomenon is the universal conductance fluctuation [1] due to the quantum interference effect caused by the scattering of electron waves by the randomly distributed scatterers (ionized impurities).

We estimate the inelastic diffusion length L_{in} of this system by comparing the magnetoconductance with the localization theory. The positive magnetoconductance of the wires showed no significant anisotropy with respect to the transverse field direction. This suggests that the system is one or three dimensional. In one dimension, the theoretical magnetoconductance [15] is given by

$$\Delta G(B)= G(B)-G(0)= \frac{e^2}{\pi \hbar L} [L_{in} - (\frac{1}{L_{in}^2}+\frac{1}{L_B^2})^{-1/2}], \quad (1)$$

where $L_B = \sqrt{3}\,\hbar/eBW$ is the magnetic length in the one-dimensional system [16], and L and W are length and width of the wire. Equation (1) is valid for $W < L_{in}$ and $B < B_m = h/2eW^2$. In Fig.6, the magnetoconductance at low field region for two thin

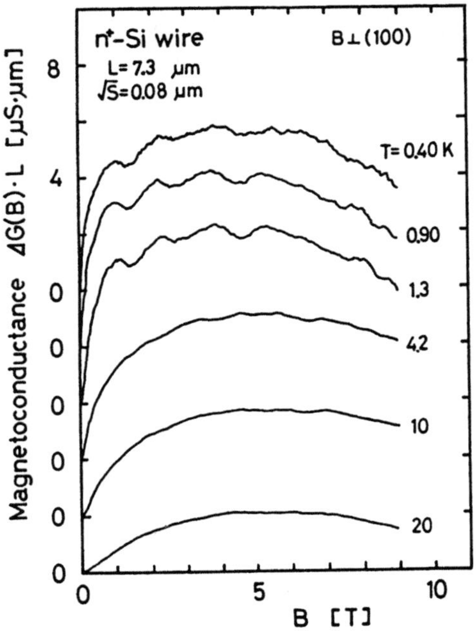

Fig. 5

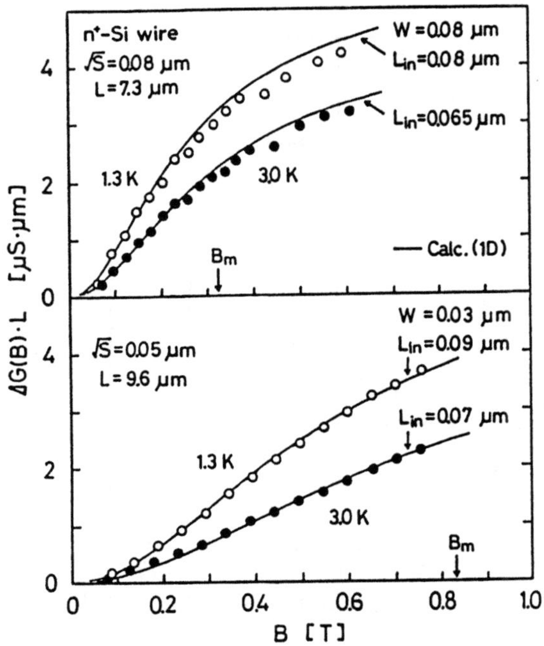

Fig. 6

Fig.5 Magnetoconductance per unit length of an n^+-Si wire at several temperatures. The wire length L is 7.3 μm and the effective thickness $\sqrt{S}$ is 0.08 μm.

Fig.6 Magnetoconductance per unit length of a wire for two thin n^+-Si wires whose effective thickness $\sqrt{S}$ are 0.05 and 0.08 μm at low field region. Solid curve indicates the calculated one using (1). The conductance per unit length of the wire at 1.3 K is 233 and 765 μS·μm for the wire $\sqrt{S}$=0.05 μm and for the wire $\sqrt{S}$=0.08 μm respectively

wires of which $\sqrt{S}$ is 0.05 and 0.08 μm are shown,together with the theoretical calculation (solid curve) using (1). Theoretical curves are fitted for low field region where $B < B_m$ for each wire with fitting parameter L_{in} and W. For the wire $\sqrt{S}$=0.05 μm, a smaller value of W=0.03 μm is obtained than $\sqrt{S}$, while for the wire $\sqrt{S}$=0.08 μm, the same value of W with $\sqrt{S}$ is obtained. An ambiguous estimation of $\sqrt{S}$ may be mainly responsible for this disagreement. The value of L_{in} obtained above fitting is 0.08 to 0.09 μm at 1.3 K for two wires and is comparable to or larger than $\sqrt{S}$ for each wire, indicating that these thin wires can be regarded as one-dimensional below 1.3 K. This value of L_{in} is about same as that obtained in the experiment on the ion-implanted and annealed n^+-Si surface layers by the authors [17]. At higher field where $B > B_m$, the system is expected to behave three-dimensionally [15]. However, the field dependence of the magnetoconductance at higher field region did not fit well to the three-dimensional theory [18].

The field spacing of the aperiodic oscillation in our n^+-Si wires was in the range of 1 to 1.5 T as is seen from Fig.5. The correlation range B_c, which characterizes the typical spacing of the magnetoconductance fluctuation, given by

$$B_c = 2.4(h/e)/(L_{in} \cdot \min(L_{in}, W)) \quad , \tag{2}$$

where W is the width of a wire perpendicular to the magnetic field direction [4,7]. Using the value of $L_{in} \sim W \sim 0.1$ μm, it is estimated to be 1.0 T. This roughly agrees with the typical spacing of the conductance fluctuation observed in our experiment.

Figure 7 shows the amplitude of the conductance fluctuation as a function of length of the wire for the samples having the length L=1.5 to 10.8 μm and effective thickness $\sqrt{S}$= 0.05 to 0.14 μm. The amplitude of the fluctuation δG is defined as a root-mean-square deviation from its average value $\langle G \rangle$, and it should be averaged over the magnetic field range much wider than the typical spacing of the fluctuation. In our case, the magnetic field range of 9 T is not so wide compared with the typical spacing of fluctuation of about 1 to 1.5 T in order to include sufficient oscillations. Furthermore, the fluctuation was superimposed on the large background of varying magnetoconductance. Therefore, for estimating the amplitude of fluctuation, we averaged the half value of the peak to peak of fluctuations over several peaks. The error bar resulting from this simple estimation of the amplitude is shown in the figure for each sample. The amplitude of conductance fluctuation varies from $5\times10^{-4}(e^2/h)$ to $1\times10^{-2}(e^2/h)$ corresponding to the wire length of 10.8 to 1.5 μm. The effective thickness $\sqrt{S}$ for each sample is smaller or comparable to the inelastic diffusion length $L_{in} \sim 0.1$ um at 1.3 K. Thus it is expected that the wires can be regarded as one dimensional with respect to the conductance fluctuation behavior. The theory predicts that the fluctuation amplitude δG is reduced from its universal value of e^2/h by the inelastic scattering of electron waves when the system length L is longer than L_{in}, and gives a relation of

$$\delta G/(e^2/h) = (L_{in}/L)^{3/2} \quad , \tag{3}$$

for one-dimensional system where $W < L_{in}$ [7,19]. The solid line in Fig.7 represents this theoretical estimate using the value of L_{in}= 0.1 μm. Considering both the approximate nature of the above theoretical prediction and the roughness of the experimental estimation of L_{in} and δG, the experimental values except for thick wires ($\sqrt{S}$= 0.13, 0.14 μm) give a good fit to the theory in the length dependence. A similar good agreement with the theory has been reported by Skocpol et al. for narrow channel Si-MOSFEF's [7]. It was predicted that when the wire was thicker than L_{in}, the amplitude of the conductance fluctuation was intensified by the classical parallel addition of the each one-dimensional amplitude of fluctuation [7,20]. The deviation of the fluctuation amplitude for thicker wires ($\sqrt{S}$= 0.14, 0.13 μm) from the length dependence of amplitude for other thinner wires ($\sqrt{S} \lesssim L_{in}$) is understood by this explanation.

Figure 8 shows the magnetoconductance at various field orientations for the wire L=2.2 μm and $\sqrt{S}$=0.08 μm. The field direction is rotated within the plane normal to

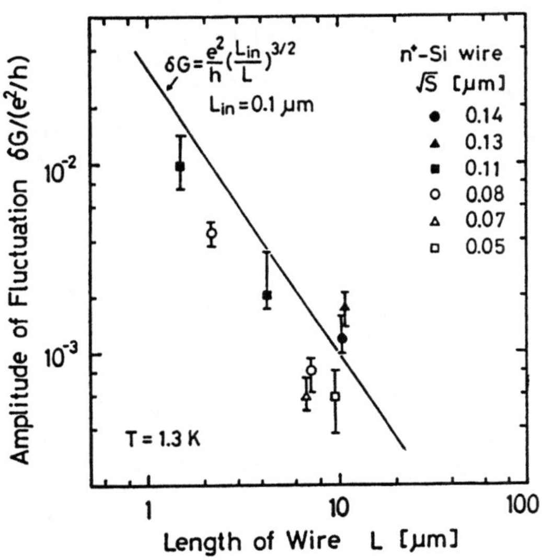

Fig.7 Length dependence of the amplitude of conductance fluctuation for n^+-Si wires L=1.5 to 10.8 μm, $\sqrt{S}$=0.05 to 0.14 μm.

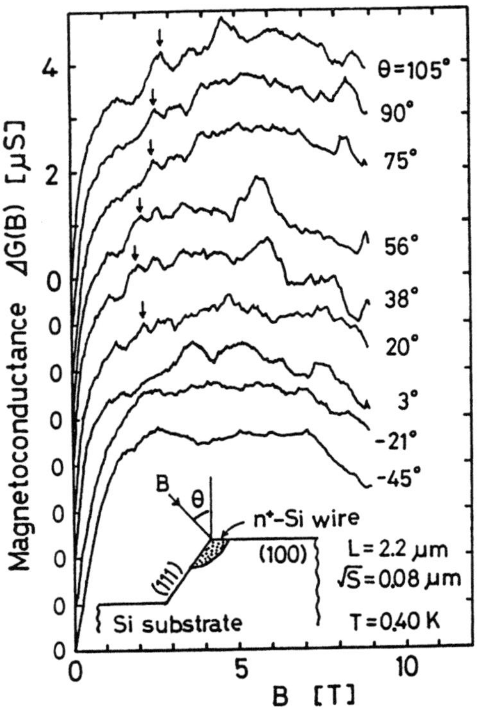

Fig.8 Magnetoconductance at various field orientations for a wire with L=2.2 µm and $\sqrt{S}$=0.08 µm at 0.4 K. The magnetic field is rotated within the plane normal to the current as shown in the insert

the current. The angle Θ is defined as shown in the insert of the figure. When the angle Θ is decreased from 105° to 38°, the fluctuation peaks shift moderately towards the lower magnetic field region, as is shown by one of the conspicuous peaks marked by arrow for each field orientation in the figure. On the other hand, the peaks shift towards the higher field region below Θ=38°. Between Θ=20° and -45°, the spacing of the peaks seems to become wider rapidly with decreasing the angle. The figure shows also several large peaks whose angular dependence cannot be readily traced. This complicated behavior may be understood qualitatively by considering the complicated cross-sectional shape of the wires. As shown in the insert of the figure, the shape of the cross section of wire may be wide along the anisotropically etched (111) plane. Thus the effective width W of the wire is largest when the field is applied normal to the inclined (111) plane (Θ=55°), and smallest when the field is applied parallel to that plane (Θ=-35°). According to (2), the spacing of fluctuation is smallest near Θ=55°, and largest near Θ=-35° as the experiment in Fig.8 shows approximately. However, there is no simple dependence on the field orientation such that the structure of the fluctuation is governed only by the normal component of the field as previously reported in narrow channel Si-MOSFET's [6] and n^+-GaAs wires [10].

4. Summary

In summary, we have fabricated mesoscopic wires of arsenic-doped n^+-Si by techniques of the selective diffusion and the anisotropic etching with no use of sophisticated microfabrication. The effective thickness $\sqrt{S}$ of the wires is estimated to be 0.05 to 0.14 µm from its conductance. Low temperature magnetoconductance shows aperiodic oscillations, superimposed on overall field dependence due to the electron localization effect. The inelastic diffusion length L_{in} was estimated to be about 0.1 µm at 1.3 K by comparing the magnetoconductance data for thinner wires with the theory of the one-dimensional localization. The amplitude of the aperiodic oscillation varied from $5 \times 10^{-4}(e^2/h)$ to $1 \times 10^{-2}(e^2/h)$ showing the length L dependence of $L^{-3/2}$ except for thicker wires. The typical spacing of the oscillation is about 1 to 1.5 T. These values of the amplitude and the spacing are consistent with the theory of the universal conductance fluctuation for the one-dimensional system ($\sqrt{S} \lesssim L_{in}$). For thin wires, the shift of the fluctuation peaks was observed by rotating the field orientation in the plane normal to the wire. This behavior suggests that the wire cross section is wide along the anisotropically etched inclined (111) plane.

Acknowledgment

The authors would like thank Dr. M. Tamura of Central Research Laboratory, Hitach Ltd. for providing ion-implanted polycrystalline Si samples.

References

1. C. P. Umbach, S. Washburn, R. B. Laibowitz, R. A. Webb: Phys. Rev. B30, 4048 (1984)
2. B. L. Al'tshuler: Pis'ma Zh. Eksp. Thor. Fiz. 41, 530 (1985) [JETP Lett. 41, 648 (1985)]
3. A. D. Stone: Phys. Rev. Lett. 54, 2692 (1985)
4. P. A. Lee, A. D. Stone, Phy. Rev. Lett. 55, 1622 (1985)
5. J. C. Licini, D. J. Bishop, M. A. Kastner, J. Melngailis: Phys. Rev. Lett. 55, 2987 (1985)
6. S. B. Kaplan, A. Hartstein: Phys. Rev. Lett. 56, 2403 (1986)
7. W. J. Skocpol, P. M. Mankiewich, R. E. Howard, L. D. Jackel, D. M. Tennanat: Phys. Rev. Lett. 56, 2865 (1986)
8. G. P. Whittington, P. C. Main, L. Evaes, R. P. Taylor, S. Thomas, S. P. Beaumont, C. D. W. Wilkinson: Int. Conf. Superlattice, Microstructures and Microdevices, Gothenburg, Aug. (1986)
9. T. J. Thornton, M. Pepper, G. J. Davis, D. Andrews: 18th Int. Conf. on the Physics in Semiconductor, Stockholm, Aug. (1986)
10. K. Ishibashi, K. Nagata, K. Gamo, S. Namba, S. Ishida, K. Murase, M. Kawabe, Y. Aoyagi: Solid State Commun. 61, 385 (1987)
11. D. P. Kennedy, R. R. O'Brien: IBM J. 9, 179 (1965)
12. M. J. Declercq, L. Gerzberg, J. D. Meindl: J. Electrochem. Soc. 122, 545 (1975)
13. R. Baron, G. A. Shifrin, O. J. Marsh, J. M. Mayer: J. Appl. Phys. 40, 3702 (1969)
14. S. M. Sze, J. C. Irvin: Solid-State Electron. 11, 599 (1968)
15. B. L. Al'tshuler, A. G. Arnov: Pis'ma Zh. Eskp. Theor. Fiz. 33, 515 (1981) [JETP Lett. 33, 499 (1981)]
16. R. Wheeler: Phys. Rev. B24, 4645 (1981)
17. K. Itoh, Y. Sasaki, S. Tanuma, Y. Oda, T. Ono, H. Nagano: J. Phys. Soc. Jpn. 54, 3679 (1985)
18. A. Kawabata: J. Phys. Soc. Jpn. 49, 628 (1980)
19. P. A. Lee, A. D. Stone, H. Fukuyama: Phys. Rev. B35, 1039 (1987)
20. Y. Imry: Europhys. Lett. 1, 249 (1986)

Aharonov-Bohm Magnetoresistance Oscillations in Selectively Doped GaAs/AlGaAs Submicron Structures

K. Ishibashi[1], *Y. Takagaki*[1], *K. Gamo*[1], *S. Namba*[1], *S. Ishida*[2], *K. Murase*[2], *Y. Aoyagi*[3], and *M. Kawabe*[4]

[1]Faculty of Engineering Science, Osaka University, Toyonaka 560, Japan
[2]Faculty of Science, Osaka University, Toyonaka 560, Japan
[3]Institute of Physical and Chemical Research, Wako, Saitama 351, Japan
[4]Institute of Materials Science, University of Tsukuba, Sakura-mura, Ibaraki 300-51, Japan

1. Introduction

Recent development in microfabrication and crystal growth techniques has made it possible to study electronic transport properties of submicron structures and opened up the new field of physics in so-called mesoscopic systems [1]. Experiments on metallic wires [2] and rings [3-5] have shown unexpected fluctuations and oscillations in magnetoconductance in spite of the presence of elastic scattering events. These phenomena are now understood as the result of quantum interference of electron waves, which preserves their phase coherence [6-8]. In the metallic system, experiments were performed at very low temperatures. In contrast, these effects are observable at 4.2K in semiconductors (Si-MOSFET [9] and GaAs system [10-13]) because of their long thermal diffusion lengths ($L_T=(\hbar D/kT)^{1/2}$).

In the case of a metallic sample, which usually has $l \ll L$ where l and L are the elastic mean free path and the sample size, respectively, the amplitude of oscillations is very small ($\Delta R/R \sim 0.1\%$). In order to achieve larger amplitudes of oscillation, which is necessary for device applications [14], we have fabricated small rings using selectively doped GaAs-AlGaAs heterostructures, where the relationship between l and L is rather close to $l \lesssim L$. However, the oscillations still have only a small amplitude ($\Delta R/R \sim 2\%$) at 1.5K.

2. Sample Preparation

The starting wafer was grown by Molecular Beam Epitaxy (MBE) and consists of a 100Å thick non-doped GaAs conductive layer sandwiched between 600Å thick Si-doped (2×10^{18} cm^{-3}) AlGaAs barrier layers with 60Å thick non-doped AlGaAs spacer layers, as shown in Fig.1. The film sample has a carrier concentration of 1.6×10^{12} cm^{-2} and a mobility of 17200 cm^2/Vs.

An Al ring pattern was first made by electron beam lithography followed by the lift-off technique, and the ring pattern of the heterostructure was made by 1keV Ar ion beam sputtering using the Al pattern as an etching mask [15]. The geometrical parameters such as diameter and width were determined by scanning electron microscopy. However, we should note that the effective line width for electrical conduction is reduced because of the presence of the inherent surface depletion layer which has a width of about 0.1um.

We made two rings (R05 & R08) with a different aspect ratio (diameter/width), and a singly connected broken ring (R02) for comparison, in which one arm of the ring was intentionally cut.

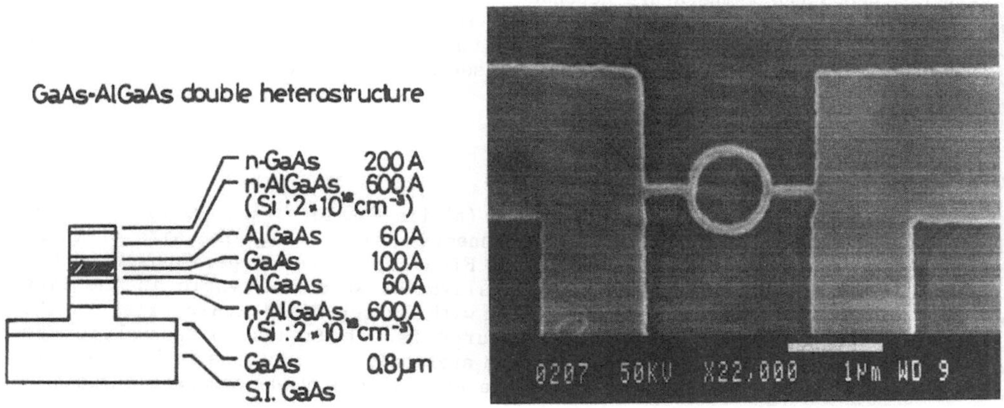

(a) (b)

Fig.1: (a) Cross-section and (b) SEM micrograph of the ring

3. Inelastic Scattering Length

The inelastic scattering length (l_{in}) is the characteristic length for mesoscopic systems. In the semiconducter sample (GaAs system and Si-MOSFET), another phase coherence length of thermal diffusion length L_T is usually longer than l_{in}. The inelastic scattering length can be determined by fitting the magnetoresistance (MR) data using some parameters. Long parallel wires were made from the same substrate, where the fluctuations were negligible due to the self-averaging effect. We used the theoretical expression for the one-dimensional case as follows [16]:

$$\frac{\Delta R}{R} = \frac{NR}{L}\frac{e^2}{\pi\hbar}\left\{ l_{in} - \left(l_{in}^{-2} + \frac{e^2W^2B^2}{3\hbar^2}\right)^{-1/2}\right\} ,$$

where L is the length. The line width W and l_{in} are the fitting parameters. The width is also used as the fitting parameter because the exact width for conduction is difficult to know [17].

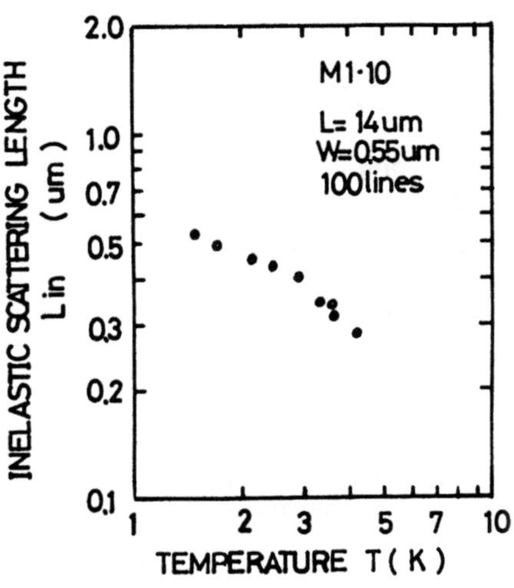

Fig.2: Temperature dependence of the inelastic scattering length of 100 parallel wires with a length of 14μm

Figure 2 shows the temperature dependence of the inelastic scattering length of 100 parallel wires with a length of 14μm. The inelastic scattering length is roughly proportional to $T^{-0.4}$ below 3K, which indicates that electron-electron interaction is dominant. Above 3K, the slope seems to become large. This may be due to electron-phonon interactions.

4. Rings

Figure 3 shows (a) the magnetoresistance and (b) its Fourier transform of two rings with different aspect ratio and a singly connected broken ring. The size of these samples is given in the caption of Fig.3. Fine structures superimposed on the background negative MR due to the weak localization were observable in the ring samples, while only aperiodic fluctuations with larger field scale exist in the broken ring. Though the field range measured is not enough to calculate the correlation function, the field scale of the aperiodic fluctuations in R02 seems to be of the order of $B_c=(h/e)/S$, where S is the area of the phase coherent subunit.

The Fourier transform shows a clear peak of the period of the h/e oscillation only in the sample with the larger aspect ratio (a.r.). Complicated structures are observed around the position corresponding to the h/e oscillation in the sample with the smaller a.r. The structures in addition to the background MR become larger with decreasing temperature. We could not identify a clear peak of the h/e oscillation above 4.2K even for the sample with the larger a.r.

The oscillations for R05 have an amplitude of ~2% in $\Delta R/R$ and $\sim 0.1e^2/h$ in absolute conductance. This value is much smaller than the values predicted by the

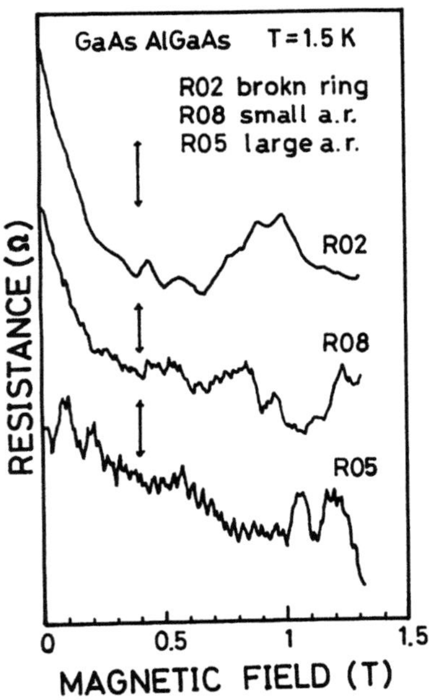

(a)

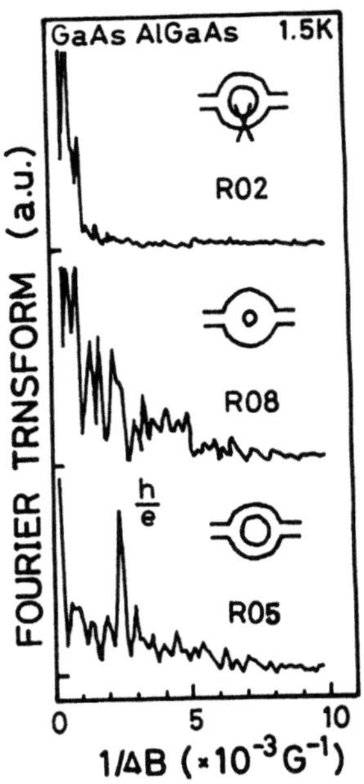

(b)

Fig.3: (a) Magnetoresistance and (b) its Fourier transform of three samples at 1.5K R02; singly connected broken ring (W=0.3μm), R05; ring (d_{out}=1μm, d_{in}=0.3μm), R08; ring (d_{out}=1μm, d_{in}=0.2μm), where d_{out} and d_{in} are the outermost and innermost diameters, respectively. The arrows in (a) show 5%xR(B=0) two theories, one for disordered systems and one for the ballistic regime.[14] The latter theory predicts 100% modulation in the magnetoresistance under postulation of the ballistic transport and a single transverse mode. Comparing two rings (R05 & R08) with different widths, we find that a large aspect ratio is important because of the reduction of the number of transverse modes.[18] The present samples have $\lambda_F \ll W$, where λ_F is the Fermi wavelength. However, oscillations may be observable in the semiconductor rings with a smaller aspect ratio than in the metallic ring because of the longer λ_F for semiconductors than for the metal. Another reason for the small amplitude of oscillations is that the inelastic scattering length is shorter than the diameter of the ring, as seen from Fig.2. The amplitude is predicted to be in proportion to $\exp(-\pi r/l_{in})$, where r is the radius of the ring.[4]

The radius which corresponds to the h/e oscillation is rather close to the inside diameter of the ring. This is probably because the width of the side wall depletion layer may not be uniform on the inside and outside for technical reasons.

The present device structure may not be optimum for the largest amplitude of oscillations. When we consider the ring structure in analogy with an optical wave guide, the ratio of the radius to the wavelength is not so large as in guide waves. A large part of the incident wave may be reflected when coming into the guide from the large electrode and being divided into two waves. DATTA et al. propose another structure to get a larger amplitude.[14]

5. Conclusion

We have observed Aharonov-Bohm magnetoresistance oscillations in submicron rings made from selectively doped GaAs-AlGaAs heterostructures. It was found that the large aspect ratio and the long inelastic scattering length are important in order to get clear oscillations. An optimum structure should be sought in order to have a much larger amplitude of oscillations, which is necessary for device applications.

Acknowledgement

We thank Dr.Hirano for providing a MBE grown substrate and T.Takaoka for his experimentally useful suggestions.

References

1. A.D.Stone: Phys. Rev. Lett. 54, 2692 (1985)
2. C.P.Umbach, S.Washburn, R.B.Laibowitz, and R.A.Webb: Phys. Rev. B30, 4048 (1984)
3. R.A.Webb, S.Washburn, C.P.Umbach, and R.B.Laibowitz: Phys. Rev. Lett. 54 2696 (1985)
4. C.V.Chandrasekhar, M.J.Rooks, S.Wind, and D.E.Prober: Phys. Rev. Lett. 55 (1985)
5. D.J.Bishop, J.C.Licini, and G,J.Dolan: Appl. Phys. Lett. 46, 1000 (1985)
6. P.A.Lee and A.D.Stone: Phys. Rev. Lett. 54, 2629 (1986)
7. B.L.Al'tshuler: JETP Lett. 41, 648 (1985)
8. For review, see Y.Imry: In Condensed Matter Physics, edited by G.Grinstein and E.Mazenko (World Publishing, Singapore 1986)
9. W.J.Skocpol, P.M.Mankiewich, R.E.Howard, L.D.Jackel, D.M.Tennant, and A.D.Stone: Phys. Rev. Lett. 56, 2865 (1986)
10. K.Ishibashi, K.Nagata, K.Gamo, S.Namba, S.Ishida, K.Murase, Y.Aoyagi, and M.Kawabe: Solid State Commun. 61, 385 (1987)
11. S.Datta, M.R.Melloch, S.Bandyopadhyay, R.Noren, M.Varizi, M.Miller, and R.Reifenberger: Phys. Rev. Lett. 55, 2344 (1985)
12. G.Timp, A.M.Chang, J.E.Cunningham, T.Y.Chang, P.Mankiewich, R.Behringer, and R.E.Howard: Phys. Rev. Lett. 58, 2814 (1987)

13. K.Ishibashi, Y.Takagaki, K.Gamo, S.Namba, S.Ishida, K.Murase, Y.Aoyagi, and M.Kawabe: to be published in Solid State Commun.
14. S.Datta, M.R.Melloch, S.Bandyopadhyay, and M.S.Lundstrum: Appl. Phys. Lett. 48, 487 (1985)
15. K.Ishibashi, K.Nagata, K.Gamo, S.Namba, S.Ishida, K.Murase, Y.Aoyagi, and M.Kawabe: Proc. of 2nd Int. Symp. of "Foundations of Quantum Mechanics", Hitachi Central Research Lab. Tokyo 1986
16. B.L.Al'tshuler and A.G.Aronov: JETP Lett. 33, 499 (1981)
17. K.K.Choi, D.C.Tsui, and K.Alavi: Appl. Phys. Lett. 50, 110 (1987)
18. M.Buttiker, Y.Imry, R.Landauer, and S.Pinhas: Phys. Rev. B31, 6207 (1984)

Conductance Fluctuations with Variations of Electron Density in Inversion Layers of Silicon

J. Kinoshita[1], *K. Inagaki*[1], *C. Yamanouchi*[1], *K. Yoshihiro*[1], *J. Wakabayashi*[2], *and S. Kawaji*[2]

[1]Electrotechnical Laboratory, Sakura-mura, Niihari-gun, Ibaraki 305, Japan

[2]Department of Physics, Gakushuin University, Mejiro, Toshima-ku, Tokyo 171, Japan

Highly reproducible aperiodic oscillations as a function of chemical potential are observed in the conductance of silicon n-inversion layers with conventional dimensions at low temperatures, where the negative magnetoresistance and a weak metallic temperature dependence of the conductance are also observed. The amplitude of the oscillations amounts to the order of e^2/h, probably resulting from a quasi one-dimensional network of electrons in the inversion layer at lowest areal concentrations.

1. Introduction

Several groups have studied ultranarrow inversion and accumulation layers (less than sub-μm wide) in silicon metal-oxide-semiconductor field effect transistors at cryogenic temperatures, and have observed reproducible aperiodic oscillations of resistance as a function of magnetic field or chemical potential [1-5]. Numerical studies carried out in this context may account for some aspects obtained in the strong localization regime [6,7]. Similar behavior has been observed also in the magnetoresistance of small metallic wires and rings [8]. Theoretical studies by STONE [9] and LEE and STONE [10] have interpreted these results based on numerical simulations on metallic samples as a consequence of quantum interference which becomes important when the inelastic diffusion length L_{in} exceeds sample dimensions. According to their prediction, conductance fluctuations of the order of e^2/h are a universal feature of quantum transport independent of sample size and degree of disorder in the low temperature limit, as long as the sample is smaller than $L_{in} \equiv \sqrt{D\tau_\varepsilon}$ and in the highly conducting diffusive regime, $k_F\ell \gg 1$. Here, $D = v_F{}^2\tau/2$ is the diffusion coefficient in a two-dimensional system, v_F the Fermi velocity, τ^{-1} and $\tau_\varepsilon{}^{-1}$ the elastic and inelastic scattering rates, respectively, k_F the Fermi wave vector, and $\ell = v_F\tau$.

In this paper, we report that similar structures are observed as a function of the gate voltage ($\propto$ chemical potential) in silicon n-inversion layers with conventional dimensions, 0.1 mm wide and 0.6 mm long, in a situation characterized by the negative-magnetoresistance, a weak metallic temperature dependence of the conductance, and $k_F\ell \gtrsim 10$. The conductance fluctuations we observed are far larger than those reported in previous works, which were performed at higher temperatures [11-13].

2. Experimental Details and Results

Traces of resistance R versus gate voltage V_g and also R versus magnetic induction B were obtained at temperatures between $T = 0.5$ and 4.2 K in magnetic fields, normal to the inversion layers, ranging from $B = 0$ to 0.5 T for both high and low mobility samples. The electron mobility at $B = 0$ ranged from $\mu = 1.0$ to 1.4 m^2/Vs and $\mu = 0.3$ to 0.4 m^2/Vs for high and low mobility samples, respectively. The concentration of two-dimensional electrons n_s was nearly proportional to V_g. The measurements were made with channel currents between $I_{SD} = 20$ and 500 nA rms at frequencies 6-20 Hz, using a lockin amplifier. The values of R were obtained using the relationship, $R = V_{PP'}/I_{SD}$, and $V_{PP'}$, the potential drop along the channel length between probes P and P' (see one of the insets in Fig. 1). The conductance is given by $G = 1/R$ as long as $\mu B \ll 1$.

During the course of this study, the effect of external noise was found to reduce the amplitude of oscillations seriously. Efforts were made not to introduce external noise into the sample, especially from the gate circuit.

Figure 1 shows typical results at 0.5 K (a) and 1.3 K (b). The noiselike random structure is highly reproducible as seen by comparing two curves, at B = 0 and B = 0.5 T. As temperature is lowered, relative amplitudes increase and finer structures become resolved, while the position of peaks and valleys remains at the same values of V_g. The typical width of oscillations resolved at 0.5 K is 10 mV, which corresponds to 0.07 K in the shift of chemical potential for uniform two-dimensional electron systems in the inversion layer. The conductance increases slightly as T is lowered, and increases also when a low magnetic field is applied. This negative magnetoresistance effect is observed in all the ranges of V_g and T studied.

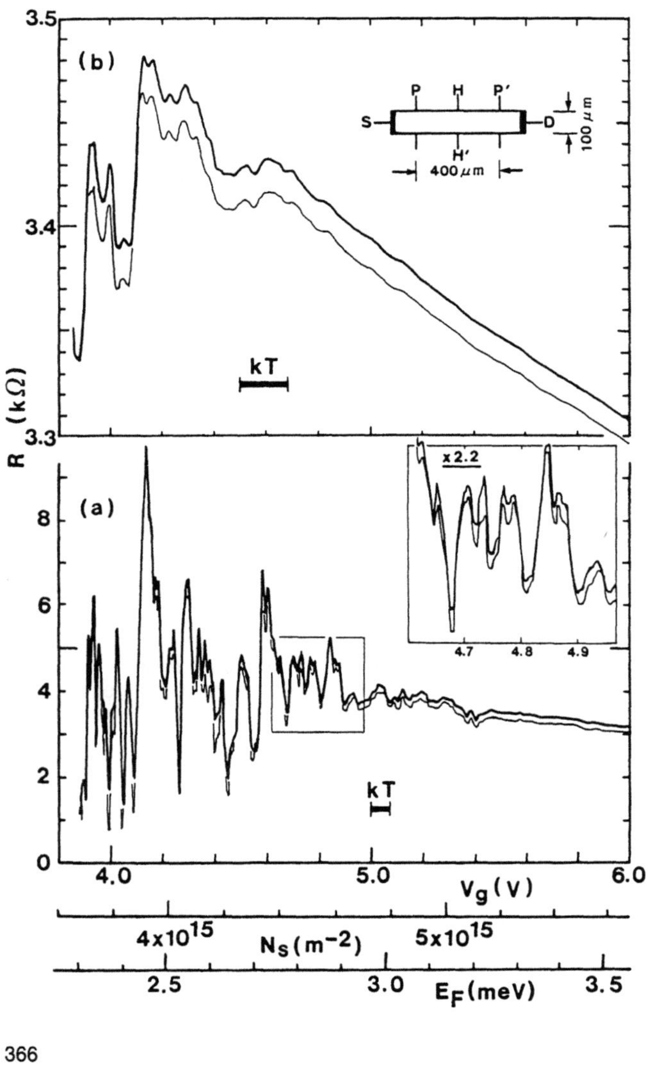

Fig. 1 Traces of resistance versus gate voltage curves at B=0 and 0.5 T (thick and thin lines, respectively). (a): T=0.51 K, I_{SD}=20 nA rms (6.0 Hz); (b): T=1.29 K, I_{SD}=20 nA rms (23 Hz): the inset in (b) illustrates the size and probe arrangement of the samples used. In the lower abscissa, $n_S \simeq 9.5 \times 10^{14} \times V_g$ m^{-2}.

Although the structure shown in Fig. 1 varies depending on the pair of potential probes selected, the typical amplitude and the spacing of peaks and valleys in V_g are almost unchanged. In order to compare the gross size of fluctuations in the R vs V_g curves obtained from different sources (probes, samples, or temperatures, e.g.), it should be appropriate to use the root mean square deviation δR, i.e., one standard deviation uncertainty around the average of $R(V_g)$ values taken in a definite small range of gate voltages ΔV_g, rather than to use the peak-to-valley amplitude.

In Fig. 2, we compare the temperature dependence of typical peak-to-valley (p-p) amplitudes, obtained at $B = 0$ and $I_{SD} < 50$ nA rms (23 Hz), with that obtained by LICINI et al. [4] from the magnetoresistance fluctuations in silicon inversion layers of mesoscopic sizes ($\sim 70 \times 50$ nm in the gate electrode area). In both cases, the amplitudes increase in proportion to T^{-2} - T^{-3} as T is lowered, and begin to saturate at the lowest temperatures. In the figure, rms data (passed through by dotted lines)

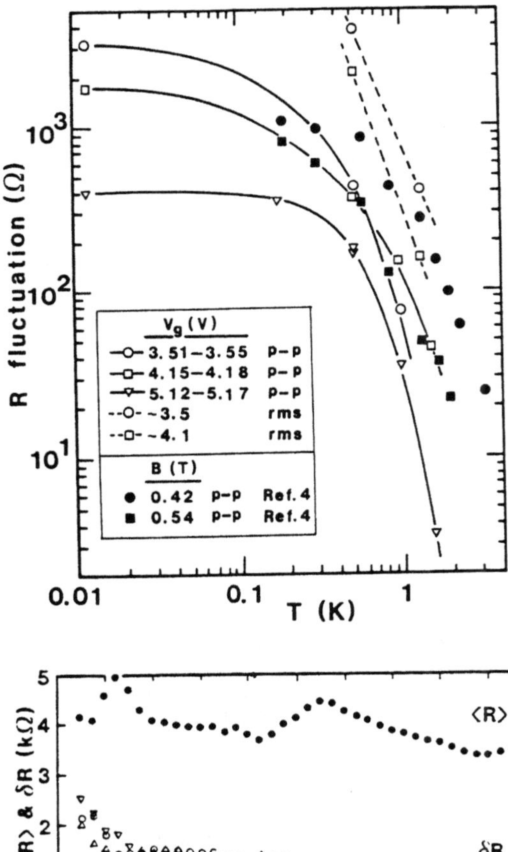

Fig. 2 Amplitude of R fluctuation as a function of T.

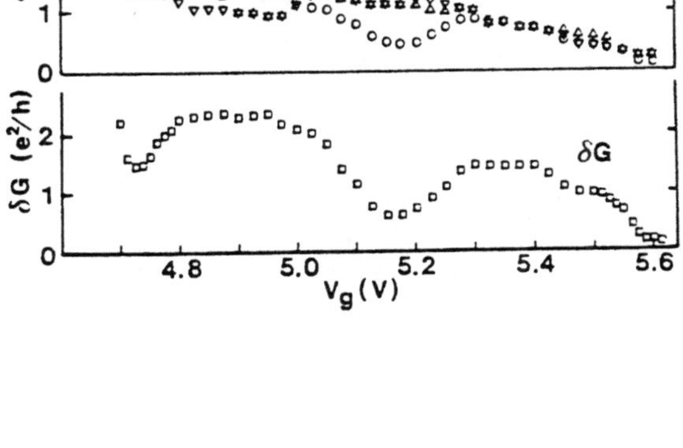

Fig. 3 <R>, δR, and δG as functions of V_g. In δR vs V_g curves, △, ▽, and ○ denote data measured via PH, HP', and PP' probes, respectively.

are also included. These were obtained recently after an improvement was made in isolating the sample from external noise. This result indicates that the effect of electrical noise suppresses the increase of the amplitude at the lowest temperatures.

Figure 3 compares the rms value of resistance fluctuations δR vs V_g curves measured at three different positions in a sample (via probes PH, HP', and PP'), at $T = 0.52$ K, $I_{SD} = 20$ nA rms (23 Hz), and the range of summation for the moving average $\Delta V_g = 0.3$ V. The average $\langle R(V_g) \rangle$ and δG, both for PP' probes, are also shown in the figure; δG is expressed in units of e^2/h. It is noted that three δR vs V_g curves are in rough agreement in spite of different distances between pairing probes, PH = H'P = PP'/2, indicating that the size of fluctuations is roughly independent of the distance of potential probes. In the range of V_g between 4.8 and 5.0 V, it is seen that $\langle R \rangle \simeq 4$ kΩ, $\delta R \simeq 1.2$ kΩ, and $\delta G \simeq 2(e^2/h)$.

The amplitude increases as V_g is decreased, i.e., $\delta R \propto \exp(-V_g/V_0)$, where $V_0 \simeq 0.7$ V (at $V_g < 7$ V) or $V_0 \simeq 2$ V (at $V_g \gtrsim 7$ V) for high mobility samples and $V_0 \simeq 0.4$ V for low mobility samples; δR is also expressed as $\delta R \propto \exp(-n_s/n_0)$, where $n_0 \simeq 6 \times 10^{14}$ m^{-2} at $n_s < 5 \times 10^{15}$ m^{-2} for both high and low mobility samples. That is, in the region where $n_s < 5 \times 10^{15}$ m^{-2}, the size of fluctuations is equal for both high and low mobility samples.

Figure 4 shows R as a function of B at $T = 0.52$ K, $I_{SD} = 100$ nA rms (23 Hz), and $V_g = 4.130$ V, at which one of the typical peak-to-valley structures arises. The curve shows a typical negative magnetoresistance and no trace of the fluctuations, such as observed in devices of mesoscopic sizes [4,5], up to $B = 0.5$ T, despite close resemblance of the present samples to those mesoscopic devices in some aspects.

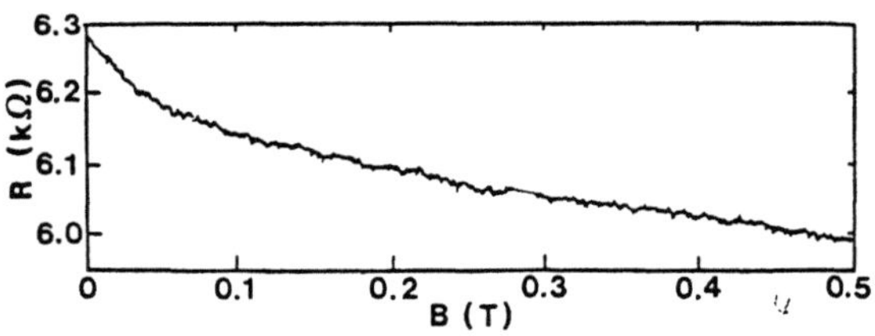

Fig. 4 R as a function of B.

3. Discussion

The negative magnetoresistance observed is considered to be the same as usually observed in the weak localization regime [14,15]. The inelastic relaxation time τ_ε can be calculated by using the present data and an expression for the magnetoconductivity given by HIKAMI et al. [16] which is valid at low magnetic fields when $\alpha\tau \ll 1$. Here, $\alpha\tau = 2\ell^2/\ell_0{}^2 = \sigma_0 \mu B/(n_v \times 1.93 \times 10^{-5})$, ℓ is the electron mean free path, $\ell_0 = (\hbar/eB)^{1/2}$ is the radius of the ground Landau orbit, σ_0 the conductivity [S] at $B = 0$, and $n_v = 2$ (the valley degeneracy factor); some other numerical constants used are given in [15].

Figure 5 shows τ_ε as a function of T at several values of V_g, and compares it with τ_ε obtained by KAWAGUCHI and KAWAJI [17] for weakly localized two-dimensional systems with n_s which is nearest to the present case. Both τ_ε values are found to be consistent. This result suggests that the present systems are of usual weakly localized electrons.

The inelastic scattering length L_{in} thus obtained for the present samples is nearly equal to or less than 1 µm. The sample dimensions are two orders of magnitude larger than L_{in}. Provided that the present samples are uniform two-dimensional free-electron systems, it is difficult to consider that the observed fluctuations come from the same origin as those observed in the mesoscopic systems. However,

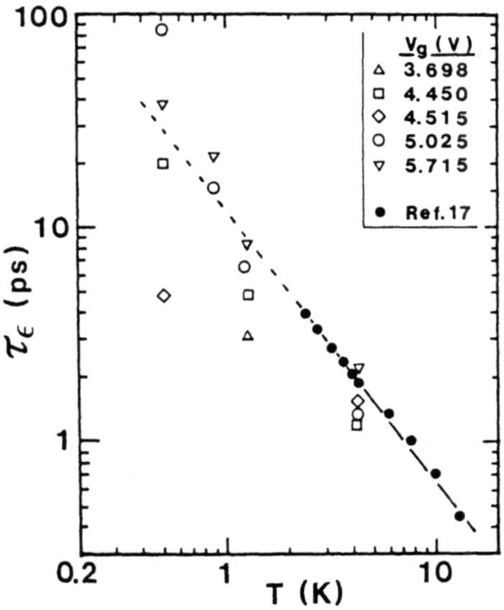

Fig. 5 τ_ε plotted against T.

the oscillations can appear when the channel current is carried via a few small regions ($\sim L_{in}$ in sizes), which are able to arise in a percolation network formed in a random mosaic structure of highly diffusive ($k_F\ell \gtrsim 10$) and strongly localized regions. A disappearance of the quantum Hall effect reported previously [18] is understood also on the basis of this picture.

References

1. A. Yagi: Jpn. J. Appl. Phys. 20, 2255 (1981)
2. A. B. Fowler, A. Hartstein, R. A. Webb: Phys. Rev. Lett. 48, 196 (1982)
3. R. F. Kwasnick, M. A. Kastner, J. Melngailis, P. A. Lee: Phys. Rev. Lett. 52, 224 (1984)
4. J. C. Licini, D. J. Bishop, M. A. Kastner, J. Melngailis: Phys. Rev. Lett. 55, 2987 (1985)
5. W. J. Skocpol, P. M. Mankiewich, R. E. Howard, L. D. Jackel, D. M. Tennant: Phys. Rev. Lett. 56, 2685 (1986); W. J. Skocpol, L. D. Jackel, R. E. Howard, P. M. Mankiewich, D. M. Tennant: Surf. Sci. 170, 1 (1986)
6. P. A. Lee: Phys. Rev. Lett. 53, 2042 (1984)
7. J. A. McInnes, P. N. Butcher: J. Phys. C: Solid State Phys. 18, L921 (1985)
8. C. P. Umbach, S. Washburn, R. B. Laibowitz, R. A. Webb: Phys. Rev. B30, 4048 (1984)
9. A. D. Stone: Phys. Rev. Lett. 54, 2692 (1985)
10. P. A. Lee, A. D. Stone: Phys. Rev. Lett. 55, 1622 (1985)
11. R. J. Tidey, R. A. Stradling, M. Pepper: J. Phys. C7, L353 (1974)
12. T. Cole, A. A. Lakhani, P. J. Stiles: Surf. Sci. 58, 56 (1976)
13. J. Wakabayashi, H. W. Myron, M. Pepper: Physica 117B & 118B, 691 (1983)
14. Y. Kawaguchi, S. Kawaji: J. Phys. Soc. Jpn. 48, 699 (1980)
15. S. Kawaji: Surf. Sci. 170, 682 (1986); S. Kawaji: Prog. Theor. Phys. Suppl. 84, 178 (1985)
16. S. Hikami, A. I. Larkin, Y. Nagaoka: Prog. Theor. Phys. 63, 707 (1980)
17. Y. Kawaguchi, S. Kawaji: Surf. Sci. 113, 505 (1982)
18. J. Kinoshita, K. Inagaki, C. Yamanouchi, K. Yoshihiro, J. Wakabayashi, S. Kawaji: in *Proc. 2nd Int. Symp. Foundations of Quantum Mechanics, Tokyo 1986*, ed. by M. Namiki et al. (Physical Society of Japan 1987)

Quantum Transport in Disordered Surfaces in Doped Silicon Near the Metal-Insulator Transition

Y. Ochiai

Institute of Materials Science, University of Tsukuba,
Ibaraki 305, Japan

1. Introduction

We have studied the low temperature transport in a disordered surface of non-metallic Sb-doped Si near the metal-insulator transition. By mechanical lapping of a Si crystal the surface of the sample has a strongly damaged layer a few micrometers thick. The conduction properties of the surface are very similar to those of amorphous Si films.[1] The low temperature transport in such a disordered layer shows Mott's $T^{\frac{1}{4}}$ type hopping conduction.[2]

Recently, we have observed several steps in the I-V characteristics of the nonmetallic Sb-doped Si just after lapping.[3,4] This nonlinear conduction can be observed even at H=0 and only at low temperatures. The steps appear at equal intervals in the bias voltage of the I-V curve and the intervals in the voltage are independent of magnetic field. In previous studies the nonlinear conduction has been explained by comparing it with the macroscopic quantum effect which has been observed in two-dimensional clusters of extremely small Josephson junctions.[5] A macroscopic quantum effect due to a new secondary quantization has been predicted theoretically in very small Josephson tunneling or single electron tunneling junctions.[6,7]

There has been growing interest in quantum effect transport phenomena in small conductors. The size of the conduction path has been able to become smaller in recent micro-fabrications of semiconductor devices and the length scale of the conduction has been nearly equal to the order of the inelastic scattering or the phase coherent length of the conduction electrons. In this paper, we focus on the relation between the quantum transport and nonlinear conduction in nonmetallic Sb-doped Si near the transition. We discuss an appropriate model for nonlinear conduction.

2. Experimental Details

We have carried out magnetoresistance measurements on nonmetallic Sb-doped Si of $2 \times 10^{18} cm^{-3}$. The critical concentration of the transition in Sb-doped Si is about $3 \times 10^{18} cm^{-3}$.[8] The samples were cut into a bridge-type shape having eight electrodes ; the dimensions were $0.5 \times 0.15 \times 0.03\ cm^3$. The surface of the sample was mechanically lapped with carborundum and Au lead wires were attached on each Sb evaporated electrode by a spot welding method. Thermal annealing was not performed after lapping.

We have made DC resistance measurements in the as-lapped sample by using a constant current source and a precise digital voltmeter. The Si sample was cooled by a one-shot-type He-3 cryostat having a 8 T superconducting magnet. The measurements of the low temperature conduction and the Fourier transformation of the magnetoresistance were carried out by a conventional microcomputer system.

3. Experimental Results

The resistance of the as-lapped sample is strongly influenced by the measurement current below about 2 K. After about 1 year such a strong dependence on the measurement current has disappeared. Figure 1 shows the magnetoresistances of the as-lapped sample at the measurement currents of 7 and 0.1 μA. For currents lower than 1 μA, we have observed a large fluctuation in the magnetoresistance, which is weak and negative, at 0.35 K as shown in the lower part of Fig. 1. At 1.2 K the magnetoresistance becomes positive but the fluctuation does not disappear. When a higher current than 1 μA is applied to the same sample, a step-like magnetoresistance can be observed as shown in the upper part of Fig. 1. The conductance fluctuation disappears and a step behavior of the resistance can be seen as well as in the I-V characteristics.[3,4] Although the whole of the magnetoresistance is positive, the horizontal part of each step is not so strongly positive. The position of the edge of the steps depends on the measurement current. After about 1 year, the conductance fluctuation and the step behavior disappear and the magnetoresistance becomes clearly positive even at 0.3 K and even at lower currents than 1 μA.

Conductance fluctuation is a kind of quantum transport phenomenon in small conductors.[9] Aging in the Fourier power spectrum of the magnetoresistance of the as-lapped sample is shown in Fig. 2. The peak in the spectrum after 5 months shifts to a higher value than that of the as-lapped. After 11 months no peak can be seen in the spectrum. The strength of the fluctuation of the magnetoresistance after 5 months becomes smaller compared with the as-lapped sample. Then we find that the fluctuation of the resistance consists of two parts of periodic and aperiodic components. The I-V characteristics corresponding to the Fourier power spectrum are shown in Fig. 3. The step becomes gradually smaller as the sample ages in air and disappears after 11 months. Since the gradient of the I-V curve means conductance, we can define the local resistance at each step. The local resistance at the first step in Fig. 3 is lower than the whole resistance of the I-V curve and gradually increases with aging. After 11 months there is no discontinuity between the local and the whole resistances. With increasing magnetic field, the whole

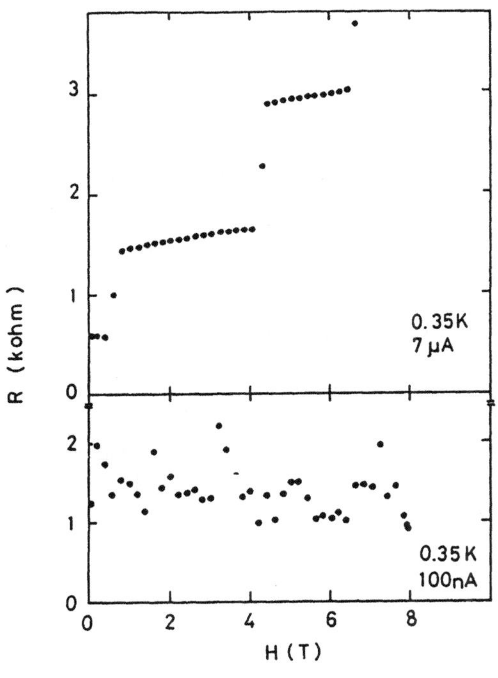

Fig. 1: Magnetoresistances of the as-lapped Sb-doped Si at 7 and 0.1 μA

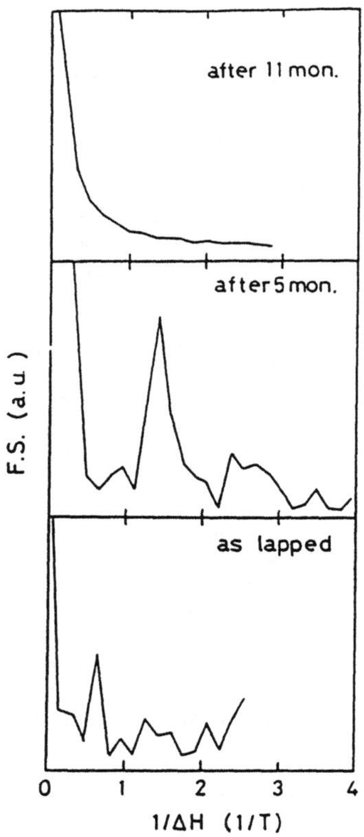

Fig. 2: Aging in the Fourier power spectra of the magnetoresistance of the as-lapped sample at 1.2 K

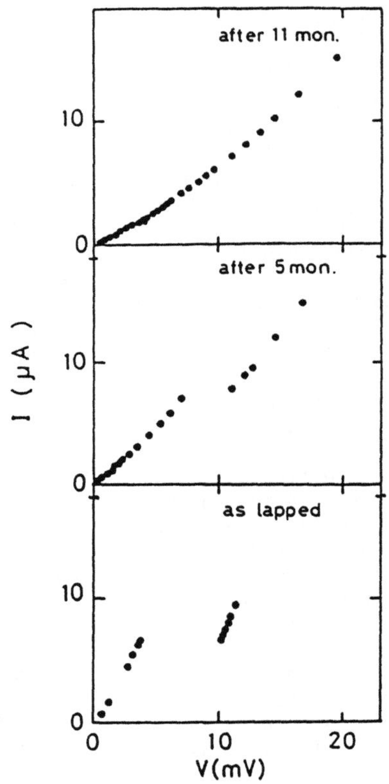

Fig. 3: Aging in the I-V characteristics of the as-lapped sample at 0.35 K and H=0

resistance increases but the local resistance decreases reversely. Then it should be noted that the higher conductive part at each step shows negative magnetoresistance. Here we consider that our fluctuating magnetoresistance comes from a quantum interference effect in coherent conduction. If the quantum interference model due to the Aharonov-Bohm effect [10] can be applied to the fluctuating magnetoresistance, the length scale of the conduction path will be estimated from the Fourier power spectrum of the resistance of the as-lapped sample. The value of $1/\Delta H$ at the peak in Fig. 2 is expected to define flux quanta of h/e. The results of h/e oscillations in the low temperature magnetoresistance show that the as-lapped and the 5 month old samples have periods of 0.88 and 1.5 T^{-1} which correspond to length scales of 60 and 80 nm, respectively. In the 11 month old sample the observable fluctuation almost disappears. Therefore the order of the length scale becomes larger than 100 nm. We can estimate the order of the inelastic scattering length to be about 30 nm from the results of ESR and the low temperature conduction in the damaged surface in Sb-doped Si.[1] Since the estimation was carried out around 1 K, the inelastic scattering length at 0.35 K should become longer than 30 nm. This longer length is close to the above values derived from the quantum transport. From the result of the 11 month old sample we find that the inelastic scattering length at 0.35 K is shorter than 100 nm.

4. Discussion and Concluding Remarks

Experimental results indicate that the low temperature conduction in the as-lapped sample consists of two types of transport. One should be an intrinsic transport in an ordinary nonmetallic sample and does not change even after 1 year or more. Another is considered to be a quasi-high-conductive transport (QHCT) which shows negative magnetoresistance and becomes gradually weakened as the sample ages in air. A similar QHCT has been reported in the nonmetallic Sb-doped Si near the transition.[11] In our as-lapped sample the conduction below 2 K is largely influenced by the measurement current. This suggests that the QHCT is due to a certain tunneling of electrons because of its field effect. Similar steps in the I-V characteristics have been observed in various transport phenomena. However, since superconducting behavior was not seen in the low temperature transport in our samples, our nonlinear conduction could not be explained by an ordinary superconducting effect. We have tried to explain the nonlinear conduction by using Likharev's single electron tunneling effect.[3,4] According to Likharev's model, [6,7] a nonlinear conduction can arise even in the absence of Josephson coupling and the physical origin of the oscillation is believed to be due to the Coulomb blockage of tunneling between normal metal electrodes. There exists correlated discrete electron tunneling of single electrons. Then, if we apply this model, a set of arrays of small tunneling junctions must be considered. This is very similar to rectangular Josephson junctions.[12]

Near the transition in doped Si, it can be considered that Anderson localization in a strongly disordered system has a certain inhomogeneity of the conduction path. We think that such an inhomogeneity is not attributed directly to donor impurity distribution, but it may be due to a large fluctuation of conductive domains coming from critical bahavior near the transition. In such a system the length scale of the nonmetallic region would be an important parameter for an explanation of the transition mechanism. Here we postulate that the conduction paths of the QHCT consist of series of many conductive domains which are joined by a weak link of localization region. When the inelastic scattering length becomes nearly equal to the mean distance between these conductive domains, then QHCT will occur. If so, the mean distance will correspond to the length obtained by quantum transport. The QHCT at higher currents can be considered to come from a nonequilibrium currnt flow through the weak links of coherent conduction. The step nature of the nonlinear conduction may be due to a breakdown of the excited nonequilibrium state in the QHCT. A similar jump in the I-V characteristics has been discussed in a chaotic state in rf and dc current driven Josephson junctions.[13]

We have observed quantum transport in the magnetoresistance of Sb-doped Si near the transition. We have found that the quantum transport is closely related to the appearance of nonlinear conduction at higher currents. The inelastic scattering length at 0.35 K has been estimated from the Fourier power spectra of the magnetoresistance.

It is our pleasure to thank Prof. Michael Pollak of the University of California, Riverside, who has taken on interest in our study, for fruitful discussions on a model of this study. Experiments were done at the Cryogenics Center of the University of Tsukuba.

References

1. Y.Ochiai, R.Yoshizaki, E.Matsuura: J.Non-Cryst.Solids 77&78, 707 (1985)
2. N.F.Mott, E.A.Davis: Electron Processes in Non-Crystalline Materials, 2nd. ed. (Oxford Univ. Press, Oxford, 1979)
3. Y.Ochiai, E.Matsuura: Jpn.J.Appl.Phys. 26, L582 (1987)
4. Y.Ochiai, R.Yoshizaki, E.Matsuura: Proc. Spring Meeting MRS Anaheim '87
5. K.Yoshihiro, J.Kinoshita, K.Inagaki, C.Yamanouchi: Proc. 2nd. ISFQM Tokyo '86
6. K.K.Likharev, A.B.Zorin: J.Low Temp.Phys. 59, 347 (1985)

7. D.A.Averin, K.K.Likharev: J.Low Temp.Phys. 62, 345 (1986)
8. Y.Ochiai, E.Matsuura: phys.stat.sol. (a) 27, K89 (1975)
9. A.D.Stone, Y.Imry: Phys.Rev.Lett. 56, 189 (1986)
10. C.P.Umbach, S.Washburn, R.B.Laibowitz, R.A.Webb: Phys.Rev. B 30, 4048 (1984)
11. A.P.Long, M.Pepper: Solid-State Electro. 28, 61 (1985)
12. T.A.Fulton, R.C.Dynes: Solid State Commun. 12, 57 (1973)
13. I.Goldhirsch, Y.Imry, G.Wasserman, E.Ben-Jacob: Phys.Rev. B 29, 1218 (1984)

Index of Contributors

MIX
Papier aus verantwortungsvollen Quellen
Paper from responsible sources
FSC® C105338

If you have any concerns about our products,
you can contact us on
ProductSafety@springernature.com

In case Publisher is established outside the EU,
the EU authorized representative is:
Springer Nature Customer Service Center GmbH
Europaplatz 3, 69115 Heidelberg, Germany

Printed by Libri Plureos GmbH
in Hamburg, Germany